U.S. OIL PIPE LINES

U.S. OIL PIPE LINES

An Examination of How Oil Pipe Lines Operate and the Current Public Policy Issues Concerning Their Ownership

by George S. Wolbert, Jr.

American Petroleum Institute
2101 L Street, Northwest
Washington, D.C. 20037

Copyright © 1979 by
American Petroleum Institute
Printed in the United States of America
FIRST PRINTING

Library of Congress
Catalogue Card #79-55344
Wolbert
U.S. Oil Pipe Lines

DC: American Petroleum Institute
7910   790906

Foreword

The latest resurgence of debate on the structure and performance of the oil pipeline industry has prompted critics to allege that shipper ownership of pipelines creates problems which only divestiture can solve. Industry defenders contend that existing common carrier laws adequately protect the interests of nonowner shippers, that the oil pipeline industry is highly efficient and only reasonably profitable, and that divestiture would be a serious mistake.

The quality of public debate has been hampered by the lack of a comprehensive and authoritative assembly of facts and analyses which bear on oil pipelines and related public policy issues. George Wolbert's book fills this need. It also provides the general reader with an interesting overview of the industry and its performance. To the serious student it presents detailed information, conveniently organized and extensively footnoted. The facts argue for themselves as the author weaves them into his discussions of the various points at issue.

Dr. Wolbert has studied the petroleum industry and its pipeline segment for more than three decades. He began writing on oil pipelines in the early 1950's. As legal counsel to a major petroleum company for many years, he dealt directly with various legal and financial issues associated with pipelines.

Dr. Wolbert's prodigious efforts have culminated in an important new book which will stand for some time as the definitive scholarly work in its field.

Charles J. DiBona
President
American Petroleum Institute
Washington, D.C.

Preface

This is a long book, but the subject matter dealt with is an important, complicated and changing one. In some respects, it is a supplement to my work of almost three decades ago, but in many significant ways, the very nature of the changes which have taken place since that time have caused the present volume to expand its horizons and treat aspects untouched by the prior approach.

Because the issues concern policy, an area in which decisions rest on judgments which necessarily must take into account imponderables, I have attempted, to a greater extent than otherwise might have been expected, to avoid stating conclusions. However, in several instances where the empirical evidence is overwhelmingly one-sided, I have made certain observations which may appear subjective to those whose deep-felt convictions run counter to them. An effort has been made to overcome this problem by laying out, in minute detail, the relevant testimony, facts and circumstances, which are in the public domain, so that the reader can reach his own conclusions. This has generated a certain amount of deliberate redundancy, but, in my judgment, the importance and controversial nature of the subject tipped the scales in favor of overinclusion rather than terseness of composition. Citations are copious, both to enable those with questions to examine the source material for themselves and for scholars to find a convenient starting point for further examinations of facets in which they have a particular interest.

In following this course, I have had to look to many sources and, hence, my acknowledgement perforce must be broad. To single out particular persons or groups for thanks would be a disservice to so many who have contributed so much. This holds true not only for those whose copyrighted material has been quoted or cited, but also to the many people who have provided information and source material. Hence I limit my acknowledgement to my wife, Winifred E. Wolbert, whose patience has rivaled that of Job, and to Sharon, George, and Jason, who must have wondered about that strange man who was in their midst, but who seemed hardly to be one of the family.

George S. Wolbert, Jr.
Houston, Texas, 1979

Table of Contents

Foreword

Preface

Chapter I A Review of Pipeline Development and Pipelines' Importance to Industry and Society 1

Chapter II Today's Pipeline System—an Overview 47

Chapter III Economic Characteristics 93

Chapter IV Pipeline Industry Structure 159

Chapter V Financing 229

Chapter VI Current Policy Issues Concerning Pipeline Ownership 253

Chapter VII Proposed Solutions to an Assumed Problem 439

Appendixes, Figures and Tables 479

Index 543

Appendixes

Appendix A	Range of Petroleum Transportation Costs	481
Appendix B	Shell Pipe Line Corporation Texas Local Gathering Tariff	482
Appendix C	Typical Gathering System	486
Appendix D	Crude Oil Gathering in Texas, December 1978	487
Appendix E	Refinery Receipts of Crude Oil and Condensate, Texas, First Half 1978	488
Appendix F	Typical Trunk System	493
Appendix G	Crude Oil Pipeline Capacities—1975	494
Appendix H	Product Pipeline Capacities—1975	495
Appendix I	Tariff Required to Earn 7% NROI	496
Appendix J	Crude Refining Areas and Crude Pipelines, December 1977	497
Appendix K	Products Pipelines, December 1977	498
Appendix L	Total Petroleum Products Carried in Domestic Transportation and Percent of Total Carried by Each Mode of Transportation	499
Appendix M	Eugene Island Pipeline System Project Development Chronology	500
Appendix N	Texoma Pipeline System Project Development Chronology	502
Appendix O	Current Capacity—Crude Oil Lines	503
Appendix P	Current Capacity—Product Lines	504
Appendix Q	Exxon Company U.S.A.—Crude Oil Sales (and Purchases)	506
Appendix R	New U.S. Refineries Built Since January 1, 1950	511

Figures

Figure 1	Pipeline Market	519
Figure 2	N.Y. Spot Cargo Prices (LOW) vs. Gulf Coast Spot Prices (LOW) Plus Transportation by Tanker and Pipeline—Premium Gasoline	520
Figure 3	N.Y. Spot Cargo Prices (LOW) vs. Gulf Coast Spot Prices (LOW) Plus Transportation by Tanker and Pipeline—Regular Gasoline	521
Figure 4	N.Y. Spot Cargo Prices (LOW) vs. Gulf Coast Spot Prices (LOW) Plus Transportation by Tanker and Pipeline—Unleaded Gasoline	522
Figure 5	N.Y. Spot Cargo Prices (LOW) vs. Gulf Coast Spot Prices (LOW) Plus Transportation by Tanker and Pipeline—No. 2 Heating Oil	523
Figure 6	Crude Oil Nominations and Purchases	524

Tables

Table I	Ratio of Throughput to Capacity for Selected Pipeline Systems	527
Table II	Ratio of Shipments to Capacity for the Plantation Pipeline, 1950–1978; Description of Prorationing on Plantation	528
Table III	Ratio of Shipments to Capacity on Colonial, 1965–1978; Description of Prorationing on Colonial	530
Table IV	"Majors" Pipeline Systems	531
Table V	Jointly Owned or Operated Pipeline Systems	535
Table VI	"Independent" Pipeline Systems	540
Table VII	Spot Price Comparisons—Cents per Gallon	542
Table VIII	Top Crude Oil Producers in Texas, 1977	437

Chapter I

A Review of Pipeline Development and Pipelines' Importance to Industry and Society

A. The Development of Petroleum Pipelines in the United States

Immediately after the initial burst of enthusiasm over the pioneer discovery oil well "brought in" by Colonel Drake near Titusville, Pennsylvania, on August 27, 1859,[1] the question arose, how do we get this material to a point of sale?[2] Because the wells were close to Oil Creek, a tributary of the Allegheny River, it was only natural that the early transportation was by water, downstream to Oil City, then on to Pittsburgh.[3] However, as the limits of the producing fields were extended, horse-and-wagon teams became the primary mode of transportation to shipping points. The charges for this service varied with the distance, season of the year,[4] and the number and depth of mud holes;[5] ranging from $1.00 to $5.00 a barrel, with the mean around $2.50.[6] Inasmuch as oil was selling at the lease for about $7.00 per barrel,[7] it was obvious that the teamsters' rate methodology was largely based on the necessities of the shipper.[8] Moreover, the service was poor, being affected by mud, wagon breakdowns, and mule and teamster problems.[9] Obviously, a less expensive and more efficient means of transporting the

---

[1] G. WOLBERT, AMERICAN PIPE LINES 5 (1952) [hereinafter cited as WOLBERT]; A. JOHNSON, THE DEVELOPMENT OF AMERICAN PETROLEUM PIPELINES 2 (1956) [hereinafter cited as JOHNSON].

[2] As stated graphically by Max Ball: "Oil in the field tanks is like a fat steer on the range; it needs to be taken thence and made into something useful." M. BALL, THIS FASCINATING OIL BUSINESS 173 (1940).

[3] HOWREY & SIMON, PIPELINES OWNED BY OIL COMPANIES PROVIDE A PRO-COMPETITIVE AND LOW COST MEANS OF ENERGY TRANSPORTATION TO THE NATION'S INDUSTRIES AND CONSUMERS, app. at 1 (1978) [hereinafter cited as HOWREY & SIMON].

[4] WOLBERT, *supra* note 1, at 6.

[5] JOHNSON, *supra* note 1, at 4.

[6] *Hearings before Temporary National Economic Committee, Pursuant to Pub. Res. 113* (75th Cong.), 76th Cong., 2d & 3d Sess., Parts 14-17A, 8584-8585 (1939-1941) [hereinafter cited as *TNEC Hearings*]; D. LEVEN, DONE IN OIL 512 (1941); P. GIDDENS, THE BIRTH OF THE OIL INDUSTRY 103 (1938) [hereinafter cited as GIDDENS].

[7] JOHNSON, *supra* note 1, at 6.

[8] A. CONE & W. JOHNS, PETROLIA 104 (1870).

[9] Burke, *Oil Pipelines' Place in the Transportation Industry*, 31 I.C.C. PRAC. J. 780, 782 (1964) [hereinafter cited as Burke]; HOWREY & SIMON, *supra* note 3, app. at 2.

crude had to be found. There had been earlier plans[10] and a successful installation in 1862 of a small diameter pilot model pipeline on the Tarr Farm in the Pithole Creek field which siphoned oil approximately 1,000 feet from a producing well to a field refinery.[11] However, an attempt one year later to scale-up the pilot model using a two-inch cast-iron pipe to pump oil from the Tarr Farm to the Humboldt Refinery below Plumer,[12] a distance of two miles, failed because of numerous leaks.[13] A more ambitious effort to build a two-inch field line from the Sherman Well to the Oil Creek Railroad Station at Miller's Farm not only leaked badly,[14] but fell prey to the teamsters who reacted to the competition by tearing up the pipe.[15] It remained for Samuel Van Syckel to complete the first commercially successful line on October 7, 1865:[16] a two-inch wrought-iron line,[17] lap-welded with screw-threaded joints.[18] The line was 32,000 feet long[19] and it transported 81 barrels an hour over the five-mile journey from Pithole City to the Oil Creek Railroad Station at Miller's Farm, Pennsylvania.[20] Van Syckel charged $1.00 per barrel and his service was reliable.[21] Notwithstanding the additional costs of maintaining armed guards to provide protection against the displaced teamsters[22] and the addition of a telegraph line to keep track of the shipments,[23] the line made enough money to return Van Syckel's investment

---

[10] As early as November 1860, G. D. Karnes had envisioned building a gathering line from a well in Burning Springs, Virginia, to the Ohio River, but the outbreak of the War Between the States interrupted his plans. D. LEVEN, DONE IN OIL 513 (1941); Finney, *Oil Pipe Line Transportation* in ELEMENTS OF THE PETROLEUM INDUSTRY 310 (1946); WOLBERT, *supra* note 1, at 6. In November 1861, a producer on the Tarr Farm, one Heman Janes, sought to promote a wood pipeline to Oil City but was deterred by the failure of a pipeline charter bill in the Pennsylvania Legislature. JOHNSON, *supra* note 1, at 5.

[11] GIDDENS, *supra* note 6, at 1421 (1938); THE HUMBLE WAY 8 (Nov.-Dec. 1949); WOLBERT, *supra* note 1, at 6.

[12] For a small scale map showing the location of these exotic places, see Map of Cornplanter Township in JOHNSON, *supra* note 1, at 6-7.

[13] JOHNSON, *supra* note 1, at 5-6; *cf.* Burke, *supra* note 9, at 782.

[14] See note 11, *supra*.

[15] JOHNSON, *supra* note 1, at 6.

[16] WOLBERT, *supra* note 1, at 6; GIDDENS, *supra* note 6, at 143 gives the date as October 9, 1865, and JOHNSON, *supra* note 1, at 8 names October 10. The differences probably represent physical completion date, commencement of operations, and receipt at the destination.

[17] WOLBERT, *supra* note 1, at 6; HOWREY & SIMON, *supra* note 3, app., at 2.

[18] JOHNSON, *supra* note 1, at 8; J. MERCHANT, *Pipeline Engineering and Construction* in AOPL EDUCATORS CONFERENCE 1 (1978) [hereinafter cited as MERCHANT].

[19] *Id.*

[20] WOLBERT, *supra* note 1, at 6; HOWREY & SIMON, *supra* note 3, app. at 2.

[21] HOWREY & SIMON, *supra* note 3, app. at 2.

[22] *Id.*

[23] JOHNSON, *supra* note 1, at 8.

[24] *Id.* at 7.

[25] Burke, *supra* note 9, at 782; HOWREY & SIMON, *supra* note 3, app. at 2.

(which, incidentally, was "leveraged" to the extent of a $30,000 loan from the First National Bank of Titusville)[24] in a very few months.[25] Van Syckel quickly followed his successful venture with a second line of similar capacity between the same points[26] and lowered his charge to $0.50 per barrel.[27] Even at these reduced rates, Van Syckel's dominant position was quickly diluted as numerous entrepreneurs, having witnessed the mechanical and economic reality of pipelines, commenced to lay multiple gathering lines from the spreading fields to local refineries and shipping points on the railroad or the Allegheny River.[28]

The railroads favored pipeline development at first because the early lines brought increasing quantities to loading racks along the railroad right-of-way[29] whence it was carried by rail tank cars to refineries on the East Coast, south to Pittsburgh, or west to Cleveland. The emerging pipeline system was caught up in a bitter struggle for trade and power between the Pennsylvania and the Erie Railroads, later joined by the New York Central, and the Baltimore and Ohio.[30] The railroad's attempts to monopolize the traffic merely created another transportation roadblock, dictating prices to producers and driving shippers from the field.[31] The resourceful pioneers met this threat by constructing trunk pipelines[32] directly from the producing fields to the refineries.[33] By 1874, a four-inch line was laid from the producing fields to Pittsburgh.[34] However, the railroads and the pipelines began to form associations and during most of the 1870's there were four such combinations which dominated the transportation of crude oil; *i.e.,* the Pennsylvania Transportation Company, associated with the Atlantic and Great Western Railroad (Erie System); the Empire Transportation Company, associated with the Pennsylvania Railroad; the Tidioute Pipe Line Company; and Vandergrift and Forman.[35] In a move to counteract these forces,

---

[26] JOHNSON, *supra* note 1, at 8; Burke, *supra* note 9, at 782.

[27] Burke, *supra* note 9, at 782.

[28] JOHNSON, *supra* note 1, at 8-14; *cf.* Burke, *supra* note 9, at 782.

[29] For an interesting picture of one of these pipeline-railroad transfer points, see JOHNSON, *supra* note 1, opp. 112.

[30] JOHNSON, *supra* note 1, at 12-54; M. DE CHAZEAU & A. KAHN, INTEGRATION AND COMPETITION IN THE PETROLEUM INDUSTRY 76 (1959) [hereinafter cited as DE CHAZEAU & KAHN].

[31] GIDDENS, *supra* note 6, at 152 (1938); WOLBERT, *supra* note 1, at 7.

[32] See definition in text at notes 373-400, *infra.*

[33] WOLBERT, *supra* note 1, at 7.

[34] This sixty mile line, having a capacity of 7,500 barrels per day, was built by Captain Vandergrift and George W. Foreman. A.P.I., PETROLEUM: THE STORY OF AN AMERICAN INDUSTRY 16 (1949); D. LEVEN, DONE IN OIL 48 (1941); THE HUMBLE WAY (Nov.-Dec. 1949).

[35] J. MCLEAN & R. HAIGH, THE GROWTH OF INTEGRATED OIL COMPANIES 60 (1954) [hereinafter cited as MCLEAN & HAIGH].

Standard Oil extended its activities into the gathering line segment of the business and, by 1877, had developed a system of crude pipeline and storage systems which fully matched that of Empire's.[36] When Empire attempted to offset the threat of Standard's pipelines by acquiring and constructing refineries on the East Coast, and entering into alliances with the few remaining independent (from Standard) refineries in Pittsburgh, Buffalo, and New York, Standard suspended all shipments over the Pennsylvania Railroad. It obtained the assistance of the New York Central, Erie, and Baltimore and Ohio Railroads and the power of this combination, together with the railroad strikes and riots of July 1877, which were directed discriminately at the Pennsylvania, caused the Pennsylvania to be receptive to a deal whereby Standard Oil would "even out" its shipments among the railroads, in exchange for which Standard was to receive a 10 percent commission for its services as "evener."[37] In addition, as part of the deal, Standard bought out Empire, including its New York and Philadelphia refineries, and its extensive crude oil gathering system.[38]

About the same time, a group of producers looking for an adequate, suitably priced outlet for their oil, formed the Tidewater Pipe Line Company and commenced the construction of a 115-mile six-inch crude-oil trunk line connecting the producing center of Coryville to Williamsport, Pennsylvania, where the oil could be loaded on tank cars and hauled by the Philadelphia and Reading Railroad to New York.[39] The Tidewater pipeline, completed in 1879, represented a major technological advance because it transported oil for a longer distance and at higher altitudes (it crossed the Allegheny Mountains) than had ever been done before.[40] Although by this time the Appalachian producing fields were served by over 1,200 miles of pipelines,[41] the joint Tidewater pipeline tariff and Reading Railroad rate made Pennsylvania grade crude oil available to Tidewater's independent refiner customers in New York at a significantly lower cost than that incurred by Standard Oil

---

[36] *Id.* at 63; HOWREY & SIMON, *supra* note 3, app. at 4.

[37] This deal was evidenced by an exchange of letters between William Rockefeller and Tom Scott, Vice President of the Pennsylvania Railroad. JOHNSON, *supra* note 1, at 66, citing I. TARBELL, STANDARD OIL COMPANY, VOL. I, 371-373 (1904).

[38] MCLEAN & HAIGH, *supra* note 35, at 64; JOHNSON, *supra* note 1, at 65; DE CHAZEAU & KAHN, *supra* note 30, at 78.

[39] Burke, *supra* note 9, at 781; *cf.* WOLBERT, *supra* note 1, at 7; HOWREY & SIMON, *supra* note 3, at 5. MCLEAN & HAIGH, *supra* note 35, at 64 report that the Reading was so eager for the traffic that it contributed about half of the capital required for the line. W. BEARD, REGULATION OF PIPELINES AS COMMON CARRIERS 13 (1941) [hereinafter cited as BEARD] reports the originating point to be Rixford, Pennsylvania.

[40] MCLEAN & HAIGH, *supra* note 35, at 64; *cf.* Reduced Pipe Line Rates and Gathering Charges, 243 I.C.C. 115, 119 (1940).

[41] WOLBERT, *supra* note 1, at 8; API, PETROLEUM FACTS AND FIGURES 138 (1947); THE HUMBLE WAY 9 (Nov.-Dec. 1949).

refineries. Standard immediately countered by building its National Transit Pipe Line System directly from the northeastern end of the Appalachian field to New York City, completing same in 1881.[42] During the same year, the Tidewater line was extended from Williamsport to the Philadelphia refining area.[43] However, the original contract between the builders of the Tidewater line and the Reading Railroad provided that Tidewater would forfeit $100,000 to the railroad if the line was built any closer to the Atlantic Seaboard during the first eight years of operation.[44] With Tidewater thus deterred from extending its line from Williamsport to the New York area and forced to rely upon railroad transportation for a substantial part of the movement, the cost advantage in the New York area switched back to Standard which had completely eliminated any reliance on rail transportation between the two points. By the time the Tidewater line was extended to Bayonne, New Jersey, in 1888, Standard had acquired the principal independent refineries which formerly had purchased from Tidewater. The latter, weakened by the loss of this business and beset by financial difficulties and internal dissension among its stockholders, entered into an arrangement with Standard whereby Standard would provide 88.5 percent and Tidewater the remaining 11.5 percent of the line's throughput.[45]

In the meantime, the Lima-Indiana fields had been discovered. At first they did not seem to be very attractive because stocks were high,[46] demand was not keeping pace,[47] and the crude was heavy,[48] black, and had a high sulphur content ("sour crude").[49] However, it was situated advantageously with respect to railroad connections from Lima to major markets such as Chicago, Toledo, and Cincinnati.[50] Standard organized the Buckeye Pipe Line Company which constructed a 50-mile six-inch pipeline from the pro-

---

[42] D. LEVEN, DONE IN OIL 513 (1941); Burke, *supra* note 9, at 783; JOHNSON, *supra* note 1, at 104 describes the route as connecting Olean, New York (near the Bradford, Pennsylvania field) to Saddle River, New Jersey, and the facilities as two side-by-side six-inch lines.

[43] Burke, *supra* note 9, at 283.

[44] MCLEAN & HAIGH, *supra* note 35, at 65 n. 15; *cf.* DE CHAZEAU & KAHN, *supra* note 30, at 78.

[45] A. NEVINS, STUDY IN POWER, Vol. I, pp. 379-380 (1953); MCLEAN & HAIGH, *supra* note 35, at 65; HOWREY & SIMON, *supra* note 3, app. at 5; DE CHAZEAU & KAHN, *supra* note 30, at 79. This event took place in 1883, at a time when production in Pennsylvania and New York had peaked out, with the possible exception of one brief "blip" in 1891. A. JOHNSON, PETROLEUM PIPELINES AND PUBLIC POLICY 4 (1967) [hereinafter cited as PETROLEUM PIPELINES]; BREAD, *supra* note 39, at 13 (1941).

[46] MCLEAN & HAIGH, *supra* note 35, at 67 noted that in 1887 there were about 31,000,000 barrels in aboveground storage as compared to the full 1886 year's production of only 28,000,000 barrels.

[47] *Id.* Russian oil was flooding into the European and Near Eastern markets.

[48] *Id.* at 67-68. The processing cost was higher than for Pennsylvania crude and the kerosene/naphtha yield was only about 57 percent compared to 75 percent for Pennsylvania crude.

ducing fields at Cygnet, North Baltimore, and Findlay[51] to Lima where Standard built a new refinery, the Solar Refining Company.[52] Buckeye had to store more than 85 percent of the total field production during 1886-1892.[53] Nevertheless, Standard increased its position in the Lima-Indiana field[54] by extending a fuel-oil pipeline from Lima to Chicago in 1888 and acquiring the field's leading producing company, the Ohio Oil Company, the following year.[55] The 205-mile eight-inch Lima to Chicago line, laid along the right-of-way of the Chicago and Atlantic Railroad,[56] represented another technological advance in pipeline construction methods.

A traction engine strung the pipe on the surface parallel to the railroad tracks; a steam-powered "screw-machine" joined the threaded pipe joints and a steam-powered ditching machine dug the trench; manual labor was employed only to remove stumps and boulders beyond the ditcher's capacity, and to lower the pipe into the trench.[57] Although the laboratories of both the Lima (Solar) and Cleveland refineries were attacking vigorously the sulphur problem, the early activity at Lima was virtually a "skimming operation," *i.e.*, removal of the naphtha and gasoline light ends from the crude and a deodorizing treatment of the skimmed crude to eliminate the bad odor before shipping it to the fuel oil market which Standard promoted aggressively.[58]

---

[49] *Id.*; PETROLEUM PIPELINES, *supra* note 45, at 12; moreover, this "skunk oil," as it was called, produced a kerosene which had a slight odor, encrusted the wicks and clouded the chimneys of lamps. MCLEAN & HAIGH, *supra* note 35, at 67-68; 8 U.S.G.S. ANN. REP. 626-627 (1889). Crude oil with a substantial sulphur content generally is referred to as "sour" crude. S. REP. No. 94-1005, 94th Cong., 2d Sess. 7 n.2 (1976) [hereinafter cited as S.2387 REPORT].

[50] H. WILLIAMSON AND A. DAUM, THE AMERICAN PETROLEUM INDUSTRY, THE AGE OF ILLUMINATION 596-597 (1959) [hereinafter cited as AGE OF ILLUMINATION].

[51] For a contemporaneous map of this area, see AGE OF ILLUMINATION, *supra* note 50, at 590.

[52] PETROLEUM PIPELINES, *supra* note 45, at 12.

[53] R. HIDY AND M. HIDY, HISTORY OF STANDARD OIL COMPANY (NEW JERSEY): PIONEERING IN BIG BUSINESS 159 (1955) [hereinafter cited as HIDY & HIDY]; PETROLEUM PIPELINES, *supra* note 45, at 12; *cf.* AGE OF ILLUMINATION, *supra* note 50, at 597-598.

[54] AGE OF ILLUMINATION, *supra* note 50, at 597; MCLEAN & HAIGH, *supra* note 35, at 68.

[55] PETROLEUM PIPELINES, *supra* note 45, at 12.

[56] The consideration for the right-of-way was a guarantee of one-third of Standard's western shipment of refined products, the C&A also benefited from its revenue obtained from hauling materials for the pipeline. AGE OF ILLUMINATION, *supra* note 50, at 603; PETROLEUM PIPELINES, *supra* note 45, at 12.

[57] AGE OF ILLUMINATION, *supra* note 50, at 603-604. The so-called "screw-machine" apparently had been tried out on the Cygnet-Lima line. A somewhat later vintage machine is shown in illustration behind page 206 of PETROLEUM PIPELINES, *supra* note 45.

[58] *Id.* at 603. The method used was the Dufur process which cost about 1-½ cents per barrel.

The persistent, systematic attack on the sulphur problem produced the Frasch process, developed between 1888 and 1890, which enabled Standard to produce an acceptable refined product from the Lima crude, whereupon Standard constructed the then largest refinery in the world at Whiting, Indiana, (forming Standard of Indiana) to refine Lima crude for the Chicago, West, and Southwest markets, following the principle that it was cheaper to move the crude by pipeline to Whiting than it would be to ship the refined product by tank car from Lima.[59] These moves resulted in expansion of the pipeline system: the Buckeye eight-inch fuel line to Chicago was supplemented by a six-inch crude line to Whiting, ownership of both being placed in the Indiana Pipeline Company, and crude lines were laid east from Buckeye's Cygnet, Ohio, Terminal to Mantau, Ohio, where it joined National Transit's western extension connecting Bear Creek, Pennsylvania, with Cleveland.[60] By 1890, Buckeye was delivering Lima crude to Standard's refineries in Cleveland[61] which were in the process of being converted to run on sour crude. Once the conversion was completed these refineries operated on it exclusively throughout the rest of the decade.[62] In order to segregate the sour Lima crude from the "sweet" Appalachian crude, Standard organized the Northern Pipe Line Company, transferred to it the National Transit lines running through Kane and Colegrove, Pennsylvania,[63] and began running the highly sulphurous sour crude at its Olean, New York, refinery in 1893.[64] After sufficient experience with refining sour crude, Standard obtained a reversal of the New York Produce Exchange's ban on exports of illuminating oil refined from highly sulphurous crude[65] and Lima crude began moving to Standard's Bayonne, New Jersey, and Philadelphia refineries in 1895.[66] Thus, Standard's willingness to put its financial and managerial resources into the Lima-Indiana fields paid off handsomely for it, but the whole industry benefited from the development of the fuel oil market and the perfection of a method of refining sulphurous crude which could readily be applied elsewhere to newly discovered crude.[67]

As to the rest of the pipeline structure in the Appalachian Region, the

---

[59] *Id.* at 608-609; PETROLEUM PIPELINES, *supra* note 45, at 13.
[60] AGE OF ILLUMINATION, *supra* note 50, at 612; PETROLEUM PIPELINES, *supra* note 45, at 13-14.
[61] PETROLEUM PIPELINES, *supra* note 45, at 14; AGE OF ILLUMINATION, *supra* note 50, at 612.
[62] AGE OF ILLUMINATION, *supra* note 50, at 612.
[63] See *Id.* for map of Standard's trunk pipeline system circa 1900.
[64] *Id.*; PETROLEUM PIPELINES, *supra* note 45, at 14.
[65] HIDY & HIDY, *supra* note 53, at 287; AGE OF ILLUMINATION, *supra* note 50, at 613.
[66] AGE OF ILLUMINATION, *supra* note 50, at 613; PETROLEUM PIPELINES, *supra* note 45, at 14.
[67] AGE OF ILLUMINATION, *supra* note 50, at 613.

independent producers in the Pennsylvania Oil Region formed the Producers Oil Company, Limited, in 1891 to market their crude to independent refineries. Their first attempt was to build a terminal at Coraopolis, Pennsylvania, connected by a gathering line from the booming McDonald field, with the expectation of shipping crude by rail to the independent Columbia Oil Company refinery at Bayonne, New Jersey. It also was hoped that the Coraopolis Terminal would be able to ship by rail to independent refineries in Titusville and Oil City. In both instances, discriminatory railroad rates aborted the scheme.[68] A second effort, which was launched with the assistance of the independent refiners (the pooled company was known as the Producers and Refiners Oil Company, Limited), achieved partial success by constructing a pipeline from the Coraopolis Terminal to Titusville and Oil City.[69]

Encouraged by this accomplishment, the company, led by an old enemy of Standard Oil, Lewis Emery, Jr., conceived the idea of overcoming the geographical disadvantage of the independents' inland refineries by its use of pipelines to transport refined oil.[70] Products pipelines had been used before[71] but only for short distances from refineries to railheads.[72] A new company, the United States Pipe Line Company, was formed on September 20, 1892, and began to plan routes to achieve the desired end. After an attempt to duplicate the earlier success of Tidewater was jettisoned by the refusal of the Reading to transport the oil from Williamsport to seaboard, Emery sought to purchase right-of-way to Hancock, New York, on the extreme northeastern border of Pennsylvania, and there to connect with the New York, Ontario, and Western Railroad. Although the Pennsylvania Free Pipeline Act of 1883 facilitated acquisition of most of the right-of-way, this effort was also turned back when the Erie reneged on its promise to permit the line to cross its rail yard at Bradford and its right-of-way at Hancock.[73] Emery simply backtracked to Athens, Pennsylvania, and projected his lines southeast to Wilkes-Barre.

By June 1893, the United States Pipe Line Company had completed two four-inch pipelines to Wilkes-Barre, one delivering crude from Bradford, 180 miles away, and the other bringing refined oil from independents' plants

---

[68] JOHNSON, *supra* note 1, at 173.
[69] AGE OF ILLUMINATION, *supra* note 50, at 570; *cf.* JOHNSON, *supra* note 1, at 173.
[70] JOHNSON, *supra* note 1, at 173; AGE OF ILLUMINATION, *supra* note 50, at 571.
[71] In 1865, John Warren and Brothers Company operated a three-mile refined oil pipeline from their Osceola Refinery at Plumer to a rail junction at McMahon's run near Oleopolis. JOHNSON, *supra* note 1, at 6, 173-174.
[72] AGE OF ILLUMINATION, *supra* note 50, at 571.
[73] *Id.* at 572-573; JOHNSON, *supra* note 1, at 174-175.
[74] *Id.*

at Titusville, Oil City, Warren and Bradford.[74] Once again, an important technological development had occurred; three grades of kerosene were moved through the line at an approximate rate of 2,000 barrels per day and with only moderate intermixture or contamination.[75] Subsequently, the United States Pipe Line secured the newly formed Pure Oil Company as a shipper. After an ill-fated attempt to build to Bayonne, it swung south through Easton, Pennsylvania, down to Marcus Hook, just below Philadelphia on the Delaware River.[76] Eventually the pipeline was acquired by Pure Oil, along with virtually the majority interest in the Producers and Refiners Oil Company and practically all of the stock of the Producers Oil Company.

The final principal segment of the Appalachian pipeline system arose out of discoveries in Washington and Greene Counties in southwestern Pennnsylvania. William Mellon, a 21 year old nephew of Andrew Mellon, utilized this crude to build an export business to Europe. When the Pennsylvania Railroad suddenly increased its tariffs, Mellon organized the Crescent Pipe Line Company to construct a five-inch line from the McDonald field to Carlisle, Pennsylvania, where the Reading Railroad, then engaged in a competitive battle with the Pennsylvania, agreed to a favorable rate to carry the oil on to seaboard. However, by the time the line reached Carlisle, the President of the Reading had been abruptly discharged and the new President refused to honor the contract; hence, the Crescent continued the line on to Marcus Hook, delivering the first oil on November 7, 1892.[77]

Little more than local attention had been paid to the activities in what was to become known as the Mid-Continent Field, although oil had been discovered in 1892 near Neodesha, Kansas, by a transplanted Pennsylvanian named W. M. Mills.[78] Lacking capital, and unable to obtain any locally, Mills sold out to James M. Guffey and John Galey, a pair of Pennsylvania wildcatters, who, after laying some short pipelines and extending the Neodesha fields down into Indian Territory, found themselves with 150,000 barrels in storage, daily output of 1,800 barrels and no markets. They in turn sold out to the Forest Oil Company, a Standard Oil producing affiliate. Forest erected a 500-barrel per day refinery in Neodesha and expanded its leasing and exploration efforts, following up with extensions of the old

---

[75] AGE OF ILLUMINATION, *supra* note 50, at 573. See note 116, *infra*, for definition of contamination.

[76] See map of the system in AGE OF ILLUMINATION, *supra* note 50, at 579. BEARD, *supra* note 39, at 19 (1941) reports that the kerosene line from the Oil Creek region to Marcus Hook, a distance of approximately 350 miles, to be the only significant product line built by the 1920's.

[77] MCLEAN & HAIGH, *supra* note 35, at 72-73; AGE OF ILLUMINATION, *supra* note 50, at 583-584; JOHNSON, *supra* note 1, at 175-176.

[78] PETROLEUM PIPELINES, *supra* note 45, at 17.

Guffey and Galey pipeline. This increased activity sparked an oil boom in Kansas, and in short order oil and gas fields were reported at Paola, Osawatomie, Fort Scott, Wyandotte, Rosedale, Greeley, Iola, LaHarpe, Chanute, Peru, Cherryvale, Coffeyville, Humboldt, and Neodesha.[79] In view of the expanding nature of the business, Forest's properties were transferred to the Prairie Oil and Gas Company which bought and gathered oil in Kansas. When oil was discovered in Bartlesville, Indian Territory, in 1903, followed by Cleveland the next year, Prairie extended the lines acquired from Guffey and Galey and laid six-inch and eight-inch lines from Red Fork, Indian Territory, to its tank farm at Neodesha. Then came the famous Glenn Pool and neighboring districts in 1905-1906 which also were connected to Neodesha. Despite a five-fold increase in capacity at its Neodesha refinery, the system fell far short of satisfying Mid-Continent output, so Prairie laid an eight-inch line in 1904 from Humboldt to Sugar Creek, near Kansas City, where Standard of Indiana built a new refinery. Even this was not enough, so a second line was constructed, half the distance an eight-inch and the remaining half a twelve-inch to which additional eight-inch lines were added.[80] The final link came in 1905 when the 460-mile eight-inch line was completed from Humboldt, Kansas, to Griffith, Indiana, a few miles south of Whiting at a junction point with Standard's trunk lines running east to the seaboard, The value of this link was shortly evidenced by the building of a parallel twelve-inch line in 1906.[81] These activities increased the total pipeline mileage, which was approximately 6,800 miles in 1900[82] to over twice that amount in 1911[83] when the Mid-Continent fields supplied 30 percent of the refinery requirements east of the Mississippi.[84]

With the exception of limited production and refining at Corsicana, Texas, in the closing years of the 19th century, Texas crude production really did not begin until the famous Spindletop discovery by Patillo Higgins and Anthony Lucas in 1901. This 75,000 to 100,000 barrel per day gusher brought wildcatters in from all over the country. Sour Lake, Batson, Dayton, and Humble were discovered and pipelines were constructed from Spindletop to Beaumont, Port Arthur, and Sabine. These lines were extended to Sour Lake, Saratoga, Batson, Dayton, and when the Humble field in north Harris County came in during 1904 and 1905, the lines were extended

---

[79] C. RISTER, OIL! TITAN OF THE SOUTHWEST 38 (1949).
[80] PETROLEUM PIPELINES, *supra* note 45, at 18.
[81] For map of major oil fields and their pipeline connections in 1906, see *Id.* at 6.
[82] *Id.* at 1.
[83] H. WILLIAMSON, R. ANDREANO, A. DAUM & G. KLOSE, THE AMERICAN PETROLEUM INDUSTRY, THE AGE OF ENERGY 66 (1963) [hereinafter cited as AGE OF ENERGY]. See Table 3:1, giving yearly mileage of crude interstate trunk lines 1901-1920. *Id.*
[84] Reduced Pipeline Rates and Gathering Charges, 243 I.C.C. 115, 119 (1940).

to there and a new line was constructed by The Texas Company south to Houston.⁸⁵ The main contribution of the lower Gulf Coast "play" in this period was the emerging of three new competitors to Standard Oil: Gulf Oil (whose predecessors had a brief play in the old McDonald, Pennsylvania, field and had built the Crescent line to Marcus Hook before selling out to Standard), Sun Oil, which had started in the Lima-Indiana field, and The Texas Company (hereafter "Texaco"). A fourth major company, the Security Oil Company, apparently was backed secretly by Standard. At any rate, the State of Texas, proceeding against Security on antitrust grounds, tied up its tank cars bringing oil in from Oklahoma and forced it into bankruptcy in 1909.⁸⁶

At the turn of the century, California likewise had important oil production. However, in 1906, most long distance crude oil movement in that area was by rail or water.⁸⁷ Because of the intrastate nature of the movements, most pipelines that were built were operated as private carriers,⁸⁸ although by 1921, the bare bones of a significant pipeline system in California were already in place.⁸⁹ From 1900 to 1919, the lines were extended, connections made between systems, and the network began filling in. Technologically, larger diameter lines began to appear. Whereas five and six-inch lines were common during 1879-1882, and Buckeye had built an eight-inch line to connect Lima to Chicago, very few companies had the 20,000 barrels per day required to load such lines to capacity.⁹⁰ By the same token, except for one 25-mile eight-inch line between Spindletop and Sabine, the 1901-1905 Gulf Coast development remained with six-inch lines.

Once the Mid-Continent fields came in, nearly 1,000 miles of eight-inch lines were laid in a single year. This size became the norm. Standard of Louisiana used it when constructing Standard's first outlet to the Gulf Coast in 1909, as did Magnolia in serving the Electra, Texas, pool in 1914. Sinclair made its initial large entry into the pipeline field with an eight-inch line from the Mid-Continent to Chicago, and the California lines constructed during this period used the same.⁹¹ Occasionally, lines of even

---

⁸⁵ PETROLEUM PIPELINES, *supra* note 45, at 17. See map of principal trunk lines in southeast Texas in 1906, *Id.* at 16. Comparing costs of laying pipeline, the longest Gulf Coast line, a six-inch line from Humble to Port Arthur, Texas, involved an estimated maximum investment of $600,000, whereas the cost of reproducing the shortest single six-inch line from Colegrove to Philadelphia was estimated in 1904 to be about $1.3 million. AGE OF ENERGY, *supra* note 83, at 88.

⁸⁶ HIDY & HIDY, *supra* note 53, at 393 trace Security's assets into Magnolia Petroleum Co. (Mobil).

⁸⁷ PETROLEUM PIPELINES, *supra* note 45, at 19.

⁸⁸ See Associated Pipe Line Company v. Railroad Commission of California, 176 Cal. 518, 169 Pac. 62 (1917).

⁸⁹ See Map 3.1 in AGE OF ENERGY, *supra* note 83, at 68.

⁹⁰ *Id.* at 70.

⁹¹ *Id.*

larger diameter began to appear. The Prairie's twelve-inch section in its looped line from Humboldt, Kansas, to Griffith, Indiana, previously has been noted.[92] California Standard employed some twelve-inch pipe on the West Coast and in 1918 Shell Oil Company laid a 428-mile ten-inch line connecting Cushing, Oklahoma, to Wood River, Illinois.[93]

The second important development of this era was the replacement of the steam pump by the diesel-engine drive pump, except in California where noncondensing steam engines were used, employing the steam exhaust from the pump to preheat the heavy, viscous oil before it entered the line.

The principal events which led to the tie-in between the Gulf Coast and the Mid-Continent Regions were the coincidental decline of the last large Southeast Texas Pool, Humble, with the fantastic outturn of the Glenn Pool in Oklahoma. Gulf and Texaco both laid eight-inch lines in 1907 from Glenn Pool to their refineries in the Sabine area. Between them they ran nearly 20 percent of Kansas–Oklahoma production in 1907.[94] Their success was a key factor in Standard's decision to build southeast from Glenn Pool to Baton Rouge, separately incorporating the Oklahoma Pipeline Company for movement through Oklahoma, Prairie Oil and Gas Company (through Arkansas) and the Louisiana Pipeline Company from the Arkansas–Louisiana border to a newly constructed refinery at Baton Rouge.[95] Then came Cushing and Healdton, Oklahoma, which, based on a comparable time span, dwarfed all other Mid-Continent fields to that date. Even with Texaco and Gulf running to capacity along with Prairie, there was more than they could take. Magnolia extended its eight-inch line running from Electra and Henrietta, Texas, where its sizeable production was stabilizing, to Corsicana, the site of the old Navarro Refining Company's refinery which Magnolia had taken over, thence to its large refinery at Beaumont. Later on, it purchased an eight-inch line which had been built by the McMann Oil Company from Cushing to Healdton. Sinclair Oil and Refining Company, formed by the merger of some of the larger field refiners and nonintegrated producers in 1916, somewhat akin to the old Producers and Refiners Oil Company in the early Pennsylvania play, began laying an eight-inch line from Drumright, near Cushing, to East Chicago, Indiana, and announced plans for new refineries at Humboldt, Kansas,

---

[92] See text at note 81, *supra*.
[93] K. BEATON, ENTERPRISE IN OIL 144-145 (1957).
[94] AGE OF ENERGY, *supra* note 83, at 92-93.
[95] This device of separately incorporating in each state failed to defeat the imposition of common carrier status. In the Matter of Pipe Lines, 24 I.C.C. 1 (1912), Prairie Oil & Gas Co. v. United States, 204 Fed. 798 (Comm. Ct. 1913); The Pipe Line Cases, 234 U.S. 548 (1914). For a history of this battle, see WOLBERT, *supra* note 1, at 117-132; BEARD, *supra* note 39, at 28-55 (1941).

and East Chicago. Shortly thereafter, Shell Oil's Yarhola Pipe Line Company placed the Healdton to Wood River, Illinois, ten-inch line in service. As noted above, it had already commenced the Cushing-Wood River leg, the additional 126 miles to Healdton was completed and the entire line was in service by August, 1918.[96]

Two exogenous events occurred in this era which deserve mention at this point, although they will be considered in more detail later. The first is the holding of the United States Supreme Court in *The Pipe Line Cases*[97] that the plain meaning of the 1906 Hepburn Act[98] amending the Interstate Commerce Act[99] was to make all pipelines engaged in the business of transporting oil in interstate commerce common carriers. Justice Holmes construed the word "transportation" in such a way as to avoid the constitutional issue of making all existing pipelines common carriers. His construction excluded from the Act the Uncle Sam Oil Company which had a refinery in Kansas, producing wells just across the Oklahoma line, and an interstate pipeline connecting the two.[100] However, the practical thrust of the decision was to make all large interstate crude lines and multiple owned or used interstate products pipelines common carriers subject to full Interstate Commerce Commission regulation, *e.g.*, to conduct their operations, make their services available to, and deal with, all shippers on a completely nondiscriminatory basis and to file just and reasonable tariff rates.[101]

The second (first, chronologically) of these events was the dissolution of the Standard Oil Company (New Jersey) ordered by the United States Supreme Court in 1911 after an explicit finding of illegal monopolization under the Sherman Act.[102] The Court affirmed the Missouri Circuit (District) Court's decree which, among other things, directed that the holding company be dissolved and that each shareholder receive a proportionate amount of stock in each of the 33 successor companies. Ten of the companies were common carrier pipeline companies and three were partially or wholly integrated companies owning pipelines.[103] It is interesting to note

---

[96] K. BEATON, ENTERPRISE IN OIL 145 (1957). For a map of the lines emanating from the Glenn Pool-Cushing area, see AGE OF ENERGY, *supra* note 83, at 101.
[97] 234 U.S. 548 (1914).
[98] 34 Stat. 584 (1906), as amended, 49 U.S.C. § 1 (1976).
[99] 24 Stat. 379 (1887), as amended, 49 U.S.C. §§ 1-27 (1976).
[100] "It would be a perversion of language, considering the sense in which it is used in the statute, to say that a man was engaged in the transportation of water whenever he pumped a pail of water from his well to his house. So as to oil when, as in this case, a company is simply drawing oil from its own wells across a state line to its own refinery for its own use, and that is all, we do not regard it as falling within the description of the act, the transportation being merely an incident to use at the end." The Pipe Line Cases, 234 U.S. 548, 562 (1914).
[101] 49 U.S.C. §§ 1(4), 3(1), and 1(5) (1976).
[102] Standard Oil Co. v. United States, 221 U.S. 1 (1911).

the dissolution was along corporate lines; [104] vertical divorcement [105] was not a target remedy. The resultant vertical separation was more by accident than design, [106] operating on those companies which were not vertically integrated but which had relied upon former members of the Standard Oil group to perform the other functional operations necessary to find and produce the crude, provide the necessary transportation to and from an affiliated refinery to the consuming public. For a period of time, these companies achieved the same result with the same companies via the market route. However, enforced managerial independence and changing conditions, not the least of which was competition from the integrated oil companies that had diminished the Standard Oil group's share of the market appreciably prior to the dissolution, [107] gradually caused the severed companies to form vertically integrated

---

[103] HOWREY & SIMON, *supra* note 3, app. at 9. The pipeline companies were: Buckeye, Crescent, Cumberland, Eureka, Indiana, National Transit, New York Transit, Northern, Southern, and South-West Pennsylvania Pipe Lines; the integrated owners were Standard Oil Company (California), Ohio Oil Company and Prairie Oil & Gas Company. PETROLEUM PIPELINES, *supra* note 45, at 65.

[104] G. WOLBERT, *The Recurring Spectre of Pipeline Divorcement*, in OIL'S FIRST CENTURY 108 (R. Hidy ed 1960) [hereinafter cited as SPECTRE]. Nothing in the opinion of the Supreme Court or that of the Circuit Court, 173 Fed. 177 (C.C.E.D. Mo. 1909) indicated disapproval of vertical integration of pipelines. HALE & HALE, MARKET POWER: SIZE AND SHAPE UNDER THE SHERMAN ACT 205 (1958) [hereinafter cited as MARKET POWER]; Hale, *Vertical Integration: Impact of the Antitrust Laws Upon Combinations of Successive Stages of Production and Distribution*, 49 Col. L. Rev. 921, 924 (1949) [hereinafter cited as Hale, *Successive Stages*].

[105] The originator of this term appears to be the colorful Attorney General of Oklahoma, Charles West, in his testimony before the House Committee on Interstate and Foreign Commerce in support of H.R. 16581, 63d Cong., 2d Sess. (1914), commonly known as the Davenport Bill. Mr. West, who was mounting a "white charger" campaign to become Governor, already had brought the so-called "sand bag" cases against The Texas Company, Prairie Oil and Gas, and Magnolia Pipe Line Company for alleged violation of Oklahoma law prohibiting a company from owning another company in the same line of business. He had also supported oil procedures in his state by pressuring the Oklahoma Corporation Commission to take action against purchasing companies. PETROLEUM PIPELINES, *supra* note 45, at 106. When the hearings on the Davenport Bill were announced, West saw yet another opportunity to spur his steed closer to the Governor's mansion. After making the point that common carrier pipeline transportation would not solve the problems of his producer constituents unless there was storage and a use for the oil at the far end of the line, he concluded that "divorcement" of pipelines from production was a *sine qua non*: "That divorce, I think, cannot be brought about by requiring directly that the pipe lines engaged in interstate commerce shall not be engaged in production. I think you will have to take the other horn of the dilemma and provide that no pipe line engaged in interstate commerce shall transport oil as to which it is interested in the production." *Id.* at 107-108.

[106] MARKET POWER, *supra* note 104, at 206; Bork, *Vertical Integration and the Sherman Act: The Legal History of an Economic Misconception*, 22 U. Chi. L. Rev. 157, 160 n.22 (1954) [hereinafter cited as Bork, *Vertical Integration — Economic Misconception*]; Hale, *Successive Stages, supra* note 104, at 924.

[107] HIDY & HIDY, *supra* note 53, at 120, 417, 474; H. LARSON, *The Rise of Big Business in the Oil Industry* in OIL'S FIRST CENTURY 38-42 (R. Hidy ed. 1960); H. WILLIAMSON & R. ANDREANO, *Competitive Structure of the American Petroleum Industry* in OIL'S FIRST CENTURY 73-84 (R. Hidy ed. 1960).

units in their own self-interest and to compete vigorously against their former affiliates.[108]

From 1919 to 1930, total pipeline mileage virtually tripled to over 115,000 miles, gathering lines more than doubled to a total of 53,600 miles and refined products lines in their first year of real operation increased to 3,499 miles.[109]

Before the 1930's, when high-strength big-inch welded-steel pipe was developed, increased capacity was obtained by laying an additional line alongside ("looping") the original line. As noted previously,[110] eight-inch pipe was the largest pipeline which could be operated at the normal operating pressures of that period. True, there were ten- and twelve-inch lines laid during the period, but they tended to split at the seams unless operated at a less-than-desirable operating pressure.[111] The economic capacity of these lines was approximately 20,000 barrels per day for the eight-inch lines and 60,000 barrels per day for the twelve-inch. The larger refineries of this period had input capacities of 80,000 to 125,000 barrels per day. Thus, the refiner of that era desperately needed the full capacity of the pipelines he designed, financed and built; there was no economic incentive to seek out non-proprietary traffic. To the contrary, movement for others would require the construction of additional pipelines, which served to increase, rather than lower, his own cost of transportation.[112]

Once again, economics and technology produced a dramatic change in the pipeline segment of the business in the late 1920's and early 1930's. The railroads, which constituted the dominant mode of delivering products to market from field-oriented refineries, had raised their rates steadily until they had reached unacceptably high levels.[113] When Mid-Continent refiners considered a cooperative gasoline pipeline venture in the 1920's to overcome this transportation cost disadvantage, they had to surmount the problems of loss of quality specifications (such as octane rating, boiling point, vapor pressure, etc.), leakage, and fire.[114] The solution to the first problem came with the issuance by the Bureau of Mines of master specifications for liquid

---

[108] A. JOHNSON, *Lessons of the Standard Oil Divestiture* in VERTICAL INTEGRATION IN THE OIL INDUSTRY 192 (E. Mitchell ed. 1976) [hereinafter cited as JOHNSON, *Lessons*].

[109] PETROLEUM PIPELINES, *supra* note 45, at 143. See map of major trunk pipelines in 1930. *Id.* at 144.

[110] See text at note 91, *supra*; *cf.* SPECTRE, *supra* note 104, at 121.

[111] Burke, *supra* note 9, at 784; R. LENNART, *Pipeline Orientation* in AOPL EDUCATORS CONFERENCE 7 (1978) [hereinafter cited as LENNART]; *cf.* U.S. Dept. of Justice, Third Report of the Attorney General Pursuant to Section 2 of the Joint Resolution of July 28, 1955, Consenting to an Interstate Compact to Conserve Oil & Gas 49 (Sept. 1, 1950) [hereinafter cited as Att'y Gen., *First Report, Second Report*, etc.].

[112] Burke, *supra* note 9, at 784; LENNART, *supra* note 111, at 8.

[113] HOWREY & SIMON, *supra* note 3, app. at 17; PETROLEUM PIPELINES, *supra* note 45,, at 251-254.

[114] PETROLEUM PIPELINES, *supra* note 45, at 254.

fuels and their testing[115] and experimentation indicated that contamination[116] was more susceptible to control than had been thought.[117] The leakage problem was made manageable by the development of improved seamless and welded pipe[118] and the introduction of electric welding of its joints.[119] Moreover, the technique of operating a "closed system[120] by the use of automatic controls had already been demonstrated by Shell in the operation of its Ventura to Wilmington, California, natural gasoline line.[121] All that remained was for someone to do it. As chance would have it, Jersey Standard's subsidiary, Tuscarora Pipe Line Company, which for years had received Prairie-originated oil at the Ohio-Pennsylvania border and carried it to Standard's Bayonne and Bayway refineries, had reached the end of its economic usefulness despite Jersey's favoring it over the then independent New York Transit Company.[122] Instead of scrapping the line, Jersey caused Tuscarora to reverse its flow and convert to the carriage of gasoline from the New Jersey refineries across the heartland of Pennsylvania to Pittsburgh.[123] This spurred Sun to build a gasoline pipeline from its Marcus Hook refinery near Philadelphia, north and west to Pittsburgh and Cleveland, with a branch running north to Syracuse.[124] Threatened by this invasion of one of its primary markets, the Atlantic Refining Company incorporated the Keystone Pipeline Company and laid 226 miles of eight-inch gasoline line from its Philadelphia refinery to Mechanicsburg, Pennsylvania,[125] across the Susquehanna River from Harrisburg. The initial system was such a financial and operating success that during 1935-1937 it was extended west to Pittsburgh and north to the New York State line and thence, by the newly-formed

---

[115] U.S BUREAU OF MINES, *United States Government Specification for Lubricants and Liquid Fuels and Methods for Testing,* Technical Paper 323A (1924).

[116] "Contamination" is the commingling or intermixing of two fluids in the region of contact as they are pumped through a pipeline. Birge, *Contamination Control in Products Pipe Lines*, OIL & GAS JOURNAL, September 20, 1947, p. 176 [hereinafter cited as Birge].

[117] PETROLEUM PIPELINES, *supra* note 45, at 254.

[118] *Id.*; HOWREY & SIMON, *supra* note 3, app. at 18; *cf.* Burke, *supra* note 9, at 784.

[119] Petroleum Rail Shippers' Ass'n v. Alton & Southern R.R., 243 I.C.C. 589, 599-600 (1941); PETROLEUM PIPELINES, *supra* note 45, at 254.

[120] A "closed" system is one in which the discharge of one station goes directly into the suction of the next as contrasted to an "open" system which operates with each station discharging into a tank at its downstream station, which takes suction therefrom and discharges into a tank further downstream. WOLBERT, *supra* note 1, at 31 n.153.

[121] PETROLEUM PIPELINES, *supra* note 45, at 254-255; HOWREY & SIMON, *supra* note 3, app. at 18.

[122] G. GIBB and E. KNOWLTON, HISTORY OF THE STANDARD OIL COMPANY (NEW JERSEY): THE RESURGENT YEARS 469-470 (1956) [hereinafter cited as GIBB & KNOWLTON].

[123] AGE OF ENERGY, *supra* note 83, at 348; PETROLEUM PIPELINES, *supra* note 45, at 255.

[124] PETROLEUM PIPELINES, *supra* note 45, at 255; AGE OF ENERGY, *supra* note 83, at 576.

[125] *Id.*

Buffalo Pipe Line Corporation, to Buffalo and Rochester.[126] The economies of the line not only enabled Atlantic to remain competitive in its "home" territory in Pennsylvania, but it also permitted it to expand in the western New York market[127] which previously it had served marginally by barge over the New York State Barge Canal.[128] In 1933, only a year after start-up of the line, Atlantic began to move kerosene and furnace oil in addition to the two grades of gasoline with which it had commenced operation.[129] Both the Keystone and its connecting Buffalo pipeline developed a profitable "outside" shipper business.[130] The economic import was unmistakable: Tuscarora's rates were only 30 to 40 percent of rail rates to comparable points. The Susquehanna-Sun Oil system moved gasoline over its line at only ⅓ the cost per ton-mile of railway delivery[131] and Keystone, based on somewhat later data, ran about 10 to 15 percent of the rail rate.[132] Socony-Vacuum, a unit formed by two former affiliates of the Standard Oil Company, whose re-combination was approved by the same court by whom they were previously divorced,[133] brought to New England its first products line in 1931 by laying an 85-mile line from its Providence plant to Springfield and Worcester, Massachusetts.[134]

The Mid-Continent refiners, who had been discussing the possibility of building a products pipeline to the Great Lakes and south to New Orleans[135] but had put their project on the back burner,[136] now had not only another decade of rising railroad rates but the encouragement of example.

Phillips Petroleum Company was the first Mid-Continent refiner to commence construction of a products pipeline. Although it was still interested in the proposed jointly-owned line, it felt that it could wait no longer and it had a somewhat different route in mind, so in 1930 it decided to connect its Borger Refinery, located in the Texas Panhandle, with its marketing

---

[126] MCLEAN & HAIGH, *supra* note 35, at 208; PETROLEUM PIPELINES, *supra* note 45, at 255.
[127] *Id.*
[128] MCLEAN & HAIGH, *supra* note 35, at 208-209.
[129] *Id.* at 209; PETROLEUM PIPELINES, *supra* note 45, at 255-256. This feat was made possible by the construction of a ½ inch pyrex glass tubing erected on uprights in such a manner as to reproduce the profile elevations of the Keystone line. See Williams, *Pumping Various Products Through the Same Pipe-Line System*, OIL & GAS JOURNAL, September 22, 1945, p. 197 The correlation between the data obtained from the model and the actual eight-inch line was extremely close. WOLBERT *supra* note 1, at 33 n.172.
[130] MCLEAN & HAIGH, *supra* note 35, at 208-209; PETROLEUM PIPELINES, *supra* note 45, at 256.
[131] AGE OF ENERGY, *supra* note 83, at 576.
[132] Calculated from data in MCLEAN & HAIGH, *supra* note 35, at 209.
[133] United States v. Standard Oil Company, 47 F.2d 288 (E.D.Mo. 1931).
[134] OIL AND GAS JOURNAL, January 29, 1931, p. 11.
[135] OIL AND GAS JOURNAL, February 27, 1920, p. 76; *Id.*, March 12, 1920, p. 56.
[136] PETROLEUM PIPELINES, *supra* note 45, at 256.

facilities in Wichita, Kansas City, and St. Louis. Accordingly, it formed the Phillips Pipe Line Company which built a 681 mile eight-inch gasoline pipeline connecting these points, coming onstream and in full operation by early 1931.[137]

The rest of the Mid-Continent refiners[138] followed shortly. Drawing from the precedents of the Texas-Empire[139] and Ajax[140] jointly-owned crude lines, they incorporated the Great Lakes Pipe Line Company in 1930, which constructed and placed in operation the first portion of the line from Ponca City, Oklahoma, to Kansas City in early 1931[141] and extended the line to Chicago by July of that year.[142] The backbone of the Great Lakes system originated in Tulsa, where branches connected refineries at Ponca City, Muskogee, Okmulgee, Tulsa, and Barnsdall, Oklahoma, to Minneapolis-St. Paul, with a branch from Osceola, Iowa, to Omaha and another from Des Moines to Chicago.[143] The line had an interconnection with the Phillips line at Paola, Kansas, just south of Kansas City, enabling Phillips to move its products north along the Great Lakes system and the "Tulsa" group to ship eastward to St. Louis.[144]

---

[137] *Id.*; HOWREY & SIMON, *supra* note 3, app. at 19. Phillips extended its line from St. Louis to Chicago in 1938. OIL AND GAS JOURNAL, October 26, 1939, p. 103.

[138] Continental Oil and Barnsdall were the original parties but they were soon joined by Skelly, Mid-Continent Petroleum, Phillips, and Pure. In the summer of 1933, Texaco and Sinclair joined the ranks of owners, PETROLEUM PIPELINES, *supra* note 45, at 256. Cities Service came in as an owner in 1938. AGE OF ENERGY, *supra* note 83, at 579.

[139] Texas-Empire, incorporated in Delaware as a common carrier in 1928, was owned 50/50 by Texaco and the Empire Gas & Fuel Company, a subsidiary of Cities Service. Its purpose was to serve the refineries of affiliated companies at Lawrenceville, Lockport and East Chicago, Illinois. In order to achieve the economies of scale, Texaco and Cities Service, with the same objectives in mind, *i.e.*, to bring crude from the flush Mid-Continent fields to market oriented refineries in the Chicago region rather than refine in the Mid-Continent and pay the high railroad rate for shipping refined products to Chicago, formed Texas-Empire which constructed a 12-inch line from Cushing, Oklahoma, to the Missouri-Illinois border and thence to Chicago via a subsidiary, Texas Empire Pipe Line Company (Illinois). The sharing of the financial burden of the $17,000,000 cost was an obvious benefit. PETROLEUM PIPELINES, *supra* note 45, at 140. One could say that this was the first jointly owned line in the sense of planning, formation, financing, and construction, although Standard of Indiana and Sinclair Consolidated Oil Corporation had achieved a similar result in 1921 when Standard of Indiana purchased a half interest in the Sinclair Pipe Line System. *Id.* at 131.

[140] In 1930, Ajax Pipe Line Corporation, a holding company, and Ajax Pipe Line Company, an operating company, were incorporated with Jersey Standard owning 53 percent, Pure 24 percent, and Sohio 23 percent. Jersey advanced the money for construction. The "takeout" loan was secured by a forerunner of today's "throughput" agreement, *viz.*, agreements by the participants to purchase or cause to be purchased specified amounts of crude oil under a five year contract with the Carter Oil Company, Jersey's Mid-Continent producing affiliate, for transmission through the Ajax system which was to be twin ten-inch lines originating at the Oklahoma Pipe Line Company station at Glenn Pool, Oklahoma, and terminating at the Wood River, Illinois, station of Illinois Pipe Line Company. PETROLEUM PIPELINES, *supra* note 45, at 140-142; MCLEAN & HAIGH, *supra* note 35, at 244-246; *see* Ajax Pipe Line, 48 I.C.C. Val. Rep. 153, 172 (Div. 1, 1939).

[141] PETROLEUM PIPELINES, *supra* note 45, at 256; HOWREY & SIMON, *supra* note 3, app. at 20.

[142] PETROLEUM PIPELINES, *supra* note 45, at 256.

[143] See map of system. *Id.* at 257.

[144] PETROLEUM PIPELINES, *supra* note 45, at 257-258.

Not to be outdone by its larger competitors, the Champlin Refining Company, located at Enid, Oklahoma, built its own 250-mile six-inch products line from its Enid refinery to Hutchinson, Kansas, and Superior, Nebraska, in 1935, subsequently extended to Rock Rapids, Iowa.[145]

Professor Arthur M. Johnson, after reviewing the factual evidence, commented that the savings that accrued to the oil companies by virtue of their ownership of gasoline lines appeared to have been passed on to the consumer.[146] He established the causal relationship by comparing the yearly average tank wagon prices for six cities in Pennsylvania served by the Tuscarora, Susquehanna, and Keystone lines against 50 representative cities in the U.S. and a similar comparison for eight cities in the Great Lakes Pipe Line territory compared to 50 representative cities; the timing ruled out accidental occurrence.[147]

From 1931 to 1940, the number of pipeline companies reporting to the ICC increased from 49 to 66. By 1940, 100,000 miles of trunk and gathering lines transported nearly 283,000,000 barrel miles of crude oil, and 23,700,000 barrel miles of refined products. Most of the gathering line was from two- to four-inch; most trunk line was eight-inch.[148] The contours and flow directions of the nation's pipeline system were well established by the end of that period, reflecting the great distances separating crude producing regions, refinery areas, and consuming markets. Crude trunk lines radiated from interior points in Texas, Louisiana, and the Mid-Continent, south to the Gulf and north to major Mid-Western refineries. But practically all of the petroleum consumed by the Eastern Seaboard was transported by tanker, with almost 90 percent having its origin in Texas.[149]

At the threshold of World War II, two major products pipeline systems were planned, designed, and making attempts to secure the necessary right-of-way. Southeastern Pipe Line (formed by Gulf and Pure) was to run 462 miles from Port St. Joe, located on the west coast of Florida, north to Tennessee with numerous branches in Georgia. Plantation (Standard of New Jersey, Standard of Kentucky and Shell) was to run 812 miles from Baton Rouge, Louisiana, to Greensboro, North Carolina, via Birmingham and Atlanta, with some 450 miles of lateral lines to intermediate points.[150]

---

[145] Champlin Refining Co. v. United States, 329 U.S. 29 (1946); *see* Petroleum Rail Shippers' Ass'n v. Alton & Southern R.R., 243 I.C.C. 589, 600 (1941).

[146] PETROLEUM PIPELINES, *supra* note 45, at 258.

[147] *Id.* at 259-260, Charts 1 and 2.

[148] *Id.* at 308. It should be noted that use of ICC reported mileage and capacity tends to understate the total measurement as it omits intrastate, nonreporting lines. THE PETROLEUM ENGINEER, THE PETROLEUM DATA BOOK H-13 (1947) gives the mileage as 125, 950 miles. However, at this point, for a mere proxy of magnitude of growth, the ICC figure is adequate.

[149] *Id.* at 307-308; HOWREY & SIMON, *supra* note 3, app. at 22-23.

[150] PETROLEUM PIPELINES, *supra* note 45, at 308-309; HOWREY & SIMON, *supra* note 3, app. at 23.

Georgia had no eminent domain law, so the railroads, which were fighting to preserve their traffic, sought to block every attempt by the lines to get across the state. After an unsuccessful attempt to get the Georgia legislature to enact an eminent domain law, the lines' backers, aided by Secretary of Interior Ickes and Navy Secretary Knox, induced the Congress to enact the Cole Act [151] on July 1, 1941, which enabled interstate pipelines to exercise the right of eminent domain when the President determined that such action was in the interest of the national defense. President Roosevelt ruled that the Southeastern and Plantation lines were necessary for the war effort, so both lines proceeded quickly to completion, Southeastern by the fall of 1941 and Plantation in the early part of 1942.[152]

A month after Pearl Harbor, German U-boats appeared off the East Coast in large numbers and by May, 1942, some 55 tankers of the Gulf Coast — eastern seaboard runs had been sunk, reducing tanker deliveries to the area to about 173,000 barrels per day, less than 1/5 of the pre-Pearl Harbor shipments.[153] Pipeline industry leaders, acting as a special subcommittee of the Petroleum Administration for War (PAW), devised a plan to bring deliveries to the eastern seaboard (District 1) up to prewar levels. Basically, the scheme utilized loops, line reversals, and new "bridge" lines to recreate the routes that had functioned before tankers had displaced the Gulf-East Coast business in the 1920's, expansion of the capacity of the newly built

---

[151] 55 Stat. 610, as amended, 15 U.S.C.A. note prec. § 715. The bill became effective on July 30, 1941. The Act swept broader than merely granting eminent domain. It provided for government aid for privately built pipelines needed for natural defense and authorized the government to undertake essential pipelines too costly for private interests.

[152] PETROLEUM PIPELINES, *supra* note 45, at 310, HOWREY & SIMON, *supra* note 3, app. at 24.

[153] AGE OF ENERGY, *supra* note 83, at 762-763.

[154] The Bayou System was the first in the United States to be organized as an undivided interest line. The distinctions between a stock ownership and an undivided interest line will be developed in greater detail later, suffice it to say here that because Bayou operated in Texas it had to comply with the provisions of Art. 1502 Tex. Rev. Civ. Stat., then in force, which forbad a corporation engaged in the oil and gas producing business from owning the stock of more than one pipeline corporation organized under the laws of Texas or of some other single state. Thus, for example, Shell Oil, which was engaged in the oil and gas producing business and owned the stock of Shell Pipe Line Corporation, could not own stock in Bayou had it been incorporated. Because most of the other participants were similarly situated, the arrangement took the form of the various pipeline subsidiary companies owning a percentage of the assets (and capacity) of the Bayou System, each underwriting its share of the construction costs and all operating costs except power and oil losses, which were apportioned on a barrel mile basis. SPECTRE, *supra* note 104, at 122. The Bayou System documents were submitted by the P.A.W. to the Department of Justice for clearance which, after several changes suggested by the Department were adopted, informed the general counsel and the Director of Transportation and Supplies of the P.A.W. that nothing in the proposed plan and contract was in violation of the antitrust laws. *Hearings on S. Con. Res. 31, et al, Consumer* Energy Act of 1974, *Before the Sen. Comm. on Commerce*, 93d Cong., 1st Sess., Ser. No. 93-63, Pts. 1-4 629 (1973-1974) (Statement of Jack Vickrey) [hereinafter cited as *Consumer Energy Act Hearings*]. For a concise differentiation between corporate form jointly owned lines and undivided ownership lines, see *Id.* at 626-627. For greater detail, see text at Section IV C, *"Jointly Owned and Jointly Operated Pipelines,"* infra.

Plantation line by the addition of pump stations and lengthening the system at both ends—adding 60,000 barrels per day by constructing the Bayou system[154] from Houston, via Beaumont, to Baton Rouge, using pipe reclaimed from unused West Texas lines, and extending Plantation from Greensboro to an inland waterway northwest of Norfolk, using secondhand eight-inch pipe.[155] There were other "jury-rigged" improvements, but the key proposal was the construction and operation of the "Big Inch," a 24-inch crude line from Longview, Texas, to Phoenixville, Pennsylvania, where the stream would be split, part going to New York and part to Philadelphia,[156] and the "Little Big Inch," a 20-inch products line from Beaumont, Texas, to Linden, New Jersey.[157]

Although pipeliners had been aware of the theoretical economies of large lines even prior to the war,[158] there was no pressure to force the technology because at that time the industry as a whole, let alone individual companies, lacked the traffic to support them.[159] However, the explosion of civilian demand for petroleum products which had been suppressed by the war effort, together with the shortage of steel, literally forced the use of the now proven "big inch" technology. More and more of the new lines were big inch.[160] The tremendous initial outlay of funds involved, and the fact that the volume of traffic required to realize fully the economies of scale exceeded what a single refiner could muster, caused more and more of these lines to be jointly owned and operated.[161] Unlike the situation referred to previously,

---

[155] PETROLEUM PIPELINES, *supra* note 45, at 315. This was subsequently extended to Richmond. *Id.* at 320.

[156] *Id.* at 323-324. The first leg was Longview to Norris City, Illinois, from whence tank cars and a tie-in to the Marathon pipeline took it the rest of the way. The pipe used was seamless, made by the National Tube Co. The first oil was received at Norris City in February 13, 1943, which represented remarkable speed inasmuch as the first pipe was not shipped from National's mills until July 18, 1942. The second segment started in December 1942, and oil arrived in Phoenixville Junction, Pa., on August 14, 1943. *Id.* at 323-324. The main line length was 1,254 miles; the design capacity was 300,000 B/D, but it actually achieved an average of 317,000 B/D. *Id.* at 325. W.E.P. was a nonprofit corporation organized with government sponsorship by 11 companies: Cities Service Co., Standard Oil Co. (N.J.), Atlantic Pipe Line Co., Gulf Refining Co., The Texas Co., Socony-Vacuum Oil Co. Inc., Sinclair Oil Corp., Shell Oil Co., Inc., Pan American Petroleum & Transit Co., Tidal Pipe Line Co., and Sun Pipe Line Co. (Texas). *Id.* at 322.

[157] *Id.* at 324-325. Again, there were two discrete construction jobs—Beaumont to Seymour, Indiana, then Seymour to Linden, New Jersey. For map of these lines, see *Id.* at 321. The main line length was 1,474 miles and design capacity was 235,000 B/D. *Id.* at 325.

[158] A technical paper given before the American Institute of Mining & Metallurgical Engineers in 1942 advocated the economies of 24-inch lines. Hill, *Engineering Economics of Long Petroleum Pipelines* in PETROLEUM DEVELOPMENT & TECHNOLOGY 231, 233-234 (1942); SPECTRE, *supra* note 104, at 121; L. COOKENBOO, CRUDE OIL PIPELINES AND COMPETITION IN THE OIL INDUSTRY 134-135 (1955) [hereinafter cited as CRUDE OIL PIPELINES].

[159] CRUDE OIL PIPELINES, *supra* note 158, at 138; SPECTRE, *supra* note 104, at 121.

[160] SPECTRE, *supra* note 104, at 121. See list of lines in *Id.* at n. 222.

when nonproprietary traffic was unwanted, by now the economic incentive had shifted strongly to seek to broaden ownership as much as possible and to solicit vigorously the traffic of those not interested in taking an ownership position.[162]

A number of technological problems and improved techniques for dealing with them had developed as a result of the war and of its aftermath. One of these arose from the fact that many refineries had been forced to develop some degree of capability to handle sour crude. Post-war demand for varied products resulted in various specialized refining processes which required the pipelines to handle crudes with unusual characteristics which required segregation from other crudes. "Batching" will be discussed in a subsequent section but suffice it to say here that the days of operating automatically on a full "common stream" basis had passed. But improvements in pipe manufacturing; the internal line-up clamp; hydraulic bending machines for large diameter thin-wall pipe; matched by similar improvements in centrifugal pumps, which are more amenable to automatic and remote controls; communications improvements such as microwave radio systems; scheduling of product movement; metering and testing; and ever-improving safety devices had enabled pipelines to resolve these problems, maintain their efficiency and provide a basis for solving the problems incurred in new frontiers.[163].

At the end of the war, the pipeline mileage approximated 150,000 miles, with the average length of hauls in 1946 for gathering lines being 18 miles, 325 miles for crude trunk lines, and 382 miles for products lines.[164]

Professor Johnson categorized post World War II pipeline growth up to 1959 into two "rounds." His first period, which runs through the Korean war, was characterized by disproportionate growth of products lines, a dramatic shift in the diameters of crude trunk lines (part of which was replacement of older looped smaller lines with the more efficient larger lines), increased tie-in of the West Texas area to both the Gulf Coast and mid-west, movement of Canadian crude to the west coasts of Canada and the United States, and southeast through Canada to upper mid-western states. One of the "independent" pipelines, Buckeye, under the progressive leadership of

---

[161] SPECTRE, *supra* note 104, at 121-122; LENNART, *supra* note 111, at 8.

[162] SPECTRE, *supra* note 104, at 122; Burke, *supra* note 9, at 784, 790; *Hearings, Consent Decree Program of the Department of Justice Before the Antitrust Subcommittee of the House Committee on the Judiciary, 85th Cong., 1st Sess., Ser. No. 9, Oil Pipelines,* pt. 1 at 1135 (Statement of J. L. Burke) [hereinafter cited as *Celler Hearings*]; *cf.* PETROLEUM PIPELINES, *supra* note 45, at 386-387, 474; AOPL, PIPELINE TRANSPORTATION 41 (rev'd ed. 1976) [hereinafter cited as PIPELINE TRANSPORTATION].

[163] PETROLEUM PIPELINES, *supra* note 45, at 353; HOWREY & SIMON, *supra* note 3, app. at 28-29.

[164] WOLBERT, *supra* note 1, at 8, citing THE PETROLEUM ENGINEER, THE PETROLEUM DATA BOOK H-13 (1947) for the total mileage and SPAL, *Oil Pipe Lines*, 15 I.C.C. Prac. J. 563, 565 (1948) for the average hauls.

its president, George Patterson and his predecessor J. Piper, absorbed the Indiana Pipe Line Company, the Northern Pipe Line Company, and the New York Transit System. It then entered the products pipeline field, first by purchasing an eight-inch line between Robinson, Illinois, and Indianapolis, Indiana, from the Ohio Oil Company (Marathon), and subsequently expanding, extending, or constructing new lines whenever it saw a niche where it could provide a service to shippers who would not find it economic to build their own products line. By imaginatively seeking out and capitalizing on these opportunities, Buckeye made a place for itself.[165]

National Transit had a more checkered career, especially after its mainstay subsidiary, the National Transit Pump & Machine Company, was spun off from it. A group of western Pennsylvania refiners, who were dependent on its lines for crude deliveries, bought a substantial interest in the company. With the stability of this affiliation, National Transit acquired the South West Pennsylvania Pipe Lines and got into the products pipeline business, transporting refined product from Philadelphia to Pittsburgh. The remaining Appalachian pipeline companies made independent by the Standard Oil decree of 1911, Eureka and Southern Pipe Line, were purchased by refiners whom they formerly had served.[166]

Technological advances made during this period as listed by Professor Johnson were:

(1) *large diameter pipe, 20- to 22-inch, giving larger volume capacity*
(2) *use of electrical or diesel power dual-fuel pump installations*
(3) *use of micro-wave radio-relay communication systems*
(4) *new methods of cutting and tapping lines*
(5) *a new mechanical sealing machine*
(6) *new methods for making hydraulic calculations*
(7) *automatic push-button operating stations*
(8) *use of radioactive tracers to direct products to proper storage tanks*
(9) *handling of a greater variety of products in the same line*
(10) *automatic devices for detecting leaks*
(11) *quicker assembly through "double-jointing" and arc welding of pipes*
(12) *use of six-stage centrifugal pumps to produce more steady pressures*
(13) *use of seamless pipe*
(14) *new longer-lasting pipe coatings*
(15) *new batching methods.*[167]

---

[165] So much so that the Pennsylvania Railroad acquired a controlling interest in its shares, Burke, *supra* note 9, at 789.
[166] PETROLEUM PIPELINES, *supra* note 45, at 351-371.
[167] *Id.* at 354.

The "second round" commenced with a crude oil pipeline battle connecting the Rocky Mountain production to Mid-Continent refineries. Service Pipe Line had an old ten-inch line laid in 1923 from Wyoming to Freeman, Missouri. It had upgraded the line in 1947 and 1948 by replacing the ten-inch line with a combination 12- and 16-inch. Platte Pipe Line built a 20-inch line from the same origin to the same destination and because of its lower operating costs was able to undercut Service by nine cents per barrel. Platte increased its capacity in 1953 and 1954 up to a total of 198,000 barrels per day. Service's calculations showed that if Platte's traffic approached its capacity, its rate to yield an 8 percent return on valuation would be so low that Service could not recover its depreciation if it published a competitive rate. If that wasn't enough, the Denver-Julesburg play in northeast Colorado caused the Arapahoe Pipe Line Company (owned by Pure and Sinclair) to build a new 18- to 20-inch line paralleling Service and Platte. Service was faced with a simple alternative: build a new line or quit. Service built a new 625 mile, 20-, 22-, and 24-inch system with the highest potential volume and lowest potential cost, but implicit in the calculated risk was the hope of securing nonproprietary traffic.[168] To that end, Service posted competitive rates when it put the line in service and attracted enough business to reduce them a year later. When the Williston Basin came in during 1953, the leading producers formed Butte Pipe Line which provided a 450 mile 16-inch outlet from the field connecting with Platte at Fort Laramie and Service at Guernsey, Wyoming. Then came the Four Corners area. The Four Corners Pipe Line Company (Shell, California Standard, Gulf, Continental, Richfield, and Superior) laid a 16-inch line west from Aneth, Utah, to Los Angeles in June, 1957, and the Texas-New Mexico Pipe Line Company (Tidewater, Texaco, Sinclair, Cities Service), which had an existing line from Jal, New Mexico, to the Gulf Coast, extended its line to the Four Corners area and commenced operations within three months of the Four Corners line. Professor Johnson comments on this development: "In the process, these pipelines built primarily to further their owners' production interests, had created a situation that benefited independent producers, encouraged production, and stimulated competition for common carrier traffic."[169]

As mentioned previously, products lines flourished. In 1954, Harbor Pipe Line System (Sinclair, Gulf, Texaco) constructed a 16-inch line from a common point near Philadelphia to Trembly Point, New Jersey, in the New York Harbor, not only reaching the largest market but, by a tie-in with Buckeye at Linden, obtaining competitive access to the upstate New York

---

[168] *Celler Hearings, supra* note 162, at 1135 (Statement of J. L. Burke).

[169] PETROLEUM PIPELINES, *supra* note 45, at 375. For map of crude oil pipelines in 1958, see *Id.* at 376.

market. The same owners built the Evangeline Pipe Line System from Port Arthur to Baton Rouge where product could enter the Plantation system or be barged to destination. In 1953, Shell, Cities, and Texaco formed Wolverine Pipe Line Company for movement from East Chicago to Detroit and Toledo. Badger Pipe Line Company (Cities, Texaco, Pure, and Sinclair) went west from the same area to Madison, Wisconsin. Wabash Pipe Line (Continental and Marathon) moved products from Wood River to Chicago. Ajax, which had suffered a severe reduction in volume for the second time in its career, was sold to the Cherokee Pipe Line Company, a new company formed by Continental and Cities Service to convert to products service. The Laurel Pipe Line Company (Gulf, Texaco, and Sinclair) paralleled the old Keystone line but the segment from Eagle Point, New Jersey, to Mechanicsburg, Pennsylvania, was 24-inch, the line telescoping down to where the last segment, from Aliquippa, Pennsylvania, to Cleveland was 14-inch.

The Little Big Inch, which had been converted to natural gas after World War II, was reconverted by Texas Eastern to product service from Beaumont, Texas, to Moundsville, West Virginia. A new 14-inch lateral was built from Seymour, Indiana, to Chicago serving Indianapolis and intermediate points.[170]

Two other developments bear mentioning. The first is the development of liquid petroleum gas (LPG) pipelines. Warren Petroleum started the movement with a 140 mile six-inch line from Liberal, Kansas, to Tulsa. Phillips, after obtaining experience in batching LPG in its products lines from Borger to Denver and Chicago, bought a ten-inch crude line of Shell's from West Texas to Houston in 1953, made obsolete by the Rancho system, and converted it gradually to full LPG use by 1960. The Missouri-Kansas-Texas (Katy) railroad, which enjoyed a substantial volume of LPG movement, determined not to be left out this time around, so it developed the Mid-America Pipe Line in 1960 delivering LPG from Texas and New Mexico to the heart of the Midwest as far north as Minneapolis-St. Paul and Janesville, Wisconsin.[171]

The other development is the entry into the pipeline business by nonrefiners. The Buckeye, Mid-America, and Little Big Inch Division of Texas Eastern have been mentioned. Southern Pacific Railroad, through its pipeline division, utilized its railroad right-of-way advantageously in certain situations to provide a pipeline service that appeared prohibitive to producers and refiners. Since 1960, "independent" pipelines have shown a greater percentage increase in growth than single company lines. They now have some $1.7 billion dollars invested.[172]

---

[170] *Id.* at 372-388. The lines have been extended to Philadelphia and Texas Eastern has constructed an LPG line into upper New York state. Burke, *supra* note 9, at 790.

[171] For maps of Products Pipelines as of 1958, see *Id.* at 385.

[172] LENNART, *supra* note 111, at 10.

The importance of pipelines in moving imported crudes to Mid-Continent and Mid-West refineries is demonstrated by Capline connecting LOOP (the deepwater unloading port off the coast of Louisiana)) to Mid-West refineries; Seaway and Texoma transport imported crudes from Nederland, Texas, and Freeport, Texas, to Cushing, Oklahoma, where Mid-Continent refiners receive part of the stream and connecting pipeline carriers transport the balance to Mid-West refineries.

This historical review closes with a brief mention of the Trans Alaska Pipeline System known as TAPS. It was announced on February 10, 1969, that a 48-inch hot oil line running approximately 800 miles from Prudhoe Bay on the north slope of Alaska to Valdez, in the Gulf of Alaska, with an initial capacity of 500,000 barrels per day (increased to 600,000 B/D in 1970) would be built. Construction was expected to be completed in 1972 and the cost of the system was expected to be about $900 million.[173] After substantial changes in the initial design of the line, which were made to increase capacity and to meet the increasingly stringent environmental protection requirements, the north slope crude, crawling at a mile per hour rate, reached Valdez on July 28, 1977,[174] after innumerable delays, frustrations, lawsuits, and political maneuvers had made the incredibly difficult job appear virtually impossible at times and had run its cost up to $9.3 billion.[175]

B. Pipelines' Importance to Industry and Society

1. A Few Statistics

Today there are over 227,000 miles of operating crude and products lines (including gathering lines) in the United States[176] exceeding by nearly 40 percent the total miles of mainline railroad right-of-way.[177] Pipelines that report to the Interstate Commerce Commission serve all states except Hawaii, Texas having the most with over 52,000 miles and the District of Columbia the least, only one mile of products line. As might be expected,

---

[173] J. ROSCOW, 800 MILES TO VALDEZ 21 (1977).

[174] *Id.* at 202; J. STONE, *Trans-Alaska Pipeline* in AOPL EDUCATORS CONFERENCE 2 (1978) [hereinafter cited as STONE].

[175] Rebuttal Testimony of Paul J. Tierney in Trans Alaska Pipeline System, F.E.R.C. Dkt. OR 78-1, October 30, 1978 (formerly I&S 9164) 4 [hereinafter cited as Tierney, *TAPS Rebuttal Testimony*].

[176] Tierney, *TAPS Rebuttal Testimony, supra* note 175, at Exh.I; M. OWINGS, *A Time Perspective of Petroleum Logistics* in AOPL EDUCATORS CONFERENCE 6 (1978) [hereinafter cited as OWINGS]; *cf.* Statement of Ulyesse J. LeGrange, I.C.C. Proceeding Ex Parte No. 308, May 23, 1977 (now F.E.R.C. Dkt. RM78-2) 3 [hereinafter cited as LeGrange, *FERC Direct Testimony*] ("Over 222,000 Miles," referring to ICC reporting lines); *Consumer Energy Act Hearings, supra* note 154, at 593 (Statement of Jack Vickrey, as of November 7, 1973 - 220,000 miles).

[177] Tierney, *TAPS Rebuttal Testimony, supra* note 175, at 17 and Exh. I; *cf.* LENNART, *supra* note 111, at 3.

refined products pipelines operate in more states (45 and the District of Columbia) than do gathering lines (23) or trunk crude lines (35).[178] Between 20 to 25 percent of intercity freight of all kinds moves by oil pipeline, which is slightly greater than that hauled by truck and is exceeded only by rail,[179] but at a cost less than 3-percent of the total intercity freight bill.[180] A partial explanation for oil pipelines' increasing share of freight movement lies in the five-fold increase in petroleum movement since 1938 coupled with the increasing share of petroleum movements enjoyed by oil pipelines.[181] About 48 percent of total tonnage and 60 percent of total ton-miles of petroleum shipments in 1976 were moved by pipelines[182] as compared to 29 percent (tons) and 35 percent (ton-miles) by the next leading mode.[183] When total petroleum movements are broken down into crude movements, pipelines predominate: The pipeline network delivers about 75 percent of crude oil intakes at domestic refineries and more than 98 percent of the crude flow that is processed at inland refineries.[184] However, on the products side, the tonnage figures are 35.6 percent for pipelines and 36.4 for motor carriers.[185]

---

[178] LeGrange, *FERC Direct Testimony, supra* note 176, at 4; PIPELINE TRANSPORTATION, *supra* note 162, at 18.

[179] LENNART, *supra* note 111, at 2-3; *Proposed Hearings Before the Subcommittee on Antitrust & Monopoly of the Senate Comm. on the Judiciary*, 95th Cong. 2d Sess. (June 28, 1978) (Proposed Statement of John H. Shenefield, p. 3) [hereinafter cited as SHENEFIELD PROPOSED STATEMENT]; LeGrange, *FERC Direct Testimony, supra* note 176, at 8; Tierney, *TAPS Rebuttal Testimony, supra* note 175, at 11; *Consumer Energy Act Hearings, supra* note 154, at 629 (Statement of Jack Vickrey).

[180] Tierney, *TAPS Rebuttal Testimony, supra* note 175, at 12; *cf.* LeGrange, *FERC Direct Testimony, supra* note 176, Exh. 29 p. 12; *Consumer Energy Act Hearings, supra* note 154, at 629 (Statement of Jack Vickrey: "At a cost of only 1.5 percent of the nation's freight dollar").

[181] LENNART, *supra* note 111, at 3.

[182] Tierney, *TAPS Rebuttal Testimony, supra* note 175, at 17; *cf.* LeGrange, *FERC Direct Testimony, supra* note 176, at 8 ("almost half" in 1972).

[183] Tierney, *TAPS Rebuttal Testimony, supra* note 175, at Exh. J; *cf.* SHENEFIELD PROPOSED STATEMENT, *supra* note 179, at 3.

[184] LENNART, *supra* note 111, at 3: *cf.* LeGrange, *FERC Direct Testimony, supra* note 176, at 9; Burke, *supra* note 9, at 783; STAFF OF SUBCOMM. ON ANTITRUST AND MONOPOLY OF SENATE COMM. ON THE JUDICIARY, 95th Cong. 2d Sess., REPORT ON OIL COMPANY OWNERSHIP OF PIPELINES (Comm. Print 1978) [hereinafter cited as KENNEDY STAFF REPORT] cites the Petroleum Industry Competition Act (PICA) hearings for an 87 percent figure. This could have been true in 1974 but increases in imports have increased tanker movements to seaboard refineries substantially. The 1976 report by the Energy Resources Council states pipeline deliveries to domestic refineries to be "over 70 percent". ENERGY RESOURCES COUNCIL, ANALYSIS OF VERTICAL DIVESTITURE 13 (May 1976) [hereinafter cited as ERC REPORT].

[185] Tierney, *TAPS Rebuttal Testimony, supra* note 175, at Exh. "K"; LENNART, *supra* note 111, at 4; If one uses ton-miles, pipeline percentage rises to 40.8 percent and motor carrier percentage drops to 5.8 but water carriers take top spot at 51 percent. This is consistent with what one might expect *a priori*. Motor carriers are used substantially in the short-haul later stages of distribution, whereas pipelines and water carriers have the long-haul carriage. ERC REPORT, *supra* note 184, at 14 uses a 50 percent figure but does not specify the quantity being measured; KENNEDY STAFF REPORT, *supra* note 184, at 34, simply echoes the ERC.

Using line capacities as a rough proxy for directional quantification, 1977-1978 crude movements would be: from the Texas/Mid-Continent area to the upper Midwest—2.0 million barrels of oil per day (MMBOD); from inland Texas, New Mexico, and Oklahoma fields to the Gulf Coast—2MMBOD; from the Rocky Mountain area to the Midwest—0.5MMBOD; from Louisiana to the Midwest—1.1MMBOD; from Western Canada to the U.S. Midwest and Eastern Canada—1.6MMBOD; from Western Canada to the Northwestern U.S.—0.4MMBOD and from the Southwestern U.S. to the West Coast—O.1MMBOD.[186]

Similar figures during the same period for products lines are as follows: from the Texas/Gulf Coast to the Southeast and the New York/Washington, D.C. area 2.4MMBOD; from the Texas/Mid-Continent/Gulf Coast area to the Upper Midwest—1.6MMBOD; from the east coast to the Middle Atlantic and Great Lakes States—0.7MMBOD; and from the Wyoming/Rocky Mountain area to the Pacific Northwest and Colorado—0.2MMBOD.[187]

2. Anti-Inflationary Rate Trends

Pipelines were not only able to resist the inflationary trend and maintain transportation rates (measured as an average cents per 100 barrel miles) but to reduce them by 40 percent from 1940 through 1968.[188] These, of course, are optimum years for selection but even after the sharp reductions that took place pursuant to the Interstate Commerce Commission's proceedings in *Reduced Pipeline Rates and Gathering Charges*[189] and the 1941 Elkins Act Consent Decree,[190] the average rate for all pipelines remained virtually constant for the 20 year period, 1945 to 1964.[191] In 1971, the average rate actually was 5 percent *lower* than in 1947[192] although since 1971, rising costs had caused the rate ratio over the 25 year period from 1947-1972 to reflect an overall increase of 1.03 percent.[193]

---

[186] LENNART, *supra* note 111, at 3-4.
[187] *Id.* at 4.
[188] *Id.* at 12.
[189] 243 I.C.C. 115 (1940), 272 I.C.C. 375 (1948).
[190] United States v. Atlantic Refining Company, Civil No. 14060, D.D.C., December 23, 1941.
[191] Burke, *supra* note 9, at 780.
[192] *Hearings on S.2387 et al, The Petroleum Industry, Before the Subcomm. on Antitrust & Monopoly of the Senate Comm. on the Judiciary*, 94th Cong., 1st Sess., Parts 1-3, 1108 (1975-1976) (Statement of Charles E. Spahr) [hereinafter cited as *Petroleum Industry Hearings*]. The slight differences between the Spahr and Steingraber figures and Tierney's are probably due to the fact that the former two were using July 1972 interim data and Tierney used July 1977 figures. At any rate, Spahr and Steingraber, if anything, understated their case.

Perhaps the most comprehensive analysis of the level of pipeline rates compared with other price indicators is set forth in the sworn testimony in the TAPS proceeding of Paul J. Tierney, former Chairman of the Interstate Commerce Commission, and currently the President of the Transportation Association of America (TAA). TAA is a national transportation policy organization made up of transport users, suppliers, investors, and carriers of all modes, including air, freight forwarders, highway, petroleum pipeline, railroad, and water carriers. Mr. Tierney's testimony stated that TAA's data clearly showed that oil pipeline rate levels, as reflected by the generally accepted Average Revenue Per Ton-Mile yardstick, had remained well below general price levels throughout the 1947-1976 period.[194] The cited figures indicate that during this period consumer prices rose 155 percent and producer (wholesale) prices 138 percent, while oil pipeline rate levels increased 40 percent—all of the latter being in recent years. While the development of the two national price measures are not developed in exactly the same manner as average revenue per ton-mile figures, the overall indices do reflect general price/cost trends and are roughly comparable. Even granting some slight inconsistencies in their makeup, the sharp differences plainly demonstrate that the real (constant dollar) cost of oil pipeline service to the nation has decreased.

Mr. Tierney attributed the principal reason for this decline in real cost to the great economies effected by the construction of high-cost but extremely efficient large diameter pipelines. He also cited advancing

---

[193] F. STEINGRABER, *Pipelines, Divestiture and Independents* in WITNESSES FOR OIL 142 (1976) [hereinafter cited as STEINGRABER]; *Petroleum Industry Hearings, supra* note 192, at 301 (Statement of E. P. Hardin). For a graphical presentation of ICC reporting pipeline revenue per 100 barrel-miles and volume of barrel-miles over the period 1955-1973, see LeGrange, *FERC Direct Testimony, supra* note 176, p. 15.

[194] Tierney, *TAPS Rebuttal Testimony, supra* note 175, at 6:

Average Revenue per Ton-Mile of Oil Pipelines
vs.
Consumer and Producer Price Indexes

YEAR	OIL PIPELINE (¢)	INDEX 1947 = 100	WHOLESALE PRICES 1947 = 100	CONSUMER PRICES 1947 = 100
1947	.292	100	100	100
1950	.315	108	106	107
1955	.322	110	114	119
1960	.315	108	123	133
1965	.279	96	126	142
1970	.271	93	143	173
1971	.285	98	148	181
1972	.285	98	155	187
1973	.291	100	175	199
1974	.315	108	208	221
1975	.368	126	227	240
1976	.409	140	238	255

technology, such as computerized scheduling and pipeline flow control, which have facilitated continuously increasing traffic volumes with progressively fewer employees. However, pipelines have begun to feel inflationary pressures in recent years, as the recent upturn indicates. Most large pipelines have about reached the limit in automation and it has become increasingly difficult to counter rising labor costs with technology. Construction costs have almost doubled since 1966. Variable costs such as fuel and power, taxes and the like, also have been rising.

As a closing observation, one notes that just prior to Van Syckel's innovative pipeline, oil priced at $7.00 per barrel at the lease was transported by teamster and railroad over roughly 350 miles from Oil City to New York Harbor for $8.35; in other words, the cost of carriage represented about 54 percent of the $15.58 selling price in New York.[195] Today, a barrel of "new" 40° API west Texas sour crude, worth $12.60 at the lease, will move from Yates in Pecos County, Texas, to Baytown, Texas, near Houston, through a 440 mile pipeline for $0.285 (seven cents gathering charge plus 21.5 cents trunk tariff), 2 percent of its value in Baytown.[196] Examined from another viewpoint, pipelines, while playing a highly important role, represent about 10 to 15 percent of the total cost for industry to transport, refine and market petroleum and their rates constitute an insignificant component of delivered petroleum prices, on the order of 2 to 3 percent.[197]

3. Efficiency Compared to Alternate Forms of Transportation

It is the general consensus that pipelines are by far the most economical means of large scale overland transportation for crude oil and products,[198] clearly superior to rail and truck transportation over competing routes, given large quantities to be moved on a regular basis.[199] This has been true for many years; figures are readily available for 1964, when the typical long haul

---

[195] LENNART, *supra* note 111, Slide 7.
[196] Exxon Pipeline Company Tariff, F.E.R.C., No. 130, effective March 1, 1978.
[197] LENNART, *supra* note 111, at 16 (percentage of cost of MTM); *Statement of Charles J. Waidelich before the Subcomm. on Monopolies and Commercial Law of the House Judiciary Committee* 3 (September 10, 1975) (Component of Final Price) [hearings not printed; hereinafter cited as Waidelich, *Rodino Testimony*].
[198] *Statement of Stewart C. Myers in Hearings Pursuant to S. Res. 45, Market Performance and Competition in the Petroleum Industry Before the Special Subcomm. on Integrated Operations of the Senate Committee on Interior and Insular Affairs*, 93d Cong., 1st Sess. 1974, page 8 of Statement [hereinafter cited as Myers]; Harmon, *Effective Public Policy to Deal with Oil Pipelines* 4 Am. Bus. L. J. 113, 118 (1966); Att'y Gen., *Third Report, supra* note 111, at 48.
[199] HOWREY & SIMON, *supra* note 3, at 132, citing U.S. CONG., CONG. RES. SERVICE, NATIONAL ENERGY TRANSPORTATION, VOL. I, CURRENT SYSTEMS AND MOVEMENTS, PUB. NO. 95-15, SENATE COMM. ON ENERGY AND NATURAL RESOURCES AND SENATE COMM. ON COMMERCE, SCIENCE, AND TRANSPORTATION, 95th Cong., 1st Sess. 213 (Comm. Print 1977) [hereinafter cited as NET-I, NET-II, and NET-III, respectively].

pipeline rate was 10 to 20 percent of the rail rate.[200] More specifically, pipeline rates average about 1/5 of rail rates and about 1/20 of truck rates and pipelines can usually compete favorably with all marine transportation except for ocean-going, long-haul supertankers.[201] There are, of course, special situations where other forms of transportation have a competitive advantage over pipelines. For example, heavier petroleum liquids or solids (*e.g.*, residual fuel oil, asphalt, and coke), while transportable by pipelines, are more economically handled by other bulk carriers.[202] Often market volume requirements make barge movement more attractive than by pipeline.[203] In large volume markets having good port facilities, tanker shipments may offer the lowest transportation costs.[204] This is especially true of crudes imported from far away foreign sources. The niche for trucking lies in situations where small volumes over short hauls with many different destinations are involved, such as gasoline movements from terminals to jobber plants and to private residences.[205] What has been stated above is more graphically displayed by the bar chart in Appendix A, taken from the testimony of U. J. LeGrange in FERC Docket No. RM78-2.[206]

4. Continuous and Reliable Operation

Because the oil business is for the most part, and unqualifiedly is at the refinery level, a 24 hour per day, 365 days per year operation (excluding scheduled maintenance "turnarounds"), there is a very high premium attached to continuity of supply and distribution.[207] This is not to say that it is absolutely impossible to operate through contracts with other modes of transportation. But it is quite obvious that a form of transportation which is itself continuous and relatively insusceptible to interruption, given the competitive rate structure described above, is the quintessential solution. The

---

[200] Burke, *supra* note 9, at 783.
[201] Appendix A; STEINGRABER, *supra* note 193, at 142, using 1971 figures reached the same ratio with respect to pipeline/rail rates but showed the pipeline/truck rates ratio to be 1/28.
[202] MARATHON PIPE LINE COMPANY, AN ANALYSIS OF CERTAIN CONSIDERATIONS RELATING TO SUGGESTIONS OF VERTICAL OIL PIPELINE DIVESTITURE WITHIN THE CONTINENTAL UNITED STATES 58 (1978) (Submission to the Department of Justice) [hereinafter cited as MARATHON]; NET-I, *supra* note 199, at 200.
[203] MARATHON, *supra* note 202, at 58.
[204] *Id.* at 59.
[205] *Id.* This is clearly demonstrated by the difference between the percentage of petroleum transported by trucks in 1975 measured on a ton-basis (28.4 percent) and on a ton-mile basis (1.7 percent). *Id.* at Table VI.
[206] For comparable pictorial presentations of this relationship, see MARATHON, *supra* note 202, at Appendix H, and LENNART, *supra* note 111, at Slide 5.
[207] Bond, *Oil Pipe Lines—Their Operation and Regulation*, 25 I.C.C. Prac. J. 730, 735 (1958) [hereinafter cited as Bond]; Att'y Gen., *Third Report, supra* note 111, at 47, 50; *Consumer Energy Act Hearings, supra* note 154, at 668 (Statement of W. J. Lamont); *cf.* Att'y Gen., *Second Report, supra* note 111, at 65-66.

great advantage of pipelines in this regard is their immobility—the commodities which they transport do the moving. Thus, they are not nearly as vulnerable to interference with movement or to attrition of equipment as are other forms of transportation. They are not only the conveyor but also the container of the materials they transport. Deadheading and back hauling are not required; and they are virtually immune to interruption by aboveground activities and buffeting from the elements.[208] By virtue of the high degree of automation achieved technologically, they are not prone to labor stoppages during the operational mode.[209] In short, they are ideally designed for continuous and uninterrupted service.

5. Ability to Operate in Remote, Hostile Environments

Pipelines have demonstrated an ability to adapt to a wide variety of environments. The "spaghetti bowl" offshore the Gulf Coast in water depths up to 360 feet; the transportation ashore from producing platforms in the treacherous North Sea; pipelines traversing the arid shifting desert sands of the Middle East; and the conquering of the challenging permafrost and the earthquake zone of Alaska each provide evidence of pipeline abilities in this area.[210] Oil supplies in western Canada, the Rockies, and offshore Louisiana would have remained undeveloped at their prevailing prices if it had not been for pipelines.[211]

6. Flexibility and Alternatives to Shippers

With very minor exceptions, largely due to locational peculiarities, most refineries are served by one or more pipelines. The comprehensive and interrelated nature of the pipeline network is such that the usual refiner has a choice of routes and schedules that greatly exceeds that made available by any other mode of transportation.[212] Despite the greatly increased capacities of the larger lines, rendered even greater by jointly owned and operated lines, it is impracticable for one pipeline to reach the scale of output necessary to

---

[208] Burke, *supra* note 9, at 791; HOWREY & SIMON, *supra* note 3, app. at 159; LENNART, *supra* note 111, at 5; Att'y Gen., *Third Report*, *supra* note 111, at 48; PIPELINE TRANSPORTATION, *supra* note 162, at 17; NET-I, *supra* note 199, at 262; *Consumer Energy Act Hearings*, *supra* note 154, at 668 (Statement of W. J. Lamont).

[209] OWINGS, *supra* note 176, at 12.

[210] Friggens, *The Great Alaska Pipeline Controversy*, in READERS DIGEST, November, 1972, at 125, 128 [hereinafter cited as Friggens]; Dadisman, *Known Problems and Unknown Effects* in NATION, October 2, 1972, at 262; OIL AND GAS JOURNAL, August 27, 1973, at 51, HOWREY & SIMON, *supra* note 3, at 154-156.

[211] PIPELINE TRANSPORTATION, *supra* note 162, at 17.

[212] LENNART, *supra* note 111, at 1,4; *cf. Consumer Energy Act Hearings*, *supra* note 154, at 40 (Statement of Jack Vickrey).

supply the entire demand of most metropolitan markets,[213] which are served from multiple origins. A trained economist would almost involuntarily translate that statement into "economese:" a pipeline's highest output generally will be less than the demand in the market served. Pragmatically, this boils down to the fact that pipelines extend from every producing region to a number of refining areas, and conversely, every refinery area is connected to a number of different producing regions.[214] Specific illustrations of these points with respect to crude oil pipelines are as follows: American Petrofina's refinery at El Dorado, Kansas, can draw upon imported crudes landed at Nederland or Freeport, Texas; West Texas and New Mexico crudes via two different combinations of carriers; and Oklahoma and North Texas crudes through a joint tariff filed by two common carrier pipelines.[215] Clark's refinery at Blue Island (Chicago) can draw on Oklahoma and Kansas crudes; West Central Texas crude; Louisiana onshore and offshore crudes via two alternative pipelines; North Texas crude; Eastern Montana and Western Wyoming crudes; West Texas and New Mexico crudes; and im-

---

[213] HOWREY & SIMON, *supra* note 3, at 19. Because the classic economic requirement of a "natural monopoly" is "an inherent tendency to decreasing costs *over the entire extent of the market*" (emphasis added), A. KAHN, THE ECONOMICS OF REGULATION: PRINCIPLES AND INSTITUTIONS II, 119 (1971), a question is raised whether pipelines are "natural monopolies."

[214] HOWREY & SIMON, *supra* note 3, at 19.

[215] **Common Carrier Crude Oil Pipeline Transportation Available to American Petrofina, El Dorado, Kansas As of April 1, 1979**

SOURCE OF CRUDE OIL PRODUCTION AVAILABLE	APPLICABLE TRUNK LINE RATES IN ¢/BBL.	TARIFF PUBLISHED BY	FERC NO.
Nederland, TX (Imports)	51.50	Texoma	304
West Texas & New Mexico-Cushing, OK (Proportional)	21.00	Shell	1881
Cushing, OK-El Dorado, KS (Proportional)	11.50	Osage	3
Freeport, TX (Imports)	58.00	Seaway	34
West Texas & New Mexico-Cushing, OK (Proportional)	23.00	Amoco	968
Cushing, OK-El Dorado, KS (Proportional)	11.50	Osage	3
Oklahoma Points-Cushing, OK (Proportional)	13.75	ARCO	1305
Cushing, OK-El Dorado, KS (Proportional)	11.50	Osage	3
North Texas Points-Cushing, OK (Proportional)	16.00	ARCO	1300
Cushing, OK-El Dorado, KS (Proportional)	11.50	Osage	3

Note: Shipper must furnish tankage at Cushing when crude is moved by Texoma. Tankage furnished by Seaway.

ported crudes landed at three different ports.[216] Ashland's refinery at Canton, Ohio, has available to it West Texas crude; Louisiana onshore and offshore crudes; Wyoming crude; Eastern Montana crude; and imported crudes landed at three different ports.[217] Gladieux's refinery at Fort Wayne, In-

[216] **Common Carrier Crude Oil Pipeline Transportation Available to Clark Refinery at Blue Island (Chicago), Illinois As of April 1, 1979**

SOURCE OF CRUDE OIL PRODUCTION AVAILABLE	APPLICABLE TRUNK LINE RATES IN ¢/BBL.	TARIFF PUBLISHED BY	FERC NO.
Oklahoma Points	40.25	ARCO	1316
Kansas Points	29.00	ARCO	1316
West Central Texas	52.00	Shell	1984
St. James, LA	34.40	Shell	2002
St. James, LA-Potoka, IL (Proportional)	22.40	Southcap	152
Patoka, IL-Blue Island, IL (Proportional)	10.40	Chicap	12
Freeport, TX-Cushing, OK (Proportional)	43.00	Seaway	2
Cushing, OK-Blue Island, IL (Proportional)	37.00	ARCO	1316
North Texas	57.00	Amoco	914
Eastern Montana-Guernsey, WY (Proportional)	35.00	Butte	492
Guernsey, WY-Salisbury, MO (Proportional)	34.20	Platte	1159
Salisbury, MO-Blue Island, IL (Proportional)	24.00	ARCO	1219
Western, WY-Salisbury, MO. (Proportional)	38.30	Platte	1156
Salisbury, MO-Blue Island, IL (Proportional)	24.00	ARCO	1219
West Texas & New Mexico-Cushing, OK (Proportional)	21.00	Shell	1881
Cushing, OK-Blue Island, IL (Proportional)	37.00	ARCO	1316
Nederland, TX (Imports)	75.00	Texoma	326

[217] **Common Carrier Crude Oil Pipeline Transportation Available to Ashland Refinery at Canton, Ohio As of April 1, 1979**

SOURCE OF CRUDE OIL PRODUCTION AVAILABLE	APPLICABLE TRUNK LINE RATES IN ¢/BBL.	TARIFF PUBLISHED BY	FERC NO.
West Texas	59.50	Shell	1925
St. James, LA*	45.45	Ashland	274
Freeport, TX (Imports)	84.45	Seaway	43
Nederland, TX (Imports)	84.45	Texoma	338
Wyoming Points	83.40	Amoco	971
Eastern Montana-Guernsey, WY (Proportional)	35.00	Butte	492
Guernsey, WY-Canton, OH (Proportional)	67.40	Platte	1186

* Movements by other Capline carriers can be made to Patoka on a proportional tariff, then to Canton on a proportional tariff via Ashland.

diana, can obtain imported crude (other than Canadian crude) landed at three different ports; Canadian crude; Eastern Montana and Western Wyoming crudes; Kansas and Oklahoma crudes; West Texas crude; and Louisiana onshore and offshore crudes.[218] Tesoro's refinery at Newcastle, Wyoming, can obtain Northeastern Montana crude; Eastern Montana crude and Wyoming crude.[219] The reader may rest assured that these are typical, not point-making-selected, examples of the manifold alternatives of pipeline routing from producing regions to refining areas. Because of the large number of interconnecting and alternative routes, origins and destinations, it is impractical to list all the combinations and permutations that evolve from

[218] **Common Carrier Crude Oil Pipeline Transportation Available to Gladieux Refinery, Ft. Wayne, Indiana As of April 1, 1979**

SOURCE OF CRUDE OIL PRODUCTION AVAILABLE	APPLICABLE TRUNK LINE RATES IN ¢/BBL.	TARIFF PUBLISHED BY	FERC NO.
Freeport, TX (Imports)	86.00	Seaway	39
Nederland, TX (Imports)	85.00	Texoma	327
Eastern Montana-Guernsey, WY (Proportional)	35.00	Butte	492
Guernsey, WY-Salisbury, MO (Proportional)	34.20	Platte	1159
Salisbury, MO-Auburn Jct., IN (Proportional)	34.00	ARCO	1220
Western, WY-Salisbury, MO (Proportional)	38.30	Platte	1156
Salisbury, MO-Auburn Jct., IN (Proportional)	34.00	ARCO	1220
Kansas-Salisbury, MO (Proportional)	17.00	ARCO	1282
Salisbury, MO-Auburn Jct., IN (Proportional)	34.00	ARCO	1220
Oklahoma-Hartsdale, IN (Proportional)	42.25	ARCO	1272
Hartsdale, IN-Auburn Jct., IN (Proportional)	10.00	Tecumseh	27
West Texas-Salisbury, MO (Proportional)	26.50	ARCO	1282
Salisbury, MO-Auburn Jct., IN (Proportional)	34.00	ARCO	1220
Edmonton, Alberta-International Boundary (Proportional)	26.40	Interprovincial	NEB 89
International Boundary-Hartsdale, IN (Proportional)	30.40	Lakehead	44
Hartsdale, IN-Auburn Jct., IN (Proportional)	10.00	Tecumseh	27
St. James, LA-Patoka, IL (Proportional)	24.25*	Texas	1523
Patoka, IL-Dyer Jct., IN (Proportional)	13.25	Texaco-C.S.	200
Dyer Jct., IN-Auburn Jct., IN (Proportional)	10.00	Tecumseh	31

* If unloaded across dock, an additional charge of 4.5 cents will be made.

the variety of sources, destinations and their alternatives.

This is not a recent development; a similar observation was made by this author in a presentation to a seminar held by the Harvard Graduate School of Business Administration in 1959 wherein it was stated: "Every refinery in America today other than those designed specifically for supply by water or local production is served by one or more pipelines."[220] This text continues: "Not only does the nonintegrated refiner have a choice of long-haul pipeline routes available to him, but he obtains thereby a wide choice of crude oil sources."[221] A typical example given was that of the Clark Refinery at Chicago (Blue Island).[222] At the time of the address, ten competing carriers offered 19 routes from 13 crude producing areas (in eight states) under tariffs

[219] **Common Carrier Crude Oil Pipeline Transportation Available to Tesoro Refinery, Newcastle, Wyoming As of April 1, 1979**

SOURCE OF CRUDE OIL PRODUCTION AVAILABLE	APPLICABLE TRUNK LINE RATES IN ¢/BBL.	TARIFF PUBLISHED BY	FERC NO.
Northeastern Montana-Mush Creek Jct., WY (Proportional)	47.00	Wesco	103
Mush Creek Jct., WY-Newcastle, WY (Proportional)	9.00	Plains	WY.PSC-6
Eastern Montana-Mush Creek Jct., WY (Proportional)	35.00	Butte	495
Mush Creek Jct., WY-Newcastle, WY (Proportional)	9.00	Plains	WY.PSC-6
Wyoming Points to Fiddler Creek, WY (Proportional)	Various (55 Points)	Belle Fourche	WY.PSC-18
Fiddler Creek, WY-Newcastle, WY (Proportional)	7.00	Plains	WY.PSC-6

[220] SPECTRE, *supra* note 104, at 112, citing SEN. REP. No. 1147, 85th Cong., 1st Sess. 124 (1959) [hereinafter cited as CELLER REPORT]; *Celler Hearings, supra* note 162, at 1136; "Map, Crude Oil Pipelines and Principal Refineries," OIL AND GAS JOURNAL, September 21, 1959.

[221] SPECTRE, *supra* note 104, at 112; *cf.* Att'y Gen., *Third Report, supra* note 111, at 79.

[222] Rostow & Sachs, *Entry Into The Oil Refinery Business: Vertical Integration Re-examined*, 61 YALE L. J. 856, 891-892 (1952) cite the case of another nonintegrated refiner, Globe Oil & Refinery Company, for the proposition that a well managed "independent" refining company had been able to thrive on common carrier pipeline service. These remarks about Clark would have applied equally to Globe; *see also Consumer Energy Act Hearings, supra* note 154, at 604 (Rock Island Refinery, located at Rock Island, Indiana, served by 14 different competing carriers offering 23 routes from 10 different producing areas) (Statement of John E. Green).

[223] See table of common carrier crude oil pipeline transportation available to Clark at that time, showing source, tariff rate, and citing the tariffs in effect by name and I.C.C. number in SPECTRE, *supra* note 104, at n.108. Fourteen years later the ten competing carriers had grown to 28 and instead of 19 routes there were 55. *Consumer Energy Act Hearings, supra* note 154, at 603 (Statement of John E. Green). There was some question whether all these routes were to Chicago because Clark had, in the interim, acquired a refinery at Wood River, Illinois, *Id.* at 671 (Statement of W. J. Lamont).

in effect on October 15, 1959.[223] Examples were also cited of Aurora (subsequently purchased by Marathon) and Naph-Sol Refiners at Muskegon, Michigan, (14 crude sources moveable through 21 routes of nine competing carriers); Aurora and Petroleum Specialties in Detroit (connected to 17 crude oil sources by 48 common carrier pipeline alternatives); and Ashland's Canton, Ohio, refinery (drawing on 13 distant crude sources by 46 possible pipeline movements).[224]

There are corresponding alternatives for petroleum products shippers as well. Plantation Pipe Line Company inaugurated service to the southeastern states in 1942. Along came Colonial in 1964 and gave independent shippers a choice of alternate pipelines.[225] Cities Service Company, which is one of the largest suppliers of independent marketers, has used Colonial to remain a viable supplier to independent marketers in Atlanta, Georgia; Charlotte, Winston-Salem, Raleigh, and Greensboro, North Carolina; and Roanoke and Richmond, Virginia.[226] Refiners in the Philadelphia-Linden, New Jersey, complex can reach the Pittsburgh or Buffalo markets by the ARCO (formerly Keystone and Buffalo) pipeline, the Laurel pipeline, or Sun's line.[227] Gulf Coast refiners shipping to the Chicago area can use Explorer or Texas Eastern Transmission[228] or combinations such as Explorer and Continental (formerly Cherokee).

This vast, interconnected, underground system has certain advantages to the nation's security. Being underground makes them less vulnerable to enemy attack than other modes of transportation which are susceptible to surface, air, or in the case of tankers, underwater attack.[229] In a previous section, mention was made of the reversals, interconnections of the existing system, and construction of the Big Inch and Little Big Inch lines during World War II.[230] In the so-called Suez Canal crisis, extraordinary steps were taken, this time to make more crude available to Gulf Coast ports for shipment to Europe.[231] There were some pipeline reversals to divert crude oil movements from other areas to the Gulf Coast. Other measures involved use

---

[224] SPECTRE, *supra* note 104, nn. 110, 111, 112; Even the Department of Justice has noted this flexibility and interconnected service. Att'y Gen., *Third Report, supra* note 111, at 73-74.

[225] HOWREY & SIMON, *supra* note 3, at 23.

[226] *Petroleum Industry Hearings, supra* note 192, at 258 (Statement of *Charles J. Waidelich).*

[227] HOWREY & SIMON, *supra* note 3, at 23.

[228] S. M. LIVINGSTON, OIL PIPELINES: INDUSTRY STRUCTURE 63 (1978) [hereinafter cited as LIVINGSTON]. Dr. Livingston's paper has been reprinted in its entirety in AEI, OIL PIPELINES AND PUBLIC POLICY 317-391 (E. Mitchell ed. 1979). Citations herein are to the mimeographed paper.

[229] HOWREY & SIMON, *supra* note 3, at 157.

[230] See text at footnotes 153 to 157, *supra.*

[231] Att'y Gen., *Second Report, supra* note 111, at 51; The text following this footnote has been derived from the Joint Trial Brief of Defendants in United States v. Arkansas Fuel Oil Corporation, Crim. No. 13318 (N.D. Okla. 1960).

of barges to move crude from inland areas to the Gulf.

Movements south to the Gulf, principally from West Texas, were increased by removing bottlenecks in lines, reactivating old lines and reversal of a ten-inch Pasotex line. Magnolia (Mobil) reactivated an old eight-inch line from Corsicana. Jersey (Exxon) expanded the Interstate Oil pipeline to increase capacity 10,000 barrels per day. Gulf reversed the flow of dual eight-inch lines which normally carried crude from East Central Texas to Tulsa, Oklahoma. Barges, which normally carried oil up the Mississippi, were turned around to carry downstream to the Gulf oil which had come from Wyoming and the Rocky Mountains to Wood River by pipeline. Substantial quantities of West Texas and New Mexico crude moved up the Basin and Ozark pipelines to Wood River and thence to the Gulf by barge (at an additional 65¢/bbl. cost).

When large long-haul lines are constructed, military considerations generally are taken into account. The Colonial system is an example. Numerous military installations, including 16 military air bases, are located along Colonial's route. At several vital points, the line can serve the Atlantic fleet. Before freezing the design, Colonial officials discussed the details of the project with the Secretaries of Defense, Interior, and Commerce, as well as the Air Force, Army, and Navy.[232]

Because our economy[233] and defense[234] are geared so closely to petroleum products, an extensive overland, uninterrupted and efficient system of pipelines, with sufficient flexibility to adapt to changing needs, is quite important. The U.S. pipeline system appears to meet those criteria.[235]

7. Noninterference with Other Activities

Because oil pipelines are mostly[236] buried and the right-of-way restored

---

[232] HOWREY & SIMON, *supra* note 3, at 160.

[233] OWINGS, *supra* note 176, at 6-7; PIPELINE TRANSPORTATION, *supra* note 162, at 2.

[234] *Statement by John J. Allen, Jr., Before Subcomm. for Transportation of House Comm. on Armed Services,* July 15, 1959; Burke, *supra* note 9, at 739; *Hearings Pursuant to H. Res. 5 and 19, Anticompetitive Impact of Oil Company Ownership of Petroleum Products Pipelines, Before the Subcomm. on Special Small Business Problems of the House Select Comm. on Small Business,* 92d Cong., 2d Sess., 144 (1972) (Statement of Fred F. Steingraber, citing Adm. Arthur Radford, Chairman of the Joint Chiefs of Staff and ODM Director Arthur S. Fleming testimony before the House Armed Services Comm.) [hereinafter cited as *H.Res.5 Subcomm. Hearings*]; *Petroleum Industry Hearings, supra* note 192, at 318 (Remarks of Peter N. Chumbris, Minority Chief Counsel).

[235] *Cf.* Att'y Gen., *Third Report, supra* note 111, at 79.

[236] The principal reason for this qualification "mostly" is obviously the Trans Alaska Pipe Line system where the permafrost and nature requirements caused large sections to be elevated on steel piles or gravel ledges. HOWREY & SIMON, *supra* note 3, at 155.

so that surface usage is preserved insofar as is consistent with safety, they offer less interference with other activities than other modes of transportation.[237] One need only consider railroad grade crossings with unit trains passing over them, high voltage transmission lines of electric power companies, highways and bridges cluttered with transport trucks and locks and dams required by barge traffic to appreciate the difference.[238]

In addition to being hidden from sight and virtually noiseless,[239] pipelines have not been subject to much spillage.[240] In 1976, the domestic industry transported over 9-½ billion barrels of crude oil while losing less than 100,000 barrels, or 0.001 percent, due to accidents.[241] The safety characteristics of oil pipelines have given impetus to the concept of underwater pipelines initially utilized to transport oil from offshore oil wells with less risk of spills than with marine vessels.[242] From this activity, their use has been extended to the connection of deepwater docking facilities with onshore connections, thereby both minimizing the environmental risks of bringing tankers through a crowded fairway into longshore docks and piers and also permitting the more economical use of very large crude carriers (VLCC's) and ultra large crude carriers (ULCC's) (700,000 d.w.t. carrying up to five million barrels) whose drafts are too great for U.S. ports.[243] Thus, the deepwater ports, as they are called, will reduce the risks of accidents in crowded harbors and the frequency of hookup and disengagement, each of which creates a possibility of spills.[244] This represents a further recognition by most pipeline operators that good ecology is good business.[245] Pipelines were concerned with control procedures before the Environmental Protection Agency came into being.[246] The early pipelines were constructed with bare pipe which was highly vulnerable to corrosion. Now it is standard practice to coat all pipe with a protective coating such as somastic to protect it externally from moisture and corrosive soils.[247] Costs are often lowered by having pipe precoated rather than on-site. The coated pipe is "jeeped" to be sure the

---

[237] PIPELINE TRANSPORTATION, *supra* note 162, at 53.
[238] *Id.;* LeGrange, *FERC Direct Testimony, supra* note 176, at 7.
[239] PIPELINE TRANSPORTATION, *supra* note 162, at 53.
[240] HOWREY & SIMON, *supra* note 3, at 152.
[241] National Transportation Safety Board Press Release 1 (May 14, 1977), cited in *Id.*
[242] HOWREY & SIMON, *supra* note 3, at 153.
[243] *Id.*
[244] Houston Chronicle, September 1, 1976 (Journal), Section 1, at 54.
[245] PIPELINE TRANSPORTATION, *supra* note 162, at 53, lists three good business reasons for taking extraordinary precautions to preclude leaks: (1) loss of product for which the lines are responsible; (2) loss in throughput while shut down for repairs; and (3) costs incurred in locating and repairing leaks.
[246] *Id.* The text immediately following this footnote is, except where specifically footnoted, taken from the treatment on pages 53-57.
[247] Burke, *supra* note 9, at 791.

coating is intact prior to its being lowered into the ditch. "Jeeping" is performed by an instrument called a holiday detector which is run along the full circumference of the coating. If the instrument passes over a break in the coating ("a holiday"), the detector sounds an alarm and the coating break is repaired. Not content to trust the whole job of protection to the coating, cathodic protection devices are used to backstop the coating. This usually takes the form of sacrificial anodes and/or rectifiers, which create a flow of electric current from the soil into the pipe to counteract the galvanic current associated with the corrosion of steel in soils. These design and construction control measures against corrosion and leaks are supplemented by operational techniques such as leak detection systems. Instrumentation is installed to flash early warning of abrupt changes in operating pressures or differences in line balances. Low flying planes have largely displaced the old time line walkers. Changes in color of the terrain or other indicia of the presence of leaks are investigated promptly. The presence of construction equipment on or near rights-of-way are special subjects of surveillance as damage by such equipment is the major cause of pipeline leaks. Internal anticorrosion measures common in products lines, although not as usual in crude lines, include the use of chemical inhibitors added in small quantities to the materials being transported so as to disperse as a thin film on the inner circumference of the pipe.[248] In addition, swabs or scrapers are run regularly to keep the pipe clean. These techniques are supplemented by periodic comparisons of the input and output of sections of the line and the transmission of instruments inside the pipe which measure the frequency and intensity of internal pitting. When severe pitting is discovered, the pipeline may be physically coated internally or bad sections replaced. In areas where there are multiple lines, a system of leak notification is installed. The first representative of a pipeline who reaches the scene is responsible for taking whatever corrective action he thinks is necessary, whether it is his company's line or that of another. The important thing is to stop the leak and he will be supported in his action by the owner.

Tank farms and terminals are protected against pollution by multiple measures such as floating roof tanks, oil-water separators to prevent oil from escaping, and fire protection measures such as tank dikes, foam injectors, and hydrants. Underwater pipeline operators employ boats which carry hoses and skimmers and containment booms, together with oil absorbent chemicals.

The substantial lengths to which pipelines have gone, voluntarily, or pursuant to exogenous forces, to protect the environment is, of course, well publicized in the TAPS line where about half the line is insulated and elevated

---

[248] *Id.*

on steel piles or gravel ledges to avoid melting the permafrost.[249] A few buried miles of line are refrigerated so as not to disturb the frozen ground.[250] River and flood plain crossings are protected by concrete coating the pipe and deep anchoring.[251] In order to make the line as earthquake-proof as possible, the pipeline is designed to withstand a 20 foot horizontal and 3 foot vertical movement without rupture.[252] In addition, remote control cutoff valves have been installed so as to seal off a leak or rupture within minutes.[253] These measures have earned TAPS the accolade of being "environmentally safe and a model for the world."[254]

8. Safety

Oil pipelines have compiled a stronger record in the area of safety than other forms of transportation and industry in general.[255] Since 1950, there has been almost a tenfold reduction in accident rates, *i.e.,* in 1950 there were 632 reported accidents while transporting almost 284 million tons (2.225 accidents per million tons) whereas by 1968, the number of accidents had been reduced to 203 while moving over 725 million tons (0.28 accidents per million tons).[256] The total reported fatalities resulting from pipeline-related accidents was 14 in 1976, which consisted of only 0.028 percent of the 49,575 transportation-related deaths reported that year.[257] When one considers that pipelines were transporting approximately 23 percent of the total intercity freight moved by all forms of transportation,[258] this record is noteworthy. These statistics evidence a continuing technological trend by oil pipelines towards the construction, maintenance and operation of ever safer pipeline systems.[259]

Design, construction, and operation of oil pipelines has been governed by standards developed by the American National Standards Institute, although since 1972, they have been supplemented by regulations promulgated by the Department of Transportation.[260] These standards incorporate the composite

---

[249] OIL AND GAS JOURNAL, August 27, 1973, p. 51.

[250] HOWREY & SIMON, *supra* note 3, on 155.

[251] Friggens, *supra* note 210, at 127-128.

[252] HOWREY & SIMON, *supra* note 3, at 156.

[253] Friggens, *supra* note 210, at 128.

[254] *Id.*, quoting Walter Hickel.

[255] *Statement of Charles E. Spahr, Before the Subcomm. on Monopolies and Commercial Law of the House Judiciary Committee* 10-11 (September 10, 1975) [hearings not printed; hereinafter cited as Spahr, *Rodino Testimony*]; HOWREY & SIMON, *supra* note 3, at 2 gives a safety ratio compared with trucks of 1,000 to 1.

[256] A.P.I., FACTS AND FIGURES 272, 581, 582.

[257] News Release of May 14, 1977, by the National Transportation Safety Board; *cf.* Spahr, *Rodino Testimony, supra* note 255, at 10. This was down from 1974 when pipelines had 34 deaths out of 50,541 transportation fatalities (.067 percent) PIPELINE TRANSPORTATION, *supra* note 162, at 50.

[258] See text at note 179, *supra.*

[259] HOWREY & SIMON, *supra* note 3, at 148.

[260] 49 C.F.R. § 195 (1972).

experience of individuals, companies, and professional societies that have been concerned with the design and operation of pipelines for many years, supplemented by research, tests, studies, and investigations.[261] Material characteristics, their susceptability to failure, and the consequences of failure were all taken into consideration. The soundness of the resulting standards has been attested to by the absence of failures attributable to inadequate design.[262] Because roughly 83 percent of the accidents that did occur were traceable to the line pipe, [263] special attention has been addressed to this aspect. The pipe used is made from steel, manufactured to standards established by the American Petroleum Institute (API), which provide for uniformity of dimension, hydrostatic testing, and grading of pipe according to the strength and composition of the steel used in its manufacture. In addition, pipelines add a safety margin by using only a portion of the lines' full strength, in conformity with the code recommendations issued by the American Standards Association (ASA).[264]

Improvements such as new welding techniques for producing the longitudinal welds, development of columbium and vanadium microalloys yielding higher pipe strength without reducing weldability, use of nondestructive testing immediately "downstream" from the welder to insure effective quality control, [265] and the employment of hydrostatic testing referred to previously virtually have eliminated pipe defects caused by material and workmanship.[266] The safety practices involved in corrosion control were discussed in the preceding subsection.[267]

Maintenance is an essential feature of pipeline operations. Large systems are divided into maintenance areas, each with its complement of personnel, equipment, and material. Included in this procedure are upkeep of the right-of-way, check of crossing markers, painting and repair of buildings, tanks, piping, and equipment, periodic checks of safety devices, and reconditioning of lines.[268]

---

[261] HOWREY & SIMON, *supra* note 3, at 148.

[262] *Hearings Before the Department of Transportation Hazardous Materials Regulation Board 53 (January 20, 1970) [hereinafter cited as DOT Hazardous Materials Hearings]* (Statement of Milton E. Holmberg).

[263] DOT Summary of Liquid Pipeline Accidents reported on DOT Form 7000-1 during 1976, at 5 with 173 out of 209 "accidents" (which includes incidents involving just property damage as well as injuries) tied to line pipe as opposed to 11 at pumping stations, 13 at tank farms, 4 at delivery points, and 8 at miscellaneous locations.

[264] Burke, *supra* note 9, at 791; for a specific example, see Exh. R-38 in Brief of Respondent, Lang. v. Colonial Pipe Line Co., Pa. Pub. Util. Comm. Dkt. 17915 where Colonial's 30" OD line had a 1352 psi tensile strength, 201 percent of the ultimate actual operating pressure and 191.5 percent of surge or maximum allowable pressure.

[265] *DOT Hazardous Materials Hearings, supra* note 262, at 38-40 (Statement of Carl E. Rawlins).

[266] HOWREY & SIMON, *supra* note 3, at 150, citing *DOT Hazardous Materials Hearings, supra* note 262, at 11.

[267] See text at notes 246-248, *supra*.

[268] PIPELINE TRANSPORTATION, *supra* note 162, at 29.

Pumping stations, which were the third most frequent cause of failure,[269] are usually started and stopped remotely by the use of matching control equipment at both the station and the control center. A signal from the dispatching center initiates a start-up procedure which must be sequentially followed with proper results or the start-up does not proceed any further and maintenance personnel are sent to the station to ascertain and correct any deficiency. A typical station has self-acting safety devices which react to (1) high or low pressure, (2) excess motor or pump bearing temperature, (3) excess pump case pressure, (4) excessive vibration, (5) overcurrent or undervoltage power supply, (6) excessive pump seal leakage, (7) high sump level, (8) low suction pressure, and (9) vapor detection. Any variation from the norm shuts down the station and signals the need for a maintenance crew.[270]

Safety regulation of pipelines commenced in 1921 when the Transportation of Explosives Act of 1909[271] was amended to include flammable liquids and solids. It conferred upon the Interstate Commerce Commission (ICC) the authority to formulate safety regulations for the transportation of explosives and other dangerous articles, including flammable liquids and compressed gases by interstate common carriers.[272]

The ICC conducted industry-wide surveys in 1930, 1935, and 1940 to determine the need for safety regulation of pipelines. It recognized their sound safety record in a report issued on February 24, 1942, which concluded that it was unnecessary to establish safety regulations for the industry.[273] An amendment to the Transportation of Explosives and Other Dangerous Articles Act in 1960[274] deleted pipelines from the coverage of the Act,[275] and from 1960 to 1965 no federal agency regulated oil pipeline safety.[276] There appeared to be no need; a survey covering the ten-year period ending in 1964 disclosed only six deaths and thirteen injuries to members of the public which had been caused by escape of liquids from oil pipelines.[277] In the absence of federal regulation, state legislatures began to consider the enactment of state safety laws which could have resulted in diverse and inconsistent requirements, hence the Transportation of Explosives Act was amended again

---

[269] See note 263, *supra*.
[270] PIPELINE TRANSPORTATION, *supra* note 162, at 29.
[271] 62 Stat. 739, 18 U.S.C. § 835 (1976).
[272] PIPELINE TRANSPORTATION, *supra* note 162, at 51.
[273] *Cf.* Bond, *supra* note 207, at 742.
[274] 74 Stat. 808, 18 U.S.C. §§ 831-835 (1976).
[275] Brief of Respondent, Lang v. Colonial Pipeline Co., Pa. Pub. Util. Comm. Dkt. 17915, p. 6 (May 15, 1964).
[276] PIPELINE TRANSPORTATION, *supra* note 162, at 51.
[277] *Id.* at 52.

on July 27, 1965, to cover oil pipelines.[278]

Before the ICC could produce a safety code, the Department of Transportation (DOT) came into being on April 1, 1967, and it was charged with the responsibility for pipeline safety. DOT began by requiring accident reports from pipelines in December 1967, and then promulgated a safety code effective April 1, 1970. It drew extensively on the industry's voluntary code, B31.4, which was sponsored by the American Society of Mechanical Engineers (ASME) and published by the American Standards Association (ASA).[279]

Miscellaneous Benefits and Summary

Because of the nature of pipeline transportation, which utilizes the pipeline as a container as well as a conveyor of the materials being transported, the stationary location of its power sources, the elimination of deadheading and backhauling,[280] the direct routing and the economies of scale involved, pipelines are relatively frugal in the consumption of energy required to do their job. Admittedly an extreme example, the movement of 1.4 million barrels per day of products handled by Colonial Pipe Line Company alone in 1974, if transported over the same distance by tank trucks, would have required 19,000 trucks running continuously (and consuming 3.3 million gallons of fuel per day).[281]

In a previous subsection, an examination was made of the cost effectiveness of pipelines vis-a-vis other forms of transportation and their relatively modest increase in costs over time.[282] There was documentation of the fact that the large jointly-owned and operated lines assisted the independent producer, encouraged production and stimulated competition for common carrier traffic.[283] These "industry lines" also made it possible for suppliers of independent marketers to sell to such customers in markets which otherwise would have been logistically nonviable for them.[284] Bringing this down to the bottom line, authority was cited for the proposition that the sav-

---

[278] 79 Stat. 285, 18 U.,S.C. §§ 831-835 (1976). *See Hearings Before the Subcomm. on Communications and Power of the House Interstate and Foreign Commerce Comm., Pipeline Safety 1969*, Ser. 91-5, 91st Cong., 1st Sess. 24 (1969) [hereinafter cited as *Pipeline Safety Hearings*].

[279] PIPELINE TRANSPORTATION, *supra* note 162, at 52. See Report Form reproduced in *Pipeline Safety Hearings, supra* note 278, at 29.

[280] See text at note 208, *supra; cf.* DE CHAZEAU & KAHN, *supra* note 30, at 335; but *cf.* NET-I, *supra* note 199, at 262 (water movement uses less energy than pipelines according to the American Waterways Operators).

[281] HOWREY & SIMON, *supra* note 3, Summ. at 2; MARATHON, *supra* note 202, at 5.

[282] See text in subsection 2, Anti-Inflationary Trends, *supra*.

[283] See text at notes 162 and 169, *supra*.

[284] See text at note 226, *supra*.

ings affected by pipeline integration had been passed on to the gasoline consumer.[285]

During the period 1960 to 1973, pipelines showed the greatest growth in productivity, as expressed in output per man-hour, among the industries surveyed by the Bureau of Labor Statistics, although the rate of growth eased off during the latter half of the period, in common with more than ⅔ of such industries.[286] Another table in the series showed that the average change in output per man-hour declined about 32 percent during the latter half,[287] perhaps reflecting the observation by Tierney that most pipelines had about exhausted the possibilities in automation.[288] This improvement has come from heavy investment; for example, there has been a 397 percent increase in net investment per employee, from $113,043 per employee in 1960 to $561,412 in 1975, a sizeable chunk of the increase in later years coming from the Trans Alaska Pipeline System.[289]

Oil pipelines have provided their own funding and have not sought government subsidies.[290] They have paid taxes at national, state, and local levels.[291]

Summing up, the nation accomplishes most of its work through the utilization of machines.[292] Oil and gas furnish approximately 74 percent of the basic energy required to operate these machines.[293] No other country has a comparable nationwide system for the transportation and distribution of

---

[285] See text at note 146, *supra; see also* Emerson, *Salient Characteristics of Petroleum Pipeline Transportation*, 26 LAND ECON. 27, 39 (1950): SPECTRE, *supra* note 104, at 111.

[286] U.S. DEPT. OF LABOR, BUREAU OF LABOR STATISTICS, REPORT 436, *Current Developments in Productivity 1973-74*, 18-20, Charts 3 and 4 (1975).

[287] *Id.* at 15, Table 6.

[288] See text at paragraph immediately preceding note 195, *supra*.

[289] Tierney, *TAPS Rebuttal Testimony, supra* note 175, at 14-15; Colonial operates with an average of one employee per million dollar investment. *Consumer Energy Act Hearings, supra* note 154, at 615 (Statement of Jack Vickrey). A question could be raised concerning the increase in the later years before TAPS was in operation but the point remains valid, because during the period 1958-1973, before the large addition to investment in TAPS, the average annual growth in interstate oil pipeline output per man-hour was twice that of the transportation sector in general. API, Productivity Change in the Petroleum Industry: A Review of Existing Estimates 14 (Discussion Paper #001R, December, 1977).

[290] Burke, *supra* note 9, at 780; *Petroleum Industry Hearings, supra* note 192, at 259 (Statement of Charles J. Waidelich), *Id.* at 306 (Statement of Vernon T. Jones); STEINGRABER, *supra* note 193, at 154; *cf.* Tierney, *TAPS Rebuttal Testimony, supra* note 175, at 19; *Consumer Energy Act Hearings, supra* note 154, at 593, 610 (Statement of Jack Vickrey).

[291] SPECTRE, *supra* note 104, at 121; *cf.* Thompson, *Recent Steps in Government Regulation of Business*, 28 Corn. L.O. 1, 14 (1942).

[292] Burke, *supra* note 9, at 791.

[293] *Id.* at 781; *cf.* Att'y Gen., *Fourth Report, supra* note 111, at 7: "Oil pervades every aspect of out national life. It is vital to what we do and how we live, to the services we need and the goods we buy."

petroleum and its related products.[294] The TAA statistics indicate that oil pipelines have been adaptable to change.[295] According to testimony by the Chairman of the Interstate Commerce Commission, oil pipelines seem to be one of the more efficient transportation systems we have.[296] They are said to be the safest, most efficient, least expensive, and most environmentally desirable overland method of transporting crude oil and refined products.[297] It would appear that the overall pipeline transportation system has physically developed into an efficient transport system, capable of serving the needs of a wide range of shippers and refiners.[298] Sellers seem able to reach more markets efficiently and at a lesser cost by these pipelines than would be possible otherwise.[299]

---

[294] National Petroleum Council, *U.S. Petroleum and Gas Transportation Capacities*, 29 (September 15, 1967); *Consumer Energy Act Hearings, supra* note 154, at 640 (Exh. C to Statement of Jack Vickrey, excerpting National Petroleum Council's Report); *cf.* SPECTRE, *supra* note 104, at 106.

[295] Tierney, *TAPS Rebuttal Testimony, supra* note 175, at 19.

[296] *Hearings Pursuant to S. Res. 45, Market Performance and Competition in the Petroleum Industry, Before the Senate Comm. on Interior and Insular Affairs*, 93d Cong., 1st Sess., ser. No. 93-24, pt.3 at 896 (1973) (Statement of I.C.C. Chairman George M. Stafford) [hereinafter cited as *S. Res. 45 Hearings*].

[297] *Petroleum Industry Hearings, supra* note 192, at 256 (Statement of Charles J. Waidelich).

[298] Att'y Gen., *Third Report, supra* note 111, at 79.

[299] STEINGRABER, *supra* note 193, at 134.

Chapter II

Today's Pipeline System — An Overview

A. Crude Oil Lines

1. Pregathering Procedures — Trucking

When an explorer is fortunate enough to become a producer, his first thought is to sell the oil so as to realize the value of his investment as rapidly as possible.[300] He first erects small tanks, known as field tanks, conveniently located on the lease; or if the field is unitized, somewhat larger tanks centrally located in the producing field,[301] where the gas is separated from the oil by a separator,[302] and Basic Sediment & Water ("BS&W")[303] are removed from the oil by gravity or sometimes by heating and/or chemicals.[304] The producer traditionally owns the smaller lines connecting the wells to the lease separators. These lines are known as "flow lines," where the well pressure is sufficient to move the oil from the wells to the tanks, or "lead lines," where the propelling force is the pump bringing the oil to the surface of the well bore.[305] Generally, these smaller lease tanks will hold no more than three or four days' production.[306] Occasional wells in "wildcat" territory[307] or wells that are too scattered[308] or distant from existing gathering systems to make it

---

[300] *Cf.* Att'y Gen., *Second Report, supra* note 111, at 87.

[301] Bond, *supra* note 207, at 732.

[302] A separator is a vertical cylinder in which the gas is permitted to expand and its velocity slowed. The reductions of velocity and pressure permit the gas to separate itself from the oil. The gas is recovered from the top of the cylinder through the "gas release line" and the oil continues into the receiving tank. If the stream were to proceed directly into the tank, the gas would escape from the vapor release at the top of the tank, carrying with it some of the lighter ends of the oil. Also, the oil might "boil" so violently that some of it would spray out of the tank. BALL, THIS FASCINATING OIL BUSINESS, 155 (1940) [hereinafter cited as BALL].

[303] This consists of sand from the producing formation, metals, water, and other undesirable materials. WOLBERT, *supra* note 1, at 27 n.142.

[304] MARATHON, *supra* note 202, at 9.

[305] *Cf.* KENNEDY STAFF REPORT, *supra* note 184, at 17.

[306] Bond, *supra* note 207, at 732; *cf.* KENNEDY STAFF REPORT, *supra* note 184, at 17 ("several days production"); Att'y Gen., *Third Report, supra* note 111, at 51 ("a relatively few days' production").

[307] Att'y Gen., *Second Report, supra* note 111, at 95 n.154: "In 1938, Commissioner Thompson of the Texas Railroad Commission stated before a congressional committee that every well in Texas, except a few isolated wildcats, had a pipeline connection."

[308] *Petroleum Industry Hearings, supra* note 192, at 399 (Statement of David Bacigalupo).

economic for the gatherer to extend its system, [309] are serviced by trucks. [310] The number of such wells, commonly known as "unconnected wells," is usually relatively small.[311] There was a brief flurry at the time of the Suez crisis when the number of unconnected wells in Texas rose to a high of just under 5 percent. The producers brought an action before the Texas Railroad Commission seeking: (1) a general order directing all common carrier pipelines to connect any well whose producer requested it; (2) an order directing all pipelines from West, West Central, and North Texas to the Gulf Coast to increase throughput capacity in an amount sufficient to carry the then current production plus a "cushion" to provide space for oil to be produced in the event of a national emergency; and (3) specific orders directing certain common carriers to extend connections in specific instances based on evidence adduced before the Commission.

The Commission held extended hearings in the spring of 1957 which disclosed that the individual complaining witnesses in fact had no purchasers for their crude and hence really were asking for an order requiring purchase and not carriage. The Commission found complainants' requests for individual connections to be unjustified and that trunk line capacity was adequate, hence it denied the specific requests for individual connections and the requested order for enlargement. It refused to enter a general order requiring connections of *all* wells in Texas, but instead established a procedure and criteria for adjudicating future specific unconnected well situations.[312] The producers took their case to the legislative forum, where they were received graciously by U.S. Senator O'Mahoney, who devoted to their cause four

---

[309] *Celler Hearings, supra* note 162, at 451 (Statement of Owen Clarke, Chairman of ICC); Section 1(4) of the Interstate Commerce Act, 49 U.S.C. § 1(4) (1976), provides that "It shall be the duty of every common carrier subject to this part to provide and furnish transportation *upon reasonable request therefor*, . . ." (emphasis added). This is a business judgment of a commerical feasibility. WOLBERT, *supra* note 1, at 25-26. "The determining factor seems to be whether the prospects of production in the new territory are sufficient reasonably to justify the expectation that an extension of the lines will result in profit to the company." Brundred Brothers v. Prairie Pipe Line Co., 68 I.C.C. 458, 463 (1922); MARATHON, *supra* note 202, at 10.

[310] MARATHON, *supra* note 202, at 9; *Consumer Energy Act Hearings, supra* note 154, at 656 (Statement of Jack Vickrey, Exhibit K); KENNEDY STAFF REPORT, *supra* note 184, at 18; *cf.* Att'y Gen., *Second Report, supra* note 111, at 71, 95, although in Louisiana this is sometimes performed by barges. Att'y Gen., *Third Report, supra* note 111, at 84.

[311] Att'y Gen., *Second Report, supra* note 111, at 155, Table 13.

[312] *Texas Railroad Comm., General Inquiry Relative to Common Carriers of Crude Oil in Texas* [hereinafter cited as *Tex. Gen. Inquiry*]; *Tex. Gen. Inquiry*, Pl. First Amended Petition 11 (1957); *Tex. Gen. Inquiry*, Pl.Br. 21; *Tex. Gen. Inquiry, Vol. I*, pp. 52-53 (George Mitchell); p. 70 (R. S. Anderson); p. 90 (E. L. Wilson); p. 156 (George Livermore); p. 214 (Harvey Gandy); *Vol. II*, pp. 8, 19 (Jack Boles) (April 1 and 2, 1957); *Texas Railroad Comm., Oil & Gas Dkt. Nos. 108, et al.,* Order No. 20-38,015 (June 4, 1958); for a commentator's view at the scene, see Rogers, *Common Purchaser, Market Demand, Pipeline Proration* in SOUTHWESTERN LEGAL FOUNDATION, NINTH ANNUAL INSTITUTE ON OIL & GAS LAW AND TAXATION 45, 75 (1958).

pages of his report on a pending bill to make ownership by an integrated oil company of a controlling interest in a pipeline a *prima facie* case of a substantial lessening of competition or of a tendency to create a monopoly under Section 7 of the Clayton (anti-merger) act. They also visited with the Justice Department which employed six pages of the Attorney General's *Second Report* to the subject, in which the author, perhaps inadvertently, created the impression that all of the "unconnected wells" belonged to "independent" producers, contrary to the fact that a very substantial percentage of such wells belonged to the large integrated oil companies. There was no finding of discrimination on the part of major oil companies.[313] The Texas Railroad commission handed down its order on June 4, 1958, in which it provided: "In any future case of request for connection, either the pipeline or the producer may bring the matter before the Commission and a hearing will be held on 10 days' notice. The Commission will consider, among other factors, the ability of the pipeline to transport the quantity of oil involved, the market or lack of market for the tendered oil, and the period required to return the investment for the connection."[314]

2. Gathering Systems

Gathering lines are those lines which serve one oil field, or a group of closely associated fields.[315] Their function is to pick up or "gather" the oil from the various field tanks and bring the oil to a central point, either a trunk line tank farm which accumulates sufficient oil to maintain the efficient rate of flow through the pipeline[316] or a direct injection point on the main trunk line.[317] These lines ideally are located in such manner as to drain the field tanks by gravity but where this is not feasible, power is supplied by small "field" or "gathering" pumps.[318] These lines, compared to trunk lines,

---

[313] S. REP. NO. 1147, 85th Cong., 1st Sess. 28-31 (1957) [hereinafter cited as O'MAHONEY REPORT]; *cf.* Att'y Gen., *Second Report, supra* note 111, at 94-101.

[314] *Texas Railroad Comm., Oil & Gas Dkt. Nos. 108, et al.,* Order No. 20-38,015 (June 4, 1958). In the 17 months following the Commission's decision, there was only a single unconnected well case filed pursuant to the order, and it was withdrawn by the complainant. SPECTRE, *supra* note 104, at 116.

[315] Testimony of W. J. Williamson in Trans Alaska Pipeline System, F.E.R.C. Docket OR 78-1, November 30, 1978 (formerly I&S 9164), 5 [hereinafter cited as W. J. Williamson].

[316] Bond, *supra* note 207, at 732. This is not a *storage* tank in the accepted sense of the word. Its function is analogous to a freight house, used to accumulate oil in sufficient quantity to permit efficient operation of the trunk line. *Id.*

[317] MARATHON, *supra* note 202, at 3; *cf.* W. J. Williamson, *supra* note 315, at 5; L. BARBE, *Pipeline Movements* in AOPL EDUCATORS CONFERENCE 2 (1978) [hereinafter cited as BARBE]; *Consumer Energy Act hearings, supra* note 154, at 610 (Statement of Jack Vickrey).

[318] Bond, *supra* note 207, at 732; BALL, *supra* note 302, at 175; BARBE, *supra* note 317, at 3; C. MATTHEWS, *Pipeline Operation and Maintenance* in AOPL EDUCATORS CONFERENCE 3 (1978) [hereinafter cited as MATTHEWS]. These small low-volume pumping units range in size from 5 to 200 horsepower. Burke, *supra* note 9, at 793.

seem small in diameter, short in length, and transport small volumes of oil.[319] However, since a gathering system consists of many small lines flowing into larger lines and, in turn, into even larger lines,[320] like a series of small streams that feed into a river,[321] their size, length, and volume transported will depend upon how extensively the gathering system develops.[322] Usually, the size runs from two inches to eight inches,[323] but it is not uncommon in large systems to find 10 or 12 inch lines.[324] Gathering lines, as one might expect, are concentrated in the leading producing states.[325] Total gathering line mileage in 1977 amounted to 67,798 miles.[326] Construction of the gathering system is commenced as early as possible in the life of an oil field, and it is expanded if and when new producing wells are brought in. This is a bit of a chicken-and-egg conundrum because the economic analysis of when and how large to build the line depends upon the gatherer's estimate of the commercial prospects for the line, which is a function of what the potential traffic will be (*i.e.*, how much oil is producable and at what rate),[327] and the producer's decision to drill additional wells and complete the development of the field is affected by how soon he can expect that his production will be served by a pipeline connection.[328] One of the advantages of vertical integration, which will be discussed in greater detail in a subsequent section, lies in the ability of the integrated company to coordinate its planning in this regard, and not having to contend with the transactional difficulty termed "opportunism" by Professor Oliver Williamson.[329]

Assume that a good-sized field is discovered. It is usual for more than one gathering system to be laid to the field because competition is very

---

[319] W. J. Williamson, *supra* note 315, at 5.
[320] Bond, *supra* note 207, at 732.
[321] BARBE, *supra* note 317, at 2-3; MATTHEWS, *supra* note 318, at 3; KENNEDY STAFF REPORT, *supra* note 184, at 18 (analogy to roots of tree).
[322] For an example of a rather extensive gathering system, see Shell Pipe Line Corporation, Texas Local Gathering Tariff No. 728, effective June 10, 1978, reprinted as Appendix B.
[323] BALL, *supra* note 302, at 175; Burke, *supra* note 9, at 793.
[324] MARATHON, *supra* note 202, at 3; MATTHEWS, *supra* note 318, at Fig. 1, reproduced herein as Appendix C; R. MILLS, THE PIPE LINE'S PLACE IN THE OIL INDUSTRY 38 (1935) [hereinafter cited as MILLS].
[325] OWINGS, *supra* note 176, at 1-2, lists these as Texas (40 percent); Louisiana (21 percent); California (11 percent); Oklahoma (5 percent); Wyoming (4 percent); and New Mexico (3 percent) in 1976. Alaska displaced California in crude oil production in 1977 although it is doubtful that the gathering systems were fully developed at that time.
[326] *Id.* at 4. These are ICC figures and do not include non-reporting intrastate systems.
[327] MARATHON, *supra* note 202, at 10. See also the discussion in note 309, *supra*, and the text corresponding therewith.
[328] W. J. Williamson, *supra* note 315, at 5.
[329] Williamson, *The Economics of Antitrust: Transactions Cost Considerations,* 122 U. Pa. L. Rev. 1439, 1445 (1974) [hereinafter cited as Oliver Williamson]; *cf.* D. TEECE, VERTICAL INTEGRATION AND VERTICAL DIVESTITURE IN THE U.S. OIL INDUSTRY 8 (1976) [hereinafter cited as TEECE].

keen.[330] Sometimes the producer,[331] but more often the purchaser, of the crude, sends to the gatherer of his selection a "request for connection" which gives the gatherer sufficient information to predicate a judgment of the commercial feasibility of making the requested connection. The gatherer then extends its line to the lease or field storage tank. The shipper gives the gatherer an estimate of the monthly volumes it expects to ship so that proper scheduling arrangements can be made. Gathering lines generally do not have minimum tenders.[332] If the system is "manual," the producer's field tanks are calibrated for volume per unit of fluid level ("strapped")[333] and when the tank is ready to be "run,"[334] the gauger samples the oil for quality of the oil ("marketability")[335] with a sampling tool commonly referred to as a

---

[330] Bond, *supra* note 207, at 732; MARATHON, *supra* note 202, at 10; W. J. Williamson, *supra* note 315, at 5-6. D. Behring, *Oral Argument on Behalf of Okie Pipe Line Company, an independent crude oil gatherer,* in F.E.R.C. Dkt. RM 78-2, Tr. 1829 (October 24, 1978).

[331] Most producers prefer to sell in the field. WOLBERT, *supra* note 1, at 47, 101; SPECTRE, *supra* note 104, at 112; Reduced Pipe Line Rates and Gathering Charges, 243 ICC 115, 122, 140 (1940); *TNEC Hearings, supra* note 6, at 7309, 8247, 8343-8344; S. REP. No. 25, 81st Cong. 1st Sess. 20 (1949); MARATHON, *supra* note 202, at 10; J. Vickery, *Oral Argument on Behalf of Belle Fourche Pipeline Company, an independent gatherer, the Independent Petroleum Association of America and the Transportation Association of America,* F.E.R.C. Dkt RM 78-2, Tr. Vol. 12 at 1927 (October 24, 1978). This is because they usually have small volumes; the average U.S. well produced approximately 17 barrels per day in 1975, and frequently production is divided among fractional interest owners under "division orders." Historically, independent producers have a ready market in the field and the last thing they want to do is to tie up money in pipeline inventory or pipelines. They are having enough trouble borrowing money to carry on the exploration business. Statement of Jack M. Allen, independent oil producer and President of IPAA in F.E.R.C. Dkt. RM 78-2, p. 7 (October 24, 1978). In fact, most independent producers are reluctant or unwilling even to invest in the gathering phase of the business. Att'y Gen., *Second Report, supra* note 111, at 99-100; AOPL, PETROLEUM PIPELINE PRIMER 5 (Jan. 8, 1979) [hereinafter cited as PIPELINE PRIMER].

[332] MARATHON, *supra* note 202, at 11. See Shell Pipe Line Corporation Tariff in Exhibit B.

[333] The process of determining accurately the volume that each tank contains at a certain level is termed "strapping." The circumference of the tanks is measured at several heights. After taking into account the tank wall thickness and the displacement of the internal bracing members, tank tables are computed which show volumes in barrels to the closest 1/100 corresponding to each 1/4 inch increment in height for small tanks and for each 1/8 inch for larger tanks. WOLBERT, *supra* note 1, at 27 n.138; BALL, *supra* note 302, at 175.

[334] The producer's employee, known as a "switcher" (in flowing well areas) or "pumper" (where the wells are on pump) may notify the gatherer's employee, commonly known as the "gauger," or if the operation is routinized the latter may just stop by on a regular basis. In either event, it is the responsibility of the gatherer to see that the tanks are emptied frequently enough that the producer always has room in his field tanks for his daily production. Bond, *supra* note 207, at 732.

[335] The minimum quality of the oil, termed "marketable" in the tariff, specifies the percentage of BS&W or other impurities above a point usually normally four to six inches below the tank outlet connected to the pipeline. See Rule 1 in the Shell Pipe Line Tariff, Appendix B (2 percent at a point six inches below the pipeline connection).

"thief,"[336] measures the quantity in the tank,[337] and takes the temperature[338] and gravity of the oil.[339] After the gauger records this information on the "opening" part of the run ticket,[340] he breaks the seal on the pipeline valve and turns it on the line and closes and seals all other connections.[341] After the oil has run from the tank, the gauger returns and makes a "closing" measurement of volume and temperature of the oil remaining in the tank below the pipeline connection. He then completes the run ticket and reverses the position and sealing of the valves.[342] If the representative of the shipper has been present, he countersigns the run ticket.[343] From the time the pipeline valve is opened on the field tanks allowing the oil to run into the gathering line, the custody and responsibility for the oil is shifted to the gatherer.[344] At the same time, if the shipper is an oil purchaser, rather than a producer, the entry into the line also constitutes a sale and title passes from the producer to the purchaser.[345] The run ticket serves as a basis for the pay-

---

[336] A thief is a cylindrically shaped instrument with a spring closing bottom plate which enables the gauger to obtain a sample from any depth in the tank. The usual practice is to take a sample from just below the surface and one from the oil adjacent to the pipeline outlet. The reason for this is that the BS&W tends to increase at the latter point due to the gravitational settling action. The samples are usually measured by a portable hand operated centrifuge equipped with two graduated test tubes. Rapid revolution produces a centrifugal force which deposits the heavier BS&W in the bottom of the test tube. In order to secure a sharper reading, the sample is usually "cut" 50 percent by gasoline or other solvent which increases the weight differential. The percentage of BS&W is obtained by adding the percentages shown on the two test tubes. If either reading exceeds the percentage specified on the tariff, the tank is "turned down" and the producer must treat the oil and bring it into specification. If both samples "pass" the test, the results are averaged.

[337] This is done by measuring the length of the surface with a graduated steel tape with a weight similar to a plumb bob. There may still be some old graduated rods in use. In either event, the measuring decree is known as a gauge—hence the term "gauger." BALL, *supra* note 302, at 175. This measurement is known as the "opening" gauge. WOLBERT, *supra* note 1, at 27 n.140.

[338] Crude oils and products expand and contract as the temperature rises and falls. In order to correlate the measurements, a common base temperature, arbitrarily selected as a standard, is 60°F. A volume adjustment is made by applying expansion coefficients to the volume determined by the gauge level and read from the tank table. See WOLBERT, *supra* note 1, at 27 n.141 for details of the adjustment.

[339] The gravity measurement, explained in detail in WOLBERT, *supra* note 1, at 27-28 n.143, is required because the price structure of crude oil hinges on API gravity with a price increment added for each degree higher gravity and subtracted for each degree lower gravity. *Id.*; Burke, *supra* note 9, at 793.

[340] See WOLBERT, *supra* note 1, at 28 n.144 for description of the "run ticket" and its use. The owner of the oil or his representative is entitled to be present while the foregoing operation is conducted. Whether or not he is present, the owner will receive a copy thereof. *Id.*; Bond, *supra* note 207, at 732.

[341] Burke, *supra* note 9, at 793; WOLBERT, *supra* note 1, at 28.

[342] *Id.*

[343] Burke, *supra* note 9, at 793; MILLS, *supra* note 324, at 38.

[344] Bond, *supra* note 207, at 732.

[345] WOLBERT, *supra* note 1, at 28 n.145.

ment to the producers and royalty owners on behalf of the purchaser [346] and for payment by the shipper to the gatherer. Trunk line delivery or transit receipts provide the basis for division of transportation charges between connecting carriers in a joint movement.[347]

In a modern field, a substantial percentage of the production from leases is transferred automatically through Lease Automatic Custody Transfer (LACT) units. These units automatically pump from a lease "surge tank"[348] when the oil reaches a predetermined level, monitor the oil for quality, measure the volume going through a meter as it is transferred, and remove a continuous sample for checking by the gauger.[349] The gauger or a meterman periodically (usually monthly) "proves" the meter by checking it against the calibrated master "prover," determines the quantity and quality of the oil moved through the meter and writes up a ticket receipting for the oil.[350] Movement through the LACT unit automatically kicks off the field pumping unit or booster station in order to maintain the continuity of the system. For a schematic diagram of an automated gathering system, see Appendix C.

Where there is a large modern unitized field, much or all of the oil goes directly from the individual wells or leases to a central treating station servicing the entire field. Under such an arrangement, the pipeline carrier may receive all the treated oil at one central, automated custody transfer point. An extreme example is the Prudhoe Bay, Alaska, field, where the total production is unitized for economy of operation and maximization of ultimate recovery. The unit operator delivers all production to the Trans Alaska Pipe Line System at Pump Station Number One. In effect, the gathering function is performed by the unit operator and not by TAPS.[351]

Another factor that bears mention in the subject of gathering is the prevalence of "common purchaser" statutes which require crude oil purchasers to purchase ratably and without discrimination.[352] The first such law was passed in Oklahoma in 1909[353] arising from the fact that the tremendous production in the Glenn Pool[354] far outstripped pipeline capacity existing

---

[346] Att'y Gen., *Second Report, supra* note 111, at 70.
[347] Burke, *supra* note 9, at 793-795.
[348] In some units there is a "bad oil" tank into which is switched any oil "flunking" the monitoring tests. Burke, *supra* note 9, at 794. Any such diversion triggers an alarm which alerts the producer to the difficulty. *Id.*
[349] MATTHEWS, *supra* note 318, at 4; BARBE, *supra* note 317, at 3.
[350] MATTHEWS, *supra* note 318, at 4; *cf.* Burke, *supra* note 9, at 794.
[351] BARBE, *supra* note 317, at 4.
[352] MARATHON, *supra* note 202, at 12; *Consumer Energy Act Hearings, supra* note 154, at 600, 610 (Statement of Jack Vickrey).
[353] OKLA. STAT. ANN. tit. 52, § 54 (West).
[354] See text at note 80, *supra*.

therefrom.[355] Texas[356] enacted such a statute in 1930,[357] followed by New Mexico[358] and Colorado.[359] In Oklahoma, a common purchaser is directed to purchase "all (or a proportionate amount) of the petroleum in the vicinity of, or which may be reasonably reached by, its pipelines, or gathering branches, without discrimination in favor of one producer or one person as against another" and such common purchaser is forbidden from discriminating in price or amount for like grades of oil or facilities as between producer or persons. Texas similarly requires that a common purchaser "shall purchase oil offered it for purchase without discrimination in favor of one producer or person as against another in the same field, and without unjust or unreasonable discrimination as between fields."

One other development of importance has been the tremendous growth of the independent gatherer (one other than a major refiner or pipeline company). The Attorney General in his Third Report on the operation of the Interstate Compact to Conserve Oil and Gas devoted seven pages to the subject.[360] The Report stated that despite the fact that published pipeline reports did not identify the ultimate source of oil movements into the interstate system, comparison of the ICC annual publications with area production figures identified a perceptible growth in gathering by firms other than the interstate pipelines.[361] Note that the Attorney General apparently attributed all non-ICC reporting activity to independents. The Report concluded that such comparison disclosed a growth of the proportion of crude oil which never reached the interstate system from 7.1 percent in 1948 to 10.1 percent in 1956; moreover, the proportion of oil introduced into the interstate network which had been gathered by non-ICC reporting lines had grown even more substantially. Thus, over the same period, 1948-1956, the amounts reported by individual companies as having been received from other gatherers indicated a growth of independent gathering from 13.9 percent in 1948 to 17.2 percent in 1956.[362] Addition of the percent of production never handled by ICC pipelines to the percent of interstate pipeline crude

---

[355] E. ZIMMERMAN, CONSERVATION IN THE PRODUCTION OF PETROLEUM, 136 (1957), quoting the saying "more oil has run down the creeks from Glenn Pool [from independent producers' earthen storage] than was ever produced in Illinois."

[356] TEX. REV. CIV. STAT. ANN, tit. 103, art. 6049a, § 8 (Vernon).

[357] E. ZIMMERMAN, CONSERVATION IN THE PRODUCTION OF PETROLEUM 146 (1957).

[358] N.M. STAT. ANN. § 65-3-15 (Allen Smith 1973).

[359] COLO. REV. STAT. § 34-60-117: a pipeline owner in "purchasing" or "taking for transportation" from a field must do so "ratably" and without discriminating in favor of his own production.

[360] Att'y Gen., *Third Report, supra* note 111, at 79-86.

[361] *Id.* at 80.

[362] *Id.* at 81.

"originating" from "outside" gatherers, produced a total of 27 percent in 1956, up from 21 percent in 1948.[363]

Examining individual oil producers' reports to state agencies, the authors opined that these data corroborated the size of the trend toward independent gathering which had been gleaned from the ICC reports. They noted that this growth could not have occurred without the active aid of the major integrated oil companies.[364] The Report mentioned "many recent instances" in which substantial pipeline gathering systems had been sold to independent gatherers by the interstate carriers, although the selling interstate carrier remained the only trunk line carrier through which the oil could be shipped.[365] The authors then asserted that the willingness of the interstate carriers to aid this process could not be wholly explained in terms of the non-profitability of the gathering operation. They cited a ten year *increase* of 4.15 percent in average "operating ratios" (ratio of operating expense to operating revenues) of gathering lines as opposed to a 16.21 percent *decrease* in trunk line "operating ratios,"[366] which would tend to lead to a conclusion that the relative unprofitability of gathering lines vis-a-vis trunk lines was precisely the reason the companies acted as they did. However, the authors concluded that "some possible benefits to the interstate carriers may be surmised," *i.e.*, insulation from state common purchaser regulation and the acquisition of a flexible means of connecting to new production without the investment involved in gathering line construction.[367]

The Report summarized its discussion of the purchase and transportation of crude oil by remarking that the past half century had seen a vast development in the physical structure of pipelines from a few isolated, low-capacity lines capable of serving only particular refineries into an interconnected pattern able to serve public convenience and necessity as true common carriers between a wide variety of shipping points in the producing fields and destinations in the refinery areas throughout the country. The function of pipelines in the marketing of crude apparently also had changed, concluded the Report, with preliminary indications that the integrated pipeline-refining companies were withdrawing from the actual gathering of crude oil in the fields.[368] McLean & Haigh, in their monumental study on *The Growth of Integrated Oil Companies*,[369] recognized the same phenomenon when they analogized the formation and alteration of integration patterns in the in-

---

[363] *Id.* at 82.
[364] *Id.* at 84.
[365] *Id.*
[366] *Id.* at 84-85.
[367] *Id.* at 86.
[368] *Id.* at 86-87.
[369] *MCLEAN & HAIGH, supra* note 35.

dustry to the process found at work throughout the entire world of living organisms, which are continually making a progressive adaption to the physical environments in which they exist. As new conditions emerge, organisms which fail to make the necessary adaptations inevitably suffer a competitive disadvantage in their struggle for survival; likewise, in the economic world, business corporations must continually alter their structures and seek new adaptations to the reality of their surroundings if they are to remain strong and vigorous and able to stand up to the relentless pressure of competition.[370] This, then, is a form of voluntary disintegration which McLain & Haigh found to be a manifestation of the working of competition in the industry.[371]

Bringing the subsection on gathering lines to a close, it should be of interest to note that the role of "independent" crude oil transporters has continued to grow. In Texas, where we have the benefit of Railroad Commission statistics, in July, 1978, the largest single gatherer in the huge East Texas field was Scurlock, which has no pipelines reporting to the ICC but a significant amount of non-ICC lines in a number of states, including 1,020 miles in Texas alone. The second largest is Matador (100 percent owned by Koch). Koch has no ICC reporting lines in its own name, although it owns a 29.5 percent interest in the 365 mile Minnesota Pipe Line Company and a 100 percent interest in the 383 mile Okie Pipe Line Company products pipeline. It has a significant number of non-ICC crude lines in a number of states, including 487 miles of such lines in Texas.

A compilation of the leading crude oil gatherers in Texas during December, 1978 is shown as Appendix D. "Independents," an amorphous grouping which has expanded and contracted over the years in a manner which appears empirically to be designed to include only those companies

---

[370] *Id.* at 672.
[371] *Id.* at 674.

that are not required for concentration statistics purposes,[372] are designated by an asterisk. Appendix E is a compilation, derived from Texas Railroad Commission Forms R-1, which lists Refinery Receipts of Crude Oil and Condensate by each refiner in Texas, during the first half of 1978, broken down by the delivering pipeline. It is intended to show that both "majors" and "independents" gather and transport crude oil for themselves and others; that there is a significant "independent" sector of transportation, and that such sector is a substantial factor in supplying some refiners. It also should demonstrate that "majors" do not "control" crude oil deliveries to the "in-

---

[372] As pointed out by W. T. Slick, Jr., in a colloquy with Charles E. Bangert, Majority General Counsel, during Slick's testimony in the *Petroleum Industry Hearings, supra* note 192, at 353, if you make the sample small enough, such as the ownership of all service stations on the northeast corner of, say, Spruce & Goose, you can get a 100 percent concentration, or if you select a number of companies that is large enough, you can also obtain as high a concentration ratio as you desire. Early on, the division was drawn between the Standard Oil Group, and all other companies, which became the "independents." *TNEC Hearings, supra* note 6, at 9932. Later the division shifted to a distinction between those companies which had interstate pipeline facilities ("majors") and those which did not ("independents"). H. LAIDLER, CONCENTRATION OF CONTROL IN AMERICAN INDUSTRY 27 n.13 (1931); Dr. Walter Splawn's report in 1933 refined this concept by ranking companies by the *size* of their investments in oil and gasoline pipeline facilities, thereby producing a list of 20 "majors" topped by Standard Oil Co. (N.J.) ("Exxon") and extending to Deep Rock Oil Corp. ("Deep Rock"), with all other companies presumably being "independents." H.R. REP. NO. 2192, 72d Cong., 2d Sess. XXVII-XXVIII (1933) [hereinafter cited as SPLAWN REPORT]. By the TNEC days of 1939, the aggregation was still 20 in number, but changed somewhat in composition. *TNEC Hearings, supra* note 6, at 7110-7111. The cut-off point appeared to be assets in excess of $25 million. *Id.* at 7111. The Department of Justice's complaint in United States v. American Petroleum Institute, Civil No. 8524, D.D.C., Sept. 30, 1940, added two more companies, Barnsdall and Standard Oil Co. (Ky.) to the "major" classification. WOLBERT, *supra* note 1, at 3. In 1973 the FTC staff, at the request of the Chairman of the Senate Interior & Insular Affairs Committee, prepared a report which classified 18 companies as "majors." *Consumer Energy Act Hearings, supra* note 154, at 666. In 1975, the FTC's Bureau of Competition and Economics released a report entitled FEDERAL TRADE COMMISSION STAFF REPORT ON THE EFFECTS OF DECONTROL ON COMPETITION IN THE PETROLEUM INDUSTRY (Sept. 5, 1975) [hereinafter cited as DECONTROL REPORT], reprinted in *Petroleum Industry Hearings*, supra note 192, at 1035-1050, which used a three-fold test of refining capacity, crude self-sufficiency and gasoline pricing policies (those which stress price versus those which place primary weight on service and image). Purporting to use those criteria, the Report came up with 16 "majors". *Id.* at 1037. In 1976, the Senate Judiciary Committee issued its report summarizing the results of its first round of hearings on S.2387, which produced a list of 18 "majors". S. REP NO. 94-1005, 94th Cong., 2d Sess. 17 (1976). Another classification in 1976, made by the staff of a subcommittee of the Senate Interior Committee, would have boosted this number back up to the mystic 20 by adding Occidental Petroleum Company and Tenneco, Inc. STAFF OF SPECIAL SUBCOMM. ON INTEGRATED OIL OPERATIONS OF SENATE COMM. ON INTERIOR AND INSULAR AFFAIRS, 94th Cong., 2d. Sess, THE STRUCTURE OF THE U.S. PETROLEUM INDUSTRY: A SUMMARY OF SURVEY DATA 109-111 (Comm. Print 1976). This number was adopted by W. K. Jones, in his report to the Department of Energy; W. K. JONES, AUTHORITY OF THE DEPARTMENT OF ENERGY TO REGULATE ANTICOMPETITIVE ASPECTS OF PETROLEUM PIPELINE OPERATIONS 8 (1978) [hereinafter cited as JONES DOE REPORT] but the Kennedy Staff declined to follow it because Occidental had no refining activity in the United States and Tenneco was a conglomerate, primarily engaged in non-petroleum activities, KENNEDY STAFF REPORT, *supra* note 184, at 30 n.93, so it chose to go with the 18 company list of S. REP NO. 94-1005 in issuing its own report. *Id.* at 30.

dependent" refiners in Texas, and , by implication, that "majors" appear to be willing to sell significant volumes of imported crude oil to the "independent" sector.

3. Trunk Lines

a. General Description

Crude oil trunk lines, which range in diameter from 8 inches to 48 inches, in the case of TAPS, commonly transport crude oil from the origin station, which receives oil from one or more gathering systems, to one or more refineries located throughout the country or to seaboard or inland waterways for transshipment by tankers or barges.[373] In times past, and to a very limited extent in the case of movements to Canada on a reciprocal basis beneficial to both sides, trunk lines have transported crude for export. Today, the flow is reversed and trunk lines have been built or expanded to carry oil received from overseas.[374] Some lines, such as Platte, were constructed on an area-to-area (*e.g.*, Producing Region to Refinery Area or to multi-interchange points with other pipelines), or logistical basis. An idea of the crude trunkline magnitude can be gleaned from the mileage which, in 1977, was 77,972 miles,[375] although this figure must be magnified mentally because the average diameters, and hence the capacities, have steadily been growing larger and larger.[376]

Trunk line components consist physically of large diameter pipe, origin stations, booster stations, delivery terminals, a communications system, and one or more control rooms.[377] A trunk line may have more than one origin station and usually has several booster stations and delivery points.[378] All of these facilities are devoted to the primary purpose of the pipeline which is to achieve the continuous movement of the fluid being transported.[379] This liquid is moved by differential fluid pressure provided by the pumps at the origin and booster stations.[380] It follows hydraulically that in order for the

---

[373] MARATHON, *supra* note 202, at 4; PIPELINE TRANSPORTATION, *supra* note 162, at 18; *Consumer Energy Act Hearings, supra* note 154, at 610 (Statement of Jack Vickrey).
[374] W. J. Williamson, *supra* note 315, at 6; NET-I, *supra* note 199, at 195, 196, 204.
[375] OWINGS, *supra* note 176, at 4.
[376] See text at notes 165-169 and 186, *supra*; *cf.* NET-I, *supra* note 199, at 202-206.
[377] MATTHEWS, *supra* note 318, at 5; *cf.* KENNEDY STAFF REPORT, *supra* note 184, at 21.
[378] MATTHEWS, *supra* note 318, at 5. See Schematic Diagram of Typical Trunk System in Appendix E.
[379] KENNEDY STAFF REPORT, *supra* note 184, at 21; *cf.* NET-I, *supra* note 199, at 198.
[380] Burke, *supra* note 9, at 792; KENNEDY STAFF REPORT, *supra* note 184, at 21-22.

fluid to achieve the desired continuous movement, which usually[381] varies from three to five miles per hour, the line must always be full of fluid ("line fill").[382] The amount of pressure required for a given flow rate ("hydraulic gradient") can be calculated with great accuracy and, with the advent of computers, rather easily; it is a function of changes in elevation along the line ("profile"), length and diameter of line, interior surface condition of the pipe, and the density and viscosity of the material being transported.[383]

An origin station is comprised of the following basic units: (a) tankage sufficient in number and capacity to segregate the different grades received and to provide working space to maintain a continuous flow to the next station or terminal, to receive the variable runs from gathering systems and from connecting carriers and to accumulate suitable quantities of different grades of crude for desired batch sizes;[384] (b) main pumping units; (c) valves and piping to connect incoming lines from connecting carriers and gathering systems to the tanks, the tanks to the main pumping units and the pumps to the outgoing trunk line; (d) booster pumps to pump the oil from the tanks to the main pumping units; (e) safety devices to protect each pumping unit and the pump station; and (f) sampling devices and meters to determine the quantity and quality of the oil handled.[385]

Booster or pumping stations are located along the line at intervals determined by the hydraulics involved.[386] Early pump stations were "floated" on the line, *i.e.*, incoming oil was received into a tank and then reinserted back into the line.[387] This system is wasteful, expensive, and introduced unnecessary hazards. It no longer is used except where special circumstances such as differential rates of flow or pressures between sections of the line require the storage known as "working" or "breakout" tankage to be used.[388] In the absence of such special circumstances, most trunk lines today are operated as a "closed" system,[389] that is to say, with continuous movement through the sequential stations without being diverted into tanks.[390] This

---

[381] The original displacement of the material used for the initial line fill and hydrostatic testing frequently is done at a lower rate in order to facilitate early detection, rapid shutdown, and repair, and to minimize leakage of the valuable, and environmentally less acceptable, petroleum. For example, TAPS started up at a flow rate of one mile per hour. See text at note 174, *supra*.

[382] KENNEDY STAFF REPORT, *supra* note 184, at 22; *cf*. NET-I, *supra* note 199, at 198.

[383] Burke, *supra* note 9, at 792.

[384] BARBE, *supra* note 317, at 4; MATTHEWS, *supra* note 318, at 5-6.

[385] MATTHEWS, *supra* note 318, at 6.

[386] Burke, *supra* note 9, at 792 (30 to 100 mile intervals); KENNEDY STAFF REPORT, *supra* note 184, at 22.

[387] KENNEDY STAFF REPORT, *supra* note 184, at 22.

[388] *Id.*; *Consumer Energy Act Hearings*, *supra* note 154, at 605 (Statement of Jack Vickrey).

[389] See note 120, *supra*.

[390] KENNEDY STAFF REPORT, *supra* note 184, at 22.

system of operation and the efficiency and economies made possible by automatic operation have caused most pump stations to be fully automated, requiring no attendant manpower.[391] Monitoring and supervisory control are maintained by a remote control room by use of communications systems, either wire or microwave.[392] This not only enables the system to operate at its maximum efficiency but also, through the use of volumetric line balances, pressure monitoring and the like, enables the supervisor to detect and control pipeline leaks, spills, etc. At the pumping stations, the discharge pressure is increased to design limits, generally ranging from 500 to 1,500 pounds per square inch ("psi"),[393] by one or more centrifugal pumps driven by electric motors, diesel engines, or gas turbines[394] comprising several thousand horsepower.[395]

A delivery terminal generally includes facilities for delivery into tankage at a refinery, a waterway dock, or a connecting carrier. Such a terminal would include valves, piping, and volume and gravity measurement facilities.[396] Again, depending upon the degree of automation, the number of operating personnel will vary from one deliveryman for each eight hour shift to only one or two deliverymen to make up delivery tickets to record the transfers and perform miscellaneous duties.[397] The degree of automation and the resulting continuity of flow also reduces the need for tankage.

A reliable communications system is the "nerve system" of pipelines. It links each pump station and terminal together to the control room and to the dispatcher.

The "brain" of the operation is the control room. Some systems have only one central room which controls the entire system whereas others utilize control rooms for designated segments of the system. A control room has a control panel which displays continuously key data such as tank levels, pressures, whether pumping units are operating or stopped, and whether valves are opened or closed. The control room operator is enabled by the control systems to operate each pump station and to regulate the flow of oil through the system. He can start and stop each pumping unit, open or close each key valve, obtain a readout of tank volumes, and determine receipts into and deliveries out of the system.[398] It is quite common to have computers

---

[391] Burke, *supra* note 9, at 792; *cf.* KENNEDY STAFF REPORT, *supra* note 184, at 22 (except for maintenance or correction of problems); MATTHEWS, *supra* note 318, at 7 (one station attendant on duty during day shift).

[392] Burke, *supra* note 9, at 792; MATTHEWS, *supra* note 318, at 77.

[393] Burke, *supra* note 9, at 792.

[394] KENNEDY STAFF REPORT, *supra* note 184, at 22.

[395] Burke, *supra* note 9, at 792.

[396] MATTHEWS, *supra* note 318, at 7.

[397] *Id.*

[398] *Id.* at 8.

in the control room and in the dispatcher's office which permit more rapid and accurate acquisition of data from the system and assist the operator in making the calculations necessary to operate the system.[399] Some pipeline systems have taken the next step and utilize computers to actually control the normal operations of the system, including the starting and stopping of pumping units, thereby freeing the dispatcher to merely monitor the operation when everything is proceeding normally and to devote full attention to any abnormalities which might occur.[400]

b. Competition

As was presaged by the previous subsection entitled "Flexibility and Alternatives to Shippers,"[401] competition is a way of life on trunk lines. If one were preparing an S-1 or S-7 registration statement for filing with the Securities and Exchange Commission ("SEC"), he could not fail to include a warning such as "This business is highly competitive" without drawing a comment from the SEC. This subject will be developed in greater detail in a subsequent section on economic characteristics of pipelines but to provide the reader with some facts of the matter, consider the following illustrative examples: There are thirteen common carrier crude trunk lines competing for the movement of crude oil from the West Texas Producing Region to the Texas Gulf Coast Refining Area;[402] nine such carriers battling for traffic from the East Texas Producing Region to the Texas Gulf Coast Refining Area;[403] ten pipeline companies provide common carriage for both indigenous and imported crudes from St. James, Louisiana, to Toledo, Ohio;[404] similar competitive routes are available to service Toledo from each of three origin points, viz., Lake Charles, Louisiana, Nederland, Texas, and Freeport, Texas;[405] and producing areas in Wyoming and the Rocky Mountains could have reached any refinery in the Midwest by three routes: Amoco Pipe Line (formerly Service Pipe Line) which runs to a junction near Kansas City, thence on to the Chicago area; Platte Pipe Line to a junction near St. Louis with other lines into Chicago; and formerly Arapahoe (now sold to Cities Service for gas transmission service) which received from other lines

---

[399] *Id.* at 8-9.
[400] PIPELINE TRANSPORTATION, *supra* note 162, at 26; *cf.* KENNEDY STAFF REPORT, *supra* note 184, at 22.
[401] See text at notes 212-235, *supra*.
[402] LeGrange, *FERC Direct Testimony, supra* note 176, at 23; HOWREY & SIMON, *supra* note 3, at 19; LIVINGSTON, *supra* note 228, at 63.
[403] LeGrange, *FERC Direct Testimony, supra* note 176, at 23; HOWREY & SIMON, *supra* note 3, at 19-20.
[404] LeGrange, *FERC Direct Testimony, supra* note 176, at 23-24, LIVINGSTON, *supra* note 228, at 63; *cf.* HOWREY & SIMON, *supra* note 3, at 20.
[405] LeGrange, *FERC Direct Testimony, supra* note 176, at 24; HOWREY & SIMON, *supra* note 3, at 20.

serving Wyoming production and in turn connected with other lines going into Chicago.[406]

c. Scramble for Traffic

In previous sections, mention was made of the competitive consequences of the movement to "big inch" lines. The tremendous outlay of funds involved, and the fact that the volume of traffic required to fully realize the economies of scale surpassed what a single refiner could provide, caused jointly owned and operated lines to become commonplace.[407] The economic incentive, which had disfavored nonproprietary traffic in the old small capacity line era,[408] was shifted strongly to broaden the base of ownership and to solicit vigorously the traffic of those not interested in taking an ownership position.[409] The battle for crude oil movements from the Wyoming/Rocky Mountain producing regions to the Kansas City/St. Louis and Chicago refinery areas was discussed.[410] These developments caused Professor Johnson to observe that "these pipelines . . . had created a situation that benefited independent producers, encouraged production, and stimulated competition for common carrier traffic."[411] They also raised a substantial doubt concerning the application of the "natural monopoly" theory to pipelines as a class. These comments were directed to the climate which existed in 1959. What has been the situation since that time? Two good indicia of a competitive pipeline system such as described above are the ratio of utilization to capacity and the degree of nonowner shipments. In the absence of a comprehensive pipeline system-by-system or segment-by-segment analysis, which would exhaust both author and reader, yearly data have been assembled on a few representative lines. Finding a selection process which minimized bias has been simplified because Assistant Attorney General Shenefield,[412] the Kennedy Staff Report[413] and Senator Kennedy's petition filed with the Federal Trade Commission (FTC)[414] have identified certain pipelines which they suggest have exhibited anticompetitive behavior.[415]

---

[406] LIVINGSTON, *supra* note 228, at 62; see text at note 168, *supra*.
[407] See text at note 161, *supra*; *see also* KENNEDY STAFF REPORT, *supra* note 184, at 29.
[408] See text at note 112, *supra*.
[409] See text at note 162; *cf.* text at note 168, *supra*.
[410] See text at note 168, *supra*.
[411] See text at note 169, *supra*.
[412] SHENEFIELD PROPOSED STATEMENT, *supra* note 179, at 28 (Colonial, Plantation).
[413] KENNEDY STAFF REPORT, *supra* note 184, at 29 (Texoma), 66-68 n.34 (Colonial, Plantation, Wolverine, Yellowstone, and Platte), 77, 79, 131 (Explorer).
[414] Before Federal Trade Commission, Petition for the Initiation of a Rulemaking Proceeding Prohibiting Ownership of Petroleum Pipelines by Petroleum Companies, Jan. 4, 1978, p. 6 (Colonial, Explorer) [hereinafter cited as Kennedy Rulemaking Petition].

Today's Pipeline System — An Overview 63

In order to take advantage of this simplification, both crude oil and products lines will be discussed together at this point because the allegations and refutations are common to both, and this discussion anticipates only slightly the more extended examination of products lines which follows in the next subsection. The lines named as suspects are: Colonial (products), Plantation (products), Wolverine (products), Yellowstone (products), Explorer (both products and crude), Platte (crude) and Texoma (crude). Discussing first the more simple issue, *i.e.*, the extent of nonowner shipments transported by these lines, we find that the percentage of nonowner shipments through Colonial during 1964, the first full year of operation, was 0.15 percent. In 1965, the year the construction of the original system was fully completed,[416] the percentage was 0.62 percent. After a brief dip in 1966 to 0.37 percent occasioned by a doubling of throughput by the addition of pumping capacity,[417] the "outside" shippers' usage rose steadily until 1970, when it increased by 150 percent due to prorationing[418] of owners' shipments to honor the requests of nonowners for space.[419] It took another quantum jump in 1971 to 19.75 percent and it has increased steadily thereafter until 1978, when nonproprietary shipments constituted an estimated 36.26 percent of its traffic.[420] In fact, from 1969, when the owners' shipments peaked out at about 1.1

---

[415] Senator Kennedy's petition virtually concedes the absence of "hard" or "solid" evidence —page 2 of the Petition states: "This method is especially appropriate where an elaborate factual record need not be developed." Page 3 contains an unsupported assertion, resting on an alleged "inherent competitive advantage" that accrues to the "oil companies."
[Presumably, this does not mean accruing to all the "independents"—as defined in the text in the last paragraph of subsection 2 *supra*—which, to a substantial degree, also own pipelines, roughly in the same percentage as their refining capacity, LIVINGSTON, *supra* note 228, at 3; PIPELINE PRIMER, *supra* note 331, at 6; *cf.* DE CHAZEAU & KAHN, *supra* note 30, at 22]. In the section of the Petition entitled "Factual Background" the Petitioner relies upon and incorporates into the Petition, the KENNEDY STAFF REPORT, *supra* note 184. This is a very slender reed upon which to rest any type of document, let alone a pleading. For a critique of this Report, *see* Section VI B 6, "Staff Report of the Antitrust and Monopoly Subcommittee of the Senate Judiciary Committee (Kennedy Staff Report)" *infra*.

[416] *H.Res.5 Subcomm. Hearings, supra* note 234, at 140 (Statement of Fred. F. Steingraber); *Id.* at A71 (Letter No. 1, dated Jan. 27, 1972 to Senator William Proxmire from Jack Vickrey, then General Counsel of Colonial).

[417] *Id.* at A71.

[418] Prorationing of line capacity is provided for in Item 90 of Colonial's Tariff, ICC 29, effective Feb. 10, 1978.

[419] Information on the owner-nonowner shipments and existence of prorationing furnished by Colonial.

[420] The year-by-year history is as follows:

1964 – 0.15	1971 – 19.75	1978 – 36.26 (estimated)
1965 – 0.62	1972 – 20.67	
1966 – 0.37	1973 – 23.64	
1967 – 1.74	1974 – 26.79	
1968 – 2.80	1975 – 27.62	
1969 – 3.25	1976 – 28.10	
1970 – 8.10	1977 – 32.11	Source: Colonial Pipeline Company

million barrels per day, their shipments have decreased to about 1.05 million barrels per day in 1978, whereas nonproprietary (*i.e.*, nonowner) movements during the same period rose from 37 thousand barrels per day to 595 thousand barrels per day. Stated another way, nonowners have received the entire increase in utilized capacity of the line. This record may underlie the Justice Department's failure to bring an action despite its investigation of Colonial from 1963 to 1976 under five different Assistants Attorney General in charge of the Antitrust Division (Wm. Orrick, Don Turner, Edwin Zimmerman, Richard McLauren, and Tom Kauper). The prime focus, at least during the latter part of the investigation, was on the question of whether all interested users were being granted access to Colonial's line on reasonable terms[421] and it utilized, among other things, numerous Civil Investigative Demands (CID's), responsive to which boxcars of information were dispatched to Washington via the Southern Railroad. However, in 1976, Justice closed its investigation. Contrary to some Congressional grumbling, the Antitrust Division was not at fault. As Deputy Assistant Attorney General Bruce Wilson testified before the House Select Small Business Committee in 1972: "To the Antitrust Division, of course, a lack of real evidence of competitive injury is critical. The Division cannot file cases simply on mere suspicion, particularly where, as here, a very large and important segment of commerce is involved. Not only must we have solid evidence in event of trial, but we cannot irresponsibly ignore the mere effect of filing."[422] The simple fact was that the Division concluded, on the basis of the evidence, that there had been no meaningful denial of access to the pipeline.

The nonowner shippers were not principally other so-called majors. Instead, they were companies such as Northville Dock Corporation, Metropolitan Petroleum, Signal Oil Company, Texas City Refining Company, Murphy Oil Corporation, Tenneco Oil Company, Crown Central Petroleum Company, Central States Marketing Company, Ashland Oil Company, Lion-Monsanto Company, and Charter International Oil Company who were shippers early in the game. Northville Dock, for example, was shipping by September 23, 1967, and Metropolitan Petroleum just two months later.[423] Plantation likewise has shown an increase in nonowner shipment percentages, starting with 9.0 percent in 1946, rising to 67 percent nonproprietary movements in 1961. It maintained approximately that percentage until 1964 when Colonial's system had commenced operations

---

[421] KENNEDY STAFF REPORT, *supra* note 184, at 125-128.

[422] *H.Res.5 Subcomm Hearings*, *supra* note 234, at 206-207.

[423] *Id.* at A84. In 1973, testimony before the Senate Committee on Commerce clearly identified 11 of the 17 non-proprietary shippers over Colonial's line at that time to be independents. *Consumer Energy Act Hearings*, *supra* note 154, at 606 (Colloquy between Senator Stevenson and Jack Vickrey and John E. Green).

and some nonproprietary shippers, mostly Colonial's owners at first, swung over to Colonial. By 1965, Plantation's "outside" traffic had dropped to 35 percent. Colonial's rates began to erode Plantation's traffic and the percentage of nonproprietary shippers slipped down each year until 1967 when the percentage bottomed out at 30 percent.

During 1967 and 1968, Plantation extended its line to the Washington National Airport, enlarged some of its lateral extensions and commenced expansion of its main line, constructing 108 miles of 30 inch line from Helena, Alabama, to Bremen, Georgia, and 216 miles of 26 inch line from Bremen to Spartenburg, South Carolina.[424] By these competitive improvements, it turned around its traffic difficulties, thereby achieving an increase to 36 percent "outside" shippers by 1970, which it matched in 1971. In 1971, Plantation completed the second step of its expansion program, which was the construction of 202 miles of 30 inch from Collins, Mississippi, to Helena, Alabama, 70 miles of 26 inch line from Spartanburg, South Carolina, to Huntersville, North Carolina, laid a 10 inch loop on the Helena to Birmingham lateral and completed the remaining 43 miles of 8 inch lateral line running from Bremen to Columbus, Georgia, removing the old 4 inch line,[425] thereby removing the need to prorate these laterals for the full year. In 1972, fire destroyed the entire manifold at the Bremen, Georgia, station and tank farm, materially reducing the system capacity until they could be rebuilt and placed in operation in the latter part of 1973. As a result, there was some prorationing east of Bremen and some "outside" shippers moved over to Colonial, reducing Plantation's percentage of outside shipments from 36 percent to 32 percent in 1972, and 30 percent in 1973. After the mainline proration was removed as a result of the reconstruction, two new nonowner shippers were acquired and the percentage climb resumed in 1974 and continued until 1976, the last year for which data were made available, when the percentage was 36 percent.[426] The difference between 65 percent nonproprietary shipments in 1963 and the mid-30's percentages which obtained thereafter is directly attributable to competition from Colonial.

---

[424] Plantation Pipe Line Company, Offering Memorandum, Aug. 25, 1971, covering $52,000,000, 7-7/8 percent Guaranteed Notes due Oct. 1, 2001, Part II, p. 2 (1971).

[425] Id.

[426] Plantation's year-by-year non-proprietary traffic percentages are as follows:

1946— 9	1957—61	1968—30
1947—17	1958—63	1969—34
1948—26	1959—63	1970—36
1949—27	1960—65	1971—36
1950—29	1961—67	1972—32
1951—31	1962—66	1973—30
1952—39	1963—65	1974—31
1953—46	1964—NA	1975—34
1954—51	1965—35	1976—36
1955—56	1966—33	Source: Plantation Pipe Line Company
1956—58	1967—30	

Texoma, 54.8 percent of whose stock is owned by non-"major" oil companies,[427] consists of a 459 mile, 30 inch line from Nederland, Texas, (near Port Arthur) to Cushing, Oklahoma,[428] with an intermediate origin and delivery point at Longview, Texas, where West Texas crudes can be received from the West Texas-Gulf Pipe Line Company and deliveries can be made to the Mid-Valley Pipe Line Company. Intermediate delivery points were established thereafter at Silsbee, Texas, (to serve South Hampton Company), Winnsboro, Texas, (to serve Dorchester Refining Company), and Wynnewood, Oklahoma. Texoma, a relatively new line, aggressively sought out nonowner shippers and the two years for which we have data show 48.1 percent nonproprietary traffic in 1977 and 51.7 percent in 1978. These include: American Petrofina*, Amoco*, Ashland, B&B, Cenex, Coastal States, CRA*, Dorchester, Derby*, Energy Coop, Gulf, Hudson, Husky, LaGloria, Sohio, South Hampton, Leonard, and Union.[429]

Explorer Pipe Line Company would seem a most unlikely choice to cite for exclusion of nonowner shippers. Explorer's mainline includes 50 miles of 12 inch line from Lake Charles, Louisiana, to Port Arthur, Texas; 546 miles of 28 inch line from Port Arthur to Pasadena (Houston), Texas, thence on to Tulsa and West Tulsa, Oklahoma, and 653 miles of 24 inch line from Tulsa, Oklahoma, to Hammond, Indiana. The system includes 85 miles of 12 inch lateral from Greenville, Texas, to Dallas-Fort Worth, and a 14 inch pipeline from Wood River, Illinois, to St. Louis, Missouri.[430] The line was designed to transport refined petroleum products. The original owners, who launched

---

[427] Memorandum dated July 19, 1976 from D. L. Jones and L. L. Gardner, Bureau of Economics, to Robert O. Forester [sic] III, Bureau of Enforcement, Interstate Commerce Commission, and filed with the Secretary of the Commission pursuant to Judge Dowell's order at a prehearing conference in Ex Parte No. 308 (Sub-No. 1), App. 1, at 10 [hereinafter cited as Jones & Gardner Memo.] lists the owners and percentages as follows:

	PERCENT
Kerr-McGee Pipeline Corp.	10.1
Lion Oil Co.	5.0
Mobil Oil Corp.	10.1
Rock Island Refinery	5.0
Skelly Pipeline Co.	10.1
Sun Oil Co.	25.0
United Refining Co.	7.0
Bickers [sic] (Vickers) Petroleum Corp.	2.7
Western Crude Oil, Inc.	20.0
Texas Eastern Transmission	5.0

[428] PIPELINE PRIMER, *supra* note 331, at 11.
[429] Data supplied by Texoma. Those companies marked with an asterisk were sent copies of Texoma's Feasibility Study dated January 30 and 31, 1973, but declined to participate, as did some 10 other companies who neither participated nor presently are shipping over Texoma's line; *e.g.*, Diamond Shamrock, Champlin Petroleum, OKC Refining and Apco.
[430] *H.Res.5 Subcomm Hearings*, *supra* note 234, at 113; LIVINGSTON, *supra* note 228, at 24.

the project in April, 1970, were Apco Oil Corporation, Cities Service Company, Gulf Oil Corporation, Phillips Investment Company, Shell Oil Company, Sun Oil Company, and Texaco, Inc. Shortly thereafter, Continental Oil Company became an owner. In 1974, Marathon Oil Company purchased part of Gulf's interest, and in 1977, Apco ceased to be an owner when it sold out to Marathon.

As most of the large jointly-owned pipelines recently have been financed, Explorer went the Throughput and Deficiency Agreement route.[431] Due to a series of events, including the Arab oil embargo which halved its throughput between November, 1973, and January, 1974,[432] the constructing of two large diameter crude lines from the Gulf Coast to refiners in the Mid-Continent Area (Texoma, described above, and Seaway, a 510 mile, 30 inch line from Freeport, Texas, to Cushing, Oklahoma)[433] and the expansion of Texas Eastern Transmission's products line from the Gulf Coast and extension into Chicago,[434] in addition to competition from barge lines, Explorer lost $9 million in the first two years[435] and almost $46 million in the first five full years of operation,[436] requiring its owners to meet cash calls of $37.3 million, which has subsequently increased to an aggregate of $42 million.[437] In an attempt to increase the volume of its movements, Explorer took the innovative step in 1974 of transporting crude oil in addition to the petroleum products for which it was designed,[438] and it has continued to provide this unique service at the current time. It also permitted 25,000 barrel minimum tenders subject to becoming a part of a joint batch of 50,000 bar-

---

[431] *H.Res.5 Subcomm Hearings, supra* note 234, at 114; HOWREY & SIMON, *supra* note 3, at 125. Financing, including Throughput and Deficiency Agreements, will be discussed in considerable detail in Chapter V, "Financing," *infra*. At this juncture, it should be enough to say that under the terms of these agreements the shipper-owner oil companies obligate themselves, in the event revenues from shipments or other revenues are insufficient, to make cash payments to the pipeline company, or to the agents of the lenders, in amounts sufficient to pay all of the pipeline company's expenses, obligations, and liabilities including the servicing of the interest and principal repayments of the pipeline company's borrowing. In Explorer's case this was $200 million initially, with an additional $50 million anticipated during the first 10 years of the project's life. The owners signed such a Throughput and Deficiency Agreement dated July 31, 1970. *H.Res.5 Subcomm Hearings, supra* note 234, at 114.

[432] *Hearings on S.1167 Before the Subcomm. on Antitrust and Monopoly of the Senate Committee on the Judiciary,* 94th Cong., 1st Sess. pt. 9 at 606 (Statement of Vernon T. Jones) [hereinafter cited as *S.1167 Hearings]; PIPELINE PRIMER, supra* note 331, at 14.

[433] PIPELINE PRIMER, *supra* note 331, at 11.

[434] LIVINGSTON, *supra* note 228, at 26.

[435] *Consumer Energy Act Hearings, supra* note 154, at 609 (Colloquy between Senator Stevenson and Jack Vickrey).

[436] PIPELINE PRIMER, *supra* note 331, at 14. For yearly loss, see LIVINGSTON, *supra* note 228, at 35.

[437] HOWREY & SIMON, *supra* note 3, at 126.

[438] LIVINGSTON, *supra* note 228, at 25; HOWREY & SIMON, *supra* note 3, at 21.

rels prior to entering the mainline. Through the use of these measures, Explorer began to build its nonproprietary traffic commencing with 0.9 percent in the start-up year of 1972[439] and reaching a 28.9 percent volume in 1977.[440] Only three of the twenty-nine "outside" shippers fall in the so-called "major" classification.[441] Why then did Senator Kennedy in his FTC Rulemaking Petition allege that Explorer clearly inhibited nonowner access,[442] and the Kennedy Staff Report zero in on Explorer as a subject of investigation by the Department of Justice, albeit an offshoot of the investigation of another pipeline?[443] If one traces the origin of their complaints, he will find it to be an Ashland internal document which came into the possession of the Department of Justice during its antitrust review, pursuant to its duties under the Deepwater Port Act of 1974,[444] of LOOP's[445] application to the Secretary of Transportation for a license under the Act. The Ashland document, while

---

[439] KENNEDY STAFF REPORT, *supra* note 184, at 131.

[440] The non-owner shipment percentages of Explorer are:

1972 – 0.9	1975 – 10.4
1973 – 11.4	1976 – 10.3
1974 – 15.2	1977 – 28.9

Source:
Explorer Pipeline Co.

[441] These shippers are:

Amoco	Kent Oil
American Petrofina	Koch Fuels
Apco	Koch Refining
Apex	Missouri Terminal
ARCO	Mobil
Braniff	Oklahoma Refining
Buckeye Petroleum	Petroleum Trad. & Transp.
Champlin	J. F. Reidy
Clark	T. P. Reidy
Coastal	Saber Petroleum
Enercham	Saber Refining
Cenex	Texas Utilities
F. S. Services	United Airlines
Gladieux	Xcel Products
Hydrocarbon Trad. & Transp.	

Source:
Explorer Pipeline Co.

[442] Kennedy Rulemaking Petition, *supra* note 414, at 6.

[443] KENNEDY STAFF REPORT, *supra* note 184, at 131.

[444] 33 U.S.C. § 1506 (1976).

[445] LOOP, Inc. is the company organized to construct and operate an oil unloading facility located 18 miles off the coast of the Gulf of Mexico, adjacent to La Fourche Parish, Louisiana, in waters of sufficient depth, 105 to 115 feet, to accommodate VLCC's, which have a draft of up to 95 feet when fully loaded; hence the name, an acronym of Louisiana Offshore Oil Port.

reflecting some musing by an Ashland employee concerning what its position might be as a nonparticipant as opposed to a participant in LOOP, gratuitously, and without any ostensible support, continued: "'Colonial' and 'Explorer,' for example, clearly inhibit nonowners [from access]."[446] This is certainly not in the category of "solid evidence" described by Deputy Assistant Attorney General Bruce Wilson, quoted above.[447] More importantly, Mr. Robert E. Yancey, President of Ashland Oil, Inc., made it unequivocally clear that the "impression" of an Ashland Pipeline Company employee "many years ago" was not, nor is today, shared by Ashland's management. Mr. Yancey's letter of May 2, 1979, expressing the Company's position is reproduced in full in Section VI C 3, "Asserted Denial of Access to Pipelines," *infra*.

The Kennedy Staff Report related an incident involving an unnamed independent marketer located in the Dallas area who desired to use Explorer's line at an unnamed date. The Report recited that this unnamed independent marketer was able to arrange for a [supply?] contract with an unnamed refiner-owner [of Explorer?] connected to Explorer. The independent marketer did not have a terminal close enough to the pipeline to make a connection "economically attractive," so the story was that he contacted several companies with terminals connected to Explorer, but "without success." There is no explanation of what type of a deal he proposed, or why the companies refused and, because of the next sentence in the Report—"Even major companies with prior relations with the marketer or friendly attitudes toward the marketer refused to terminaling [sic] space" — the Report raises the inference that some of the companies approached were non-majors; it is also not unreasonable to suspect that space was tight except in the one instance cited—that of a shipper-owner who had excess terminal capacity available for sale. Although the putative seller made an arrangement with the excess space terminal owner, the latter is purported to have retained a veto over any marketing arrangement not to its satisfaction, which it allegedly exercised when the putative seller attempted to arrange deliveries to the unnamed independent marketer.[448] In addition to the extreme vagueness of the claim the bare bones of the allegation do not impugn Explorer, which, as has been shown, needed all the traffic it could get. Vernon T. Jones, then President of Explorer, testified at the *Energy Industry Hearings* in January, 1970, that Explorer had never been approached by any independent and asked to

---

[446] Attorney General of the United States, Report Transmitting Antitrust Advice on the Applications of LOOP, Inc. and SEADOCK, Inc. for Licenses to Construct, Own and Operate Deepwater Ports in the Louisiana and Texas Coastal Areas, respectively, of the Gulf of Mexico, dated November 5, 1976, pp. 80-81 [hereinafter cited as A.G.'s Deepwater Port Report].

[447] *See* text at note 422, *supra*.

[448] KENNEDY STAFF REPORT, *supra* note 184, at 79.

provide terminal facilities.[449]

The only other complaint against Explorer of which this author is aware is found in the Neal Smith-Conte hearings.[450] Mr. Meyer Kopolow, President of Marine Petroleum Company of St. Louis, Missouri, stated that he attempted to arrange a connection between Explorer and his terminal at a time when the line was still in the planning stage; that despite his follow-up contacts with Explorer, he never heard from it until he received a proposed pipeage contract on May 1, 1972, when the main line was already constructed and in existence. According to Mr. Kopolow, Explorer told him that the proposed spur to Marine would cost approximately $1.7 million and that Marine would have to execute a "ship or pay" contract which would guarantee Explorer approximately $5 million in revenues over a five-year period, which Mr. Kopolow said would be a three-year payout. Mr. Kopolow stated also that none of the "owners" of the pipeline had been asked to sign an agreement guaranteeing the payment of the spur into St. Louis (although he conceded they had to guarantee the overall profits of the entire line). Moreover, he objected to the surmised fact that Humble, Arco, Mobil, Amoco and J. R. Street, nonowners of Explorer, had not been asked to sign an agreement guaranteeing their St. Louis connections to Explorer. However, some five pages later in the printed hearings, there appears a reply by Vernon T. Jones,[451] then President of Explorer, now President of Williams Pipe Line, which stated that Explorer had no record of any contact from Marine or Mr. Kopolow prior to April, 1970, when he first called concerning a possible connection, later confirmed by a May 6, 1970, letter expressing an interest in a connection with the line. Mr. Jones stated that the letter arrived after the initial routing had been planned and at a time when actual construction was proceeding.

Explorer communicated with Mr. Kopolow in an effort to ascertain what Marine Petroleum's proposed movements would be and requested information concerning Marine's financial responsibility. According to Mr. Jones, no financial statements were supplied, although Mr. Kopolow represented that Marine's volume over the first five years would be approximately 18.6 million barrels. The Marine terminal is located in a very congested area of St. Louis and the construction of a spur to Marine's terminal along the West Bank of the Mississippi, approximately 2.7 miles in length, would be extremely difficult to accomplish. Studies were made to see

---

[449] *S.1167 Hearings, supra* note 432, pt. 9 at 609 (Statement of Vernon T. Jones).
[450] *H.Res.5 Subcomm Hearings, supra* note 234, at 105-109.
[451] *Id.* at 112-125.

whether the more feasible route would be along the West Bank or along the East Bank, with a river crossing. Explorer knew from other river crossings in this area that there were a number of obstacles to construction, making accurate cost estimates difficult. Preparing cost estimates which were reliable took some time, but Explorer finalized its proposal within 60 days of becoming operational in the St. Louis area, at which time the proposed pipeage contract was sent to Mr. Kopolow.[452] As Jones pointed out in his letter, Kopolow's own testimony revealed that barge rates at the time were favorable.[453] In addition to the nine barges that Kopolow admitted owning,[454] it appears from Jones' statement that Marine owned several large modern towboats, which, according to Jones' statement, provided Marine with an alternative means of transporting its petroleum products at a cost which ". . . should be significantly lower than Explorer's tariff into St. Louis from Mid-Gulf origins,"[455] There appears to be a flat contradiction between Kopolow's assertion that nonowner shippers were not required to amortize the cost of their individual connection and Jones' statement that they were.[456] One of the persons named by Kopolow as potential corroborators of his story[457] verified Jones' version that Kopolow was not the only one required to make throughput commitments.

Jones also stated that the amortization period, as specified in the contract, was 5 years, not 3 years as stated by Kopolow. If Marine only shipped the specified minimum over the first five years, Explorer would realize no profit in the connection. According to Jones, if Marine did not ship during the first five years but used the advance transportation credits during the succeeding five years, Explorer would not have realized any profit at the end of 10 years.[458] Of the three companies referred to by Mr. Kopolow as being in the St. Louis area and having expressed an interest in being connected to Explorer,[459] Martin Oil replied that it had never made a "formal" request to Explorer because its supplier was not in a position to put material into Explorer's line. While Martin's spokesman admitted that he had no direct experience with Explorer concerning his business, he gratuitously volunteered that he "knew" from "conversations with others who have tried to get on

---

[452] *Id.* at 119-120 (V. T. Jones' April 28, 1972, letter to Kopolow transmitting proposed pipeage agreement and forms of alternative Security Bond and Letter of Credit); *Id.* at 120-123 (proposed pipeage contract); *Id.* at 123 (Exhibit to V. T. Jones' letter spelling out the economics involved); *Id.* at 124 (Surety Bond and Letter of Credit forms); *Id.* at 125 (Map showing proposed connection).
[453] *Id.* at 107.
[454] *Id.* at 108.
[455] *Id.* at 116.
[456] Compare *Id.* at 107 (Kopolow) *with Id.* at 116.
[457] *Id.* at 109. These were Martin Oil Co., Apex Oil Co. and Triangle Refineries.
[458] *Id.* at 118.
[459] *Id.* at 109.

the Explorer Pipeline, [that] they are not interested in business other than their own." But Martin, by the spokesman's own admission, never even considered pipelines [including Williams Pipe Line] until Explorer was built "next door," due to the fact that it had Ohio and Mississippi River terminals served by barge.[460] If it was receiving by barge up the Mississippi, then the reason its supplier was "not in a position to put material into Explorer's line" could well be that it was nowhere near Explorer. The second reply was from J. H. Barksdale, of Triangle Refineries, which directly contradicted two of Mr. Kopolow's statements, *i.e.*, Mr. Barksdale stated that Explorer *was* interested in shipping product for Triangle's account provided it had a destination to receive it, and Barksdale also negated Kopolow's statement that no one else was being asked for throughput guarantees because Triangle would have had to execute one for a St. Louis tie-in and Barksdale's plans for his St. Louis terminal were not definite enough at that time to pursue the matter with Explorer. The third company, Apex, declined to reply to the Subcommittee despite two follow-up telephone calls. Mr. Samuel Goldstein, President of Apex, did confirm that his company was interested in connection to and shipping over Explorer's line but nothing came of discussions with Explorer, not because of a lack of good faith on Explorer's part but because of Apex's inability to secure supplies. A quick look at footnote 441 reveals another interesting fact. Apex is presently shipping over Explorer. To sum up, complaints about Explorer appear to unsubstantiated and contratictory to other evidence in the record.

Turning to a broader treatment of nonowner shipments we find that as early as 1957 there is a statement in the Dirksen Views filed with the O'Mahoney Report that approximately 40 percent of all crude oil movement ultimately reaching refineries was for consignees other than the affiliates of the pipeline.[461] Unfortunately, there were no hard data cited in support of the statement so it must be viewed with some caution. Some hard data are at hand for 1968. Aggregate numbers[462] and shippers' names are available yearly from 1968 through 1972 for Badger Pipeline Company,[463] in which the

---

[460] *Id.* at A56.
[461] Reproduced in *Consumer Energy Act Hearings, supra* note 154, at 655.
[462] Because of the restrictions on disclosure of individual shipments imposed on pipelines by 49 U.S.C. § 15(13) (1976), data are available only in aggregate form and are not broken down by individual shippers.
[463] Badger is a 332-mile, 6 to 16 inch products line running from East Chicago, Indiana, to Lemont and Peru, Illinois, thence on to Madison, Wisconsin. It is owned by ARCO - 34%; Cities Service - 32%; Texaco - 22%, and Union - 12%. Jones & Gardner Memo., *supra* note 427, App. 1 at 2.

volume and revenues represented by nonowner shipments rose from 25 percent and 34 percent, respectively, in 1968 to 39 percent and 44 percent in 1972.[464] Nonowner shippers were: Ashland, F. S. Services, Smith Oil, Badger Petroleum, Mobil, Shell, Koch, Marathon, Murphy, Midland Coop., Continental, Hydrocarbon Transportation Inc., Kerr-McGee, Sun, Humble, American Airlines, Williams Bros., Great Northern, Gustofson, United Air Lines and Skelly.[465] Note that non-"majors" clearly outnumber "major" nonowner shippers. The same data are available for Kaw Pipe Line Company,[466] whose volume and revenue percentages represented by nonowners was 53 and 51 percent, respectively, in 1968 and 55 and 51 percent in 1972;[467] Kaw's "outside" shippers were: American Petroleum of Texas, Clerk [sic] Creek, Continental, CRA Coop., Derby, Mobil, NCRA, Amoco, Rock Island, Skelly, Sohio, Tenneco, Permian and Koch.[468] Again, there is a predominance of "independents." Texaco-Cities Service[469] had nonproprietary shipments accounting for one-third the volume and one-fourth

---

[464] *Consumer Energy Act Hearings, supra* note 154, at 634 (Statement of Jack Vickrey) the numbers are:

YEAR	NO. OF NONOWNER SHIPPERS	VOLUME PERCENTAGE	REVENUE PERCENTAGE
1968	16	25	34
1969	15	27	33
1970	15	30	35
1971	16	35	38
1972	13	39	44

[465] *Id.* at 634 n.95.

[466] Kaw consists of 1,395 miles of 6 to 12 inch crude lines in North Central Kansas. It is owned by Texaco - ⅓, Cities Service - ⅓ and Phillips - ⅓. Jones & Gardner Memo., *supra* note 427, App. 1 at 4.

[467] *Consumer Energy Act Hearings, supra* note 154, at 634. The yearly numbers are:

YEAR	NO. OF NONOWNER SHIPPERS	VOLUME PERCENTAGE	REVENUE PERCENTAGE
1968	15	53	51
1969	15	55	52
1970	15	55	51
1971	15	57	52
1972	15	55	51

[468] *Id.* at 634 n.96.

[469] Texaco-Cities Service during 1968-1972 operated 1,966 miles of crude trunk lines and 206 miles of product lines in Illinois, Indiana, Kansas, Michigan, Missouri and Oklahoma. *Id.* at 634. It is owned 50-50 by Texaco and Cities Service. Jones & Gardner Memo., *supra* note 427, App. 1 at 6.

the revenues during the 1968-1972 period.[470] Texaco-Cities Service's nonproprietary shippers were: Amoco, Ashland, Arco, Clark, Continental, Gulf, Mobil, Phillips, Skelly, Sohio, Sun, Union, United Refining, Marathon, Permian, Leonard, and Suntide, about one-third "independents."[471] Texas-New Mexico[472] had 32 nonowner shippers which accounted for one-third of the barrels and one-fourth of the revenues.[473] These shippers are: Tenneco, Marathon, Admiral, Amoco, Charter, Famariss, Permian, Scurlock, Sun, Gulf, Humble, Mobil, Phillips, True, Petrolia, Tesoro, Chevron, Crown Central, Shell, Sohio, Union, Western Crude, Champlin, Douglas Oil, Amarada, Coastal States, South Western Crude, Nickers [sic], Cardinal Petroleum, and Signal, almost 60 percent "independents."[474]

Senator Stevenson, who was presiding at the *Consumer Energy Act Hearings* at which the foregoing statistics were adduced, requested the Association of Oil Pipelines (AOPL) to furnish similar data for other products lines in the country.[475] In response to this request, AOPL requested similar information from the 55 interstate products pipelines which operated in the United States. It received answers from 51 of these pipelines and submitted the results to Senator Warren Magnuson, Chairman of the parent Commerce Committee, on February 19, 1974. The figures, which covered

---

[470] *Consumer Energy Act Hearings, supra* note 154, at 634. The yearly breakdown is (*Id.* at 635):

YEAR	NO. OF NONOWNER SHIPPERS	VOLUME PERCENTAGE	REVENUE PERCENTAGE
1968	11	30	23
1969	9	28	19
1970	9	27	18
1971	10	29	23
1972	12	35	25

[471] *Id.* at 634 n.97.

[472] Texas-New Mexico during 1968-1972 operated 4,296 miles of crude gathering and trunk lines from the Southwest producing region to the Houston, Texas refinery area. *Id.* at 635. It is owned by Texaco (45%), ARCO (35%), Cities Service (10%) and Getty (10%). Jones & Gardner Memo., *supra* note 427, App. 1 at 6.

[473] *Consumer Energy Act Hearings, supra* note 154, at 635. The yearly breakdown is:

YEAR	VOLUME PERCENTAGE	REVENUE PERCENTAGE
1968	40	28
1969	36	26
1970	30	23
1971	31	23
1972	34	26

[474] *Id.* at 635 n.98.
[475] *Id.* at 606.

the latest year available, 1973, showed that 806 of the total number of shippers (909), or 89 percent, were nonowner shippers, and of these 806 nonowner shippers 573, or 71 percent, were "independents" as classified by the FTC staff report prepared at the request of the Chairman of the Senate Interior and Insular Affairs Committee and released by it in July 1973.[476]

A similar analysis for *all* pipelines—crude gathering systems, crude trunk lines and products pipelines—was later presented to the ICC. It showed that in 1973 there were 1,759 shippers using these lines, of which number 1,549, or 88 percent, were nonowner shippers, and of the 1,549 nonowner shippers, 1,011, or 65 percent, were "independents."[477] Nor has the trend subsided. The corresponding figures for Badger during 1973 through 1976, was 21 nonowner shippers (13 non-majors)[478] moving 42 to 49 percent of the volume and 52 to 56 percent of the revenues.[479] Kaw's statistics for the same period were 17 nonowner shippers (10 non-majors)[480] whose traffic declined from 59 to 51 percent of volume and from 56 to 50 percent of revenues.[481] Texaco-Cities Service, in the 1973 through 1976 period, had 29 nonowner

---

[476] *Id.* at 656-666. The submission also contained the interesting information that of the shipper-owners, 17 were "independents" and 86 were "majors," using the FTC Staff Report classification. *Id.* at 666.

[477] *S.1167 Hearings, supra* note 432, pt. 9 at 596 (Statement of Fred F. Steingraber); MARATHON, *supra* note 202, at 56.

[478] HOWREY & SIMON, *supra* note 3, at 52 n.1 lists the non-owner shippers as: Ashland, F. S. Services, Mobil, Shell, Koch, Amoco, Go-Tane Service, Inc., Marathon, Murphy, Midland Coop., Continental, Hydrocarbon Transportation, Inc., Kerr-McGee, Champlin, Martin Oil, Sun, Exxon, American Airlines, Gustafson Oil, United Airlines and Williams Energy Co.

[479] *Id.* at 52; *cf.* MARATHON, *supra* note 202, at 55. The yearly figures are:

YEAR	VOLUME PERCENTAGE	REVENUE PERCENTAGE
1973	42	52
1974	44	51
1975	44	53
1976	49	56

[480] HOWREY & SIMON, *supra* note 3, at 53 n.1 lists the non-owner shippers as: American Petrofina of Texas, Clear Creek, Continental, CRA Corp., Derby, Marathon, Apco, Mobil, Amoco, Union, Clark Oil, DOMA Corp., Skelly, Sohio, Koch, General Energy Corp. and National Coop. Ref.

[481] *Id.* at 53; *cf.* MARATHON, *supra* note 202, at 55. The yearly figures are:

YEAR	VOLUME PERCENTAGE	REVENUE PERCENTAGE
1973	59	56
1974	59	57
1975	55	54
1976	51	50

shippers (17 "independents")[482] shipping 47 to 60 percent of the volume and 60 to 67 percent of the revenues.[483] Texas-New Mexico's updated statistics are 33 nonowner shippers (19 non-majors)[484] whose movements represented 34 percent of the volume and 25 percent of the revenues.[485] There were some additional lines for which data were available during the 1973-1976 period. Amoco Pipeline Company,[486] had 51 nonowner shippers (34 non-majors)[487] using its lines comprising about 18 percent of the volume and about 14 percent of the revenues.[488] West Shore[489] had 13 nonowner shippers (11 non-

---

[482] HOWREY & SIMON, *supra* note 3, at 54 n.1 lists the non-owners as: American Petrofina, Amoco, Ashland, Apco, Clark, ARCO, Continental, Gulf, Mobil, Phillips, CRA, Inc., OKC Corp., Skelly, Sohio, Sun, Union, Rock Isle Ref., National Ref., Murphy, Marathon, Lakeside Ref., Crown Central, Western Crude Oil, Herndon Oil & Gas, Champlin, Derby, General Energy, Gladieux Ref. and Koch.

[483] *Id.* at 53-54; *cf.* MARATHON, *supra* note 202, at 55 [both of these sources say there are 30 nonowner shippers but data available to this author only show the 29 listed in note 482, *supra*]. The breakdown by years is as follows:

YEAR	VOLUME PERCENTAGE	REVENUE PERCENTAGE
1973	47	60
1974	55	64
1975	52	59
1976	60	67

[484] HOWREY & SIMON, *supra* note 3, at 55 n.1 lists these shippers to be: Navajo Ref., American Petrofina, Marathon, Amoco, Continental, Famariss, Permian, Apco, Skelly, Lion Oil, Diamond-Shamrock, Sun, Tesoro, Chevron, Crown Central, Shell, Union, Charter, Herndon Oil, Clark, Western Crude, Gulf, Exxon, Mobil, SoCal, Amerada Hess, Basin, Inc., Summit Gas Co., OKC, Vickers, Phillips, B&B Trading Co. and United Ref.

[485] *Id.* at 54-55; *cf.* MARATHON, *supra* note 202, at 55.

[486] Owned 100 percent by Standard Oil Company (Indiana). Jones & Gardner Memo., *supra* note 427, App. 1 at 2. During 1973-1976 Amoco operated 2,162 miles of 3 to 12 inch products lines in Illinois, Iowa, Indiana, Michigan, Minnesota, Missouri, North Dakota, South Dakota and Texas. HOWREY & SIMON, *supra* note 3, at 55.

[487] HOWREY & SIMON, *supra* note 3, at 56 n.1 names the following shippers: Amerada Hess, American Petrofina, Apco, ARCO, Basin, Inc., Burlington Northern, Caribou Four Corners, Inc., Champlin, Charter, Chevron, Cities Service, Clark, Continental, CRA, Inc., Crown Central, Derby, Diamond-Shamrock, Dow Chemical, Famariss, Getty, Jefferson Chemical, Kerr-McGee, Gulf, Husky, Laketon Asphalt Ref., Little America Ref., Marathon, Mobil, Nat. Coop., Navajo, Pasco, Inc., Phillips, Premium Oil Co., Rock Isle. Ref., Shell, Sinclair, Skelly, Sohio, Summit Gas, Sun, Texaco, Texas City Ref., Permian, Union Carbide, Union, Union Texas Petroleum, United Refining, Vickers Petroleum, Western Crude Oil, Western Ref., Wing Corp.

[488] *Id.* at 56. The yearly breakdown is as follows:

YEAR	VOLUME PERCENTAGE	REVENUE PERCENTAGE
1973	19	14
1974	20	17
1975	18	15
1976	17	12

Today's Pipeline System — An Overview 77

majors)[490] whose shipments declined from 23 to 12 percent of the volume and from 21 to 14 percent of the line's revenues.[491] Olympic[492] transported for 10 nonowner shippers (7 of which were "independents")[493] whose shipment accounted for 38 to 42 percent of the line's volume and 40 to 47 percent of its revenues.[494] A survey conducted by the AOPL in respect to calendar year 1976 of 74 multiple-owned and single-owned crude oil and products pipelines disclosed that of the 2,194 shippers[495] on these lines, 1,958, or 89.24 percent, were nonowner shippers. More importantly, of the 1,958 nonowner shippers, 1,408, or 64 percent, of the *total* shippers were "independents."[496]

---

[489] West Shore is owned by Shell Pipe Line Corp. (20%), Mobil Pipe Line Co. (14%), Texaco, Inc. (9%), Amoco Pipeline Co. (16.5%), Exxon Pipeline Co. (3.5%), Union Oil Co. (5.5%), Clark Oil & Refining Corp. (8%); Cities Service Co. (8%), Marathon Oil Co. (9%) and Continental Pipe Line Co. (6.5%). Jones & Gardner Memo., *supra* note 427, App. 1 at 6. During the years in question, it operated 326 miles of 10 to 16 inch products lines running from East Chicago, Indiana, area to various terminals including Des Plains, Illinois, and Milwaukee, Granville and Green Bay, Wisconsin. HOWREY & SIMON, *supra* note 3, at 57.

[490] HOWREY & SIMON, *supra* note 3, at 57 n.1 enumerates the following: American Airlines, ARCO, E. L. Bride, Farmers Union Central, Gustafson Oil, Industrial Fuel, Jacobus Oil, Midland Coop., Murphy, Phillips, United Airlines, U.S. Oil Co. and Wisconsin Electric Power.

[491] *Id.* at 57; *cf.* MARATHON, *supra* note 202, at 55. Year-by-year figures are given as:

YEAR	VOLUME PERCENTAGE	REVENUE PERCENTAGE
1973	23	21
1974	9	11
1975	10	12
1976	12	14

[492] Olympic is owned by Shell Pipe Line Co. (43.5%), Mobil Pipe Line Co. (29.5%) and Texaco, Inc. (27%). Jones & Gardner Memo., *supra* note 427, App. 1 at 4. During 1973-1976, it operated 406 miles of 6 to 20 inch products lines from the Ferndale and Anacortes Refineries in Northern Washington to Renton (Seattle), Tacoma and Olympia, Washington, and on to Portland, Oregon. HOWREY & SIMON, *supra* note 3, at 58.

[493] HOWREY & SIMON, *supra* note 3, at 58 n.1 names: Douglas Oil, Powerine Oil, Toscopetro Oil, Burmah Oil, Aminol Oil, Lion Oil, SoCal, Union, ARCO and Signal.

[494] *Id.* at 58; *cf.* MARATHON, *supra* note 202, at 55. By years, the figures are:

YEAR	VOLUME PERCENTAGE	REVENUE PERCENTAGE
1973	38	40
1974	39	42
1975	41	45
1976	42	47

[495] There may be some duplication in the figures because each pipeline company reported the total number of individual shippers on their lines for 1976, and some shippers undoubtedly tender shipments on more than one line. However, it is not believed for the purposes used herein that there is any significant distortion.

[496] PIPELINE PRIMER, *supra* note 331, at 24; Rebuttal Testimony of Jack M. Allen Before F.E.R.C. Dkt. OR78-1, Exh. JMA-5 (Oct. 30, 1978) [hereinafter cited as Allen, *TAPS Rebuttal Testimony*] gives a breakdown of these figures.

Reviewing these figures from 1968 to 1976, it seems difficult to come to any conclusion other than that nonowner shippers, including "independents," have, and are taking advantage of, access to the nation's pipelines.

A second index of the continuation of a competitive pipeline system is the ratio of throughput to capacity. A numerical figure of 1.00 for the ratio is used to indicate when throughput equals rated capacity under normal conditions, although actual capacity at a given time could be less due to unexpected shutdowns, due to accidents, for example, or throughputs could slightly exceed rated capacity due to operations going more smoothly than normal. Despite these variations, the 1.00 is a reasonable proxy for times when prorationing might be expected, and numbers under 1.00 usually indicate the absence of prorationing. With those qualifications, and the additional caveat that seasonal swings and particular line segments may cause peculiarities, it can be stated that Explorer, Platte, Wolverine and Yellowstone, as shown in Table I, ran well under capacity in the years for which data are available. Table II portrays the history of Plantation, and Table III depicts Colonial's situation. The subject of prorationing is given a detailed examination in Chapter VI. A more complete overall picture of prorationing on integrated pipelines was provided recently by survey data collected by the Association of Oil Pipe Lines (AOPL). A total of 46 integrated pipeline companies, including those wholly-owned by 14 of the 16 largest oil and natural gas liquids producers in 1977, provided information on the extent and duration of prorationing on their systems during the three-year period 1976-1978. These companies accounted for approximately 85 percent of total barrel miles shipped on trunklines in 1977, so the coverage of the U.S. pipeline system is fairly complete. Of the 46 integrated systems, 29 had no prorationing whatsoever during the period under review. Of the 17 which had to proration capacity, only two had substantial prorationing; on the remaining 15 systems prorationing was limited in duration and/or confined to a relatively small part of the system. Thus, the sweep of the data make it difficult to reach a conclusion that integrated pipelines have been noncompetitively (under)sized.

d. Tenders

There always has been some confusion attached to the word "tender." One reason for this is that different pipelines give the word different meanings. Some consider tenders to be strictly accounting records against which daily receipts and deliveries are charged;[497] others consider a tender to be an offer by the shipper to ship a certain amount and acceptance by the carrier to effect a contract of carriage; still others treat the tender as a device to permit

---

[497] *Cf.* BEARD, *supra* note 39, at 97 (1941), citing Brundred Brothers v. Prairie Pipe Line Co., 68 I.C.C. 458, 462 (1922); WOLBERT, *supra* note 1, at 29, and authorities cited therein at n.149.

scheduling through the line.[498] Obviously the first and third descriptions render controversy over the size of tender more or less meaningless. Critics of the industry take it to be a requirement that the shipper must be prepared to deliver the "tender" in such way as to permit continuous input to the pipeline, and in this sense the amount could present a problem to a small shipper who might have to construct costly storage to accumulate the required tender. This is the meaning under which the ICC proceedings in *Brundred Brothers v. Prairie Pipe Line Company*,[499] *Petroleum Shippers' Ass'n v. Alton & Southern R.R.*,[500] and *Reduced Pipe Line Rates* and *Gathering Charges*,[501] considered the question of what were reasonable tenders for interstate crude and products lines. The ICC found that minimum tenders for interstate crude lines in excess of 10,000 barrels for crude lines and 25,000 barrels for products lines would be unreasonable. The Texas Railroad Commission's Rules for the less complicated intrastate movements provide for 500 barrels and Texas intrastate tariffs so provide.[502] Other states vary.[503] "Petroleum" is not a single homogenous substance but rather is a mixture of hydrocarbons which are found in varying degrees of volatility, impurities and usefulness.[504] Different fields produce oil of different characteristics[505] and even crudes from different horizons in the same field can have different physical and chemical characteristics.[506] These differences and the fact that refinery requirements have become so complex, together with longer lines and more refineries served per line, have forced many (especially the interstate) crude lines to abandon[507] the old "common-stream" method of handling.[508] This situation presents two problems, one being the need to segregate, fully or partially, one batch from another, and the consequent requirement of larger tenders in order to prevent the interfaces or admixtures between successive "batches" or "slugs" from spoiling

---

[498] WOLBERT, *supra* note 1, at 25. See Eureka Pipe Line Co. v. Hallanan, 257 U.S. 265, 271, 274 (1921) for a discussion of tenders.

[499] 68 I.C.C. 458 (1922).

[500] 243 I.C.C. 589, 665 (1941).

[501] 243 I.C.C. 115, 136-137 (1940); 272 I.C.C. 375, 383 (1948).

[502] See Rule 6 of Shell Pipe Tariff, Exh. B; Rule 7 of Gulf R.C.T. No. 541, eff 8/1/78.

[503] Louisiana - 10,000 barrels, Marathon Pipeline Co. Tariff L.P.S.C. No. 4, effective 10/5/78 (Rule 2); Oklahoma - 10,000 barrels, Shell Pipe Line Corp. Tariff O.C.C. No. 16, effective 10/1/78 (Item 30 - but received as currently available provided the total is achieved within 30 days); New Mexico, same as Oklahoma, Shell Pipe Line Corp. Tariff N.M.S.C.C. No. 32, eff. 9/21/78 (Item 30); Michigan, same as Oklahoma, Shell Pipe Line Corp. Mich. Rate. Sheet No. 31, eff. 10/1/77.

[504] DE CHAZEAU & KAHN, *supra* note 30, at 62; S.2387 REPORT, *supra* note 49, at 7 n.2 (1976); SPLAWN REPORT, *supra* note 372, at XV.

[505] MILLS, *supra* note 324, at 39, Bond, *supra* note 207, at 733; Burke, *supra* note 9, at 795.

[506] *Id.*

[507] See text at note 162, *supra*; Bond, *supra* note 207, at 734; Burke, *supra* note 9, at 795.

[508] WOLBERT, *supra* note 1, at 30; *cf.* Att'y Gen., *Second Report*, *supra* note 111, at 70 n.94.

the quality of the higher grade. This accounts in part for differences in practice. A simple gathering system, gathering one grade of crude may not require any minimum tender;[509] however, a system carrying, say, 15 grades with as high as 100 batches currently moving through the line,[510] may well require a 10,000 barrel minimum tender for shipments by one shipper to one consignee, or perhaps permit aggregation of shipments by several shippers of like crudes to be shipped as a joint batch.[511] One does not need to be an old oil field hand to recognize the problem faced by a refiner with limited ability to handle sour crudes if his shipment of sweet Ellenberger was delivered to him common-streamed with West Texas sour. The second problem is an offshoot of the first, *i.e.*, provision of sufficient storage to accumulate sufficient quantities of the different quality crudes to make up a "batch" sufficiently large to permit proper scheduling and at the same time preserve the integrity of the different batches. Storage will be discussed in Chapter VI, "Current Policy Issues," *infra*.

e. Batching and Quality Banks

As discussed in the preceding paragraph, because of the complexity of refining operations and the variety of crudes produced, it frequently is necessary to segregate crudes by grades so that the refineries to which the oils are consigned can secure the grades which their respective programs specify. The process by which the integrity of their shipments is preserved is termed "batching."[512] Batching is not a simple problem of "slugging" a large volume of one kind of oil, and then merely displacing it with another grade.[513] It is a complex technical problem of fluid mechanics and physical properties of different fluids transported through a pipeline, discussed in greater detail in subsection B 3 of this chapter, "Technical Aspects," *infra*. In the early days, when pipeline systems were not as advanced technologically as they now are, batches as large as 300,000 barrels were used.[514] This was reduced, on an empirical basis to 100,000 barrels, which was common until the *Brundred Brothers* decision in 1922 which, balancing the shippers' needs

---

[509] See text at note 332, *supra*.

[510] Burke, *supra* note 9, at 795.

[511] PIPELINE PRIMER, *supra* note 331, at 25; *cf.* JONES DOE REPORT, *supra* note 372, at 53, citing Denver Oil Co. v. Platte Pipe Line Co., 316 I.C.C. 599 (Div. 2, 1962), 319 I.C.C. 725 (Div. 2, 1963) (pipeline should include specifications for different "streams" in its tariff).

[512] MILLS, *supra* note 324, at 39; BARBE, *supra* note 317, at 4; Bond, *supra* note 207, at 734.

[513] MILLS, *supra* note 324, at 39.

[514] *Id.* at 40.

Today's Pipeline System — An Overview

and the technological requirements of the pipeline operation,[515] lowered the tenders (which might or might not necessarily coincide with the batch size) to 10,000 barrels.[516] While this decision legally was applicable only to the carriers before the ICC in the *Brundred Brothers* case, it was extended to all ICC crude pipelines by the *Reduced Pipe Line Rates and Gathering Charges* decisions of 1940 and 1948.[517] Where the difference in grade is relatively minor (as for example, when the crudes are of the same "base," sulphur content, but vary only in gravity) or the operational problems in batching are so grave as to outweigh the benefits from segregation (such as in TAPS), "common stream" is used. In order to compensate shippers of higher quality crudes for any loss which might result from any downgrading of their quality, "Quality Banks" are becoming more common. At first, these were simply "gravity banks" under which the stream is sampled and the shipper is credited/charged for each degree of API gravity above/below a set standard gravity. The money is distributed among the shippers who receive crude that has been downgraded in proportion to the degradation. The carrier or gravity bank operator retains none of the money but simply acts as a collecting/paying agent. The more recent versions take both gravity and sulphur content into account. Cook Inlet, Seaway and Texoma have led the way; in fact, a gravity-sulphur bank was a part of Texoma's first tariff bulletin. TAPS and Eugene Island presently are developing mandatory quality banks. Capline set up a separate organization called "Gravcap" to act as the "banker" for the adjustment.

---

[515] The 10,000 figure was a "Solomon Compromise" by the Commission, faced with a request by the shipper for a 2,000-barrel minimum tender and 100,000 barrels urged by Prairie Pipe Line and its connecting carriers, supported by certain Pennsylvania refiners in Warren, Pennsylvania, which were receiving sweet Pennsylvania crude through the Pennsylvania carrier and did not want it contaminated by the sour Lima-Indiana crude. The Commission's reasoning went this way: The minimum must not be too low because transportation of oil by pipeline "is essentially a bulk business" which cannot be successfully operated "on a driblet basis"; on the other hand, it must not be too high, as the Commission felt the 100,000 barrels to be, because it "reserves the pipelines to a few large shippers and essentially deprives the lines of the common carrier status with which they were impressed by the interstate commerce act." The Commission selected 10,000 barrels as being a figure that would neither be so low as to produce unduly aggravating operating difficulties nor so high as to prevent the Brundred Brothers and others from using the line on a common carrier basis. The Commission forthrightly admitted that there was nothing precise about its figure saying: "We are practically without precedent upon which to base our determination of a reasonable minimum, and the reasonableness of any minimum can only be verified by actual experience." Brundred Brothers v. Prairie Pipe Line Co., 68 I.C.C. 458, 465-466 (1922); *see* BEARD, *supra* note 39, at 99; JONES DOE REPORT, *supra* note 372, at 51.

[516] Brundred Brothers v. Prairie Pipe Line Company, 68 I.C.C. 458, 466 (1922).

[517] See text at note 501, *supra*.

B. Products Lines
1. General Description

Functionally, products pipelines provide the transportation link between refineries and market.[518] The usual linkage is from a refinery to bulk terminals, from whence distribution is made to the wholesale and retail markets, generally by tank trucks.[519] Sometimes, spur lines run from pipeline terminals to individual company terminals serving a particular marketing area[520] and there has been an increase of spur lines directly serving large consumer installations such as airports. Products pipelines commonly carry gasoline, kerosene, jet fuel, home heating oil ("No. 2") and other refined products[521] on the "light" end of the barrel, leaving to other forms of transportation movement of heavier petroleum liquids or solids such as residual fuel oil, asphalt and coke.[522] Products pipeline mileage[523] and traffic[524] have grown substantially in recent years. The products mix has also changed considerably: gasoline has dropped from 61 percent in 1964 to 55 percent of the total in 1974; jet fuel has tripled its share of the total and natural gas liquids have increased their share of the traffic by more than 50 percent.[525]

Products pipelines generally vary from 6 inches[526] to 36 inches in diameter in the case of Colonial's largest segment,[527] with the more recent lines skewed toward the larger size.

2. Competition

Products pipelines, like their crude oil counterparts, require a high percentage use of capacity in order to achieve maximum operating efficiency. Thus, as has been mentioned previously,[528] the economic incentive since

---

[518] W. J. Williamson, *supra* note 315, at 8.
[519] MARATHON, *supra* note 202, at 4.
[520] KENNEDY STAFF REPORT, *supra* note 184, at 20.
[521] MARATHON, *supra* note 202, at 4.
[522] See text at note 202, *supra*.
[523] In 1959, there were approximately 40,000 miles of products pipelines in the United States, SPECTRE, *supra* note 104, at 105. In 1977, there were 81,296. OWINGS, *supra* note 176, at 4.
[524] Traffic had grown from 1.5 billion barrels in 1964 to more than 3.2 billion barrels in 1974. PIPELINE TRANSPORTATION, *supra* note 162, at 15.
[525] *Id.*
[526] There are still some 4-inch and even a small number of 2- and 3-inch products lines around as a holdover from earlier times. Apco had a 4-inch line from Cyril to Duncan, Oklahoma, and there may be some "tap" lines of that diameter but as a general proposition, 6-inch is about as small as economics will let you go.
[527] NET-I *supra* note 199, at 198. This is the Houston, Texas, to Greensboro, North Carolina segment. The line "telescopes" down to 32 inches then down to 30 inches on the last segment, Dorsey Jct., Md., to Linden, N.J. Colonial's recent expansions have used 40-inch line.
[528] See text at note 162, *supra*.

the advent of large diameter pipelines has been to seek nonproprietary business to achieve these efficiencies. We have just witnessed the head-to-head struggle for the Gulf Coast refinery area to Southeastern and East Coast consumer markets.[529] Repeating one key fact to epitomize the reality of the Colonial-Plantation struggle, in 1963, Plantation transported 352,000 barrels per day (BPD). In 1964, the first full year of Colonial's operation, Plantation carried only 242,000 BPD, a decline which resulted in a 38 percent reduction in its revenue [and more like 50 percent reduction in net income].[530] We likewise saw Explorer's projected traffic eroded by competitive moves by Texas Eastern to expand its mainline capacity out of Houston and to extend its line from Seymour, Indiana, to Chicago and by the construction of two large diameter crude lines carrying locally produced and imported crudes to refineries in Explorer's Mid-Continent market area.[531] Refiners in the Philadelphia/Linden, New Jersey, area seeking to market their products in the Pittsburgh/Buffalo area can select between the ARCO line, the Laurel pipeline, the Buckeye pipeline, the Mobil pipeline or the Sun pipeline; thus we have a five line direct competitive race for business.[532] Williams Pipeline Company (formerly Williams Bros.) faces competition along segments of its line from ARCO, Explorer, Kaneb, Continental (formerly Cherokee), Phillips, Badger, Mobil, Amoco, National Coop Ref. Association and Koch.[533] A more detailed discussion on the subject of competition will be found in a subsequent chapter entitled "Economic Characteristics [of Pipelines]," but the foregoing should provide a rough idea of the competitive nature of the products line business.

3. Technical Aspects
a. Preservation of Product Integrity

The problems involved in transporting different grades of material through the same line have already been alluded to in the "Batching and Quality Banks" subsection on crude lines.[534] This difficulty is significantly increased in the case of products pipelines. Some products pipelines operate on a "quasi-common-stream" basis, *i.e.*, all products of a particular grade, say, for example, motor gasoline having the same general specifications, are commingled together and the carrier is not required to segregate different

---

[529] See text at notes 424 to 426, *supra*.

[530] LeGrange, *FERC Direct Testimony, supra* note 176, at 26; M. CANES & D. NORMAN, PIPELINES & ANTITRUST 10 n.1 (API, Nov. 1978) [hereinafter cited as PIPELINES & ANTITRUST]; O'Donnel, *Pipelines Pumping to Meet Southeast Products Demand*, OIL AND GAS JOURNAL, Feb. 13, 1967, pp. 74-75.

[531] See text at notes 433 & 434, *supra*; *cf.* PIPELINE PRIMER, *supra* note 331, at 22. Vernon T. Jones, then President of Explorer, testified before the Senate Judiciary Committee: "We actively solicit outside business." *S.1167 Hearings, supra* note 432, at 606.

[532] PIPELINE PRIMER, *supra* note 331, at 22; *cf.* MARATHON, *supra* note 202, at 20.

[533] MARATHON, *supra* note 202, at 20-21; *cf.* PIPELINE PRIMER, *supra* note 331, at 22.

[534] See text at notes 509 through 516, *supra*.

shippers' movements of such products from those of other shippers.[535] However, most products pipelines follow the complete segregation method whereby the shipments of an individual shipper are maintained separate and apart from the shipments of others even where two products have generally similar characteristics.[536] An example of the "common stream within product batches" operation is Williams Pipeline;[537] an example of a segregated system is Colonial.[538] The former method of operation requires less total tankage because all shipments of like product (meeting tariff specifications) can be commonly stored. Smaller tenders per individual shipper can be accepted because they can be combined, thus making it easier for small shippers to get on the line, and the shipper can frequently obtain immediate receipt at the far end of the line of a like quantity of the same product that he delivers to the pipeline at the originating location, thus eliminating transit time.[539] The latter method is used where individual specification product is important—take for example Shell's unleaded gasoline, "Super Regular" (now "Super Shell Unleaded"). Not only must there be no contamination from adjacent leaded products which might cause it to exceed the Environmental Protection Agency's (EPA) grams per gallon limit,[540] but the quality (among other characteristics the octane rating) vis-á-vis other unleaded products requires protection. The storage situation at destination usually is not a significant problem, as the shippers, or their consignees, already have terminals built to receive shipments from the line and from which to deliver product, usually by tank truck, to their various reseller, retail and consumer outlets. Storage at origin is either tankage at the shipper's refinery, or in some cases, storage provided by an independent storage firm such as GATX in the Houston area.[541] If small shippers are willing to commingle their products in the GATX terminal and, by combining their shipments, meet the batch requirements, Colonial has no objection.[542] It should be noted that, even in the quasi-common-stream operation, batching is required[453] because although all shipments of a given product may be

---

[535] Williams Pipe Line Co. (formerly Williams Bros.) utilizes this method of operation. *H.Res.5 Subcomm Hearings, supra* note 234, at 16 (statement of D. W. Calvert, Exec. V.P., The Williams Cos.).
[536] *Id.*
[537] *Id.*
[538] *Consumer Energy Act Hearings, supra* note 154, at 607-609 (colloquy between Senator Stevenson and Jack Vickrey and John Green).
[539] *H.Res.5 Subcomm Hearings, supra* note 234, at 17; *cf.* KENNEDY STAFF REPORT, *supra* note 184, at 23.
[540] *Consumer Energy Act Hearings, supra* note 154, at 608.
[541] *Id.* at 605. About one-third of the products going through Colonial's system enter the system through the GATX Terminal. *Id.* at 606.
[542] *Id.* at 607. This not uncommon in the industry. MARATHON, *supra* note 202, at 83.
[543] KENNEDY STAFF REPORT, *supra* note 184, at 23.

treated as fungible, the pipeline must still segregate such shipments from those of a different type of product.

Minimum batch size depends upon the amount of interface or mixed product that can be tolerated as the batches move down the line.[544] If batches that are too small are accepted for long-distance transportation, the amount of "contamination spread"[545] may be too large to permit appropriate blending with lower grade shipments.[546] In older systems, involving unregulated velocity in line loops,[547] station bypasses and equipment,[548] and tanks "floating" on the line,[549] the contamination problem was severe because the design work had been done prior to realization of the effect that these "dead fluid pockets" had on intermixing.[550] Present day lines have eliminated most of these now identified line configuration sources of contamination and adopted operational techniques designed to reduce contamination, but the "normal" contamination continues. Normal contamination appears to vary as a function of velocity (rate of flow), density and viscosity differentials between the two products, pipe friction factor (affected by length, diameter and condition of the line), and other factors not yet identified.[551] The incremental effect of velocity is somewhat difficult to pin down but it must be sufficient to maintain whirling or eddy type ("turbulent") flow[552] in order to prevent excessive contamination.[553] Turbulent flow begins when the

---

[544] NET-I, *supra* note 199, at 261.

[545] Birge, *supra* note 116, at 176 defines this as the quantity of commingled material extending from pure product of the leading fluid to pure product of the following fluid. J. D. Durand defines this as "interface." *Consumer Energy Act Hearings, supra* note 154, at 664.

[546] NET-I, *supra* note 199, at 261; PIPELINE PRIMER, *supra* note 331, at 24-25; *cf.* MARATHON, *supra* note 202, at 84.

[547] Birge, *supra* note 116, at 177, 179, 273; Roach, *Factors Influencing Commingling in a Products Pipe Line,* WORLD OIL, Jan. 1948, pp. 172-174 [hereinafter cited as Roach]; Neptune, *Operation of Partial Loops,* OIL AND GAS JOURNAL, Sept. 23, 1944, pp. 204-207.

[548] Roach, *supra* note 547, at 172, 174; Birge, *supra* note 116, at 178, 273. Items included are station bypasses, scraper traps, hay tanks, sump pits, and pumps not operating.

[549] WOLBERT, *supra* note 1, at 31 n.153.

[550] Roach, *supra* note 547, at 172.

[551] Birge, *supra* note 116, at 278; WOLBERT, *supra* note 1, at 32; *cf.* BARBE, *supra* note 317, at 5; *Consumer Energy Act Hearings, supra* note 154, at 664 (Letter to Senator Magnuson from J. Donald Durand, dated Jan. 23, 1974, responding to questions asked by Senator Stevenson).

[552] Turbulent flow is characterized by whirls and eddies, thus having a greater tendency to wash off liquid which might otherwise cling to the surface of the pipe, as contrasted to streamline flow in which all the liquid particles follow paths generally parallel to the walls of the pipe. This produces an effect analogous to an unfolding telescope or a collapsible drinking cup, which tends to leave behind the particles on, or close to, the surface of the pipe to mix with the following fluid.

[553] Birge, *supra* note 116, at 178; Williams, *Pumping Various Products through the Same Pipe Line System,* OIL AND GAS JOURNAL, Sept. 22, 1945, p. 197 [hereinafter cited as Williams]; Fowler & Brown, *Contamination by Successive Flow in Pipe Lines,* PETROLEUM ENGINEER, Aug. 1944, p. 121; WOLBERT, *supra* note 1, at 32; KENNEDY STAFF REPORT, *supra* note 184, at 24.

Reynolds number[554] is between 2,000 and 2,500.[555] The early observers in the field believed that once turbulent flow had been achieved, a further increase in the Reynolds number caused a decrease in contamination.[556] A variety of this belief has persisted, and statements are made with respect to the pumping rate that as the velocity of the product in a pipeline increases, the growth rate of the interface decreases.[557] Based on the early work, a widely accepted rule of thumb was that pipeline flow should remain above two and one-half feet per second in order to avoid excessive contamination.[558] However, further experimentation tended to discount this theory. Kerosene-gasoline and kerosene-diesel fuel contacts were transported in a commercial line at velocities as low as 0.5 feet per second (Reynolds Number 20,000) without appreciable abnormal spread of contamination.[559] It would appear that once a Reynolds number of 20,000 is reached, the incremental velocity does not affect appreciably the spread of the interface.[560] Another important factor is the effect of length of line on contamination. Here again, we must distinguish between the *rate* of spread, which *decreases* with the distance traveled,[561] from the *amount* or length of the contamination spread, which *increases* with the distance traveled.[562] Birge's observations of the Plantation Pipe Line indicate that normal contamination increases exponentially with

---

[554] Named after Professor Osborn Reynolds who was the pioneer investigator in the field. Reynolds injected colored liquids into streams of water flowing through glass tubing and observed that under controlled conditions, the flow assumed a viscous or streamline character at certain velocities and became turbulent at others. He empirically determined that the nature of the flow was a function of the relation $DV\delta/\mu$, where D is the diameter of the pipe, V is the velocity of the fluid, δ is the density of the fluid and μ is the viscosity of the fluid (in any consistent absolute system of units. Williams, *supra* note 553, at 197.

[555] J. ZABA & W. DOHERTY, PRACTICAL PETROLEUM ENGINEERS' HANDBOOK 445 (2d ed. 1939); Williams, *supra* note 553, at 197. Velocities giving a Reynolds number of 2,000 to 2,500 are said to be "critical velocities."

[556] Fowler & Brown, *Contamination by Successive Flow in Pipe Lines*, PETROLEUM ENGINEER, Aug. 1944, p. 127.

[557] *Consumer Energy Act Hearings, supra* note 154, at 664. Note this is not the same as Fowler & Brown's theorem because it addresses the *rate* of growth which is the first differential of the increase.

[558] WOLBERT, *supra* note 1, at 32.

[559] Birge, *supra* note 116, at 178.

[560] *Id.* at 179, 280. Fowler & Brown's conclusions do not necessarily conflict with Birge's because they were experimenting with mixtures moving at velocities associated with Reynolds numbers under 20,000 and Birge was concerned with Reynolds numbers above 20,000. Williams, *supra* note 553, at 198, plotted relative velocities against the logarithms of Reynolds numbers and his graph indicates that there would be a decrease in contamination spread from Reynolds number 3,160 to 20,000 but little change thereafter. Thus both Birge's and Fowler & Brown's conclusions could be valid. However, commercial pipelines are operated in the range examined by Birge; hence for our purposes, his findings are more significant than those of Fowler & Brown.

[561] Birge, *supra* note 116, at 276.

[562] WOLBERT, *supra* note 1, at 33; *Consumer Energy Act Hearings, supra* note 154, at 664.

the length of the line.[563] Although maintaining a constant rate of flow and pressure will minimize the contamination spread,[564] line shut-downs under pressure with gasoline-kerosene interfaces in transit have been found not to cause excessive spread of contamination.[565]

We have mentioned operational techniques which are utilized to minimize the effects of contamination. In certain situations, mechanical devices are used to inhibit the spread of contamination.[566] These are generally referred to by pipeliners as "batch pigs"[567] or, more genteelly, as "batching spheres."[568] A batching sphere is an inflatable, water-filled rubber sphere conforming to the inside diameter of the pipeline which is inserted between dissimilar products to aid in their separation.[569] Sometimes a buffering material is used to separate two products whose properties render difficult disposal of their admixture.[570] Another technique is making intermediate deliveries out of "heart-cuts."[571] There are a number of other techniques used, depending upon the configuration of the particular

---

[563] Birge, *supra* note 116, at 278. For example his empirically derived equation for a gasoline-kerosene contamination was $Y = 1.93\ X^{0.529}$, where Y is the contamination spread in feet and X is the line distance travelled in feet. *See* graph of this relationship, *Id.* at 177. SMITH & SCHULZE, INTERFACIAL MIXING CHARACTERISTICS OF PRODUCTS PIPE LINES 63, 64 (1947) [hereinafter cited as SMITH & SCHULZE] also found a logarithmic ratio.

[564] *Consumer Energy Act Hearings, supra* note 154, at 664; *cf.* BARBE, *supra* note 317, at 5.

[565] Birge, *supra* note 116, at 282.

[566] BARBE, *supra* note 317, at 5.

[567] MARATHON, *supra* note 202, at 18.

[568] NET-I, *supra* note 199, at 199.

[569] *Id.*; *cf.* MARATHON, *supra* note 202, at 18.

[570] WOLBERT, *supra* note 1, at 34 n.179. For example Product X has a high flash and a low end point and Product Y has a low flash and high end point. Very small amounts (0.25%) of Product Y will reduce the flash of Product X by as much as 25° or 30°F., which is extremely undesirable because it renders use of Product X hazardous. At the same time, contaminating Product Y by small amounts of Product X will lower disproportionately the end point of Product Y, an important specification of Y. Under these facts, we could not cut sharply at either end lest we spoil one product, nor at the middle because to do so would ruin both. A solution is to insert a buffer slug of Product Z which has a high flash and a high end point. The contamination buffer can be cut both ways because the resultant contamination between X and Z will not affect the high flash point of X nor will the contamination between Y and Z lower Y's high end point. *Id.*

[571] WOLBERT, *supra* note 1, at 32; KENNEDY STAFF REPORT, *supra* note 184, at 24. We have already noted that the *rate* of spread decreases with the distance travelled. Text at note 561, *supra*. Therefore, instead of making intermediate deliveries from the ends of the batch and suffering an iteration of the more rapid spread by starting over again, it makes sense to allow the two contaminated ends to proceed undisturbed to their last terminal and take intermediate deliveries out of the pure product in the "heart" of the batch. This is called a "heart cut." Birge, *supra* note 116, at 276; see BARBE, *supra* note 317, at 6-7.

system.[572] Obviously, the more sophisticated and accurate methods one has to identify and to track the location of the ends of the contamination spread and to measure the differential change of the stream from one product to another, the better he is able to cope with the problem.[573]

Disposition of the interfacial mixture depends upon the materials involved. Ideally, it may be split equally between the two grades by making a "flying switch"[574] at the middle of the interface,[575] which requires optimum scheduling procedures and minimum tenders large enough to absorb the cut without pulling the pure product off specification.[576] This is a matter which is determined by the characteristics of the pipeline system concerned and will vary with the configuration and contamination characteristics of the line involved. Thus, the example which was given in footnote 576 of Plantation's World War II line with respect to a kerosene-diesel fuel interface will not be applicable even to a kerosene-diesel movement through Plantation's current line, and would definitely not be a guide for large diameter lines such as Ex-

---

[572] See WOLBERT, *supra* note 1, at 31-32 for a discussion of some of these. Another technique is to "tight-line" with connecting carriers, which is the equivalent of operating both systems on a "closed basis," *i.e.*, direct discharge from the last station of the first carrier directly into the first station of the succeeding carrier, without the use of tankage. MARATHON, *supra* note 202, at 18.

[573] The evolution has come from taking samples at one minute intervals for testing by an ordinary field hydrometer, WOLBERT, *supra* note 1, at 31 n.156, to continuously recording gravimeter and special sight glass installations, Lundberg, *Magnolia Pipeline's Hebert Terminal Designed and Planned for Minimum Interruption and Product Commingling*, OIL AND GAS JOURNAL, Oct. 6, 1949, pp. 208, 211, 213, to the insertion of radioactive isotopes picked up by a Geiger Counter. *Bond, supra* note 207, at 734-735; *cf.* KENNEDY STAFF REPORT, *supra* note 184, at 24.

[574] This is a sharp (as possible) cut on the interface. Because it is made across a finite time interval, it must be accomplished in as short a time span as possible to avoid introducing an appreciable admixture of product into the receiving tank. Petroleum Rail Shippers' Ass'n v. Alton & Southern R.R., 243 I.C.C. 589, 657 (1941); SMITH & SCHULZE, *supra* note 563, at 68.

[575] BARBE, *supra* note 317, at 5.

[576] WOLBERT, *supra* note 1, at 35 illustrates this point with an example taken from data contained in Birge, *supra* note 116, at 288-291. The data came from Plantation's World War II operations, where the kerosene-diesel contamination spread over Plantation's line was approximately 800 barrels each way. In this case the entire cut was downgraded into the diesel so that approximately 800 barrels of kerosene had to be absorbed by the diesel. At that time, 3 percent kerosene was the limit that specifications would permit to be blended with the diesel. The minimum batch or tender of diesel that would accommodate this amount of kerosene and remain on specification was 800 times 100/3, or 26,667 barrels of diesel in the batch, hence a requirement of 25,000 barrels.

plorer and Colonial.[577] Typical interfacial mixtures in today's lines, depending on the characteristics of the line and the products moved, vary from 50 to 2,000 barrels, relating to segregated batches of 10,000 to 300,000 barrels.[578]

Minimum tenders (and product specifications) are an integral part of each company's tariffs which it files with the appropriate regulatory agency, which in the case of interstate pipelines, is now the FERC.[579] They are in the public domain and the FERC has, and its predecessor, the ICC, has exercised, adequate power to correct or eliminate any unreasonable requirements or specifications.[580]

b. Scheduling and Dispatching

In addition to the batch size, contamination disposal problems can be mitigated to some degree by proper sequencing of batches within the cycle.[581] The basic rule is to place products adjacent to each other which will be disposable into one or the other product or into some other product being

---

[577] In the *Consumer Energy Act Hearings, supra* note 154, held in November 1973, Jack Vickrey, then in private law practice but still quite familiar with Colonial because of his having served as its General Counsel for 11 years, explained to Senator Stevenson that Colonial's operating conditions, *i.e.*, moving 1.4 million barrels a day through essentially 1,900 miles of 36 inch line, with 27 front-end shippers moving 6 or 7 different products each in 10 day cycles, caused the minimum figures to come out to 75,000 barrels. The Company did permit, and Vickrey said it was not unusual for, two or more shippers to get together and pool their 25,000 barrel tenders so as to make up a workable batch. A 75,000 barrel batch on Colonial's 36 inch line would occupy about 12 miles in length with an interface of one mile on either side. A 25,000 barrel batch would be 4 miles long with an interface of one mile on either side of the segregated product, thus making the proportional size of the interface intolerable. PIPELINE PRIMER, *supra* note 331, at 25; *see also* HOWREY & SIMON, *supra* note 3, at 46-47 where the authors give comparable figures for a 28 inch line, *i.e.*, when the ICC set a 25,000 barrel limit for a 6 to 8 inch product line, the slug would extend 71.63 miles and require 20 hours of pumping time, whereas in a modern 28 inch line, this same size batch would only occupy 6.48 miles and require only 69 minutes of pumping time; *accord,* MARATHON, *supra* note 202, at 85. When you consider that some interfacial mixtures of refined products must be run into a "slop" tank and be returned to a refinery for reprocessing, BARBE, *supra* note 317, at 6, you begin to realize how important batch size is. Actually, the usual slug going through Colonial's line is substantially larger than 75,000 barrels; otherwise Colonial would be in a "bind." *Consumer Energy Act Hearings, supra* note 154, at 609.

[578] BARBE, *supra* note 317, at 6.

[579] See, *e.g.*, Exxon Pipeline Co., F.E.R.C. No. 129, eff. 3/1/78, Items 15 & 20; Crown-Rancho Pipe Line Corp. Texas Joint Tariff No. 43, eff 8/1/78, Rules 1 & 7; Shell Pipe Line Corp. O.C.C. No. 16, eff. 10/1/77, Items 20 & 30; Shell Pipe Line Corp. N.M.S.C.C. No. 32, eff. 9/21/78, Items 20 & 30; Marathon Pipe Line Co., Ill.C.C. No. 54, eff. 1/1/76, Rules 2 & 5; Shell Pipe Line Corp. L.P.S.C. No. 74, eff. 12/1/77, Items 20 & 30.

[580] See text at notes 499-501, *supra*; HOWREY & SIMON, *supra* note 3, at 46.

[581] WOLBERT, *supra* note 1, at 34; *cf.* BARBE, *supra* note 317, at 6; NET-I, *supra* note 199, at 199; PIPELINE TRANSPORTATION, *supra* note 162, at 26.

handled currently.[582] Basically, this means attempting to abut products whose properties or uses are similar;[583] the approach is to group gasolines in one series of batches and distillate fuels in another series.[584] A typical sequence is regular gasoline, premium, regular gasoline, No. 1 fuel oil, No. 2 fuel oil or diesel, jet fuel or kerosene and regular gasoline (unleaded gasoline follows an unleaded product).[585] This sequencing takes place within the framework of a "product cycle" which is drawn up from the needs of the shippers, the capacity of the line, the delivery points involved and the contamination characteristics discussed above. It used to be a rule of thumb that product cycles were multiples of five days,[586] but the practice now seems to be seven, ten,[587] or fourteen day cycles.[588] Each shipper notifies the pipeline company in advance of the type, quantity, source, destination, and time he can make available his shipments and the dates and places he desires deliveries to be made.[589] The composite of the information forms the basis for the pipeline scheduler to make up his schedule. In crude oil systems this is usually done on a monthly basis, so as to synchronize with refinery run schedules and, in times past, with production allowables.[590] From the monthly schedules and the line's operational requirements and limitations, the scheduler breaks down the monthly schedules to a daily basis. Products lines similarly derive a monthly schedule but emphasis is placed on repetitive cycles which observe the batch sequencing described above.[591] The final result is a detailed operating plan which has dates and hours of movements, batch sizes, stations and pumping units to be operated and the tanks which

---

[582] Dreyer, *Scheduling and Dispatching: Description of the Procedures of Plantation Pipe Line*, OIL AND GAS JOURNAL, Oct. 6, 1949, p. 178. For example, an interface mixture between premium and regular gasoline would be cut into the regular tank. BARBE, *supra* note 317, at 6, or in the pre-unleaded days, when intermediate grades were being sold, the mixture itself became a saleable product. cf. *H.Res.5 Subcomm Hearings*, *supra* note 234, at 17 (Statement of D. W. Calvert).

[583] NET-I, *supra* note 199, at 199; BARBE, *supra* note 317, at 6.

[584] BARBE, *supra* note 317, at 6. For example, Colonial is said to run a gasoline sequence ranging from unleaded regular through leaded regular and leaded premium, followed by a sequence of kerosene, aviation kerosene, diesel fuel, various grades of light heating oils, aviation kerosene, and kerosene. NET-I, *supra* note 199, at 200.

[585] PIPELINE TRANSPORTATION, *supra* note 162, at 26; KENNEDY STAFF REPORT, *supra* note 184, at 24. Another sequence might be naptha, gasoline, kerosene, gas oil, kerosene, gasoline and naptha. NET-I, *supra* note 199, at 199.

[586] WOLBERT, *supra* note 1, at 34.

[587] Colonial is reputed to be on a 10 day cycle. *Consumer Energy Act Hearings*, *supra* note 154, at 607 (Statement of Jack Vickery); KENNEDY STAFF REPORT, *supra* note 184, at 24.

[588] BARBE, *supra* note 317, at 9.

[589] PIPELINE TRANSPORTATION, *supra* note 162, at 25; KENNEDY STAFF REPORT, *supra* note 184, at 23.

[590] BARBE, *supra* note 317, at 8.

[591] *Id.* at 9.

will be involved in the movements.[592]. The detailed daily execution of these schedules is done by dispatchers. Because a batched product may require days or weeks of in-transit time, may be increased or decreased in volume as it passes through the system and is only one of many separate batches in the system,[593] the dispatchers must keep track of each batch on an hour-by-hour and day-by-day basis,[594] so that the receiving consignees can be advised when their shipments will arrive[595] and the necessary valves, pumps and tankage operations performed properly to accomplish the movement, either by remote control or by the issuance of operating orders to field personnel.[596] Movement schedules are in a constant state of flux: shippers change plans, receiving refineries or terminals have equipment failures, pipeline equipment can have breakdowns or be shut down for maintenance, extreme weather can cause malfunctioning, require adjustments in the batches (say, more home heating oil in a time of shortage), or a number of other causes.[597] These changes are fed into the computer so that the schedule constantly is kept up to date.[598]

C. Petroleum Pipelines Viewed as a System

In previous subsections, mention has been made of pipeline mileage of each type of system as it was being discussed. It might assist the reader to have these figures brought together, take a brief look at historical trends and translate mileage, which was the only recorded indicium of growth in the formative years, into the more meaningful barrel-mile figure, which reflects more accurately the effect of the adoption of "big-inch" technology after World War II. Giving the mileage first, ICC data on total system mileage have been available since 1926, when the mileage was 90,170 miles as compared to 227,066 miles in 1977.[599] Breakdown by crude gathering, crude trunk lines and product lines is available from 1950, when the respective mileages were: 60,560 miles, 71,373 miles and 20,881, with a total system mileage of 152,814 miles, as compared to 1977's figures of 67,798 miles,

---

[592] *Id.*
[593] Burke, *supra* note 9, at 795.
[594] KENNEDY STAFF REPORT, *supra* note 184, at 23.
[595] Burke, *supra* note 9, at 795.
[596] BARBE, *supra* note 317, at 10.
[597] *Id.*
[598] *Cf.* KENNEDY STAFF REPORT, *supra* note 184, at 23.
[599] OWINGS, *supra* note 176, at 6.

77,972 miles and 81,296 miles, for total system mileage of 227,066.[600] Barrel miles of capacity increased in the following magnitudes: crude lines from 839 million barrel miles in 1955 to 1,609 million barrel miles in 1976; products lines from 205 million barrel miles in 1955 to 1,307 million barrel miles in 1976.[601] Although the absence of data on non-ICC reporting lines causes the total figures to be understated, the trends are believed to be reliable. Thus, the following observations appear to be appropriate: crude gathering line mileage peaked in 1959, which slightly antedated the cresting of domestic crude oil production. Gathering line mileage can be expected to decline as production declines and consolidations take place at a rate greater than new discoveries generate additional mileage; crude trunk line mileage has continued to increase as imported crude has increased; however the growth of products line mileage has expanded about 400 percent, reflecting big-inch technology displacing coast-wise tanker movement and increased crude imports into coastal refineries, with resultant products pipeline transportation to inland markets.

A more graphic illustration of pipeline movements can be derived from the maps of crude oil movements and products movements which are included herein as Appendices G and H. These maps not only give the location of the various pipelines but they add a "feel" for the volume of the movements and illustrate the interconnections and possibilities for alternate routes which were described in Section I B 6, "Flexibility and Alternatives to Shippers,"[602] provide the competition which was set forth in Sections II A 3 b, "Competition,"[603] and II B 2, "Competition,"[604] and account for the growth of non-proprietary shipments which was discussed in Section II A 3 c, "Scramble for Traffic."[605]

---

[600] *Id.* The data are as follows:

YEAR	CRUDE GATHERING MILEAGE	CRUDE TRUNK MILEAGE	PRODUCT MILEAGE	SYSTEM MILEAGE
1926	N.A.	N.A.	N.A.	90,170
1927	N.A.	N.A.	N.A.	110,580
1950	60,560	71,373	20,881	152,814
1959	75,182	70,317	44,483	189,982
1971	71,132	75,143	72,396	218,671
1977	67,798	77,972	81,296	227,066

[601] *Id.*

[602] See text at notes 212-228, *supra*.

[603] See text at notes 401-406, *supra*.

[604] See text at notes 528-533, *supra*.

[605] See text at notes 407-496, *supra*.

Chapter III

Economic Characteristics

A. General Observations

In order for a petroleum pipeline to be successful financially, there must be a need for continuous movements of a rather large quantity of one basic class of materials[606] in one direction [607] from one fixed point to another fixed point[608] over a long period of time, usually 30 to 40 years.[609] Unlike a railroad, which can haul products different from those which caused it to be built initially, a pipeline must look to a limited number of potential customers for its traffic.[610] Once in place, it cannot be adjusted readily to shifting sources of supply or changing markets[611] although it does have the ability to make limited extensions and expansions.[612] Pipelines are extremely

---

[606] *Petroleum Industry Hearings, supra* note 192, at 257 (Statement of Charles J. Waidelich); *Consumer Energy Act Hearings, supra* note 154, at 595 (Statement of Jack Vickrey); Spahr, *Rodino Testimony, supra* note 255, at 8; SPLAWN REPORT, *supra* note 372, at LXXVIII; WOLBERT, *supra* note 1, at 56.

[607] *Consumer Energy Act Hearings, supra* note 154, at 595 (Statement of Jack Vickrey); Waidelich, *Rodino Testimony, supra* note 197, at 8; LeGrange, *FERC Direct Testimony, supra* note 176, at 27; SPLAWN REPORT, *supra* note 372, at LXXVIII; WOLBERT, *supra* note 1, at 56; PIPELINE PRIMER, *supra* note 331, at 14.

[608] Waidelich, *Rodino Testimony, supra* note 197, at 8; LENNART, *supra* note 111, at 5-6; PIPELINE TRANSPORTATION, *supra* note 162, at 37; LeGrange, *FERC Direct Testimony, supra* note 176, at 27; *cf. Petroleum Industry Hearings, supra* note 192, at 257 (Statement of Charles J. Waidelich) (inflexible in their routes); WOLBERT, *supra* note 1, at 55; PIPELINE PRIMER, *supra* note 331, at 14.

[609] *Consumer Energy Act Hearings, supra* note 154, at 595, (Statement of Jack Vickrey); *cf.* Waidelich, *Rodino Testimony, supra* note 197, at 9; STEINGRABER, *supra* note 197, at 146 (35-40 years).

[610] *Petroleum Industry Hearings, supra* note 192, at 257 (Statement of Charles J. Waidelich); *H.Res.5 Subcomm Hearings, supra* note 234, at 142 (Statement of Fred F. Steingraber); WOLBERT, *supra* note 1, at 55. This fact can create a "small numbers bargaining problem" in the absence of integration. *See* TEECE, *supra* note 329, at 8.

[611] LENNART, *supra* note 111, at 6; *cf.* Spahr, *Rodino Testimony, supra supra* 255, at 8-9 (line cannot be moved like a garden hose to a new location).

[612] See text at notes 424-426 (Plantation Pipe Line expansions and extension to Washington, D.C. area) and note 434 (Texas Eastern expansion of Houston-Seymour, Indiana, main line and extension from Seymour to Chicago), *supra*.

capital intensive,[613] requiring eight times as much net plant investment as does the average industrial company[614] and eight and one-half times as much investment per employee as the railroad industry, which itself is capital intensive.[615] Buried pipe and right-of-way costs constitute 75-80 percent of total pipeline capital investment.[616] Fixed costs consume about two-thirds of every dollar of revenue.[617] Because of pipelines' high proportion of fixed costs, profitability is extremely sensitive to volume.[618] For example, if a 20-inch pipeline can earn a fair return carrying 100,000 barrels per day at a 20 cents per barrel tariff rate, an additional 50,000 barrels per day throughput will produce the same return at about 16 cents per barrel.[619] Conversely, however, if the throughput drops even slightly below the "break-even"

---

[613] *Consumer Energy Act Hearings, supra* note 154, at 595 (Statement of Jack Vickrey); Statement of Raymond B. Gary, Managing Director of Morgan Stanley & Co., Inc., in Ex Parte No. 308 (now FERC Dkt. RM 78-2) 5 (May 27, 1977) [hereinafter cited as Gary, *FERC Direct Testimony*]; PIPELINE TRANSPORTATION, *supra* note 162, at 35; MARATHON, *supra* note 202, at 5 gives three examples: Explorer $219 million, Colonial $378 million [original cost—then $540 million, *Petroleum Industry Hearings, supra* note 192, at 256 (statement of Charles J. Waidelich) and later, $675 million] MARATHON, *supra* note 202, at 42, Aleyeska (TAPS) $9 billion.

[614] Gary, *FERC Direct Testimony, supra* note 613, at 13 cites government figures of $4.00 net plant investment in pipelines per $1.00 revenue compared to average industrial company net plant investment of $0.50 per $1.00 revenue.

[615] Tierney, *TAPS Rebuttal Testimony, supra* note 175, at 14 ($561,412 per pipeline employee, compared to $65,872 per railroad employee, in year 1975). This figure is higher for more recent lines. Explorer's is $1.3 million per employee. *S.1167 Hearings, supra* note 432, pt. 9 at 615 (Statement of Vernon T. Jones, then President of Explorer).

[616] *Petroleum Industry Hearings, supra* note 192, at 257 (Statement of Charles J. Waidelich); PIPELINE TRANSPORTATION, *supra* note 162, at 35; *cf.* LeGrange, *FERC Direct Testimony, supra* note 176, at 27 (80%); SPECTRE, *supra* note 104, at 117 (75%); Burke, *supra* note 9, at 787 (buried pipe and right of way represent 80-90 percent of a pipeline's total investment). For a breakdown of petroleum pipeline investment by line item, see NET-I, *supra* note 199, at 212.

[617] LENNART, *supra* note 111, at 14; *cf. Consumer Energy Act Hearings, supra* note 154, at 595 (Statement of Jack Vickrey: about two-thirds of every dollar of revenue is required to service debt, pay taxes and cover fixed operating costs); Burke, *supra* note 9, at 787 (approximately 70 percent); PIPELINE TRANSPORTATION, *supra* note 162, at 36; STEINGRABER, *supra* note 193, at 147: Colonial's 1974 fixed costs were approximately $80 million.

[618] *Petroleum Industry Hearings, supra* note 192, at 257 (Statement of Charles J. Waidelich); PIPELINE TRANSPORTATION, *supra* note 162, at 36; *cf.* Gary, *FERC Direct Testimony, supra* note 613, at 14; *Consumer Energy Act Hearings, supra* note 154, at 595 (Statement of Jack Vickrey). Note, however, that the *degree* of sensitivity decreases as pipeline diameters approach 24 inches because the intermediate-run cost curves begin to flatten out and the line is close to its optimal cost over a wider range of throughputs adjacent to its design capacity. See *CRUDE OIL PIPELINES, supra* note 158, at 26, Chart 6; Appendix I; PIPELINE TRANSPORTATION, *supra* note 162, at 40; *cf.* KENNEDY STAFF REPORT, *supra* note 184, at 99.

[619] PIPELINE TRANSPORTATION, *supra* note 162, at 36. See Appendix I, "Tariff Required to Earn 7% Net Return on Investment."

Economic Characteristics 95

volume, the line will operate at a substantial loss.[620]

The risk and rewards to a pipeline owner are asymmetrical because, as illustrated by the Explorer example, the owners have no "downside risk safety net" and can incur substantial losses, but there is no way that they can earn more than the amount permitted by regulation.[621] Because of the foregoing, and the risks and uncertainties discussed in sections F and G of this Chapter, pipelines are regarded as high risk investments by members of the industry,[622] investment bankers,[623] and institutional investors[624] These circumstances,

---

[620] *Consumer Energy Act Hearings, supra* note 154, at 595 (Statement of Jack Vickrey); Waidelich, *Rodino Testimony, supra* note 197, at 9; Gary, *FERC Direct Testimony, supra* note 613, at 14 cites Explorer as a case in point, losing $35 million between 1972 and 1975 primarily because of insufficient throughput; *cf.* KENNEDY STAFF REPORT, *supra* note 184, at 99; MARATHON, *supra* note 202, at 32.

[621] Rebuttal Testimony of Professor Ezra Solomon Before FERC in Trans Alaska Pipeline System, Dkt. OR78-1, 27 (Oct. 30, 1978) [hereinafter cited as Solomon, *TAPS Rebuttal Testimony*]; *cf. Consumer Energy Act Hearings, supra* note 154, at 595 (Statement of Jack Vickrey) ("A [shipper-owner] pipeline investor is shooting craps with a 7-percent limit on what he can earn, but no limit on what he can lose."); Gary, *FERC Direct Testimony, supra* note 613, at 5, 9.

[622] LeGrange, *FERC Direct Testimony, supra* note 176, at 26; PIPELINE TRANSPORTATION, *supra* note 162, at 37; *cf. Consumer Energy Act Hearings, supra* note 154, at 595 (Statement of Jack Vickrey) (extremely high risk).

[623] Gary, *FERC Direct Testimony, supra* note 613, at 12.

[624] *Cf.* Solomon, *TAPS Rebuttal Testimony, supra* note 621, at 28. Apparently, some critics of the industry have difficulty in perceiving the risk. For example, Senator Tunney, in questioning witness Richard Hulbert (President of Kaneb Pipe Line Company, which is unaffiliated with an oil company) in August, 1974, could not see the risk in Colonial Pipeline Company's venture because during its first year of operation it paid dividends of $36 million to its shipper-owners on an equity of $25 million. *S.1167 Hearings, supra* note 432, pt. 8 at 5946. With all due respect to the Senator, he has compounded several analytical errors. First, he is assessing risk at a time more than 10 years *after the correct assessment date, i.e.,* in March, 1962, when the shipper-owners committed themselves to expenditures of at least $378 million (which grew to $675 million), in the face of questions whether: (a) right of way could be acquired in Alabama, North Carolina and Maryland, which did not have eminent domain laws [remember Plantation's and Southeastern's World War II difficulties which required the Cole Act to get the lines built—see text at notes 150-152, *supra*]; (b) supertankers then being built would make the pipeline non-competitive (Vickrey, *Proxmire Hearings, supra* note 179, at 18); (c) there would be a substantial change in the relative size of refining centers; (d) there would be substantial imports of foreign or Caribbean refiner products directly into the New York Harbor Area (*S.1167 Hearings, supra* note 432, pt. 8 at 5946, Testimony of Richard C. Hulbert); and (e) there would be "grass-roots" refineries constructed on the east coast (*Petroleum Industry Hearings, supra* note 192, at 314, Testimony of E. P. Hardin). Second, Senator Tunney confused equity with investment; the shipper-owner's credit was "on the line" for the *full $378 million*, not just the $25 million *cash* which they put up. See the comment from the bench by Judge Richmond B. Keech, of the United States District Court for the District of Columbia, during the trial of the *Arapahoe* case; United States v. Atlantic Refining Co., Civil No. 14360, D.D.C., Mar. 24, 1958: "Actually as a practical proposition, if I preempt my credit and obligate myself to draw dollars from some other sources, it is to that extent my money, isn't it, because I have a credit rating of a given amount of dollars and

the need of oil companies for the lines,[625] and their willingness to underwrite the pipelines' risks have been the reasons why, absent special situations,[626] oil companies, large and small,[627] have owned and operated the nation's pipelines.[628]

B. Cost Components

By far the largest component of cost is capital. As noted in the preceding paragraph, pipelines are extremely capital intensive.[629] The amounts are significant, from $219 million in the case of a line such as Explorer to over $9 billion in the case of TAPS.[630] Once the line is lowered in the ground and backfilled, these dollars are sunk.[631] The shipper-owners, by virtue of their

---

if I should say place an easement on them, why I am sort of preempting my source of supply? Tr. p. 34, excerpted in *H.Res.5 Subcomm Hearings, supra* note 234, at A81. Third, the Senator engaged in a bit of circuitous reasoning when he said "Where's the risk involved" when "The consortium [Colonial's owners] had a throughput agreement at the very beginning." *S.1167 Hearings,* at 5946. For the same theme, see *Petroleum Industry Hearings, supra* note 192, at 229 (Statement of Jesse Calhoon). What the non-perceivers fail to recognize is *who* is the risk-taker. All that has happened is that the risk has been shifted from the pipeline company to the shipper-owner parent. For an instance where cross-examination brought an understanding to a non-perceiver, see Attorney Howard Trienens' questioning of David L. Jones, Acting Chief, Section of Motor, Pipeline and Water Carrier Analysis, Bureau of Economics, Interstate Commerce Commission. F.E.R.C., Valuation of Common Carrier Pipe lines, Dkt. RM 78-2, Tr. Vol. 3, p. 489-491 (Nov. 3, 1977).

[625] *Petroleum Industry Hearings, supra* note 192, at 257 (Statement of Charles J. Waidelich); Spahr, *Rodino Testimony, supra* note 255 at 12; Waidelich, *Rodino Testimony, supra* note 197, at 3; LIVINGSTON, *supra* note 228, at 12.

[626] See text at IV A, "Ownership," *infra.*

[627] DE CHAZEAU & KAHN, *supra* note 30, at 22 ("The most striking fact disclosed by [McLean & Haigh's] inquiry is how widespread is the integrated form of organization among refiners, *small as well as large") (emphasis supplied);* LIVINGSTON, *supra* note 228, at 3 ("All of the 42 largest refining companies, accounting for 94% of total U.S. crude running capacity in January 1974, own crude oil pipelines, and all but three own refined products lines. *Even among the smaller refiners, 23 own crude lines and 11 own product lines, and the smaller refiners as a group have roughly the same share of pipeline ownership as of refining capacity.") (emphasis supplied);* PIPELINE PRIMER, *supra* note 331, at 6 ("The smaller refiners, as a group, own about the same percentage of the pipelines as their share of total refining capacity . . . The fifty-eight smallest refiners owned 1.9 percent of the refining capacity compared with 2.0 percent ownership of the crude oil lines and 0.9 percent of the refined product lines.").

[628] Gary, *FERC Direct Testimony, supra* note 613, at 5; *Consumer Energy Act Hearings, supra* note 154, at 595 (Statement of Jack Vickrey); *Petroleum Industry Hearings, supra* note 192, at 25 (Statement of Charles J. Waidelich); Waidelich, *Rodino Testimony, supra* note 197, at 6; *H.Res.5 Subcomm Hearings, supra* note 234, at 142 (Statement of Fred F. Steingraber); *cf.* LENNART, *supra* note 111, at 6. This is not a new phenomenon; it was so as far back as 1933. SPLAWN REPORT, *supra* note 372, at LXXVIII.

[629] See text at note 613, *supra.*

[630] See note 613, *supra.*

[631] LeGrange, *FERC Direct Testimony, supra* note 176, at 27; Myers, *supra* note 198, at 9; LIVINGSTON, *supra* note 228, at 18; PIPELINE PRIMER, *supra* note 331, at 14.

debt guarantees and throughput and deficiency agreements, are locked into an investment that cannot readily respond to changing patterns of supply or distribution.[632] Moreover, the salvage value of the line is negligible at best[633]

---

[632] Gary, *FERC Direct Testimony, supra* note 613, at 13; MARATHON, *supra* note 202, at 35-36; *cf.* HOWREY & SIMON, *supra* note 3, at 10-11.

[633] *Consumer Energy Act Hearings, supra* note 154, at 595 (Statement of Jack Vickrey); Myers, *supra* note 198, at 9 ("minimal"); MARATHON, *supra* note 202, at 36 ("very little"); Spahr, *Rodino Testimony, supra* note 255, at 9 ("frequently, it has no scrap value"); PIPELINE PRIMER, *supra* note 331, at 15, lists Gulf's Northern Pipeline System from Tulsa to Dublin, Indiana, shut down in 1970 and sold for scrap in 1972 and West Texas Gulf Pipe Line, connecting the West Texas production region to Port Arthur, Texas, salvaged for "minimum scrap value" in 1972-1974. In a few cases, new uses have been found for oil lines such as reversals of flow, shifts from crude to products transportation, vice versa, or both. Myers, *supra* note 198, at 9 n.4. Examples that come to mind are the reversal of flow and conversion to products service of the Tuscarora Pipe Line, see text at notes 121-122, *supra*; sale of Ajax to the Cherokee Pipe Line Company (now Continental) for conversion to products service, SPECTRE, *supra* note 104, at 117 ("at bargain basement prices"); sale to Phillips of a Shell Oil 10-inch crude line from West Texas to Houston, rendered obsolete by the Rancho system, for LPG use, see text at note 171, *supra*; sale of the Four Corners Pipe Line to ARCO for reversal of flow, now to bring North Slope Crude from California to Red Mesa, Arizona, and thence through the Texas-New Mexico system to West Texas where a number of alternate routes will be available, see REPORT OF THE ANTITRUST DIVISION, DEPARTMENT OF JUSTICE, ON THE COMPETITIVE IMPLICATIONS OF THE OWNERSHIP AND OPERATION BY STANDARD OIL COMPANY OF OHIO OF A LONG BEACH, CALIFORNIA-MIDLAND, TEXAS CRUDE OIL PIPELINE 27-28 [hereinafter cited as D. J., SOHIO REPORT]. Sohio is attempting to construct a new tanker terminal at Long Beach, California; then after diverting 200,000 B/D to Los Angeles area refineries, to move about 500,000 B/D to Midland, Texas, by constructing 227 miles of new 42-inch line, reversing and converting to crude transportation some 125 miles of Southern California Gas Company's line and approximately 675 miles of El Paso Natural Gas Company's line into Midland, Texas. NET-III, *supra* note 199, at 211. These salvage operations have been sporadic and fortuitous, and constitute exceptions, rather than the rule. MARATHON, *supra* note 202, at 36; Myers, *supra* note 198, at 9 n.4 ("by no means generally available"). The idea advanced by Acting Chief, Section of Motor, Pipeline and Water Carrier Analysis, Bureau of Economics, ICC (now F.E.R.C.) David L. Jones in his prepared direct testimony, pp. 25-27, that oil pipelines may be converted to carrying nonoil commodities such as coal, capsulized "grain, chemicals, minerals, machine parts, urban waste and bulk mail" evaporated on cross-examination as he admitted that he had no back-up studies on the cost of conversion or market potential for hauling such materials to or from areas now served by petroleum pipelines. F.E.R.C. Dkt. RM 78-2, Tr., Vol. 3, at 496-498 (Nov. 3, 1977). LeGrange, *FERC Direct Testimony, supra* note 176, at 28 flatly states that, contrary to Jones' suggestion, there is little likelihood of oil pipeline systems being converted to such uses because they are either not located properly, sized suitably or the technologies are not compatible.

and can be highly negative at worst.[634]

Once the original design has been fixed, the principal operational costs are those of labor and horsepower. There are, of course, related costs such as site improvements, water supply, service tanks, auxiliary buildings, employee accommodations, heating plants and miscellaneous machinery and equipment such as communications, fire protection equipment, office furniture, machines, tools, etc.[635] When the line does not achieve its design throughput, some reduction of cost can be realized due to less power required to move fewer barrels,[636] but the savings will always be far less than the loss in income occasioned by the shortfall in throughput.[637]

C. *Economies of Scale*

There is no dispute that pipelines have substantial economies of scale.[638] The basic reason is that as pipeline diameters are increased, pipe costs (up to

---

[634] Direct Testimony of Raymond B. Gary, Before F.E.R.C. in Trans Alaska Pipeline System, Dkt. OR 78-1, 25 (Nov. 29, 1977) (TAPS owners required to dismantle and remove system after operations ceased and to be responsible for all damages caused by construction, operation maintenance or dismantling of the system) [hereinafter cited as Gary, *Direct TAPS Testimony*]; *cf.* LeGrange, *FERC Direct Testimony, supra* note 176, at 28.

[635] For a detailed, although somewhat outdated description of these costs, see Cookenboo, *Costs of Operating Crude Oil Pipe Lines*, Rice Institute Pamphlet, Vol. XLI, No. 1, 71-99 (April, 1954) [hereinafter cited as Cookenboo, *Crude Line Operating Costs*].

[636] DE CHAZEAU & KAHN, *supra* note 30, at 69; actually the reduction in electric power is disproportionate to the volume decrease in throughput. CRUDE OIL PIPELINES, *supra* note 158, at 31. A rule of thumb is that if throughput is decreased by a factor of 0.3, horsepower (hence electrical demand) will decrease by a factor of 0.4. *Cf.* PIPELINE TRANSPORTATION, *supra* note 162, at 36.

[637] *Cf.* CRUDE OIL PIPELINES, *supra* note 158, at 30, and *see* Figure 1 on page 9. Although a decrease in throughput means a more than proportionate decrease in electric power, the capital cost of the now "oversized" pipeline and pumps must be serviced. The applicable short run cost is derived by subtracting from the "intermediate-run" cost of the designed throughput the cost of the applicable power which is saved when throughput drops from the design level. This figure is always higher than the "intermediate-run" cost of the lower throughput. *Id.* at 30. Sometimes, temporary relief is obtained by pulling stations off the line, especially if they are needed elsewhere. The cents per barrel cost curve will shift downward and, at a fixed volume of throughput, significant reductions in cost occur, but the slope of the new cost curve is steeper and unit costs increase at a greater gradient than before as throughput declines. See MCLEAN & HAIGH, *supra* note 35, at 189.

[638] JONES DOE REPORT, *supra* note 372, at 4; PIPELINE TRANSPORTATION, *supra* note 162, at 39; S.2387 REPORT, *supra* note 49, at 12; KENNEDY STAFF REPORT, *supra* note 184, at 34.

Economic Characteristics 99

existing mill facilities) increase somewhat less than proportionately,[639] construction costs increase linearly,[640] but capacity increases exponentially.[641] At design capacity, unit operating costs drop as line size increases.[642] Illustrating each of these propositions, the cost of line pipe for a 20-inch line is only 4⅓ times that of a 4-inch line, yet the capacity of the larger line is 48 times that of the smaller line.[643] Stated another way, the cost of pipe for one 20-inch line is about 28 percent less[644] than the equivalent number (8) of 8-inch lines required to transport the same volume.[645] The construction cost proposition has been expressed as follows: one 36-inch line is equal in capacity to seven-

---

[639] JONES DOE REPORT, *supra* note 372, at 4; Tierney, *TAPS Rebuttal Testimony, supra* note 175, at 5; *Consumer Energy Act Hearings, supra* note 154, at 626. The most widely quoted diameter-pipe cost and diameter-capacity figures are found in *Petroleum Industry Hearings, supra* note 192, at 1100 (Statement of Charles E. Spahr). This table was submitted previously in Spahr, *Rodino Testimony, supra* note 255, at 6; its original source appears to be D. Hendricks, "Economics of Pipeline Transportation," (API School of Technology, 1966). Apparently these figures represent the most economical combination of grade, wall thickness, etc., in each size. If one used the same grade of standard wall thickness, the cost factor would rise much more rapidly than the table shows.

DIAMETER (INCHES)	PIPE COST FACTOR	CAPACITY FACTOR
4	1.0	1
8	1.9	6
12	2.4	15
16	3.5	28
20	4.3	48

[640] PIPELINE TRANSPORTATION, *supra* note 162, at 39; NET-I, *supra* note 199, at 209; PIPELINE PRIMER, *supra* note 331, at 10; KENNEDY STAFF REPORT, *supra* note 184, at 99; *cf.* SHENEFIELD PROPOSED STATEMENT, *supra* note 179, at 6; SHENEFIELD, PIPELINE POLICY—A LIGHT AT THE END OF THE TUNNEL 8 (Remarks before Corporate Counsel Section, Tulsa County Bar Association, November 8, 1978) [hereinafter cited as SHENEFIELD, PIPELINE POLICY].

[641] PIPELINE TRANSPORTATION, *supra* note 162, at 39; SPECTRE, *supra* note 104, at 121; NET-I, *supra* note 199, at 209; KENNEDY STAFF REPORT, *supra* note 184, at 99; Kaplan, Statement of the U.S. Department of Justice Before F.E.R.C. Valuation of Common Carrier Pipelines, Dkt. RM 78-2, at 6 (October 23, 1978) [hereinafter cited as Kaplan, RM 78-2 Statement]; *cf. Celler Hearings, supra* note 162, at 1135 (Statement of J. L. Burke) (cube of diameter); JONES DOE REPORT, *supra* note 372, at 4 (square of radius); SHENEFIELD PROPOSED STATEMENT, *supra* note 179, at 6. Because of the friction factor, capacity does not vary as the square of the radius but this change will never be less than that of the square of the radius. The exponent is between 2.7 and 3.

[642] PIPELINE TRANSPORTATION, *supra* note 162, at 39; Burke, *supra* note 9, at 783.

[643] See bottom line on chart in note 639, *supra.*

[644] This percentage is derived by first determining how many 8-inch lines it would take to equate to the 20-inch line capacity. From the chart in note 639, *supra*, the capacity factor of an 8-inch line is 6 whereas that of the 20-inch is 48, hence 48/6 or eight of the smaller lines are required. If each of the eight 8-inch lines has a cost factor of 1.9, the total cost factor for equivalent capacity is 8 x 1.9 or 15.2. The cost factor of the 20-inch is 4.3 so the comparison is 4.3/15.2 or 0.28, which expressed in percentage is 28 percent.

[645] JONES DOE REPORT, *supra* note 372, at 4; *Petroleum Industry Hearings, supra* note 192, at 1100 (Statement by Charles E. Spahr); Spahr, *Rodino Testimony, supra* note 255, at 7.

teen 12-inch lines,[646] but its construction cost is less than 3½ times that of one 12-inch line.[647] The operating comparison is illustrated by the fact that the per barrel cost of operating a 36-inch line is about ⅓ the cost of operating a 12-inch line.[648] The basic reasons are that certain capital costs such as surveying, right of way, damages and communications do not vary with line diameter.[649] The big ticket item is the cost of steel which will decrease per unit of carrying capacity as the size increases.[650] The second most important item is the friction factor which is influenced principally by the inner surface area of the pipe.[651] Because the volume increases more than does the surface area, it follows that in the larger pipe, a smaller proportion of the oil touches pipe surface area.[652] Less friction per barrel is created and hence the energy required for pushing the fluid will increase at a rate significantly less than the increase in throughput.[653] In addition to reducing operating costs, this factor also affects capital expenditure because for a given throughput, the optimal sized line will require the least horsepower capital investment.[654] The combined effect of all these component factors is that the cost of transporting a barrel of oil generally decreases about ⅓ each time the design pipeline throughput is doubled.[655]

---

[646] S.2387 REPORT, *supra* note 49, at 12 n.20; *Petroleum Industry Hearings, supra* note 192, at 256 (Statement of Charles J. Waidelich); PIPELINE PRIMER, *supra* note 331, at 10; *H.Res.5 Subcomm Hearings, supra* note 234, at 142 (Statement of Fred F. Steingraber); KENNEDY STAFF REPORT, *supra* note 184, at 98; *Proposed Hearings Before the Subcommittee on Antitrust and Monopoly of the Senate Comm. on the Judiciary*, 95th Cong., 2d Sess. (June 28, 1978) (Proposed Statement by Alvin L. Alm, p. 1) [hereinafter cited as ALM PROPOSED STATEMENT].

[647] S.2387 REPORT, *supra* note 49, at 12 n.20; *H.Res.5 Subcomm Hearings, supra* note 234, at 142 (Statement of Fred F. Steingraber); *Petroleum Industry Hearings, supra* note 192, at 256 (Statement of Charles J. Waidelich); HOWREY & SIMON, *supra* note 3, at 119; ALM PROPOSED STATEMENT, *supra* note 646, at 1; KENNEDY STAFF REPORT, *supra* note 184, at 98.

[648] *S.1167 Hearings, supra* note 432, pt. 9 at 615 (Testimony of Fred F. Steingraber); Waidelich, *Rodino Testimony, supra* note 197, at 5; HOWREY & SIMON, *supra* note 3, at 119-120.

[649] Cookenboo, *Crude Line Operating Costs, supra* note 635, at 92.

[650] LENNART, *supra* note 111, at 8; KENNEDY STAFF REPORT, *supra* note 184, at 34 (the ratio of the circumference to the area of a circle is a decreasing function of the radius). Spelled out, the circumference of a circle is: πd and the area is πr^2, or $0.7854d^2$. Every inch increase in diameter incrementally increases the area substantially more than the circumference; *cf.* J. ATWOOD & P. KOBRIN, INTEGRATION AND JOINT VENTURES IN PIPELINES 3 (API September 8, 1977) [hereinafter cited as Atwood & Kobrin, Res. Study 005].

[651] Cookenboo, *Crude Line Operating Costs, supra* note 635, at 45.

[652] *Id.*; Atwood & Kobrin, Res. Study 005, *supra* note 650, at 3.

[653] LENNART, *supra* note 111, at 8.

[654] Cookenboo, *Crude Line Operating Costs, supra* note 635, at 45; *cf.* KENNEDY STAFF REPORT, *supra* note 184, at 34.

[655] LENNART, *supra* note 111, at 8; Burke, *supra* note 9, at 783 (about 30 percent). This, of course, presupposes that the utilization remains near the optimal design capacity.

D. Similarities and Dissimilarities to a Natural Monopoly

The foregoing economies of scale have led enforcement agencies,[656] Congressional staffers,[657] Senators,[658] and those working for such agencies[659] to fasten the inappropriate label of "natural monopoly" on pipelines. The seminal statement appeared in the Department of Justice's Deepwater Port Report,[660] where it was said that deepwater ports were "natural monopolies" in their respective markets and "in economic terms this means that they have increasing returns to scale;" and, finally, that "pipelines have

---

[656] *Department of Justice; e.g.,* A.G.'s Deepwater Port Report, *supra* note 446, at 3-4; Statement of Views and Arguments by the U.S. Department of Justice Before the ICC, Ex Parte No. 308, Valuation of Common Carrier Pipelines 20 (May 27, 1977) [hereinafter cited as D.J. Views and Arguments]; SHENEFIELD PROPOSED STATEMENT, *supra* note 179, at 6-7; SHENEFIELD, PIPELINE POLICY, *supra* note 640, at 8; D. FLEXNER, OIL PIPELINES: THE CASE FOR DIVESTITURE 2-3 (Remarks before the American Enterprise Institute, March 1, 1979) [hereinafter cited as FLEXNER, OIL PIPELINES]. *Federal Trade Commission:* REPORT OF THE FEDERAL TRADE COMMISSION TO THE HONORABLE WILLIAM T. COLEMAN, JR. SECRETARY OF TRANSPORTATION ON THE APPLICATIONS OF LOOP, INC. AND SEADOCK, INC. TO OWN, CONSTRUCT, AND OPERATE DEEPWATER PORTS UNDER THE DEEPWATER PORTS ACT OF 1974 30 (November 5, 1976) [hereinafter cited as FTC, DEEPWATER PORT REPORT]; *Proposed Hearings Before the Subcommittee on Antitrust & Monopoly of the Senate Comm. on the Judiciary,* 95th Cong., 2d Sess. (June 28, 1978) (Proposed Statement of Alfred F. Dougherty - Rev'd ed. July 10, 1978, p. 3) [hereinafter cited as DOUGHERTY, REVISED PROPOSED STATEMENT]. *Department of Transportation: Proposed Hearings Before the Subcommittee on Antitrust and Monopoly of the Senate Comm. on the Judiciary,* 95th Cong., 2d Sess. (June 28, 1978) (Proposed Statement of John G. Wofford, p. 2) [hereinafter cited as WOFFORD PROPOSED STATEMENT].

[657] KENNEDY STAFF REPORT, *supra* note 184, at 99 (but adds qualifications of "limited competition from other pipelines and other modes of transportation").

[658] *Proposed Hearings Before the Subcommittee on Antitrust and Monopoly of the Senate Comm. on the Judiciary,* 95th Cong., 2d Sess. (June 28, 1978) (Proposed Opening Statement of Senator Edward Kennedy—flat conclusionary statement: "Pipelines . . . are natural monopolies") [hereinafter cited as Kennedy, Proposed Opening Statement]; *cf.* Remarks Prepared for Delivery by Senator Howard M. Metzenbaum Before the American Enterprise Institute (March 1, 1979) [hereinafter cited as Metzenbaum, Remarks in Absentia].

[659] JONES DOE REPORT, *supra* note 372, at 5-6 (hedged with qualifications but "main point" the same for "most routings").

[660] A.G.'s Deepwater Port Report, *supra* note 446, at 3-4.

the same natural monopoly characteristics as deepwater ports."[661] In the "Statement of the Views and Arguments of the Department of Justice" in the *Ex Parte No. 308* proceedings[662] after stating the "definitive characteristic of a natural monopoly" to be that "average cost, including the cost of capital, falls throughout the relevant range of market demand" (this rephrasing departs subtly from the classic economic requirement that this must extend over the *entire extent of the market*,[663] *i.e.*, to the saturation point).[664] That is, the transportation demand between two points may be so great that pipeline costs no longer are declining in that range. Furthermore, the law abounds with subtle distinctions. The restatement injected the concept of "relevant market," which has a distinct legal (antitrust) connotation. Applying the economic definition of "natural monopoly" to oil pipelines without a factual examination of what is the extent of, and whether or not there are economic alternative sources of satisfying, the "market" (demand) *assumes* the applicability of the label without making the requisite factual inquiry. The Department later appeared to revert to the classic economic concept by saying "a single firm can serve the *entire market* at a lower cost than two or three firms," (emphasis added) and, furthermore, that "any other firm is unlikely to enter the market because the original firm could undersell it by further exploiting its economies of scale." The Department's attorneys making this argument thus remained reasonably close to the classical definition of "natural monopoly" and implied that pipelines somehow fit into the mold. Significantly, they did not come out and say pipelines were natural monopolies. They contented themselves with a suggestive heading entitled "Natural Monopoly Characteristics of Pipelines."

---

[661] It is interesting to note that the FTC, DEEPWATER PORT REPORT, *supra* note 656, at 37, states flatly that merely because port facilities include pipelines does not convert such ports into pipelines, citing Brooks Gas Corp. v. F.P.C., 383 F.2d 503 (D.C. Cir. 1967), and indicating that the very fact of passage of the Deepwater Ports Act reflects the Congress' determination that deepwater ports are different from pipelines and should be subject to a different regulatory scheme. See also statement before DOT Secretary Coleman on November 12, 1976, Tr. p. 73, where the same position is asserted. On a more analytical basis, one can understand, although not agree with, the Department's analogizing a deepwater port (especially if there is to be only one because of the unattractiveness of the additional regulatory burdens contained in the deepwater port licenses which were imposed on an already hazardous venture) to the *Terminal Railroad*, United States v. Terminal Railroad Ass'n, 224 U.S. 383 (1912), type of situation, but to extend this "bottleneck theory" to pipelines in general by simply adding a parenthetical phrase "(pipelines have the same natural monopoly characteristics as deepwater ports)" without providing any supporting fact or reasoning leaves the reader with a feeling of disquietude. When this assertion is thereafter used repeatedly as a fundamental truth upon which a plea for pipeline divorcement is based, uneasiness turns into alarm.

[662] D.J. Views and Arguments, *supra* note 656, at 20.

[663] A. KAHN, THE ECONOMICS OF REGULATION: PRINCIPLES AND INSTITUTIONS, Vol. 2, at 119 (1971).

[664] F. SCHERER, INDUSTRIAL MARKET STRUCTURE AND ECONOMIC PERFORMANCE 520 (1970).

The next public pronouncement by the Department of Justice contained the denouement. Through the use of an arguable analogy between deepwater ports and pipelines, and a blurring of the distinction between saturation of the entire market and saturation of "the entire existing range of technologically feasible pipeline sizes," a declaration that pipelines are "natural monopolies" was silently slipped into Assistant Attorney General Shenefield's proposed testimony before the Kennedy Subcommittee. The technique made use of the following syllogism: *Major premise*—"as to competition with other modes of transport, pipelines are the most efficient (*i.e.*, cheapest) overland transport mode for the movement of significant, sustained quantities of petroleum between any two given points; [665] *Minor premise*—"as to competition within the pipeline transport mode itself, pipelines exhibit great economies of scale. . . . This phenomenon of declining unit costs has been observed to take place over the entire existing range of technologically feasible pipeline sizes which are now approaching 4 feet.[666] As a result, in almost all circumstances, one properly sized pipeline will be more efficient than two or more in satisfying the available transportation demand between any two given points;" *Conclusion*—"These two basic economic facts—premier cost efficiency and tremendous economies of scale—are usually referred to as the 'natural monopoly' characteristics of pipelines."[667] This was repeated virtually verbatim in Mr. Shenefield's speech to the Tulsa County Bar Association on November 8, 1978.[668] A recent restatement on the subject appeared in Deputy Assistant Attorney General (Antitrust Division) Donald Flexner's paper before the American Enterprise Institute's Conference on Oil Pipelines and Public Policy held in Washington on March 1, and 2, 1979.[669] Mr. Flexner repeated the premise that, compared with other modes of transportation, pipelines are the most efficient overland transport mode for the movement of significant amounts of petroleum between any two given points [he even included waterborne competition except for all but the largest tankers]. He made a subtle shift and

---

[665] SHENEFIELD PROPOSED STATEMENT, *supra* note 179, at 5-6, citing NET-I, *supra* note 199, at 213-214. While probably unobjectional as a rule-of-thumb statement, the *any* two given points is a bit too broad as Figure 32 on cited page 214 demonstrates with respect to barge traffic (see also Appendix A.) There are significant areas where barge rates are lower than pipeline tariffs. *H.Res.5 Subcomm Hearings*, *supra* note 234, at 107 (Statement of Meyer Kopolow, hardly a friend of pipelines); *cf.* Statement of Glenn A. Welsh, Before the ICC in Ex Parte No. 308, Valuation of Common Carrier Pipelines, May 19, 1977, at 3.

[666] SHENEFIELD PROPOSED STATEMENT, *supra* note 179, at 6, citing his own Deepwater Port Report, *supra* note 446, at 28, which only dealt with the size of pipe available rather than the economics of scale.

[667] *Id.* at 7.

[668] SHENEFIELD, PIPELINE POLICY, *supra* note 640, at 8.

[669] FLEXNER, OIL PIPELINES, *supra* note 656, at 2-3.

instead of talking about the declining unit costs throughout the market, he cited the declining unit costs throughout the range of technologically feasible pipelines for the point that it is more efficient to build a single pipeline of sufficient size to satisfy the entire demand for transportation between two points than it would be to build several smaller pipelines *at the same time* (emphasis added). These two "economic facts," said Mr. Flexner, constitute the "natural monopoly" characteristics of pipelines.[670] In all of these statements the Justice Department spokesmen have been frank to admit that merely because pipelines have "natural monopoly characteristics," it does not necessarily follow that they have monopoly power.[671]

The first counterpoint is that the critical characteristic of an economic monopoly is an inherent tendency to decreasing unit costs over the *entire extent of the market*[672]—not just over current "technologically feasible pipelines." This presupposes that the market demand is sufficiently small that it *can* be satisfied by a single firm which is operating in an area of decreasing costs.[673] First, any pipeline that could even approach such size, accepting *arguendo* that it would run from the Justice Department's hypothetical Point A to hypothetical Point B, generally will be jointly owned and operated by no less than three but more likely eight to ten owners. Joint ownership is not always an economic mandate. However, as will be demonstrated in the following subsection, a typically large line will haul more material than any one shipper will have available, and the size of the investment is too large for a prudent investor to place in "one basket." Moreover, the risks are such that it is good business to spread them among several owners, and the rate of return regulation imposed on the venture makes it unattractive for a shipper-owner to put up more than his share of the required capital. There has been no charge of overt collusion between such owners and, considering the numbers of participants and their conflicts in position both upstream and downstream of the pipeline, implicit collusion seems quite implausible.[674]

---

[670] *Id.* at 3.

[671] *E.g.,* SHENEFIELD PROPOSED STATEMENT, *supra* note 179, at 7 (second sentence); FLEXNER, OIL PIPELINES, *supra* note 656, at 3-4. This seems prudent in light of Demsetz, "*Why Regulate Utilities?*" XI J. Law & Econ., 55-65 (March, 1968). The market power argument reappears under a new name, "undersizing," which will be discussed in Chapter VI, "Current Policy Issues Concerning Pipeline Ownership," *infra*.

[672] A. KAHN, THE ECONOMICS OF REGULATION: PRINCIPLES AND INSTITUTIONS, Vol. 2, at 119 (1971); HOWREY & SIMON, *supra* note 3, at 15.

[673] MCGRAW-HILL, DICTIONARY OF MODERN ECONOMICS 394 (2d ed., 1973); HOWREY & SIMON, *supra* note 3, at 15-16.

[674] TEECE, *supra* note 329, at 100-102; R. MANCKE, *Competition in the Oil Industry* in VERTICAL INTEGRATION IN THE OIL INDUSTRY 38 (E. Mitchell ed., 1976) [hereinafter cited as MANCKE, *Oil Industry Competition*]; PIPELINES AND ANTITRUST, *supra* note 530, at 11; M. CANES & D. NORMAN, PIPELINES AND PUBLIC POLICY 16 (Remarks before the American Enterprise Institute March 2, 1979) [hereinafter cited as PIPELINES AND PUBLIC POLICY].

Second, even markets such as hypothetical Point B don't remain static, demand grows over time[675] and this growth may result in more than one pipeline serving the two points.[676] As Tom Spavins, an economist with the Justice Department, has observed, once a pipeline is built, many of the possibilities for economies of scale are exhausted.[677] This is exactly what has happened as has been demonstrated in earlier sections where Amoco Pipeline (then Service) was overtaken by growth in the Montana/Rocky Mountains production region, and Platte and Arapahoe were built.[678] Where there were one or two sole proprietor lines from West Texas to the Texas Gulf Coast, now there are 13 lines,[679] some of which are jointly owned and operated by both large and small companies. Nine common carrier pipelines now transport crude from the East Texas producing region to the Texas Gulf refining area.[680] The same is true where the consuming market grows. Plantation, when built at the start of World War II, served inland areas not accessible by other means and hence would have come within the Justice Department's broadened definition of "natural monopoly," but what happened to its traffic in 1967 when Colonial came on-stream? It suffered a 31 percent reduction in its revenue.[681] One couldn't blame Plantation Directors at the time for feeling that theirs was not a natural monopoly.[682]

The McGraw-Hill *Dictionary of Modern Economics* states the economic theory: "Economic growth can destroy natural monopolies, because demand also grows, and the natural monopoly is broken when the monopolist no longer operates in the limited area of decreasing costs."[683] More to the point, however, is that in the real world of oil pipelines, there is no large scale movement from fixed Point A to fixed Point B. There are crude oil pipelines from the West Texas-New Mexico producing region to the Gulf Coast refinery area and to the Mid-Continent and Chicago refinery areas and products pipelines from the Gulf Coast refinery area serving the southeastern United States and the New York market. No single line could

---

[675] JONES DOE REPORT, *supra* note 372, at 5.
[676] T. SPAVINS, THE PRESENT SYSTEM OF OIL PIPELINE REGULATION 5 (paper presented before the American Enterprise Institute on March 1-2, 1979) [hereinafter cited as SPAVINS, PIPELINE REGULATION].
[677] *Id.* at 5 n.8; see also CRUDE OIL PIPELINES, *supra* note 158, at 18.
[678] See text at notes 168 and 410, *supra*.
[679] See text at note 402, *supra*.
[680] See text at note 403, *supra*.
[681] See text at note 530, *supra*.
[682] See Statement of Ezra Solomon Before the ICC in Ex Parte No. 308 (now FERC Dkt. 78-2), Valuation of Common Carrier Pipelines 13 (May 27, 1977) [hereinafter cited as Solomon, *FERC Direct Testimony*].
[683] MCGRAW-HILL, DICTIONARY OF MODERN ECONOMICS 394 (2d ed., 1973).

begin to saturate these markets—it fairly boggles the mind even to contemplate one—TAPS would look like a garden hose compared to it.[684]

Wholly aside from the fact that the "natural monopoly" theory itself is not immune from the criticism that it is illogical,[685] the definition used by economic writers speaks in terms of *markets* which are capable of being saturated[686] by one or, at most, a very few firms,[687] so that the very size and diversity of real world petroleum markets rule out the application of "natural monopoly" to today's pipelines.[688] It would seem that the economies of scale inherent in petroleum pipelines are not lastingly *internal*

---

[684] See remarks of G. S. Wolbert, Jr., Discussant, before American Enterprise Institute Conference on Oil Pipelines and Public Policy (March 1, 1979), pp. 57-58 [hereinafter cited as AEI, *Oil Pipelines and Public Policy*]. There are some lines that perhaps might meet the Department's "fixed point A" to "fixed point B" description. However, these are true "plant facilities" of a single refiner constructed to obtain low cost transportation of crude oil to its refinery or to realize low cost movement to the markets for its refined products. Usually, the lines are located in places where they would be of little or no use to any other shipper, although the additional traffic might be welcomed by the owner in order to realize the cost savings of large volumes. LIVINGSTON, *supra* note 228, in Appendix A, gives a number of examples: *Navajo Refining* owns a 423 mile crude oil gathering system connecting producing fields in southeastern New Mexico to its Artesia, New Mexico, refinery. It also owns a refined products line which runs from the refinery about 40 miles north to Roswell, New Mexico, and 150 miles west to El Paso; *Champlin Petroleum* owns and operates approximately 500 miles of crude gathering system bringing crude to its Enid, Oklahoma, refinery and 600 miles of products lines running north from Enid through Kansas and Nebraska into Iowa and South Dakota. This was the subject of litigation in Champlin Refining Co. v. United States, 329 U.S. 29 (1946) and 341 U.S. 290 (1951); *Apco Pipe Line* owns and operates 468 miles of crude oil gathering and trunk lines gathering oil in Kansas and Oklahoma for use in its parent's Arkansas City refinery (recently sold to Total Petroleum); *Mid-Continent Petroleum* owns and operates a network of crude oil lines serving its refinery at West Tulsa, Oklahoma. See Livingston's Exhibit 3 which shows the system in 1949 prior to merger with Sunray Oil and later with Sun. Livingston also discusses *Pasco, Indiana Farm Bureau Coop., OKC, United Refining, Farmer's Union Central Exchange, Allied Materials Corp., Diamond-Shamrock, Esmark* (Vickers Petroleum), *National Cooperative Refinery Ass'n, Derby Refining* (Coastal States), and *Total Petroleum*, showing, in most instances, maps of their systems. It stretches credulity to believe that the Justice Department had these lines in mind as supportive of its "natural monopoly"- market power thesis.

[685] Demsetz, *Why Regulate Utilities?*, XI J. Law & Econ. 56, at 59 (April, 1968).

[686] F. SCHERER, INDUSTRIAL MARKET STRUCTURE AND ECONOMIC PERFORMANCE 520 (1970).

[687] *Id.*; A. KAHN, THE ECONOMICS OF REGULATION: PRINCIPLES AND INSTITUTIONS 119 (1971); C. F. PHILLIPS, JR., THE ECONOMICS OF REGULATION 22-23 (1965); MCGRAW-HILL, DICTIONARY OF MODERN ECONOMICS 394 (2d ed., 1973); Waverman, *Regulation of Intercity Telecommunications* in PROMOTING COMPETITION IN REGULATED MARKETS 203 (Almarin Phillips ed. The Brookings Institution 1974). The Justice Department apparently recognized this point in its D.J. Views and Arguments, *supra* note 656, at 20, wherein it is stated: "Thus, a single firm can serve the *entire market* at a lower total cost than two or more firms. Furthermore, any other firm is unlikely to enter the market because the original firm could undersell it by further exploiting its economies of scale." (emphasis added).

[688] MARATHON, *supra* note 202, at 63-64; *cf.* C. F. PHILLIPS, JR., THE ECONOMICS OF REGULATION 23 (1965).

to the individual firm but rather tend to become *external* to the firm and are permanently internal only to the *industry*. Increasing returns of this kind are fully compatible with a competitive organization of the industry. All firms can benefit equally from these economies no matter what the scale of their individual outputs.[689] This is reflected in the multiline competition mentioned in the preceding subsections on "Competition" in crude and product lines. In this regard, Congress' repeated refusals to require Certificates of Convenience and Necessity to be granted by the Commission prior to new entry of petroleum pipelines (as is the case in natural gas lines) have been wise decisions. It is not unknown for government regulation to lessen competition and even to foster monopolies.[690] The much maligned[691] regulation by the ICC has permitted a degree of competition in oil pipelines rare in regulated industries. By observing the National Transportation Policy's mandate "to promote safe, adequate, economical, and efficient service and foster sound economic conditions in transportation and among the several carriers,"[692] the ICC has encouraged a system whereby the consumer has received the benefit of receiving the *lesser* of the regulatory-permitted rate of return or a competitive rate set by newer lines with better technology[693] or alternative modes of transportation.[694] The usual subsidization accorded to franchised utilities[695] is absent because the pipeline (and its shipper owners by reason of

---

[689] A. KAHN, THE ECONOMICS OF REGULATION: PRINCIPLES AND INSTITUTIONS, Vol. 2, 119 n.15 (1971); see Ellis & Fellner, *External Economies and Diseconomies* in READINGS IN PRICE THEORY 242, 263 (Amer. Econ. Ass'n, 1952).

[690] HOWREY & SIMON, *supra* note 3, at 18; *cf.* WAVERMAN, *Regulation of Intercity Telecommunications* in PROMOTING COMPETITION IN REGULATED MARKETS 205 (Almarin Phillips ed. The Brookings Institution, 1974). This is exactly what has taken place in natural gas pipelines, which are shielded from competitive entries, do not carry for others and can only survive by expanding their rate base or entering other lines of business. *Cf.* Averch & Johnson, *Behavior of the Firm Under Regulatory Constraint*, 52 Amer. Econ. Rev., 1052-1069 (1962).

[691] *E.g.*, Kaplan, RM. 78-2 Statement, *supra* note 641, at 8-9 ("hear no evil—see no evil" approach); SHENEFIELD PROPOSED STATEMENT, *supra* note 179, at 14-15 (same); DOUGHERTY, REVISED PROPOSED STATEMENT, *supra* note 656, at 3 ("singularly ineffective"); Kennedy, Proposed Opening Statement, *supra* note 658, at 2 ("It has never been seriously contended that the ICC did a good job of regulating pipeline tariffs"). KENNEDY STAFF REPORT, *supra* note 184, *passim*; Metzenbaum, Remarks in absentia, *supra* note 658, at 2 ("regulatory laxity on the part of the ICC"). For a serious contention that the ICC did a good job in regulating pipeline tariffs, see Solomon, *TAPS Rebuttal Testimony*, *supra* note 621, at 60-64.

[692] 49 U.S.C. preceeding §§ 1, 301, 901, and 1,001 (1976).

[693] The Colonial-Plantation and Amoco-Platte-Arapahoe situations are illustrative of this point. See text at notes 168 (Amoco-Platte-Arapahoe) and 432-426 (Colonial-Plantation), *supra*.

[694] MANCKE, *Oil Industry Competition*, *supra* note 674, at 45. For a noticeable trend, see note 184 which gives an idea of recent inroads on pipeline carriage made by tankers.

[695] Gary, *FERC Direct Testimony*, *supra* note 613, at 14-15.

their debt guarantees and throughput and deficiency agreements) take the losses.[696]

Summarizing, because no one pipeline can saturate today's markets, free entry, plentiful competition from other pipelines, both crude and products (access to which readily is available on a nondiscriminatory basis to all shippers who desire to use them);[697] intermodal competition;[698] plus the fact that the charge for transportation is an insignificant component of final market price;[699] the case for pipelines as natural monopolies seems extremely attenuated,[700] leading one knowledgeable observer to state flatly that oil pipelines are *not* natural monopolies, and another to state that in a period of changing technology and rapidly shifting patterns of energy source and distribution, the very size of the scale factor works in reverse to expose pipeline investors to potentially huge losses if and when a shift does occur in the pattern or mode of energy movement.[701]

E. *"Big-Inch" Technology and Jointly Owned and Jointly Operated Pipelines*

Prior to the 1930's, high-strength welded steel pipe was not available and the line capacities were such that most pipeline companies owned their own lines individually.[702] After the advent of seamless pipe in the 1930's made larger diameter lines physically possible, the required technology was developed during World War II because of the need to transport large quantities of crude oil and products to the East Coast by pipeline due to submarine-induced tanker shortages[703] and the inability of railroads to handle the traffic in addition to their other demands. The "Big-Inch" (24-inch crude line with design capacity of 300,000 barrels per day)[704] and "Little Big-

---

[696] *E.g.,* Ajax Pipe Line, Four Corners, West Texas Gulf, Laurel, and Explorer.
[697] See Section II A 3 c "Scramble for Traffic" at notes 407-496, *supra.*
[698] See text at, and note 694, *supra.*
[699] See text at note 197, *supra.*
[700] Gary, *FERC Direct Testimony, supra* note 613, at 14-15; *cf.* MARATHON, *supra* note 202, at 62-65.
[701] Gary, *FERC Direct Testimony, supra* note 613, at 14 ("oil pipelines are not natural monopolies"); Solomon *FERC Direct Testimony, supra* note 682, at 13 (size of scale works in reverse to expose investors to potentially huge losses).
[702] Spahr, *Rodino Testimony, supra* note 255, at 5; Burke, *supra* note 9, at 793.
[703] B. BLEDSOE, *Corporate Structure and Finance, Common Carrier Pipelines* in AOPL EDUCATORS CONFERENCE 3 (1978) [hereinafter cited as BLEDSOE]. Today, almost all of the large diameter, high tensile pipe is butt-welded.
[704] See text at, and details of the line in, note 156, *supra.*

Inch" (20-inch line products line with design capacity of 235,000 barrels per day)[705] demonstrated conclusively the operational and economic feasibility of large diameter lines.[706] Also, as a result of the war and its aftermath, a number of technological improvements, such as advances in pipe manufacturing, internal line-up clamps, hydraulic bending machines for large diameter thin-wall pipe, multistage centrifugal pumps, sophisticated communications and control mechanisms, were in place.[707] All that remained was for demand to provide the economic basis. This was not long in developing. Contrary to general expectations of a post World War II recession, the nation experienced an expansion of unprecedented proportions and the pent-up demand for petroleum, formerly suppressed by the war effort, required large volumes of oil to be transported.[708] Large diameter lines became commonplace.[709]

Although such pipelines were not always necessarily beyond the *absolute* financial capability of a single company, they generally exceeded its *practical* financial capability.[710] Not only were the risks too great to prudently concentrate the bulk of an oil company's financial resources in one project, especially when the company had a wide range of alternative demands for capital,[711] most of which envisioned a greater expected rate of return, but also most single company shipper-owners were not able to muster the throughput volumes necessary to achieve the economies of scale of the large diameter lines.[712] Therefore, in order to raise the capital, secure the necessary throughput volume, and spread the business risk,[713] the industry turned to jointly-owned and operated pipelines in ever increasing numbers.[714] In major producing areas, the situation arises because of the intense competition for

---

[705] See text at, and details of the line in, note 157, *supra*.
[706] SHENEFIELD PROPOSED STATEMENT, *supra* note 179, at 4; MARATHON, *supra* note 202, at 12-13.
[707] See text at note 163, *supra*.
[708] W. J. Williamson, *supra* note 315, at 8.
[709] NET-I, *supra* note 199, at 181; SPECTRE, *supra* note 104, at 121 and note 222; see text at note 160, *supra*.
[710] MARATHON, *supra* note 202, at 5; W. J. Williamson, *supra* note 315, at 9. For an example of a situation where one line was well beyond the financial capability of one company, Marathon Pipe Line Company's asset value in 1960 was $47.8 million which could have represented only 12.7 percent of the inital $378 million cost of Colonial and only 7.1 percent of its eventual cost of $675 million. MARATHON, *supra* note 202, at 42. For an example of one where virtual survival of shipper-owner itself depended upon the pipeline's success there is the case of Sohio and its *share* of TAPS. Gary, *Direct TAPS Testimony*, *supra* note 634, at 55.
[711] MARATHON, *supra* note 202, at 5.
[712] W. J. Williamson, *supra* note 315, at 9; SPECTRE, *supra* note 104, at 121-122.
[713] Gary, *FERC Direct Testimony*, *supra* note 613, at 6.
[714] See text at notes 159-161, *supra; cf.* Broden & Scanlan, *The Legal Status of Joint Venture Corporations*, 11 Vand. L. Rev. 673, 689 (1958).

leases, the multiplicity of ownership of undeveloped acreage and the leasing of federal lands. No one company can justify a large diameter efficient line by itself and, even if it could, it would stand a good chance that a still larger and more efficient line could be put together by a cooperative effort of other parties. Thus producers, refiners, and other interested parties are brought together by a common need for efficient and economic pipeline service which generally has resulted in minimizing the final cost of product to the consumer.[715]

Charles Spahr, Chairman of the Board and Chief Executive Officer of Sohio, in his statement before the Rodino Committee, related how Sohio, which had relied principally on Illinois crude to operate its refineries, was forced by the decline of Illinois production to purchase more and more oil in the Southwest, which is transported by barge to Ohio. Although it was the largest single shipper of crude oil in the inland waterways, its movements were not sufficient to support a pipeline that would be competitive with barge rates. As it turned out, Sun Oil Company was faced with a similar situation with respect to its shipments to its Toledo refinery. The net result was that Sohio and Sun joined to build the Mid-Valley System, a 1,053-mile 20–22-inch line from Longview, Texas, to Lima, Ohio, where it connected with the Buckeye system serving Toledo. Neither Sun nor Sohio alone had the traffic for a pipeline the size of Mid-Valley. It took the throughput of both companies to justify the large line. The result was that both Sun and Sohio (and any other shipper, *e.g.,* Gulf, needing transportation in that direction) were able to move their oil at a lower cost than either could have achieved on their own.[716] Similar conditions exist for the development of economic large diameter pipelines from refinery areas to consumer markets. When the need arises, study groups of all those who might be interested are formed and various alternative routes and destinations are drawn up for examination by the prospective participants.[717]

Case studies of actual situations will be discussed in a subsequent subsection, but it should be noted at this point that as a practical matter, generally the projects jointly owned or operated by oil companies, large and small, are the ones that attract the necessary traffic, economic and financial resources as well as the technological expertise needed to design, finance, construct, and operate large diameter, long distance crude and products pipelines.[718] Thus, the incentive has turned strongly to obtaining the broadest

---

[715] BLEDSOE, *supra* note 703, at 6.
[716] Spahr, *Rodino Testimony, supra* note 255, at 7-8.
[717] BLEDSOE, *supra* note 703, at 6.
[718] HOWREY & SIMON, *supra* note 3, at 118, 129-130; *S.1167 Hearings, supra* note 432, pt. 8 at 5929 (Statement of Robert E. Yancey); *cf.* Waidelich, *Rodino Testimony, supra* note 197, at 4; Direct Testimony of Joe L. Cooper, Before F.E.R.C. in Trans Alaska Pipeline System, Dkt. OR78-1, 11-12 (Nov. 30, 1977) [hereinafter cited as Cooper, *Direct TAPS Testimony*].

Economic Characteristics 111

possible ownership and to seek aggressively the traffic of those not prepared to incur the financial risk that accompanies such ownership.[719]

F. Risks Involved in Pipeline Ventures

Although some have suggested[720] that the risks of pipeline investment are "minimal,"[721] the record clearly shows that the investment is one of high risk;[722] moreover, this risk is asymmetrical, *i.e.*, the investor is limited by rate regulation and/or the *Atlantic Refining Company* Consent Decree on the upside potential but there is no equivalent limitation on the amount that he can lose.[723] Before examining in detail the various elements of risk and uncertainties faced by pipelines, it should be noted that the proper time and perspective[724] to measure risk is at the moment the investors commit themselves[725] to these large 30-40 year investments and not many years after

---

[719] See text at, and authorities cited in, note 162, *supra*.

[720] *S.1167 Hearings, supra* note 432, pt. 8 at 5,946: Senator Tunney speaking. "It is my understanding that this money [cash equity of roughly 10 percent of the investment] that the owners put up was $36 million, and that they had dividends the first year of $25 million. That is a fairly profitable venture. Where is the risk there, where they had these long-term throughput commitments [from themselves]?" For a brief analysis of Senator Tunney's misconceptions, see note 624, *supra*.

[721] SHENEFIELD, THE NEW RATIONALISM AND OLD VERITIES: DECIDING WHO GETS WHAT AND WHY (Remarks before the Graduate School of Business, Invitation Lecture Series, University of Chicago, October 25, 1978) ("Divestiture of [pipeline] assets is not a popular solution for [the] owners, especially since our studies show that ownership of pipelines is often profitable for oil companies to a degree inconsistent with *the minimal risks involved*") (*emphasis added*). The Assistant Attorney General stated he believed that the "risk is minimal" in deepwater ports. Supplemental Memorandum of the Department of Justice to the Secretary of Transportation on the Deepwater Port Applications of LOOP, Inc. and SEADOCK, Inc. (Nov. 19, 1976) at p. 24.

[722] See text at notes 622-624, *supra*.

[723] See text at note 621, *supra*.

[724] In the words of witness L. Dale Wooddy in his Rebuttal Testimony Before F.E.R.C. in Trans Alaska Pipeline System, Dkt. OR 78-1 (October 30, 1978), at 5: "With due respect to the [three Commission's] witnesses you have mentioned, I think that their assessment of the risks associated with TAPS would have been quite different if, during the 1969-1977 period, it had been their money or their companies' money at stake." [hereinafter cited as Wooddy, *TAPS Rebuttal Testimony*].

[725] MANCKE, *Oil Industry Competition, supra* note 674, at 97; Gary, *Direct TAPS Testimony supra* note 634, at 21; *cf.* Rebuttal Testimony of Dr. John M. Ryan in Trans Alaska Pipeline System, F.E.R.C. Dkt. OR 78-1, 18, 20 (October 30, 1978) [hereinafter cited as Ryan, *TAPS Rebuttal Testimony*]; Cooper, *Direct TAPS Testimony, supra* note 718, at 31.

the event, when everything that has succeeded seems easy in retrospect and that which has failed seldom is mentioned.[726]

1. Incorrect Forecast of Traffic

Putting aside for the moment the difficulties of correctly assessing the supply (reserve) that provides the traffic for a crude line[727] or in calculating the market to be served by a products line,[728] and disregarding the effects of competition and governmental actions, the economics of a pipeline venture can be adversely affected to a significant degree by the timing of the throughput achieved. Stated as baldly as possible, even if the expected total throughput is achieved, delays in accomplishing that result reduce the value of the revenues by the time value (*i.e.*, the discounted rate) of money.[729] For example, in planning its participation in TAPS, Exxon based its budget estimates on an ultimate throughput figure which assumed 20 billion barrels of Alaskan reserves, at a time when the proven recoverable reserves were only 9.6 billion barrels.[730] This, of course, was just a projection and was therefore quite uncertain. Even if every barrel of the additional reserves is found (and in the ten years of exploration activity following the Prudhoe Bay discovery, no additional commercial reserves have been announced),[731] these new discoveries must be found at locations such that TAPS will be the most economic means of transporting the newly discovered oil to market and, moreover, they must be found at times that will maintain the constant throughput needed for TAPS to realize optimal pipeline operation.[732]

---

[726] The risks *borne* will always cover a broader spectrum of possible events than the risks which *eventuate* with harmful effects (*e.g.,* higher costs). In the case of TAPS the very completion of the system indicates that all possible risks did not eventuate. At several points in time there was a very real risk either that the line would not be constructed or, if constructed, would be delayed so long and cost so much that capital recovery would have been an impossibility. See Rebuttal Testimony of Raymond B. Gary, Before F.E.R.C. in Trans Alaska Pipeline System, Dkt. OR 78-1 10, 17 (October 30, 1978) [hereinafter cited as Gary, *TAPS Rebuttal Testimony*]; Direct Testimony of Frank P. Moolin, Jr., Before F.E.R.C. in Trans Alaska Pipeline System, Dkt. OR 78-1, 2-3 (Nov. 30, 1977) [hereinafter cited as Moolin, *Direct TAPS Testimony*,]; Direct Testimony of Edward L. Patton, Before F.E.R.C. in Trans Alaska Pipeline System, Dkt. OR 78-1, 51 [hereinafter cited as Patton, *Direct TAPS Testimony*]; Rebuttal Testimony of Edward L. Patton, Before F.E.R.C. in Trans Alaska Pipeline System, Dkt. OR 78-1, 3 (Oct. 27, 1978) [hereinafter cited as Patton, *TAPS Rebuttal Testimony*].

[727] This will be discussed in the following Subsection 2 "Faulty Supply (Reserve) Estimates."

[728] See discussion in Subsection 4, "Market Miscalculation," *infra*.

[729] Wooddy, *TAPS Rebuttal Testimony, supra* note 724, at 13; Gary, *TAPS Rebuttal Testimony, supra* note 726, at 10; *cf.* Burke, *supra* note 9, at 787 (demand grows more slowly than anticipated.)

[730] Wooddy, *TAPS Rebuttal Testimony, supra* note 724, at 13-14.

[731] *Id.* at 15; Gary, *TAPS Rebuttal Testimony, supra* note 726, at 11-12. There may be a possible exception to this statement if ARCO's Kuparak field develops favorably.

[732] Wooddy, *TAPS Rebuttal Testimony, supra* note 724, at 15; see text at note 620, *supra*.

An example on the products side of the business is Explorer. Explorer was built on the basis of a projected need for a large diameter refined petroleum products pipeline from the Gulf Coast to the Chicago area. In order to be competitive with barge rates, the system had to be of large diameter (28-inch line from Port Arthur to Houston to Tulsa and 24-inch on to the Chicago area). As we have seen, Explorer's projected usage was in error or *premature*,[733] and its failure timely to achieve the projected throughput caused it to incur losses amounting to almost $46 million in the first five full years of operation.[734] Although it has improved its throughput somewhat since that time it has never realized its projected traffic.

Pipelines are a high risk investment because of the high ratio of fixed costs to total costs, inflexibility of the routing, limitation of the products that can be transported, the limited number of potential customers and sensitivity to fluctuations in throughput.[735] It follows that the economic evaluation of a proposed pipeline is extremely important. Among the factors to be considered are: estimates of reserves, refining pattern trends, market growth, alternate modes of supplying that market, both at the time of construction and in the foreseeable future, and possible governmental actions.[736] Thus, when it comes to sizing the line, it is most important that the very best data available be assembled, but it must be recognized that the capacity of the human mind (even aided by computers) for formulating and solving complex problems is very small compared to the size and complexity of the problems that must be solved in correctly sizing a pipeline.[737] This "bounded rationality" disability arises from the fact that there simply is no way to generate *a priori* the entire "decision tree."[738] Sizing decisions are at best imperfect business judgments made by human beings with limited knowledge and are subject to the subsequent occurence of events which could not have been perceived. The requirements are exacting: pipelines must be properly sized to meet long-term (30-40 years) future requirements and yet be attractive in the short term.[739] Typically, a crude line must be designed to be large enough to meet the requirements of a new field, and yet not so large as to be economically inoperable as production declines.[740] A products line must be economically sized to serve the demand which generated its being but it must

---

[733] HOWREY & SIMON, *supra* note 3, at 124-125.
[734] See text at notes 435 and 436, *supra*.
[735] See text at notes 606 to 624, *supra*.
[736] PIPELINE TRANSPORTATION, *supra* note 162, at 37.
[737] *Cf.* Oliver Williamson, *supra* note 329, at 1444.
[738] *Cf.* TEECE, *supra* note 329, at 123.
[739] PIPELINE PRIMER, *supra* note 331, at 19.
[740] Bond, *supra* note 207, at 733.

not be so small that it cannot handle the increased demands that market growth may place upon it.⁷⁴¹

The penalties for "guessing wrong" are substantial. The essence of the dilemma is that if the pipeline is too small, operating costs are higher than they would have been had the line been sized "correctly" as measured by hindsight (*i.e.*, the economies of scale are not realized), thus placing the shipper-owners at a competitive disadvantage against any current competition and risking the loss of outside shipments to such competition. If no competition currently is present, the line may not be able to handle all shipments, thereby forcing the shipper-owners to "prorate" their shipments and probably encourage the construction of a larger sized—lower cost line.⁷⁴² On the other hand, if the pipeline turns out to be too large, the "overinvested" capital costs will result in higher unit cost of the oil transported through the line,⁷⁴³ which will result in reduced earnings, possible losses⁷⁴⁴ and, if the situation continues, deficiency cash calls upon the shipper-owners.⁷⁴⁵ The difficulty in steering a fine line between the Charybdis of an "oversized" line and the Scylla of an "undersized" pipeline has caused one (independent)

---

⁷⁴¹ One needs only refer back to the Plantation experience for support of this point. See text at notes 424 to 426, *supra*.

⁷⁴² PIPELINE TRANSPORTATION, *supra* note 162, at 37; MARATHON, *supra note 202, at 6-8; PIPELINE PRIMER, supra* note 331, at 19. An example of this is the West Texas Gulf Pipe Line from the West Texas production region to the Port Arthur, Texas, refinery area where West Texas Gulf's 6" to 12" lines became unprofitable to operate after a 26-inch line was constructed over the same roude. *Id.* at 15. Another example is Amoco's (then Service) 10-inch line from Wyoming to Freeman, Missouri, being rendered obsolete by Platte's 20-inch line. See text at note 168, *supra*. This is usually handled in markets that are expected to grow by building in some "surplus" capacity which the owners hope will be used as the market expands or as non-owners seek the increased efficiency offered by the system. See text at note 168, *supra*.

⁷⁴³ *Id.*

⁷⁴⁴ MARATHON, *supra* note 202, at 35. ICC Statistics for 1975 reveal that twelve pipeline companies reported losses to the Commission. HOWREY & SIMON, *supra* note 3, at 9. According to 1976 ICC Statistics, nine pipeline companies—Eureka, Four Corners, Getty, Gulf Central, Kenai, Mobil, Eugene Island, National Transit and Seaway, operated at a loss for that year. In Rebuttal Testimony of Rodney A. Harrill Before F.E.R.C. in Trans Alaska Pipeline System, Dkt. OR 78-1, 18-20 (Oct. 30, 1978) [hereinafter cited as Harrill, *TAPS Rebuttal Testimony*], witness Harrill produced figures derived from data obtained from Standard and Poor's Compustat Services, Inc. and from pipeline company ICC Form P reports which showed that out of the 79 oil pipeline companies in the sample (he excluded those with 2.5 million dollars or less annual revenue to achieve comparability with the Compustat data), 10 had 5-year average returns on total capital of less than 4 percent whereas out of the 171 utility companies in the Compustat sample *none* had less than 4 percent; and the 79 pipeline companies had a variance of 23.7, while the 171 utility companies had a variance of only 2.0.

⁷⁴⁵ The Explorer example fits this description nicely. See text at notes 430 to 437, *supra*.

Economic Characteristics

pipeline official to remark wryly: "There are only two kinds of oil pipelines. One is too small and the other is the kind that never should have been built."[746]

Any pipeline owner—whether an oil company or one without any affiliation with the business—has every incentive to avoid excess capacity. Typically, the owners are committed for 30 to 40 years[747] and the economic success of the venture depends upon the accuracy of their forecast of supply and demand for that period. As will be seen from the immediately following subsections, many elements, some foreseeable, others impossible to predict or sometimes even to conceive at the time of decision, can alter the economics substantially.

Looking at the cost curves in Leslie Cookenboo's classic study,[748] it can be seen that the degree of cost sensitivity to volume tends to decrease substantially with increase in line diameter and the intermediate-run cost curves tend to flatten out around the 24-inch line, hence a 24-inch or larger diameter line is close to its optimum cost over a large range of throughput volumes.[749] It follows that throughput can be expanded appreciably over the design throughput before higher per-barrel costs than the original costs will be incurred.[750] This fact creates the ability to achieve a limited degree of flexibility by installing the minimum number of pumping stations until the operator can see how much throughput develops[751] and then expanding to optimal design capacity, and beyond, if the traffic materializes.[752]

---

[746] LENNART, *supra* note 111, at 2 citing an unnamed MAPCO official. Lennart translates the remark as follows: ". . . if the entrepreneurs who provided the funds correctly forecast demand, the line will attract volumes and be fully used. But if their forecasts were incorrect, the volumes will not appear and the owners or shippers which guaranteed repayment of the pipeline debt through tariff prepayments will take a bath."

[747] See text at, and note 609, *supra*.

[748] CRUDE OIL PIPELINES, *supra* note 158, at 26, Chart 6; for the same information in tabular form, see Cookenboo, *Crude Line Operating Costs*, *supra* note 635, at 106-107 (Table 19). The same pattern can be discerned from the curves in Appendix I.

[749] *Cf.* PIPELINE TRANSPORTATION, *supra* note 162, at 40; KENNEDY STAFF REPORT, *supra* note 184, at 99.

[750] CRUDE OIL PIPELINES, *supra* note 158, at 28-29. Actually, for a while, there would be a decline in per-barrel costs. *Id.* at 29.

[751] Of course, you are back on the short-run cost curve and the per-barrel cost is greater than it would be had you been able to immediately start off with design throughput (see text at notes 636-637, *supra*) but because this is an imperfect world, if you started out with design throughput, you would probably end up requiring expansion. In the meantime, you will realize substantial savings in power costs. See note 636, *supra*.

[752] CRUDE OIL PIPELINES, *supra* note 158, at 28-29 points out that because of the slope of the cost curve, a 24-inch line constructed to transport 200,000 barrels per day could increase the throughput to 300,000 barrels per day (after adding the required number of stations) without incurring increased costs per barrel.

2. Faulty Supply (Reserve) Estimates

One of the risks faced by crude lines is that the volume to be transported from a producing oil field diminishes as the field is depleted.[753] Despite considerable advance in petroleum reservoir engineering, the unavoidable fact is that estimating ultimate production from a field is not a science but an art, and not a precise one at that; there is a risk in relying on any such estimate.[754] Moreover, even if compensating errors operate to produce an accurate estimate, there is no assurance that those reserves will be sufficient to provide traffic for all the lines that may be constructed.[755] As a matter of historical interest, the first commercial pipeliner, Samuel Van Syckel, who constructed the pioneer line from the Pithole City field in Pennsylvania, fell victim to depletion of the reservoir (and competition from other pipelines) and although his company recovered its investment in the first line in good order, it defaulted on its bank obligations two years later and went into bankruptcy.[756] Another early example where the decline of a producing field was unexpectedly rapid, leaving the pipeline's costs unamortized, is the Prairie Pipe Line Company's extension of its line to the Ranger field in Texas.[757] A crude line built from the Wyoming producing fields to Freeman, Missouri, and purchased by a predecessor of Amoco Pipeline Company, remained idle as a crude carrier from 1928 to 1931 due to the decline in Wyoming production from 44.8 million barrels in 1923 to 11.2 million barrels in 1933.[758]

In the same era, Sohio built a line in the 1930's from producing areas in central Michigan to the parent's refinery at Toledo, Ohio. Contrary to estimates, production peaked in 1939 at around 23 million barrels and declined so substantially that by 1950 it was down to around 16 million barrels. This was less than the requirements of local refineries. As a result Sohio's line was "unloaded," and Sohio was forced to settle for recovery of a small portion of its investment by selling the line in 1950 to a local refinery, Mid-West

---

[753] STEINGRABER, *supra* note 193, at 151.

[754] W. J. Williamson, *supra* note 315, at 6; Burke, *supra* note 9, at 787; Cooper, *Direct TAPS Testimony*, *supra* note 718, at 16; F.E.R.C., Valuation of Common Carrier Pipe Lines, Dkt. RM 78-2, Tr. Vol. 12, pp. 1833-1834 (Oct. 24, 1978) (Oral argument of Darell Behring on behalf of Okie Pipe Line Company, an independent crude gatherer.) [hereinafter cited as *RM 78-2 Hearings Transcript*].

[755] W. J. Williamson, *supra* note 315, at 7. This has been expressed as the "possibility of shipments drying up." *S.1167 Hearings*, *supra* note 415, pt. 9 at 704 (Statement of Gary L. Swenson) [hereinafter cited as Swenson, *S.1167 Testimony*].

[756] Burke, *supra* note 9, at 782.

[757] WOLBERT, *supra* note 1, at 10.

[758] LIVINGSTON, *supra* note 228, at 20. The line had never operated at more than 50 percent capacity and was operating at about 15 percent when closed down. It was not until 1936 that the line was returned to crude service. PETROLEUM PIPELINES, *supra* note 45, at 133.

Refineries. Mid-West reversed the flow in part of the line to supply its refinery at Alma, Michigan, with Mid-Continent crude through a connection with another line at Toledo.[759]

Neither the Four Corners Pipe Line,[760] the Texas-New Mexico Pipe Line nor the Cook Inlet Pipe Line ever reached expectations because the crude oil reservoirs did not contain the recoverable reserves that the estimators had predicted.[761] All that remains of the Arapahoe Pipe Line is the old Sterling system; the big line from Gurley, Nebraska, to Humboldt, Kansas, was sold to Cities Service for conversion to gas service.[762] Another line that threatens to take a "bath" (no pun intended) is the Eugene Island Pipeline System which was forced to proceed with construction of a 20-inch line on the basis of tentative estimates because of the delay in finalizing the ownership nominations due to prospective participants changing their minds. The design capacity was 173,000 barrels per day. It is reported that it never realized much more than 73,000 and even that throughput has diminished.

3. Shifts in Logistical Patterns or Plans of Potential Shippers

Because pipelines are merely a transportation link, they are dependent on other segments of the petroleum industry for their survival.[763] Even the most accurate forecasts can be rendered incorrect by conditions completely beyond the control of the pipeline company, such as discovery of new crude

---

[759] *Id.* at 19-20.

[760] LIVINGSTON, *supra* note 228, at 20-21; MARATHON, *supra* note 202, at 35; HOWREY & SIMON, *supra* note 3, at 9. Four Corners, completed late in 1957, was built by a coalition of owners of Los Angeles area refineries and local crude oil producers to transport crude oil from the producing area where Utah, Colorado, Arizona and New Mexico meet to refineries near Los Angeles. Prior to construction, the local production was limited to only a few thousand barrels per day above the limits of the local refiners. There were believed to be very sizeable reserves in the area: the OIL AND GAS JOURNAL in 1957 published an estimate of upwards of 500 million barrels of undeveloped reserves and predicted that exploration induced by the construction of a low cost line to the Los Angeles market would boost production to 110,000 barrels per day within two years. Traffic through the Four Corners line peaked in the second year of operation at just 97,000 barrels per day. It steadily declined to the point that when ARCO bought out the other owners in 1976, see note 633, *supra*, the line was transporting only about 3,500 barrels per day. LIVINGSTON, *supra* note 228, at 21.

[761] *S.1167 Hearings, aupra* note 432, pt. 9 at 607 (Statement of Vernon T. Jones: Texas-New Mexico's line from the same producing area to Jal, New Mexico, was in witness Vernon Jones' words, "far from a roaring success."); HOWREY & SIMON, *supra* note 3, at 9; MARATHON, *supra* note 202, at 35; see also LeGrange, *FERC Direct Testimony, supra* note 176, at 28. "Recoverable reserves" can be affected drastically by the well-head price because unless such price provides for a return on, and of, capital, as well as out-of-pocket expenses, future exploration and production investments will not be made and this will cause "recoverable reserves" to shrink and pipeline throughputs will fall short of projections; Ryan, *TAPS Rebuttal Testimony, supra* note 725, at 23.

[762] *Cf.* LeGrange, *FERC Direct Testimony, supra* note 176, at 28.

[763] Gary, *FERC Direct Testimony, supra* note 613, at 17.

reserves,[764] shifts in demand or distribution patterns,[765] and exogenous events which change the plans of potential shippers.[766] An early example of this type of event was the misfortune of Prairie Pipe Line's Arkansas Division, which was part of a link in the movement of crude from the Mid-Continent field to Baton Rouge, Louisiana. Louisiana Standard had built a line into the El Dorado, Arkansas, field and later added a connection to the booming Smackover field. This diversion of crude source caused a sharp decline in Prairie's Arkansas Division traffic. Prairie was "not unhappy" to sell the line to Standard[767] which used it to form an unbroken, solely-owned pipeline system from the Glenn Pool area to its Baton Rouge refinery.[768]

Another instance was the "drying up" of Tuscarora pipeline due to the decline in overland Mid-Continent crude shipments to Jersey Standard's East Coast refineries.[769] A third example is Ajax, which was a jointly owned and operated line built in 1930 to carry oil from Oklahoma Pipe Line's Glenn Pool station to Wood River, Illinois, where it connected with the Illinois Pipe Line Company. The Illinois field subsequently was discovered and boomed from 4-1/2 million barrels in 1936 to 147.6 million barrels in 1940, providing a much less expensive alternate feed for Ohio refineries.[770] Sohio, which accounted for half of Ajax's traffic, moved aggressively into the Illinois field, becoming a producer and operating its own pipeline system to its refineries, thereby reducing its Mid-Continent crude offtakes from 88 percent of its refinery throughput in 1936 to 2.4 percent in 1939;[771] and in the process, Ajax's throughput-to-capacity ratio dropped from 99 percent in 1936 to 25 percent in 1938.[772] The volume continued to decline, dropping from 60,500 barrels per day in 1937 to 7,400 barrels per day in 1940, making Ajax financially inoperable.[773]

The Illinois boom also made uneconomic the western portion of Sun's products pipeline from Philadelphia to Cleveland. From a business standpoint, it was cheaper to build an Illinois crude-fed refinery in Ohio to serve

---

[764] *Consumer Energy Act Hearings, supra* note 154, at 595 (Statement of Jack Vickrey); HOWREY & SIMON, *supra* note 3, at 8.

[765] Spahr, *Rodino Testimony, supra* note 255, at 10; MARATHON, *supra* note 202, at 34; *cf.* Burke, *supra* note 9, at 787 (refinery shutdown).

[766] Gary, *FERC Direct Testimony, supra* note 613, at 17; Myers, *supra* note 198, at 9 (shift in refinery locations); HOWREY & SIMON, *supra* note 3, at 8 (construction of deepwater ports).

[767] PETROLEUM PIPELINES, *supra* note 45, at 149.

[768] See text at note 95, *supra.*

[769] PETROLEUM PIPELINES, *supra* note 45, at 150.

[770] LIVINGSTON, *supra* note 228, at 19.

[771] PETROLEUM PIPELINES, *supra* note 45, at 142.

[772] WOLBERT, *supra* note 1, at 10-11.

[773] LIVINGSTON, *supra* note 228, at 19.

the market formerly served by the pipeline. As J. Howard Pew, President of Sun Oil Company, ruefully expressed it: "[the new field in Illinois] completely upset the whole economic balance, and so we have to go out in that western territory and build ourselves a new refinery into which we put an investment well over $6,000,000 in order to make obsolete the $6,000,000 that we earlier put into the pipeline."[774] With the decline of the flush production in Illinois and the advent of World War II, Ajax was revived and once again did a satisfactory business. But newly discovered Canadian crude production in the western provinces and the construction of the Interprovincial Pipeline from Western Canada to Sarnia, Ontario, which had been supplied through Ajax and connecting lines, together with the competition from Ozark (a 22-inch line as compared to Ajax's twin 10-inch lines) and the new Mid-Valley 22-inch line from Longview, Texas, to Lima, Ohio, which brought Texas-Louisiana and Mississippi crudes to Ajax's former Lake Erie customers, caused Ajax's throughput virtually to vanish.[775] "No longer able to defy the lightning, Ajax hit the floor, this time to stay. It was sold on September 1, 1954, at bargain basement prices to the Cherokee Pipe Line Company for conversion to products service."[776]

In the 1950's Laurel Pipe Line System was built by Philadelphia area refiners across southern Pennsylvania to the Pittsburgh market. It had barely been completed when the refinery plans of the shipper-owners were changed by the imposition of the mandatory crude oil imports program. Because of the abrupt effect on the Philadelphia area refinery runs, and the alternate sources of supplies for the market, Laurel lost money for years and has never been particularly profitable.[777] The Tecumseh Pipe Line, which was constructed in 1957, from Griffith, Indiana, to Cygnet, Ohio, has suffered a similar decline in throughput.[778] Finally, although only a small risk, a diversion of substantial crude volumes from Valdez to an alternate intersecting point on the TAPS line could have devastating results, particularly to the smaller owners, due to the decrease in revenues.[779]

---

[774] *TNEC Hearings, supra* note 6, at 7184 (Statement of J. Howard Pew).

[775] LIVINGSTON, *supra* note 228, at 19; SPECTRE, *supra* note 104, at 117.

[776] SPECTRE, *supra* note 104, at 117.

[777] *S.1167 Hearings, supra* note 432, pt. 9 at 607 (Statement of Vernon T. Jones); HOWREY & SIMON, *supra* note 3, at 9; *see also* LeGrange, *FERC Direct Testimony, supra* note 176, at 28; *Consumer Energy Act Hearings, supra* note 154, at 595 (Statement of Jack Vickrey).

[778] LeGrange, *FERC Direct Testimony, supra* note 176, at 28; PIPELINE PRIMER, *supra* note 331, at 14.

[779] Gary, *Direct TAPS Testimony, supra* note 634, at 35-36.

4. Market Miscalculation

One of the business risks in the construction of products lines arises from erroneous estimates of market potentials[780] or shifts or changes in market conditions or locations,[781] including changes in fuel requirements.[782] Crude oil pipelines face a similar problem; there may be shifts in refining locations,[783] or changes in market conditions. For example, the natural market for North Slope crude is the West Coast. But due to logistical and quality-of-crude factors, this market will not absorb the full North Slope production. There have been estimates of surpluses in the order of 300,000 to 600,000 barrels per day which will have to find their way to Midwestern markets.[784] In short, petroleum pipeline investors, integrated or "independent," must concern themselves with the possibility of loss of markets at the delivery point.[785]

5. Fluctuation in Throughput

The high capital-intensive nature of pipelines[786] and the associated high proportion of fixed costs[787] results in a pattern of profitability that is extremely sensitive to volume changes.[788] Thus, the maintenance of constant throughputs at projected levels is a prerequisite for profitable operations and even slight variations in the volume of throughput can adversely affect the economics of the pipeline,[789] and if sufficient in magnitude, can result in losses.[790] Such fluctuations can be caused by unfavorable wellhead pricing which retards field development,[791] prorationing of production,[792] partial[793]

---

[780] W. J. Williamson, *supra* note 315, at 7.

[781] *Id.* (changes in market conditions); MARATHON, *supra* note 202, at 34 (shifts in marketing locations).

[782] *Consumer Energy Act Hearings, supra* note 154, at 595 (Statement of Jack Vickrey); HOWREY & SIMON, *supra* note 3, at 8.

[783] MARATHON, *supra* note 202, at 34; *but cf.* Burke, *supra* note 9, at 787 (refineries and populations do not make major shifts or disappear). The difference in opinion perhaps arises from the difference in dates of observation. Domestic crude oil shortages not present when Burke made his observation have affected refineries and population shifts, for example, from the Northeast to "sunbelt" locations have become quite pronounced in recent years.

[784] Gary, *Direct TAPS Testimony, supra* note 634, at 32-33.

[785] Swenson, *S.1167 Testimony, supra* note 755, at 704; Cooper, *Direct TAPS Testimony, supra* note 718, at 16, and on cross-examination, he reiterated that substantially higher than anticipated [downstream] transportation costs could result in excess capacity [in TAPS]. Hearings before F.E.R.C. in Trans Alaska Pipeline System, Dkt. OR 78-1, Tr. Vol. 12, p. 1989 [hereinafter cited as *TAPS Hearings Transcript*].

[786] See text at notes 613 to 615, *supra*.

[787] See text at note 616, *supra*.

[788] See text at notes 618 to 680, *supra*.

[789] Ryan, *TAPS Rebuttal Testimony, supra* note 724, at 10.

[790] Waidelich, *Rodino Testimony, supra* note 197, at 8-9.

[791] Ryan, *TAPS Rebuttal Testimony, supra* note 725, at 23.

or complete shutdowns of the line[794] or disadvantageous market conditions downstream of the pipeline.[795] In a previous subsection, Sohio's difficulties in moving its North Slope crude were mentioned.[796]

Sohio's plan to reverse the flow of, and convert to crude carriage, the natural gas lines of Southern California Gas Company and of El Paso Natural Gas Company so as to reach Midland, Texas, where connecting crude lines are in place to transport the crude to the Midwest refining area[797] has been thwarted by California environmentalists. However, there have been efforts from time to time to revive the project. The only present alternative is movement of the surplus (to California refiners' needs) by U.S. flag tankers through the Panama Canal and thence to other refining centers. This may result in a reduction by Sohio of its throughput through TAPS until more satisfactory arrangements can be made.[798] However, judging from the fact that TAPS' owners have agreed to expand the line from 1.2 to 1.36 million barrels per day and there is talk of a further expansion, Sohio's cutting back on its shipments appears to be unlikely.

---

[792] *Consumer Energy Act Hearings, supra* note 154, at 595 (Statement of Jack Vickrey); HOWREY & SIMON, *supra* note 3, at 8. This risk, unfortunately for the nation, has become rather rare.

[793] Gary, *Direct TAPS Testimony, supra* note 634, at 30; Wooddy, *TAPS Rebuttal Testimony, supra* note 724, at 21. On July 8, 1977, an explosion and fire occurred at TAPS Pump Station 8 which resulted in extensive damage to the station's main pump building and pump equipment. Until the station was repaired, which was not until the following year, the pipeline throughput was reduced by about 40 percent (1.2 million barrels per day lowered to 730,000 barrels per day) causing an estimated loss in gross revenues of $2.9 million dollars per day—based on tariffs in effect at the time.

[794] STEINGRABER, *supra* note 193, at 147 (for Colonial in 1974 every day the line was shut down completely, the loss, measured solely in terms of fixed costs and not taking into account anticipated "loss of business," was approximately $219,000 per day).

[795] Gary, *Direct TAPS Testimony, supra* note 634, at 33-34; Gary, *TAPS Rebuttal Testimony, supra* note 726, at 12.

[796] See text at note 784, *supra*.

[797] See note 633, *supra*.

[798] Gary, *Direct TAPS Testimony, supra* note 634, at 33-34. On cross-examination, Gary said he was of the impression that the cost of transporting crude, at the time of his testimony, to the Gulf Coast area was almost equal to the delivered price of the oil, if not in excess of it. *TAPS Hearings Transcript, supra* note 785, Vol. 12, pp. 2169-2172 (Feb. 16, 1978). The cheapest way to deal with the West Coast surplus is to exchange oil with the Japanese so that Alaska oil could go to Japan and in return Japan would deliver oil which it purchased in, say, Mexico or South America, to U.S. Gulf Coast or East Coast ports thus saving both parties the cross-haulage. But such exchange is expressly prohibited by law, 30 U.S.C. § 185 (1976), without the permission of the President, which he has expressly withheld. Direct Testimony of Alton W. Whitehouse, Before F.E.R.C. in Trans Alaska Pipeline System, Dkt. OR 78-1, 5-6 (Nov. 30, 1977) [hereinafter cited as Whitehouse, *Direct TAPS Testimony*]. See Houston Chronicle, Mar. 19, 1979, Sec. 1 at 17, Col. 14 for details of Sohio's cancellation of its proposed line to Midland, Texas, due to the inaction of the California South Coast Air Quality Management District and the Air Resources Board. Governor Brown of California has raised a question concerning the validity of Sohio's claim that the environmental constraints are the true reason for the abandonment. Houston Chronicle, Mar. 25, 1979, Sec. 3, p. 21, Cols. 3-5.

Perhaps the most graphic method of illustrating how pipeline throughputs can fluctuate, and the effect on the pipeline concerned, is to cite an actual case history, that of Platte Pipe Line. Platte was formed in 1950 by five oil companies as a "logistical" or area-to-area line tying the Rocky Mountain/Wyoming crude producing region to the Midwest refinery area.[799] At the time Platte was formed, the area was served by Amoco Pipeline (then Service) which had a combination 12- and 16-inch line with a capacity of 50,000 barrels per day. Platte's line was designed to provide a competitive service for the area's producers to Midwest oil refiners. Platte's line was 20 inches in diameter and had a 100,000 barrels per day capacity. Area reserves were then estimated to be about 1½ billion barrels and it was expected that the new line would stimulate further exploration and development, which it did. After considerable delay due to the Korean War steel shortage, the line was placed in operation and oil reached Wood River, Illinois, on December 8, 1952. Platte's initial tariffs were 9 cents per barrel lower than Amoco's, so Platte rapidly achieved initial capacity, whereupon it installed additional booster stations and pumping equipment, increasing its capacity to 171,000 barrels per day in 1953. Platte planned further construction to increase capacity to 192,000 barrels per day.

While this new expansion was in the planning stage. Amoco replaced its old 12-16-inch line with a modern 20-22-24-inch system, which was completed in 1954, having a capacity between 125,000 and 130,000 barrels per day, and it lowered its tariffs. At the same time, Arapahoe built an 18-inch line from southwest Nebraska and northeast Colorado to Humbolt, Kansas, which came on-stream in 1955 with an estimated capacity of 80,000 barrels per day. The effect on Platte was to decrease significantly its throughput, so the planned expansion was temporarily shelved. Then several smaller lines tied into Platte and increased its volume by approximately 10,000 barrels per day.

In 1961, the Glacier Pipe Line from Cut Bank, Montana, to Byron, Wyoming, was connected to Platte, thereby adding a Canadian stream of about 7,000 barrels per day which raised Platte to an average of 180,000 barrels per day in 1963. From that high, Platte's volume steadily declined from 1963 to mid-1977 when the combination of new oil discoveries in southeast Montana and eastern Wyoming and increases in the Canadian stream built Platte's volume up to 231,000 barrels per day by May 1974. Then came the curtailment of Canadian exports and Platte's throughput again suffered a sharp decline. 1977 volumes averaged around 125,000 barrels per day (which was only slightly above what they had been in 1953). This actual case history shows real world pipelining "like it is."

---

[799] Platte's case history is drawn principally from MARATHON, *supra* note 202, at 37-40. Some supplementary data was derived from J. L. Burke's prepared Statement in the *Celler Hearings, supra* note 162, at 1239-1241, and his testimony, *Id.* at 1135-1136.

6. Construction and Technical Problems Now Take the Form of Environmental and Regulatory Difficulties

In the "good old days" that everyone talks about, right-of-way was considered difficult if it took two visits and a dollar-a-rod. Weather,[800] terrain,[801] scarcity of essential materials and equipment[802] and labor problems affecting either material sources or the construction of the line itself[803] were the main problems. As population increased and newer lines were forced to cross improved property, or were constructed in states which either lacked eminent domain laws or had antiquated laws originally drawn for railroads, containing language difficult to construe as covering crude oil lines let alone products lines,[804] or across federal or state lands, obtaining the necessary permits, licenses, and rights-of-way became an increasingly difficult problem.[805]

A more recent and potentially greater obstacle is environmental protection legislation begun with the National Environmental Policy Act of 1969[806] ("NEPA"), and enforced not only by governmental agencies but by

---

[800] Burke, *supra* note 9, at 787; HOWREY & SIMON, *supra* note 3, at 8; *Consumer Energy Act Hearings, supra* note 154, at 595 (Statement of Jack Vickrey); Gary, *Direct TAPS Testimony, supra* note 634, at 29, Cooper, *Direct TAPS Testimony, supra* note 718, at 32.

[801] Gary, *Direct TAPS Testimony, supra* note 634, at 29; Gary, *TAPS Rebuttal Testimony, supra* note 726, at 10; Cooper, *Direct TAPS Testimony, supra* note 718, at 32.

[802] HOWREY & SIMON, *supra* note 3, at 8; Gary, *Direct TAPS Testimony, supra* note 634, at 29.

[803] HOWREY & SIMON, *supra* note 3, at 8.

[804] See text at notes 150 to 152, *supra*, for difficulties faced by Plantation Pipe Line and Southeastern Pipe Line companies at the start of World War II; Colonial's troubles with Frederic Lang and associates are well known, see *e.g., Frederic A. Lang v. Colonial Pipeline Company of Pennsylvania,* Pa. P.U.C. Dkt. 17915 (1964), 41 Pa. P.U.C. Rep. 651 (Dec. 9, 1964), 41 Pa. P.U.C. Rep. 680 (Aug. 7, 1967), 43 Pa. P.U.C. Rep. 243; Procedural Order, 207 Pa. Super. Ct. 312, 217 A.2d 750 (1966); order, dated Feb. 21, 1964, by Del. River Basin Comm., Dkt. D-64-1, *In the Matter of the Application of Colonial Pipeline Company; Colonial Pipeline Co. v. Frederic A. Lang,* Equity No. 1632, Ct. of Common Pleas, Chester County, Pa. (Decision of Judge Thomas Riley, Mar. 4, 1964); *Ott v. Colonial Pipeline Co. of Pa.,* Equity No. 1636, Ct. of Common Pleas, Chester County, Pa., 36 D&C.2d 76 (Order of dismissal, Sept. 24, 1969); *Frederic A. Lang v. Colonial Pipeline Co. and Colonial Pipeline Co. of Pa.,* Equity No. 1938, Ct. of Common Pleas, Chester County, Pa., and *Colonial Pipeline Co. v. Luther C. Peery* (Decisions of Judge Riley on Sept. 5 and 6, 1968, respectively); aff'd., *Colonial Pipeline Co. v. Frederic A. Lang,* 252 A. 2d 697 (Pa. Sup. Ct. 1969); *James A. Hudson v. Colonial Pipeline Co. of Del.,* Civil No. 2978 (D.Del. 1966); *Frederic A. Lang v. Colonial Pipeline Co.,* 266 F.Supp. 552 (E.D.Pa. 1967), aff'd., 383 F.2d 986 (3d Cir. 1967).

[805] For example, TAPS had to get both State of Alaska and Federal permits, the terms of which gave the "Authorized Officer" (Federal) and the "Pipeline Coordinator" (Alaska) the authority to approve or disapprove of construction plans and enforce (according to their respective interpretations of the various stipulations and conditions contained in the two Permits). Wooddy, *TAPS Rebuttal Testimony, supra* note 724, at 8-10; Gary, *Direct TAPS Testimony, supra* note 634, at 25-27. See also Swenson, *S.1167 Testimony, supra* note 755, at 704 (Condemnation); Burke, *supra* note 9, at 787 (landowner resistance).

[806] 42 U.S.C. § 4321, et seq. (1976).

"Private Attorneys-General."[807] As a result, pipelines have encountered longer than anticipated construction periods.[808] In the case of TAPS, there was doubt until the oil commenced flowing into the Valdez terminal whether there was ever going to be an operating pipeline.[809] It is significant that one of the key documents required by lenders is a Completion Agreement, signed by the shipper-owner, which guarantees that the owners will advance sufficient funds to the pipeline company to enable it to construct the pipeline and to satisfy all indebtedness until the Throughput and Deficiency Agreements take over.[810]

7. Possible Negative Salvage Value

Pipeliners have become accustomed to the prospect that once the pipeline has been constructed, the possibility for salvaging the line is negligible.[811] But with the advent of the environmental movement and the increasing involvement of Federal and State agencies in "supervising" the line,[812]

---

[807] The difficulties with environmental groups reached such proportions that after 3 years' delay in even getting right-of-way permits (total delay was 4 years) Congress had to enact the Trans Alaska Pipeline Authorization Act in 1973 and amend the Mineral Leasing Act in order to break the logjam. See Gary, *Direct TAPS Testimony, supra* note 634, at 23-24; Direct Testimony of Richard M. Donaldson, Before F.E.R.C. in Trans Alaska Pipeline System, Dkt. OR 78-1, 7-8 (Nov. 30, 1977) [hereinafter cited as Donaldson, *Direct TAPS Testimony*]; Patton, *Direct TAPS Testimony, supra* note 726, at 12-40; HOWREY & SIMON, *supra* note 3, at 168-169.

[808] Burke, *supra* note 9, at 787; *cf.* MARATHON, *supra* note 202, at 49 ("Long construction delays, increased regulatory involvement, social and environmental concerns, all contributing to misestimates of cost and high financing requirements, have caused the investor to perceive financial risks that threaten the ability of even the petroleum companies to guarantee.").

[809] *Petroleum Industry Hearings, supra* note 192, at 1962 (Statement of Raymond B. Gary); Gary, *TAPS Rebuttal Testimony, supra* note 726, at 10; Wooddy, *TAPS Rebuttal Testimony, supra* note 724, at 6, 12. At the end of 1973, Mobil could not be sure that the line could be built at all or if it could be, whether it would be economically feasible. It became clear that a 600,000 barrel per day line would be uneconomic and even at the changed 1,200,000 barrels per day, it could prove uneconomic. Mobil considered withdrawal until the Arab oil embargo and OPEC price doubling saved the project from collapse. Cooper, *Direct TAPS Testimony, supra* note 718, at 28-29. Mobil still wasn't sure that increasing the throughput to 1.2 MMBD and the higher crude oil price could make the line economic, so it "hedged its bet" by reducing its interest to 5 percent. *Id.* at 31.

[810] *Petroleum Industry Hearings, supra* note 192, at 1931 (Statement of Peter Bator); *Id.* at 1962 (Statement of Raymond B. Gary).

[811] See text at notes 633 and 634, *supra*.

[812] Donaldson, *Direct TAPS Testimony, supra* note 807, at 15-16. The "Right-of-Way" Agreement was 26¼, 7½ × 10 inch pages of single-spaced, double-columned clauses plus 5 exhibits from 4 to 18 pages each. It is very sobering to read them. They have been described as "requiring the pipeline to be double and triple built for reasons that did not make economic or technical sense." *Id.* at 14. A more homey expression was employed by E. L. Patton in answering Judge Kane's question whether the pipeline would have been no better or worse because of the Federal Government's intervention: "Where the government got involved was the belt or suspenders area, and the aesthetics, where a great deal of money went. And I think from the structural integrity standpoint, industry would have built just as good a pipeline as we ended up with governmentwise." *TAPS Hearing Transcript, supra* note 785, Vol. 5, at 537-540.

dismantling and restoration obligations imposed on pipelines has become a substantial contingent liability, and qualifies as a bona fide "risk" to be considered in evaluating the economic viability of new ventures.[813]

8. Competition

At the outset, it must be borne in mind that not all refineries are located in a position where they could use the large diameter crude lines developed in recent years, which have tended to be area-to-area or "logistical" lines, as might reasonably be expected from the volumes required to justify their capacities. Hence concentration statistics couched in terms of a nationwide pipeline "market" have no economic or legal significance.[814]

As early as 1955, Leslie Cookenboo noted that about 80 percent of independent refining capacity was located in areas where there were contiguous crude supplies, hence were not potential shippers over "major" company common carrier crude lines.[815] In addition, there is a degree of natural or geographic market separation.[816] Thus a crude line from New Mexico to California is not in competition with a line running from Louisiana to Illinois,[817] any more than a products line from West Texas to Albuquerque would be in competition with one from Chicago to Detroit.[818] When the focus of attention is directed to those lines which *are* potential competitors, it becomes apparent that one of the reasons that a number of pipelines in operation today are being run at a very low profit or at a loss is because their volumes have been siphoned off by competition.[819] Much of this competition

---

[813] TAPS' owners are required to dismantle and remove the system after it ceases to operate and are liable for damage to United States property or natural resources resulting from the construction, operation, maintenance or dismantling of the system. Gary, *Direct TAPS Testimony, supra* note 634, at 25. See Section 14, "Indemnification of the United States," of the Agreement and Grant of Right-of-Way for Trans-Alaska Pipeline between the United States of America and Amerada Hess Corporation, ARCO Pipe Line Company, Exxon Pipeline Company, Mobil Alaska Pipeline Company, Phillips Petroleum Company, Sohio Pipe Line Company and Union Alaska Pipeline Company, dated January 23, 1974. A harbinger of things to come can be found in the LOOP license which provides that upon termination or revocation of the license (with the possible exception of a novation by a creditworthy party in a transfer situation) the licensees must, at their expense, remove all components of the installation or, failing to do so, must pay for the government doing it. Art. 23, United States, Dept. of Trans., License to Own, Construct and Operate a Deepwater Port issued to LOOP, Inc.

[814] SPECTRE, *supra* note 104, at 118; CRUDE OIL PIPELINES, *supra* note 158, at 37.

[815] CRUDE OIL PIPELINES, *supra* note 158, at 52-53.

[816] *Id.* at 37; SPECTRE, *supra* note 104, at 118; LIVINGSTON, *supra* note 228, at 37-41; Myers, *supra* note 198, at 9.

[817] Myers, *supra* note 198, at 9.

[818] SPECTRE, *supra* note 104, at 118.

[819] LeGrange, *FERC Direct Testimony, supra* note 176, at 28; *cf.* W. J. Williamson, *supra* note 315, at 6-7.

has been mentioned previously in a different context, but in this section the competitive risk is re-examined by classifying the form in which the competition arises.

a. Point to Point

On the crude oil side, Ajax was knocked out, after reviving from an earlier loss, by competition from Ozark and Mid-Valley.[820] There is side-by-side competition among 13 common carriers for the movement of West Texas Area product to the Texas Gulf Coast refining area.[821] Nine common carrier crude lines strive for the crude traffic between the East Texas producing region to the Texas Gulf Coast refining area.[822] There are ten pipeline companies vying for the carriage of both indigenous and imported crudes from St. James, Louisiana, to Toledo, Ohio.[823] Seaway and Texoma are "cheek-to-jowl" between Texas Gulf Coast ports and Cushing, Oklahoma, carrying both domestic and imported crudes and, since 1974, Explorer has been "batching" crude in competition with them.[824] Mobil, Shell, Texas, and Amoco pipelines are directly competitive for crude oil shipments from Texas and Oklahoma points to Illinois refineries.[825] Mention has been made in several contexts about the three-way struggle between Amoco (Service), Platte, and Arapahoe for the Wyoming/Rocky Mountain to Kansas City/St. Louis and Midwest refinery areas crude transportation business.[826] Gulf Refining Company's Northern Pipeline 10-inch system from Tulsa, Oklahoma, to Dublin, Indiana, was wiped out by Mid-Valley's more modern 20-22-inch line.[827] Several different witnesses in the TAPS proceeding noted the competitive posture of the eight TAPS owners in competing for shipments from the Prudhoe Bay field to Valdez,[828] which could

---

[820] LIVINGSTON, *supra* note 228, at 19.
[821] See text at notes 402-406, *supra* LeGrange, *FERC Direct Testimony, supra* note 176, at 23-24; HOWREY & SIMON, *supra* note 3, at 19; LIVINGSTON, *supra* note 228, at 63.
[822] See text at note 403, *supra;* LeGrange, *FERC Direct Testimony, supra* note 176, at 23; HOWREY & SIMON, *supra* note 3, at 19-20; see also MARATHON, *supra* note 202, at 15.
[823] See text at note 404, *supra*; LeGrange, *FERC Direct Testimony, supra* note 176, at 23-24; LIVINGSTON, *supra* note 228, at 63; *cf.* HOWREY & SIMON, *supra* note 3, at 20; MARATHON, *supra* note 202, at 15-16.
[824] MARATHON, *supra* note 202, at 16; HOWREY & SIMON, *supra* note 3, at 20-21; *cf.* LeGrange, *FERC Direct Testimony, supra* note 176, at 24.
[825] MARATHON, *supra* note 202, at 16-17; HOWREY & SIMON, *supra* note 3, at 21.
[826] See text at notes 168 and 406, *supra*; LIVINGSTON, *supra* note 228, at 62.
[827] PIPELINE PRIMER, *supra* note 331, at 15.
[828] *TAPS Hearings Transcript, supra* note 785, Vol. 12 at 2187 (cross-examination by Department of Justice Attorney Donald Kaplan of witness Raymond B. Gary); Ryan, *TAPS Rebuttal Testimony, supra* note 725, at 49.

break out into an open tariff war,[829] in about eight or nine years from now when the estimated production of the Prudhoe Bay field begins to decline.[830]

Products lines also are exposed to the risk of point-to-point competition. Colonial and Plantation run virtually side-by-side from Baton Rouge, Louisiana, to Washington, D.C., and until Plantation modernized its line, the effect on its traffic was traumatic.[831] ARCO, Laurel, Buckeye, Mobil and Sun are in direct competition for movements between the Philadelphia/Linden, N.J., refinery area to the Pittsburgh/Buffalo markets.[832] Evangeline Products System, constructed in 1953 by The Texas Pipe Line Co., Gulf, and Sinclair, to transport refined products from the Port Arthur, Texas/Lake Charles, Louisiana, refinery area to a Baton Rouge connection with Plantation, was rendered uncompetitive by the Colonial system which diverted Evangeline's entire 170,000 barrels per day throughput, causing it to be inactivated and converted to natural gas use in 1965.[833] Explorer and Texas Eastern compete directly for refined products traffic from the Houston refinery area to the Chicago marketing area, although, or perhaps because, their routes to Chicago differ.[834]

b. Area

In addition to point-to-point competition, there is competition between different producing areas (via different pipelines) to a single refinery area;[835] for example, refineries in the greater Chicago area are accessed by crude from virtually every producing area in the lower 48 states except California.[836] Conversely, from most producing regions, crude pipelines extend to a number of refining areas;[837] for example, crude oil from West Texas moves by pipeline to practically every refining center in the lower 48 states except the West Coast[838] and the Rocky Mountains.[839] Crude

---

[829] Cooper, *Direct TAPS Testimony*, *supra* note 718, at 24.
[830] Whitehouse, *Direct TAPS Testimony*, *supra* note 798, at 6-7.
[831] See text at notes 424-426 and 529, *supra*.
[832] See text at note 532; PIPELINE PRIMER, *supra* note 331, at 22; *cf.* MARATHON, *supra* note 202, at 20; HOWREY & SIMON, *supra* note 3, at 23.
[833] Information on Evangeline furnished by Texaco, Inc., parent of line's operator.
[834] See text at notes 434 and 531; PIPELINE PRIMER, *supra* note 331, at 22; LIVINGSTON, *supra* note 228, at 26.
[835] MARATHON, *supra* note 202, at 14-15. See map of crude oil movements included herein as Appendix G.
[836] *Id.*; LeGrange, *FERC Direct Testimony*, *supra* note 176, at 25; HOWREY & SIMON, *supra* note 3, at 21; PIPELINE PRIMER, *supra* note 331, at 22.
[837] PIPELINE PRIMER, *supra* note 331, at 22; HOWREY & SIMON, *supra* note 3, at 19.
[838] LeGrange, *FERC Direct Testimony*, *supra* note 176, at 25; HOWREY & SIMON, *supra* note 3, at 21.
[839] PIPELINE PRIMER, *supra* note 331, at 22.

movements to Toledo, Ohio, from St. James, Louisiana, via four routes with permutations, respectively, of 10, 5, 8, and 10 pipeline companies,[840] and from Lake Charles, Louisiana, Nederland, Texas, and Freeport, Texas, have been previously described.[841] The Texas Gulf Coast refinery area is served by 13 lines from the West Texas producing region and 9 from the East Texas Field[842] plus some local gathering lines.

Prudhoe Bay crude enters the West Coast refining centers above Seattle, Washington, in the San Francisco Bay area and Los Angeles, California, in competition with California crudes, Canadian crude and foreign production. The excess will enter the Midwest refinery area (Petroleum Administration for Defense District II, known as "PAD II") or PAD III[843] via U.S. flag tankers through the Panama Canal, in competition with domestic and imported crude through the pipelines described above.

Many market areas are supplied by a number of products lines with differing origins. For example, the Chicago area is served by Amoco, Explorer, ARCO, Texas Eastern, Shell, Phillips, Williams,[844] Wabash (now owned by Marathon) and Buckeye. Smaller and more remote markets naturally have fewer pipelines bringing product in to supplement local refineries. Denver, for example, is served by Phillips, Diamond-Shamrock, Chase, Wyco, and Medicine Bow. Even the relatively small community of Spokane, Washington, is served by the Yellowstone pipeline from Billings, Montana, and the Chevron pipeline originating in Salt Lake City, Utah.[845] A graphic demonstration of the area competition is shown by Appendices G and H and the schematic, non-scale Appendices J and K which show the entry and exit from major Midwest refining areas by multiple crude lines and volumes of products moving in and out of Midwest destinations by multiple products lines.

c. En Route to Other Markets

Another variety of pipeline competition arises when another pipeline of comparable efficiency is constructed to serve basically a different market from the first line but its location is such that it passes through the market, or one of the markets, served by the first line.[846] The Williams Pipe Line Com-

---

[840] See text immediately preceding and immediately following note 418, and text at 404, *supra*.

[841] See text at note 405, *supra*.

[842] LeGrange, *FERC Direct Testimony, supra* note 176, at 23; HOWREY & SIMON, *supra* note 3, at 19-20. See text at notes 402-403, *supra*.

[843] Consisting of the States of New Mexico, Texas, Louisiana, Arkansas, Mississippi and Alabama.

[844] HOWREY & SIMON, *supra* note 3, at 24.

[845] *Id.*

[846] MARATHON, *supra* note 202, at 17.

pany's main line runs from Tulsa, Oklahoma, via Paola, Kansas, to Kansas City, thence to Des Moines, Iowa on to Minneapolis-St. Paul, with branch lines from Osceola, Iowa, to Omaha, Nebraska, and another from Des Moines to Chicago.

In 1977, Vernon Jones, President of Williams, testified before the Interstate Commerce Commission that Williams faced competition from ARCO, Explorer, Kaneb, Continental, Phillips, Badger, Mobil, Amoco, National Coop. Refinery Association, and Koch.[847] ARCO has two products lines coming into the Kansas City marketing area. Its competition with Williams could be classified as "area" competition in the sense that Kansas City is one of Williams' main drop-off points but it could also be placed in the "passing through" class because Williams is en route to Minneapolis-St. Paul. Explorer's origin (Texas and Louisiana Gulf Coast) is different from Williams', but from Tulsa to the Chicago area there is competition with Williams' movements from Tulsa to Chicago. Kaneb, originally created to serve seven land-locked refineries in southern Kansas, originates at Arkansas City, Kansas, Proceeds through the Wichita, Kansas, area, and competes with Williams' line in the Lincoln, Nebraska, marketing area, again in the Sioux Falls, South Dakota, area, and finally in the Fargo, North Dakota, area.

Continental also originates near the Tulsa area (Ponca City), and has a branch competing point-to-point against Williams up to Wichita, Kansas, and also (via the former Cherokee line) serves the St. Louis market, which competes with Williams' line from Tulsa to St. Louis. Phillips' line, which originates in Borger, in the Texas Panhandle, competes with Williams in the Wichita, Kansas, Kansas City, St. Louis, and Chicago areas. Mobil's line originating at Augusta, Kansas, (near Wichita) virtually competes point-to-point with Williams' line at Kansas City, Omaha, Nebraska, and Sioux Falls, South Dakota. Amoco competes with Williams in the Omaha, Cedar Rapids, and Dubuque, Iowa, and Minneapolis/St. Paul markets. National Coop Refinery Association competes with Williams at Omaha, Nebraska, with its line from McPherson, Kansas, to Council Bluffs, Iowa. Koch's line runs from its refinery in Minneapolis to Stevens Point, bringing products into several Wisconsin markets in competition with shippers over Williams' line, as does Badger, which also has a delivery point in Des Plains, Illinois, (Chicago's O'Hare airport).

---

[847] Statement of Vernon T. Jones, President of Williams Pipe Line Co., Before the I.C.C. in Ex Parte No. 308 (now F.E.R.C. Dkt. RM78-2) 20 (May 27, 1977) [hereinafter cited as Jones, *FERC Direct Testimony*]. The last three lines are private carriers.

d. Intermarket

Competition is not limited to crude oil line versus crude oil lines or products lines versus products lines. The more important refined products markets are supplied partially by refineries located in proximity to such markets, which receive their crude oil by low cost pipelines (or tankers) and partially by refineries located in proximity to crude oil production or on a seaboard receiving foreign crude by tankers and penetrating these markets by low cost products lines.[848] Therefore, a crude oil pipeline supplying a refinery near one of these markets competes on a comparable cost basis with refined products lines from distant refineries.[849]

In Chapter I, mention was made of the pioneer products lines, such as Tuscarora, Sun, and ARCO, penetrating inland markets from seaboard refineries enjoying low cost crude,[850] and the Mid-Continent refiners, such as Phillips, Great Lakes' owners, and Champlin, competing in the Midwest markets with market oriented refineries.[851] As the lines became larger, the competition became even more vigorous. Explorer, a products line transporting refined products from Texas/Louisiana Gulf refineries destined for the Chicago market, ran head on into two newly-constructed large diameter crude lines carrying crude from Texas Gulf Coast origins (Seaway from Freeport, Texas, and Texoma from Nederland, Texas) to refineries which competed in Explorer's market destinations, in addition to direct point-to-point refined products line competition from Texas Eastern.[852]

DeChazeau and Kahn observed that pipelines had become so essential to the cheap long-distance overland movement of oil that they created traffic rather than displacing other modes of transportation [that was in 1959—the larger diameter pipelines have eroded the traffic of some other modes since they made their comment] but more apt, and equally true today, was their observation that crude oil pipelines were more significantly locked in competition with products lines than with other forms of transportation.[853]

e. New Conformations of Existing Facilities

There have been several instances where existing facilities have been reconformed and have become competitive influences in their new usage. One instance that come to mind is the Tuscarora Pipe Line, which had reached the end of its economic usefulness as a crude line, but upon reversal

---

[848] PIPELINE PRIMER, *supra* note 331, at 22-23; HOWREY & SIMON, *supra* note 3, at 164.
[849] MARATHON, *supra* note 202, at 8; HOWREY & SIMON, *supra* note 3, at 164.
[850] See text at notes 122 to 132, *supra*.
[851] See text at notes 137 to 145, *supra*.
[852] See text at notes 433 and 434, *supra*; LIVINGSTON, *supra* note 228, at 26.
[853] DeCHAZEAU & KAHN, *supra* note 30, at 334.

Economic Characteristics   **131**

of flow and conversion to products service triggered a wave of products lines from Eastern Seaboard refineries to markets as far west as Cleveland and northwest to Buffalo and Rochester, New York.[854] Although the primary thrust of the new lines had been to lower the cost of transporting products which had previously been moved by rail, the new levels, which ranged from one-third to one-tenth of rail, brought the new products lines in competition with crude lines serving refineries in destination market areas.

Ajax, which has been discussed previously in the contexts of loss of crude throughput due to the newly discovered Illinois field[855] and a second drastic reduction in throughput caused by diversion of traffic from its feeder connection and competition from other crude producing areas to its destination area by new large diameter crude oil lines,[856] itself became a new source of competition when it was converted to products service.[857]

Shell Oil converted its eight-inch crude oil line from Wood River, Illinois, to its East Chicago, Indiana, refinery into a products line, thereby enabling it to save the cost of modernizing the refinery, turning it into a distribution terminal and expanding the capacity of its Wood River refinery and the variety of its outturn, achieving both economies of scale and flexibility. This was accomplished between 1938 and 1940. It created an additional benefit for the nation because it enabled the Wood River refinery to produce substantial quantities of aviation gasoline for use in World War II.[858] In 1953, the line was replaced with a 14-inch pipeline and its competition was felt not only in the Chicago area[859] but, through its connection with Wolverine, in the Toledo and Detroit markets.

A final example is the "Little Big Inch," constructed during World War II to carry petroleum products from Beaumont, Texas, to Linden, New Jersey; converted to natural gas service after the war but reconformed to products service by Texas Eastern, which increased its competitive force by constructing an extension from Seymour, Indiana, to Chicago,[860] and the Northeast.

---

[854] See text at notes 122 to 132, *supra*.
[855] See text at notes 770 to 773, *supra*.
[856] See text at notes 775-776, *supra*.
[857] See text following note 846, *supra*.
[858] K. BEATON, ENTERPRISE IN OIL 444-448 (1957).
[859] See text at note 844, *supra*.
[860] See text at note 434, *supra*.

f. From Other Forms of Competition

Oil pipelines must compete not only with other pipelines but also many pipelines must face significant competition from alternate modes of transportation,[861] such as tankers, barges, trucks and railroads.[862] Each mode has its own peculiar advantages and it fulfills needs for which it is unique with respect to the others,[863] thus attesting to the wisdom of the National Transportation Policy adopted in 1940, which calls for a balanced transportation system.[864]

In the movement of crude oil to refineries, where the movement is long distance, overland, and in large quantities, pipelines are the favored mode.[865] On the products side, pipelines generally handle the lighter, less viscous fluids, such as gasolines and distillates, whereas the heavier petroleum liquids and solids, such as residual fuel oil, asphalt, and coke, generally are the province of the other bulk carriers.[866] All modes but trucks can make long haul shipments economically,[867] but in the short haul, small volume area, the flexibility of trucks gives them the competitive edge.[868] This is demonstrated by the fact that motor carriers account for the largest share of total tonnage but are the lowest share, except for railroads, when it comes to ton-miles.[869]

---

[861] MANCKE, *Oil Industry Competition, supra* note 674, at 45; HOWREY & SIMON, *supra* note 3, at 41; Swenson, *S.1167 Testimony, supra* note 755, at 704; MARATHON, *supra* note 202, at 21.

[862] HOWREY & SIMON, *supra* note 3, at 41; Tierney, *TAPS Rebuttal Testimony, supra* note 175, at 17-18; *Rm 78-2 Hearings Transcript, supra* note 754, Vol. 12 at 1927 (Oral argument of Jack Vickery on behalf of Belle Fourche Pipeline Company, an independent crude oil gatherer); *cf. Id.* at 1829 (oral argument of Darell Behring on behalf of Okie Pipe Line Company, an independent crude oil gathering system—truck and barges); Att'y Gen., *Second Report, supra* note 111, at 72 (barge and tankers).

[863] HOWREY & SIMON, *supra* note 3, at 163.

[864] See text at note 692, *supra*.

[865] Tierney, *TAPS Rebuttal Testimony, supra* note 175, at 18; HOWREY & SIMON, *supra* note 3, at 162.

[866] MARATHON, *supra* note 202, at 58.

[867] *Id.*

[868] *Id.* at 59; PIPELINE TRANSPORTATION, *supra* note 162, at 32; *cf.* HOWREY & SIMON, *supra* note 3, at 162-163; LIVINGSTON, *supra* note 228, at 48; Trucks also handle temporary high density movements. PIPELINE TRANSPORTATION, *supra* note 162, at 32.

[869] In 1976, motor carriers carried 36.41% (next highest competitor was pipelines at 35.58%) of domestic petroleum transportation in tons, but next to last in ton-miles with 5.8% (next lowest competitor was also pipelines with 40.8%). Tierney, *TAPS Rebuttal Testimony, supra* note 175, at Exhibit K, reproduced herein as Appendix L. For a description of truck systems and operations, see NET-I, *supra* note 199, at 240-246; for economics and delivery patterns of trucks, see *Id.* at 246-250.

Economic Characteristics

Barging is a highly competitive form of domestic transportation.[870] Looking at the bar chart in Appendix A, it can be seen that while the lowest cost per 100 barrel miles in the range of barge costs (4¢) is not as low as the cheapest point in the pipeline range of costs (2½¢), barges are competitive with pipelines over the range 4¢ to 12¢ per hundred barrel miles.[871]

Barges are able to make use of many river systems, including the Mississippi River as far north as Minneapolis, the Illinois River to Chicago, the Ohio River to Pittsburgh, the Missouri River into Nebraska, the Arkansas River into Oklahoma, the Columbia and Snake rivers into Washington, Oregon, and Idaho; the Tennessee River into Chattanooga and Knoxville, the Warrior and Alabama rivers throughout Alabama and the Hudson River (and canals) into upstate New York and Vermont.[872] Barge operators are aggressively attacking the problem of costs. They have managed to reduce the average charge for barge transportation by 25 percent during recent years.[873] This has been accomplished by keeping their fleets modern, increasing the size of the unit movement (the individual flotilla of barges) and scheduling their crews so as to maintain as near continuous operation as possible.[874]

Large foreign-flag, oceangoing tankers are the lowest cost means of transporting petroleum.[875] Reference to the cost bar charts in Appendix A and NET-I, clearly shows the advantage. However the 1¢ per 100 barrel mile minimum applies to the large foreign flag vessels (VLCC's and ULCC's) which cannot be accommodated in United States ports because of their drafts. When LOOP is in operation, VLCC's will be able to proceed directly from their foreign loading ports to discharge at LOOP. The highest rates apply to United States flag vessels in coastwise trade.[876] This is due to the Jones

---

[870] PIPELINE TRANSPORTATION, *supra* note 162, at 34; HOWREY & SIMON, *supra* note 3, at 26; *cf.* PIPELINE PRIMER, *supra* note 331, at 23. In 1972, Meyer Kopolow, President of Marine Petroleum Co., testifying in the Smith-Conte *Hearings* before the House Small Business Committee against Explorer, conceded that [vis-a-vis Explorer] barge rates were "favorable." *H.Res.5 Subcomm Hearings, supra* note 234, at 107.

[871] Appendix A. These barge costs do not include construction, operation and maintenance costs. PIPELINE TRANSPORTATION, *supra* note 162, at 34. A similar chart supporting the same conclusion can be found in NET-I, *supra* note 199, at 214. Appendix A represents 1975 figures. LeGrange, *FERC Direct Testimony, supra* note 176, at 13. For more recent developments in barge rates, see note 873, *infra*.

[872] MARATHON, *supra* note 202, at 22; HOWREY & SIMON, *supra* note 3, at 26; *cf.* PIPELINE PRIMER, *supra* note 331, at 23.

[873] NET-I, *supra* note 199, at 234: For example the 4.0 cents per 100 barrel mile lower range figure cited above has been shaved to 3.0 cents on particular high speed tows engaged in continuous movement, contracted on a long-term basis, or shipper-owned barges. PIPELINE TRANSPORTATION, *supra* note 162, at 34.

[874] NET-I, *supra* note 199, at 234-235. For details of barge operations see *Id.* at 217-229; for details on barge economics, see *Id.* 234-237.

[875] PIPELINE TRANSPORTATION, *supra* note 162, at 35.

[876] *Id.*

Act, which restricts coastwise trade to U.S. flag vessels, and the current labor and operating costs, which are higher than those of the foreign flag operators.[877]

Even with these disadvantages, ocean vessels have been competitive at times with certain pipelines such as Colonial,[878] and, at the time Colonial was being built, were so competitive that in order for Colonial to be a viable proposition it had to build the largest products line in the world.[879] Indeed, among Colonial's important business risks when it was being formed was that there could either be foreign crude imports by large sized tankers at destination either to, or in addition to, newly constructed "grass-roots" large scale refineries,[880] substantial imports of products directly into the New York harbor area,[881] or that the current coastwise tankers would find ways of reducing their costs and reduce their tariffs and thereby divert substantial traffic from Colonial's line.[882] Dr. John Ryan, in his rebuttal testimony in the TAPS proceeding before FERC, noted the potential competition to the TAPS line from oceangoing ice-breaking tankers.[883] The design, technology and economics of operations of oceangoing tankers are described in the Report by the Congressional Research Service prepared for the Committee on Energy and Natural Resources and on Commerce, Science and Transportation of the United States Senate.[884]

Rail transportation costs are competitive only in a very narrow range of pipeline costs. Their competition with pipelines is in the area of low volume products service under special conditions.[885] However, in areas where

---

[877] NET-I, *supra* note 199, at 237.

[878] PIPELINE TRANSPORTATION, *supra* note 162, at 35; HOWREY & SIMON, *supra* note 3, at 26; MARATHON, *supra* note 202, at 21. Exxon used tankers consistently that way until it was forced to divert its Jones Act tankers to the Alaskan-West Coast run.

[879] *Consumer Energy Act Hearings, supra* note 154, at 614 (Statement of Jack Vickrey); *cf.* PIPELINE PRIMER, *supra* note 331, at 23.

[880] *Cf. Petroleum Industry Hearings, supra* note 192, at 314 (Testimony of E. P. Hardin), *Id.* at 265 (answer by Charles J. Waidelich to question by Senator Bayh).

[881] *S.1167 Hearings, supra* note 432, pt. 8 at 5946 (Answer of witness Richard C. Hulbert, President of Kaneb Pipe Line Co., to Senator Tunney); *Petroleum Industry Hearings, supra* note 192, at 265 (Answer of Charles J. Waidelich to question by Senator Bayh).

[882] This was one risk which did not mature. Colonial's initial tariffs were designed to compete with the then T-2 tanker rates of 35¢ per barrel for the long haul from the Gulf Coast to the New York Harbor. *H.Res.5 Subcomm Hearings, supra* note 234, at A73 (supplemental submission by Jack Vickrey to Senator Proxmire, Chairman of the Subcommittee on Priorities and Economy in Government, Joint Economic Committee). Since that time Colonial's rates have been reduced and the coastwise tanker rates have gone up.

[883] Ryan, *TAPS Rebuttal Testimony, supra* note 724, at 49-52.

[884] NET-I, *supra* note 199, at 229-234 (design and technology), 237-240 (economics).

[885] PIPELINE TRANSPORTATION, *supra* note 162, at 33; *cf.* LIVINGSTON, *supra* note 228, at 46-49. For example, rail rates on propane in 30,000 gallon cars moving long distances are about 17¢ per 100 barrel miles. Pipeline rates are lower but the tank car can go right to the customer's bulk plant while the pipeline shipper must pay for transportation between the pipeline terminal and the bulk plant. PIPELINE TRANSPORTATION, *supra* note 162, at 33.

pipelines cannot be justified, rail transportation plays a part, being more economical than tank trucks on movements above 200 miles because of the larger volume capabilities.[886] Rail transportation of petroleum has declined substantially since the late 1950's, although innovations such as the Tank Train—which consists of 20 to 80 specially built rail tank cars linked by a unique flexible hose, which allows the loading and unloading of a string of cars through a single connection at flow rates up to 3,000 gallons per minute—holds promise of boosting rail's share of total petroleum traffic in the future.[887] A graphic picture of the relative quantities of petroleum products carried by each mode of domestic transportation over a period from 1968 to 1976 (measured in tons) and 1972 to 1976 (measured in billions of ton miles) can be found in Appendix L.

9. New Frontier-High Cost Locations Bring Greater Risk

While oil pipelines have always involved risk, the degree of risk in today's pipelines has risen substantially[888] because most new oil is being discovered currently in new frontier and/or high cost locations.[889] Working in a new hostile environment makes it extremely difficult,[890] if not impossible, to forecast accurately at an early stage of project development what will be the ultimate cost of construction.[891] In projects the size of TAPS, the risks become prodigious;[892] the most serious risk of all was that the line would not be completed.[893] The hazards were considered to be so great that four different financial experts[894] advised officials of the State of Alaska that the line

---

[886] PIPELINE TRANSPORTATION, *supra* note 162, at 33.

[887] Tierney, *TAPS Rebuttal Testimony, supra* note 175, at 18; NET-I, *supra* note 199, at 255. According to GATX, a tank train, consisting of 90 cars of 23,150-gallon capacity each, in five 18-car strings could load or unload 49,607 barrels in less than six hours and may be competitive with pipelines of capacities up to 350,000 barrels per day, depending on each individual situation. *Id.* at 255. For description of rail systems and quantities see NET-I, *supra* note 199, at 250-256; for rail's economic and delivery pattern, see *Id.* at 256-259.

[888] Gary, *FERC Direct Testimony, supra* note 613, at 17-18.

[889] *Id.* at 9.

[890] *Petroleum Industry Hearings, supra* note 192, at 333 (Statement of W. T. Slick, Jr.).

[891] Wooddy, *TAPS Rebuttal Testimony, supra* note 724 at 6; *cf.* Gary, *TAPS Rebuttal Testimony, supra* note 726, at 10. Among the factors that made the ultimate cost inestimable were the uniqueness of the project, inflation, remoteness of location, uncertainty regarding labor contracts, critical material delivery schedules, litigation and pervasive government regulation. Wooddy, *TAPS Rebuttal Testimony, supra* note 724, at 6.

[892] *Petroleum Industry Hearings, supra* note 192, at 1962 (testimony of Raymond B. Gary).

[893] *Id.*; Gary, *TAPS Rebuttal Testimony, supra* note 726, at 10. See text at, and note 809, *supra*.

[894] Kuhn, Loeb & Co. letter to then Governor William Egan on Dec. 3, 1971; Salomon Brothers letter to Governor Egan on same date; Bank of America letter of Dec. 15, 1971, to Eric E. Wohlforth, then Commissioner of Alaska's Department of Revenue, and Merrill Lynch, Pierce, Fenner & Smith letter of December 3, 1971, to Governor Egan. Significantly, the State of Alaska not only turned away from 100 percent ownership, but also it did not even pick up its 12.5 percent fair share of the risk.

could not be financed by the State itself but would require the unconditional guarantee of revenues by the oil companies.[895]

In preparing the loan agreement under which TAPS was financed, the lenders negotiated the tightest agreement possible, including, in Sohio Pipe Line Company's case, negative covenants preventing any disposal of Prudhoe Bay reserves, or of shares in the subsidiary pipeline company, and required the owner companies to prevent any acceleration of any other indebtedness they had.[896] The lenders required completion agreements, throughput and deficiency agreements[897] and guarantees which were drawn so to give them recourse against the project in addition to the going credit of the oil companies themselves.[898]

The risk thus placed squarely upon the oil company owners of TAPS, by reason of the nature of the investment itself, were greatly increased over that which had always been present in pipeline ownership because: (1) the huge cost increased the scale of the investment in the pipeline relative to their respective total wealths inasmuch as it concentrated their exposure to loss in exactly the same way that diversification is understood to reduce exposure to risk (in homey parlance, they were putting a lot of their eggs in one basket),[899] and (2) the escalation of construction cost raised the total cost of transporting Alaskan oil by pipeline, and the higher the cost, the greater became the risk that the pipeline would turn out to be economically obsolete before the investment made in it was recovered.[900]

Although the Alaskan North Slope discoveries are unique in many ways, they are believed to presage the future trend, either in Alaska or in deeper offshore waters.[901] Unlike the crude trunk lines coming out of West Texas, which maintained reasonably full capacity over a number of years because of the discovery of additional fields on the route, pipelines transporting crude oil produced from the Outer Continental Shelf to adjoining coastal states face production declines so steep that only during the first 5 or 6 years can they expect to achieve full capacity and thereafter throughput falls off very rapidly.[902]

---

[895] Gary, *TAPS Rebuttal Testimony, supra* note 726, at 18-21.
[896] *Petroleum Industry Hearings, supra* note 192, at 1930 (Statement of Peter A. Bator).
[897] *Id.* at 1931; *Id.* at 1962 (Statement of Raymond B. Gary).
[898] *Petroleum Industry Hearings, supra* note 192, at 1962 (Statement of Raymond B. Gary).
[899] *TAPS Hearings Transcript, supra* note 785, at 9785 (Department of Justice's witness, Herman G. Roseman: ". . . when the investment is all in one bundle and the return is all in one bundle, . . . there is an element of extra risk involved . . . ."); Ryan, *TAPS Rebuttal Testimony, supra* note 725, at 45.
[900] Solomon, *TAPS Rebuttal Testimony, supra* note 621, at 30.
[901] *Petroleum Industry Hearings, supra* note 192, at 333 (Statement of W. T. Slick, Jr.).
[902] *Id.* at 312 (Statement of E. P. Hardin).

Turning back to TAPS for the moment to expand on the discussion of the increased risk, there was the hostile climate which required special metallurgy, decreased the productivity of workers, affected the transportation of men, materials and equipment,[903] and required special construction techniques.[904] As a result, a risk was created that untested design technology, which was required to solve the unique problems, would prove to be inadequate under actual operating conditions.[905] The combination of environmental, weather and terrain conditions required not only new design, construction techniques and prototype equipment to be developed,[906] but because of the combination of hot oil in the pipeline and the effect of heat on the permafrost, some 423 of the 798 miles of line had to be constructed aboveground, requiring Vertical Support Members (VSM's), which in turn required holes to be dug with specially designed drills, which differed according to the soil. Soil conditions changed so dramatically that drills had to be changed constantly. The holes had to be located precisely by surveyors and the depths had to be determined by a soil engineer. The upright steel members were held in place by sand-water mixtures which had to harden, and if the soil conditions were such that it was feared that the sand would soften, a heat exchange device was installed within the steel pile.[907] This meant that special equipment, large construction crews and new technology were required.[908] The problem was aggravated greatly by the fact that, due to the unpredictable variety of soil conditions, plans often had to be changed on the spot and line that had been planned to be laid beneath the ground had to be changed to aboveground, and vice versa, thereby not only interrupting the construction sequence ("cadence"), but frequently calling for on-the-spot design.[909]

Even without a change in mode from below ground to above ground, the geotechnical conditions often required design change. There were five or six different types of VSM's (thermal, slurry, adfreeze, friction, end bearing, thermal end bearing, etc.) and until the VSM hole was drilled and the soil

---

[903] Gary, *Direct TAPS Testimony, supra* note 634, at 27. For example, in the summer of 1975 a sea lift of major materials and equipment to Prudhoe bay was jeopardized by severe weather and ice conditions. Certain barges carrying equipment had to be rerouted to Southern Alaska ports and then transshipped by land, causing extra cost and delay. *Id.* at 29.

[904] Patton, *Direct TAPS Testimony, supra* note 726, at 54.

[905] Cooper, *Direct TAPS Testimony, supra* note 718, at 27; *cf.* Gary, *Direct TAPS Testimony, supra* note 634, at 27.

[906] Gary, *Direct TAPS Testimony, supra* note 634, at 27.

[907] Moolin, *Direct TAPS Testimony, supra* note 726, at 21.

[908] Gary, *Direct TAPS Testimony, supra* note 634, at 27-28.

[909] Moolin, *Direct TAPS Testimony, supra* note 726, at 22-23.

analysis was made, the length of the member and the type of setting material was not known. VSM's no more than 15 feet apart would require radically different designs.[910]

Even the line laid underground had special problems in unstable permafrost requiring parts of the line to be "refrigerated" by use of refrigeration pipes installed next to the line and brine pumped through these pipes similar to an ice rink.[911] The greatest problems appear to have been: *First*, there was a total lack of any infrastructure and the line was located in a remote, virtually uninhabited area. This meant starting from the beginning, building a 70-mile road north of Livengood to the Yukon River and then a 360-mile road from the Yukon River to the North Slope. Airfields had to be constructed, communications facilities, utilities, warehouse facilities, supply dumps, material yards, transportation facilities and camps for employees all had to be self-provided.[912] *Second*, it was impossible to set up and maintain an orderly sequence of work. There were constant costly "cadence" breaks, which involved shifting engineers, specialists and crews around to solve problems on the spot, which resulted in duplication of effort, facilities and transportation. *Third*, the construction was complicated by the pervasive extent and the potentially arbitrary nature of governmental regulation which was ever present.[913] Every initiation of construction of any segment of the pipeline had to be authorized by a "Notice to Proceed" granted by the "Authorized Officer" (Federal) and the "Pipeline Coordinator" (Alaska) upon whatever terms they (or "he," as they sometimes differed) deemed to be required.[914] *Fourth*, the very size of the project raised substantial coordination problems.[915] *Fifth*, the uniqueness of everything from above ground supports, below ground refrigeration, hydrostatic testing in terribly cold ground or permafrost, construction through the Atigun and Thompson passes, Keystone Canyon and river crossings during winter environmental "windows" (when the fish could not be disturbed) taken together made the

---

[910] *Id.* at 23.

[911] *Id.* at 36.

[912] *Id.* at 8-10.

[913] This item was mentioned by at least four witnesses: Wooddy, *TAPS Rebuttal Testimony, supra* note 724, at 6, 7-9; Patton, *Direct TAPS Testimony, supra* note 726, *passim*; Moolin, *Direct TAPS Testimony, supra* note 726, at 2, 5; Donaldson, *Direct TAPS Testimony, supra* note 807, at 10-11. In negotiating the Agreements for Federal Right-of Way, Donaldson had to deal with 26 different agencies. *Id.* at 15-16. Patton, in his rebuttal testimony, Patton, *TAPS Rebuttal Testimony, supra* note 726, at 16, stated that governmental regulation was perhaps the single most significant uncertainty confronting the owners in late 1973 and 1974. He devoted the next 47 pages to elucidating the point. On pages 23-26 he named "more than 50 governmental agencies which were 'getting into the act.'"

[914] See note 805, *supra*.

[915] Patton, *Direct TAPS Testimony, supra* note 726, at 53.

Economic Characteristics 139

task incredibly difficult and costly.[916] In short, to use witness Moolin's words: "TAPS was the largest private construction project in history. It was constructed along an 800-mile front. Because of the numerous geotechnical, climatic, and logistic problems involved, it was more akin to a civil construction project, which is concerned principally with the movement of earth, than to traditional pipeline construction. Moreover, unlike a building, which is constructed along traditional lines regardless of where built once it rises above the foundation level, the construction of TAPS was all 'foundation-type,' with the uncertainties inherent in such construction, magnified by little previous experience with permafrost and arctic soils."[917]

G. Uncertainties

1. Governmental Action

a. In-Place Regulation — Change in Methodology

Pipeline revenues are affected not only by utilization (throughput) but also by the regulatory treatment of tariffs. While all regulated firms face uncertainties associated with fluctuations in regulatory policy,[918] the only rational basis for pipeline owners, such as TAPS's, to plan their economics was the current valuation methodology which had been employed by the ICC for over 30 years. Being subject to the limitations of the *Atlantic Refining Company* Consent Decree, TAPS's owners, just as they had been doing ever since the decree in 1941, assumed a tariff structure based on this methodology. Although in the TAPS proceedings protestants' witness Dunn described this approach as being "naive,"[919] TAPS's owners and other pipeline owners had filed literally thousands of tariffs over a 30-year period on that basis without challenge by shippers or the regulatory body.[920] The ICC precedents in the area,[921] the language of the Valuation Act,[922] the unbroken chain of Valuation Reports from *Ajax Pipe Line Corporation*[923] to the date of construction, the incorporation of ICC valuation methodology by the Consent

---

[916] Moolin, *Direct TAPS Testimony, supra* note 726, *passim.*
[917] *Id.* at 4.
[918] Gary, *TAPS Rebuttal Testimony, supra* note 726, at 15.
[919] *TAPS Hearings Transcript, supra* note 785, at 10, 719 (Testimony of John C. Dunn).
[920] Wooddy, *TAPS Rebuttal Testimony, supra* note 724, at 17.
[921] Reduced Pipe Line Rates and Gathering Charges, 243 I.C.C. 115 (1940), 272 I.C.C. 375 (1948); Petroleum Rail Shippers' Ass'n v. Alton & Southern R.R., 243 I.C.C. 589 (1941); and Minnelusa Oil Corp. v. Continental Pipe Line Co., 258 I.C.C. 41 (1944).
[922] 37 Stat. 701 (1913), 49 U.S.C. § 19a (1976).
[923] 50 I.C.C. Val. Rep. 1 (1949).

Decree,[924] reaffirmations by the ICC of its valuation rate base by the adoptions of several versions of the Uniform System of Accounts for Pipeline Companies, the latest version available when TAPS was being planned being 1970,[925] all pointed the same way; as TAPS rebuttal witness Wooddy stated, it would have been naive to file the TAPS tariffs on any other basis.[926]

Nor was TAPS alone in its view. The State of Alaska and potential debt securities investors followed the same line of reasoning.[927] It was only *after* the TAPS project was under construction that it became known that a small group of shippers were attempting to convince the ICC that it should adopt a different methodology of rate regulation.[928] The possibility of such a change then became a substantial risk to TAPS's owners and the parent oil companies and weighed very heavily on the attitudes of the investment and financial community.[929] Clearly the risk was increased by the dictum of the Court of Appeals for the District of Columbia in the *Farmers Union Central Exchange*[930] case which, despite inferring a congressional intent to permit a freer play of competitive forces among oil pipeline companies than in other common carrier industries and professing an unwillingness uncritically to import public utility concepts into oil pipeline regulation without taking note of the degree of regulation and of the nature of the business,[931] proceeded to speculate on just that basis. It relied upon the *Hope Natural Gas* case[932] to muse that the historical link between ratemaking theory and the valuation-computation authority was broken,[933] and ICC precedents cited in note 921 to be of "little additional guidance,"[934] and the fact that the Commission's accounting procedures were geared to the use of a valuation rate base was outmoded as an "artifact of a bygone era."[935]

While not expressly basing its reasoning on reliance upon the transfer of jurisdiction over oil pipeline rates from the ICC to FERC by virtue of the

---

[924] United States v. Atlantic Refining Co., Civil No. 14060, D.D.C., Dec. 23, 1941, Final Judgment, Paragraphs III(a) and (b).

[925] Uniform System of Accounts for Pipeline Companies, 337 I.C.C. 518, 523 (1970).

[926] Wooddy, *TAPS Rebuttal Testimony, supra* note 724, at 17.

[927] *Id.* at 18.

[928] *Id.*

[929] Gary, *TAPS Rebuttal Testimony, supra* note 726, at 15-16; See Research Note by Kenneth Hollister, CFA, First Vice President, DEAN WITTER REYNOLDS INC., reproduced as Exhibit D to Statement of Jack M. Allen, Before F.E.R.C. Dkt. RM78-2 (Oct. 27, 1978) [hereinafter cited as Allen, *FERC Direct Testimony*].

[930] Farmers Union Central Exchange v. F.E.R.C., 584 F.2d 408 (D.C. Cir. 1978), *cert. den.*, 99 S.Ct. 596 (1979).

[931] *Id.* at 413.

[932] FPC v. Hope Natural Gas Co., 320 U.S. 591 (1944).

[933] Farmers Union Central Exchange v. F.E.R.C., 584 F.2d 408, 413 n.8 (D.C. Cir. 1978), *cert. den.*, 99 S.Ct. 596 (1979).

[934] *Id.* at 413.

[935] *Id.* at 418, referring back to 416.

Department of Energy Organization Act,[936] the court indicated its inclination toward FPC-type regulation by referring to the "fact [that] it was FERC (in its previous incarnation as the Federal Power Commission) that, by deviating from 'fair value' rate making, inspired the Supreme Court's holding that valuation is not the *sine qua non* of 'just and reasonable' ratemaking" in the *Hope Natural Gas* case.[937] Before sending the case to FERC to try its hand at fashioning a ratemaking method which would satisfy *Hope Natural Gas'* test (laid down in a case under the Natural Gas Act where the transmission company had "locked in" gas purchase and gas sales contracts) that investors would be enabled to cover operating expenses and capital costs without burdening consumers with exorbitant rates,[938] the Court declared itself "not persuaded by the Commission's conclusion that 'consistency and fairness' dictate resurrection of the 'fair value' method last used thirty years ago."[939]

If this was not enough to make the risk seem real, the Court of Appeals for the Fifth Circuit in *Mobil Alaska Pipeline Company v. United States*[940] bluntly stated in a dictum that "we do not accept the 1941 consent decree as a standard of reasonableness under the Interstate Commerce Act."[941] Unless

---

[936] Pub. L. No. 95-91, § 402(b), 91 Stat. 584 (1977), *effectuated*, Exec. Order No. 12609, 42 Fed. Reg. 46,267 (Sept. 15, 1977), *implemented*, 42 Fed. Reg. 55,534 (Oct. 17, 1977). The Court's unwillingness to make this an articulate premise was advised because Congress had full opportunity to mandate such a change in methodology when it enacted the legislation transferring jurisdiction to FERC but chose deliberately not to do so. To the contrary, when the Department of Energy (DOE) Organization Act was passed, Senator Ribicoff, who managed the bill, specifically assured Senator Hansen that the transition from the ICC to FERC would be accomplished with maximum consistency and continuity, and with minimum disruption for all concerned parties. 123 Cong. Rec. S13,282-S13,283 (daily ed. Aug. 2, 1977). It is difficult to conceive of anything that would be more disruptive than the application of utility controls to oil pipelines. The mere threat that this could happen already has put a chill on pipeline investors. Because Congress always has relied upon competition working in tandem with ICC methodology to protect consumers insofar as oil pipeline rates are concerned—and Section I B 2, "Anti-inflationary Rate Trends," see text at notes 188-197, *supra*, bears testimony to the wisdom of this reliance—it would seem that changes in this concept should be made by the Congress, not the regulatory agencies or the courts. *RM78-2 Hearings Transcript, supra* note 754, Vol. 12, at 1930-1931, 1955 (Oral Argument of Jack Vickrey on behalf of Belle Fourche Pipeline Company, an independent crude oil gatherer).

[937] Farmers Union Central Exchange v. F.E.R.C., 584 F.2d 408, 416 n.21 (D.C. Cir. 1978), *cert. den.*, 99 S.Ct. 596 (1979).

[938] FPC v. Hope Natural Gas Co. 320 U.S. 591, 603 (1944). The court did not discuss the Supreme Court's language, on page 603, that the return to the equity owner should be commensurate with returns in other enterprises having corresponding risks; and that return, moreover, should be sufficient to assure confidence in the financial integrity of the enterprise, so as to maintain its credit and to attract capital.

[939] Farmers Union Central Exchange v. F.E.R.C., 584 F.2d 408, 418 (D.C. Cir. 1978), *cert. den.*, 99 S.Ct. 596 (1979).

[940] 557 F.2d 775 (5th Cir. 1977), *aff'd sub. nom.*, Trans Alaska Pipeline Rate Cases, 436 U.S. 631 (1978).

[941] *Id.* at 786.

the testimony by the TAPS witnesses and the differences between the two modes of transportation discussed herein convince the FERC that the only commonality between oil pipeline and natural gas pipelines is that both transmit articles of commerce through pipelines,[942] and there is no such thing as a single pipeline industry that includes both the oil pipelines and the natural gas pipelines,[943] there appears to be a possibility that FERC might adopt the natural gas regulatory philosophy for use in regulating oil pipelines.[944]

According to testimony in the cases pending before FERC, adoption of this new methodology in oil pipeline regulation cases would not be considered desirable by independent producers,[945] especially the "stripper-well" operators,[946] who remember all too keenly the depressive effect of such regulation on independent producers' prices and cash flows in the natural gas area.[947] The financial community is concerned by the prospect of producers of petroleum products being subjected to "another act" with possibly similar results to that resulting from the 1960-1965 period of regulation which produced disastrous results.[948] Witnesses in the TAPS proceedings for the Department of Justice,[949] independent producers,[950] the investment community,[951] the leading transportation trade association,[952] and, of course,

---

[942] *RM78-2 Hearings Transcript, supra* note 754, Vol. 12, at 1926 (Oral argument by Jack Vickery on behalf of Belle Fourch Pipeline Company, an independent crude oil gatherer).

[943] W. J. Williamson, *supra* note 315, at 1-2. There is ample documentation on the important differences between oil pipelines and natural gas pipelines and other public utilities, which indicate the reason, and need, for a difference in approaching the regulatory function in their respective areas. *E.g.,* NET-III, *supra* note 199, at 136-140; Gary, *FERC Direct Testimony, supra* note 613, at 18-20; W. J. Williamson, *supra* note 315, at 1-2, 13-21; *Rm 78-2 Hearings Transcript, supra* note 754, Vol. 12 at 1928-1930 (Oral argument by Jack Vickrey on behalf of Belle Fourche Pipeline Company, an independent crude gatherer); Solomon, *FERC Direct Testimony, supra* note 682, at 12-13.

[944] NET-III, *supra* note 199, at 141; K. HOLLISTER, OIL PIPELINES - NEW METHOD OF REGULATION REARS ITS HEAD (Research Note, July 28, 1978), reproduced as Exhibit D in Allen, *FERC Direct Testimony, supra* note 929 [hereinafter cited as HOLLISTER].

[945] Allen, *FERC Direct Testimony, supra* note 929, at 8; " . . . any drastic changes in pipeline regulations at this time . . . will deter exploration and production activities, particularly by independents."

[946] *Id.* at 5. "If pipelines are regulated on a cost of service basis, [continued service to stripper well producers] would likely cease and producers would be forced to shut in their wells or utilize more expensive tank truck transportation."

[947] Allen, *TAPS Rebuttal Testimony, supra* note 496, at 14-15.

[948] HOLLISTER, *supra* note 944, at 2; *cf.* Gary, *Direct TAPS Testimony, supra* note 634, at 38.

[949] *TAPS Hearings Transcript, supra* note 785, at 9839-9843 (Herman G. Rosemann).

[950] Allen, *TAPS Rebuttal Testimony, supra* note 496, at 5, 11, 13.

[951] *Cf.* Gary, *Direct TAPS Testimony, supra* note 634, at 38, 40; Gary, *TAPS Rebuttal Testimony, supra* note 726, at 16.

[952] Tierney, *TAPS Rebuttal Testimony, supra* note 175, at 4.

Economic Characteristics

for TAPS' owners,[953] all expressed the concern that changes in the "rules of the game" for the owners of pipelines *after the lines have been completed* would not only be unfair but would be a substantial disincentive to future investments in oil pipelines.

Perhaps the overall tenor of the proponents' rationale was expressed by IPAA President Jack Allen when he stated: "The oil pipeline industry, under existing methodology, has provided immediate well connections to all producers without discrimination to permit transportation without delay at reasonable cost. . . . The public interest demands that this be continued. Why change something that has worked so well in the past?"[954] When Congress brought oil pipelines under the Interstate Commerce Act, it did not regulate them as utilities but as a special type of *competitive* common carrier.[955] This is obvious from the fact that Congress did not give ICC the usual tools for utility control.[956] Under existing law, oil pipelines not only compete with other pipelines but must compete with tankers, barges, railroads, and trucks, many of which are unregulated.[957] In many cases, the independent shipper has several competing pipeline routes to choose from[958] and can change to other modes of transport at will,[959] and frequently does so when it is to his advantage. Under this system, the real cost of oil pipeline service has decreased[960] and oil pipeline levels under ICC regulatory procedures have been quite reasonable,[961] and a flexible, interconnected, efficient transport system has developed with continuing technological advances.[962]

Despite the declared intention of Congress to transfer the regulation of oil pipelines from ICC to FERC "with maximum consistency and continuity, and with minimum disruption for all concerned parties,"[963] the protestants in the pending FERC cases seek to import utility-type, or at least natural gas

---

[953] Wooddy, *TAPS Rebuttal Testimony, supra* note 724, at 34.

[954] Allen, *FERC Direct Testimony, supra* note 929, at 9.

[955] *Rm 78-2 Hearings Transcript, supra* note 754, Vol. 12 at 1926-1927 (Oral Argument by Jack Vickery on behalf of Belle Fourche Pipeline Company, an independent crude gatherer).

[956] *Id.; cf* Farmers Union Central Exchange v. F.E.R.C., 584 F.2d 408, 413 (D.C. Cir. 1978), cert. den., 99 S.Ct 596 (1979).

[957] See text at notes 819-887, *supra*.

[958] See text at notes 216-228 and 401-441, *supra; cf* Allen, *FERC Direct Testimony, supra* note 929, at 7; Allen, *TAPS Rebuttal Testimony, supra* note 496, at 9.

[959] See text at notes 861-887, *supra*.

[960] See text at notes 188-194, *supra*.

[961] Tierney, *TAPS Rebuttal Testimony, supra* note 175, at 4; *cf.* Allen, *TAPS Rebuttal Testimony, supra* note 496, at 4; Allen, *FERC Direct Testimony, supra* note 929, at 3.

[962] See text at notes 212-235, *supra*.

[963] See the Ribicoff-Hansen colloquy referred to in note 936, *supra*.

cost-of-service, regulation,[964] which in the natural gas area generally has been acknowledged to have resulted in misallocation of gas, provided a strong disincentive for exploration resulting in, or at least contributing substantially to, sharply lowered exploration effort which appears to be a principal element in the current natural gas shortage.[965]

Moreover, to apply such regulation to oil pipelines would, according to Jack Allen, President of IPAA, be like trying to put a square peg in a round hole.[966]

b. In-Place Enforcement Agency Changes in Antitrust Theory

Another uncertainty that faces prospective investors in oil pipelines, especially those who are shipper-owners, is a change in theory by enforcement agencies having antitrust jurisdiction. The first significant change took place in 1940, when the Antitrust Division of the Department of Justice brought four cases against the industry. The principal case, *United States v. American Petroleum Institute*,[967] commonly referred to as the "Mother Hubbard" case,[968] named as defendants 22 "major" oil companies, together with 379 subsidiary or affiliated companies, as well as the largest industry trade association, the API. The principal descriptive paragraphs dealing with oil pipelines were paragraphs 20-22 which claimed that the defendant "majors" owned or controlled 57 percent of the crude oil gathering lines, 85 percent of the crude oil trunk lines, 87 percent of the dead weight tonnage oil tankers and 96 percent of the gasoline pipeline mileage;[969] that the dominance of the defendant "majors" had been steadily increasing and the position of the "independent" segment of the industry had been steadily decreasing;[970] and that this control was aggravated by exchange arrangements, which were rarely made with independent refiners.[971] The charging paragraphs which are germane to pipelines alleged that defendants jointly used their control of oil pipeline transportation for the purpose of compelling "independent" producers to sell "at the well" rather than permitting them to use defendants' pipelines on a common carrier basis.[972]

---

[964] See, e.g., *TAPS Hearings Transcript, supra* note 785, at 9273 (Herman G. Roseman); *Id.* at 10,433-10,437 (John C. Dunn); *Id.* at 12,270 (Michael J. Ileo).
[965] *S.1167 Hearings, supra* note 432, pt. 9 at 609 (Statement of Vernon T. Jones).
[966] Allen, *TAPS Rebuttal Testimony, supra* note 496, at 15.
[967] Civil No. 8524, D.D.C. Sept. 30, 1940. The complaint is reproduced in *Celler Hearings, supra* note 162, pt. I, Vol. 1, pp. 125-156.
[968] Opinions vary whether the name came from the style of ladies' dresses which were so voluminous that they covered everything or from the nursery rhyme in which Mother Hubbard, seeking to get her poor doggie a bone, found the cupboard to be empty.
[969] *Celler Hearings, supra* note 162, at 144-145 (pars. 20-21).
[970] *Id.* at 145 (par. 21).
[971] *Id.* (par. 22).
[972] *Id.* at 147 (par. 32).

Economic Characteristics   **145**

The alleged means were the concerted adoption of tariff rules and regulations which required unreasonably large "minimum tenders" and the maintenance of "uniform, noncompetitive, onerous and oppressive" rates for oil pipeline transportation, which, the government claimed, had the purpose and effect of reserving defendants' pipelines for their own use and precluding their use by the "independent" sector.[973] This asserted excessive rates collusion was charged to be a violation of Sections 2 and 6 of the Interstate Commerce Act[974] and Section 1 of the Elkins Act,[975] because defendants received back as "refunds" or "rebates" [in the form of dividends] a substantial part of the revenue, which was used to "subsidize" operations in other branches of the industry, thereby enabling defendants to enjoy an unfair competitive advantage over their "independent" competitors in such branches.[976]

The prayer for relief, as originally drafted, requested divestiture of ownership in transportation and marketing, but this was deleted on the advice of the Council of National Defense whose report observed that divorcement of oil pipelines could hinder the national defense effort.[977] While the parties were negotiating, World War II broke out and the case was suspended.[978] After the war, there were many changes in corporate relationships and the new antitrust administration decided that the case was unmanageable and that the public interest would be served by the institution of segment suits.[979] The case was dismissed on June 6, 1951.[980]

---

[973] *Id.* at 147-148 (par. 33).

[974] 49 U.S.C. §§ 2 & 6 (1976).

[975] 49 U.S.C. § 41 (1976).

[976] *Celler Hearings, supra* note 162, at 148 (par. 34). This canard has been thoroughly discredited. Dean Rostow, who hardly can be accused of being in favor of the industry, long ago forthrightly stated that the theory would not stand up under inquiry. ROSTOW; A NATIONAL POLICY FOR THE OIL INDUSTRY 65 (1948); it has not fared any better at the hands of present day academicians. *Cf.* W. LIEBELER, *Integration and Competition* in VERTICAL INTEGRATION IN THE OIL INDUSTRY 32 (E. Mitchell ed. 1976) [hereinafter cited as LIEBELER, *Integration and Competition*]; nor has it been received favorably by the legal commentators. SPECTRE, *supra* note 104, at 110; Hale, *Successive Stages, supra* note 104, at 938-939; Hale, *Size and Shape; The Individual Enterprise as a Monopoly,* 1940 Law Forum 515, 533 (1950); Bork, *Vertical Integration - Misconception, supra* note 106, at 200; Comment, *Vertical Forestalling Under the Antitrust Laws,* 19 U. Chi. L. Rev. 583, 612 (1952). The theory, at least in the past, has had its supporters. See J. BAIN, THE ECONOMICS OF THE PACIFIC COAST PETROLEUM INDUSTRY, Vol. 3 at 5-8 (1947); R. COOK, CONTROL OF THE PETROLEUM INDUSTRY BY MAJOR OIL COMPANIES 22-23, 28 (TNEC Monograph 39, 1941); Prewitt, *Operation and Regulation of Crude Oil and Gasoline Pipelines,* 56 Q. J. Econ. 177, 199-201 (1942).

[977] CELLER REPORT, *supra* note 220, at 142.

[978] PETROLEUM PIPELINES, *supra* note 45, at 425.

[979] *Celler Hearings, supra* note 162, at 202.

[980] CELLER REPORT, *supra* note 220, at 147.

The Attorney General had filed, on the same day as the *Mother Hubbard* case, three cases in the pipeline area, alleging violations of the Elkins Act. The cases were designed to cover the various forms of pipeline operation then employed, *i.e.,* a pipeline operated as a department of the shipper-owner,[981] a jointly-owned pipeline,[982] and a single shipper owner and its wholly-owned pipeline subsidiary.[983] The original complaints were grounded upon the novel[984] theory that the payment of dividends to the shipper-owners amounted to a rebate,[985] and the relief requested was that the defendant pipelines and shipper owners be enjoined from granting/receiving the alleged rebates and that the individual defendant shipper-owners be required to pay to the plaintiff United States the authorized forfeiture[986] of three times the total amount of money and "other valuable considerations" found by the court to have been received in contravention of the Elkins Act. After a preliminary joust over a venue question[987] and a period of negotiation,[988] the three pipelines cases were dismissed after the parties entered into a consent decree on December 23, 1941, in a new proceeding entitled *United States v. Atlantic Refining Company* which was filed in the District Court in the District of Columbia concurrently with the decree.[989]

The second significant change on the part of the Antitrust Division was the commencement, virtually on the eve of the Cellar Committee Consent Decree Hearings, of four proceedings in the District Court for the District of Columbia. Three were entitled "Motion for Order for Carrying Out Final Judgment" and were served upon: (1) Arapahoe Pipe Line Company; (2)

---

[981] United States v. Phillips Petroleum Company and Phillips Pipe Line Company, Civil No. 182, D.Del. Sept. 30, 1940. See complaint reprinted in its entirety in *Celler Hearings, supra* note 162, at 156-164.

[982] United States v. Great Lakes Pipe Line Company, Civil No. 183, D.Del. Sept. 30, 1940. See reprint of complaint in *Celler Hearings, supra* note 162, at 161-166.

[983] United States v. Standard Oil Company (Indiana), Civil No. 201, N.D.Ind. Sept. 30, 1940. A copy of the complaint can be found in *Celler Hearings, supra* note 162, at 166-169.

[984] *Cf.* 9 U. Chi. L. Rev. 503 (1942).

[985] For the distinction between a rebate and a dividend, see WOLBERT, *supra* note 1, at 145 n.217.

[986] Section 1(3) of the Elkins Act, *supra* note 975.

[987] United States v. Phillips Petroleum Company, 36 F.Supp. 480 (D.Del., 1941); United States v. Great Lakes Pipe Line Company, 36 F.Supp. 486 (D.Del. 1941).

[988] The particular facts surrounding the negotiations between the Justice Department and the defendant oil companies are rather obscure. The threat of damages amounting to approximately 2½ billion dollars, together with the agitation raised by certain legislators for pipeline divorcement, at a time when the defendants were being called upon by the Petroleum Coordinator, Interior Secretary Ickes, to undertake a substantial cooperative effort to solve defense issues (see text at notes 153-157, *supra*) caused the parties to enter into a compromise, which, like most compromises, was not fully satisfactory to either. See WOLBERT, *supra* note 1, at 146. Subsequently, Congressman Emanuel Celler held hearings on two consent decrees, the *AT&T* decree and the *Atlantic Refining Company* decree. See CELLER REPORT, *supra* note 220, at 149-170 for his staff's version of the negotiations.

[989] CELLER REPORT, *supra* note 220, at 148-149.

Service Pipe Line Company and Standard Oil Company (Indiana); and (3) Tidal Pipe Line Company and Tidewater Oil Company. The fourth proceeding, which was settled on consent, was a petition for civil contempt addressed to and served only upon The Texas Pipe Line Company.[990] The principal issue was raised in the *Arapahoe* case wherein the Government, after sixteen years of acquiescence, decided that what it then thought it should have said back in 1941, should be used by Judge Keech to "construe" the decree in a way that clearly violated the plain meaning[991] of the words, *i.e.,* to limit payment of dividends by pipelines to shipper-owners to seven percentum of their *equity,* in other words, to forbid oil companies from making money on borrowed money if the borrowing was done by the pipeline company.[992] Judge Keech ruled that the decree was clear upon its face, and being clear, he had no right to reword the agreement reached between two parties in 1941 after due deliberation and approved by Judge Pine in 1941 and again in 1942 by the supplemental *Great Lakes* Order.[993]

While the Government had clearly disavowed that it was requesting abandonment of the decree *in toto,* even if it had done so, said Judge Keech, "I would not hold that the decree as it has been interpreted by the parties over a period of sixteen years violates the Elkins Act." Moreover, although he had found no ambiguity in the decree—to the contrary, that it was clear on its face—even if there had been an ambiguity, Judge Keech said he "certainly would be constrained to hold that ambiguity had been resolved through the practice of the defendants, acquiesced in by the government after full disclosure, throughout the sixteen years."[994] The United States Supreme Court affirmed Judge Keech's order.[995] Mr. Justice Black, who wrote the opinion stated: "If the decree had meant to limit dividends to 7% of the current value of a parent company's actual investment in a subsidiary, as the government claims, one can hardly think of less appropriate language in which to couch the restriction."[996]

---

[990] United States v. Atlantic Refining Company, No. 210, Supreme Court of the United States, Motion of Interstate Oil Pipe Line Company and Tuscarora Pipe Line Company, Limited to Dismiss the Appeal From, or Affirm, the Order of March 25, 1958, pp. 7-8 (Aug. 22, 1958).

[991] "[T]he plain, obvious and rational meaning is always to be preferred to any curious, narrow, hidden sense that nothing but the exigency of a hard case and in the ingenuity and study of an acute and powerful intellect would discover." Lynch v. Alworth-Stephens Co., 267 U.S. 364, 370 (1925); *cf.* Old Colony R.R. v. Comm'r, 284 U.S. 552, 560 (1932); Caminetti v. United States, 242 U.S. 470, 485 (1917).

[992] United States v. Atlantic Refining Co., Civil No. 14060, D.D.C., Mar. 24, 1958, Tr. 132-133.

[993] *Id.* at 134.

[994] *Id.* at 135.

[995] United States v. Atlantic Refining Co., 360 U.S. 19 (1959).

[996] *Id.* at 22.

The last of the trilogy of significant enforcement agency changes in antitrust theory is the current Federal Trade Commission proceeding against the eight largest oil companies.[997] Although the doctrine of "conscious parallelism" has been expressly distinguished from conspiracy,[998] the FTC's approach in the Big Eight case appears to seek a new frontier of law in terms of an amorphous concept of "group monopoly."

c. Environmental Complications

Mention has been made previously of the National Environmental Policy Act of 1969 (NEPA) and its complications,[999] not only in the lower 48 states, but especially in Alaska.[1000] As time goes on, the effects of these complications will produce even more important uncertainties, as the EPA; other governmental agencies which are bound to take NEPA into account; environmental groups, such as the Wilderness Society, Friends of the Earth, Natural Resources Defense Council, the Sierra Club, etc.; and the industry strive to achieve an appropriate balance between energy requirements and environmental goals.

d. Legislation Authorizing New Forms of Regulation

There has been a pronounced inclination for the Congress (and some state legislatures) to enact legislation which authorizes and requires the appropriate Executive Department Secretary or Agency Head to issue regulations and to subject permittees, licensees or regulatees to such terms and conditions as the Secretary or Agency Head shall deem necessary or appropriate to execute the purpose or legislative intent expressed in the authorization act. While not purporting to cover even a substantial portion of such acts as they relate to pipelines, perhaps a few examples will illustrate the point. The Trans-Alaska Pipeline Authorization Act,[1001] enacted in November, 1973, while curing the width of right-of-way problem which had enabled the

---

[997] In the matter of Exxon, et al., D. 8934.
[998] Triangle Conduit and Cable Co. v. Federal Trade Commission, 168 F.2d 175 (7th Cir. 1948), *aff'd by equally divided court, sub. nom.*, Clayton Mark & Co., 336 U.S. 956 (1949).
[999] See text at notes 806 and 807, *supra*. Perhaps it could be said that NEPA's prologue was Earth Day. In NEPA's wake came the Air Quality Act of 1967, 42 U.S.C. §§ 1857-1858a (1976); the Clean Air Act of 1970, 42 U.S.C. § 7401, *et. seq.* (1976), under which emissions from pipelines are controlled; the Water Pollution Prevention and Control Act of 1972, 33 U.S.C.A. §§ 1251-1376, regulating discharges from pipelines to surface waters, and the Coastal Zone Management Act of 1973, 16 U.S.C.A. §1456(a), dealing with coastal zone energy development. *See* P. CORCORAN, *Oil Pipelines-Law and Regulation* in AOPL EDUCATORS CONFERENCE 14-15 (1978).
[1000] See text at 906-911, *supra*.
[1001] 43 U.S.C. §1651-1654 (new provisions), and amendment of the rights-of-way provisions of the Mineral Leasing Act of 1920, 30 U.S.C. § 185 (1976).

Wilderness Society to prevail in its litigation and providing that no further action under NEPA should be taken in the case of TAPS, simultaneously imposed on TAPS's owners unprecedented requirements, liabilities and restrictions.[1002]

The Interior Secretary was required by the Act to review a *total* plan of construction, operation, and rehabilitation of the line and its ripple effects. The Secretary was authorized to require provision for restoration, revegetation and erosion prevention and to impose requirements to control or prevent damage to the environment, to protect the interests of those living in the area and to ensure safety. The entire cost of monitoring the construction, operation, maintenance, and termination of any pipeline or related facilities was to be charged to the permittees, and the Secretary could suspend forthwith all activities within the right of way if, in his sole discretion and without any proceedings of any nature, he deemed it advisable for the protection of public health or safety, or the environment. He could impose strict liability for damage caused by all activities within the right of way except those resulting from an act of war or negligence of the United States. The Act also prohibited export of any domestically produced crude oil without a specific and express finding by the President that such exports were in the national interest and in accordance with the Export Administration Act of 1969. Acting under this authority, the Secretary "negotiated" an "Agreement and Grant of Right-of-Way for Trans-Alaska Pipeline" between the United States and the seven owners.[1003] The Agreement carried along with it, among other items, an Exhibit D, "Stipulations for the Agreement and Grant of Right-of-Way for the Trans-Alaska Pipeline" (18 pages of single-spaced, double-columned uncertainties for TAPS's owners).[1004] Nothing less than a careful reading of these documents can convey the pervasiveness of the Government in the business of constructing, maintaining, and operating TAPS. This is a perfect illustration of what Professor J. W. McKie has termed the "tar-baby effect"[1005] which occurs when regulators extend themselves further and further into the network of enterprise decisions.[1006]

[1002] Donaldson, *Direct TAPS Testimony, supra* note 807, at 9-10. Most of the text following this note was derived from Donaldson's testimony.

[1003] The right-of-way agreement is described in note 812, *supra*.

[1004] The Agreement and all of its Exhibits are reproduced in Patton, *Direct TAPS Testimony, supra* note 726, as Exhibit ELP-1.

[1005] The analogy is to Joel Chandler Harris' stories told by Uncle Remus.

[1006] McKie, *Regulation and the Free Market: The Problem of Boundaries,* Bell J. Econ. Vol. I, Spring 1970, p. 9.

Several of the requirements were obviously addressed solely to aesthetics regardless of cost or loss of efficiency; in fact one section of the stipulations was so entitled.[1007] Ironically, the Government was among the protestants in the *Trans Alaska Pipeline Rates Cases* who argued that "much of $9 billion [in TAPS's cost on which tariffs were calculated] represented waste and mismanagement on the part of the TAPS owners and could not, therefore, be included in the TAPS rate base."[1008]

Another example is the Deepwater Port Act of 1974.[1009] Among other things, the Act called upon the Attorney General and the Federal Trade Commission to advise the Secretary of Transportation whether, in their respective opinions, the issuance of a license to own, construct, and operate a deepwater port would create a situation in contravention of the antitrust laws and to submit reports assessing the competitive effects that may result from the issuance of such a license. The Secretary was entrusted with broad authority (obligation?) to enforce common carrier operation of the port(s) and to prescribe license conditions necessary to implement the enforcement agencies' recommendations and was to retain authority to impose whatever conditions of service he determined in his discretion to be necessary. Lurking in the background was the retained power of the Attorney General to challenge "any anticompetitive situation involved in the ownership, construction, or operation of a deepwater port," after the license had been issued and construction and operations had commenced.[1010]

Out of these circumstances came the Justice Department's "Competitive Rules" which, simply stated, are: (1) Deepwater Ports must provide open and nondiscriminatory access to all shippers, owners and nonowners alike; (2) Any deepwater port owner or shipper providing adequate throughput guarantees at the standard tariff can unilaterally request and obtain expansion of capacity; (3) Deepwater ports must provide open ownership to all shippers at a price equivalent to replacement cost less economic depreciation; and (4) the ownership shares of the deepwater port's owner's share must be revised frequently (annually) so that each owner's share equals his share of average throughput.[1011] In addition to the "Competitive Rules" there was a myriad of detailed suggestions for changes in the Shareholders

---

[1007] Stipulations, Sec's. 2.3.2.1; 2.10 ("Aesthetics"); 3.6.1.1.1.1. Professor McKie also had presaged this tendency in his article: "If the primary purpose were to prevent excessive returns in a [deemed] protected monopoly situation, a simple limit on rate of return would be enough. Some complications arise because regulatory authorities actually apply a variety of welfare criteria, which do not always reconcile with each other." McKie, *Regulation and the Free Market: The Problem of Boundaries*, Bell J. Econ. Vol. I, Spring, 1970, p. 8.

[1008] Trans Alaska Pipeline Rates Cases, 436 U.S. 631 (1978).

[1009] 33 U.S.C.A. §§ 1501-1524.

[1010] Section 7(b)(2) of the Act, 33 U.S.C.A. § 1507 (1976).

[1011] AG's Deepwater Port Report, *supra* note 446, at 106.

Agreement, in the design, operation, mooring and unloading facilities, onshore storage and a proposal to extend the deepwater port regulation downstream to connecting pipeline facilities. The Secretary did not accept all these suggestions but the license conditions he did issue were so restrictive and placed such heavy burdens on the licensees that several of the participants in SEADOCK, Inc., withdrew from the project, stating that the tough license conditions were a material factor in their decision. LOOP, Inc., went forward with the project, however it remains to be seen if it will be able to carry the extra load and still be a good investment. All that it will take for it to go under will be large scale importation of crude from Mexico where VLCC's will not be a factor as they are in liftings from the Persian (Arabian) Gulf, or large scale embargoes imposed by OPEC, Middle East countries or the United States Government.

While the foregoing statement is subject to debate by the proponents of the license provisions and seemingly the discussion is limited solely to Deepwater Ports and not to pipelines as such, it seems undisputable that the "bottom line" of this regulation has been the addition of a substantial uncertainty to those already facing the operators of new pipeline and related ventures because the authorities already have indicated a tendency to carry over the principles to pipelines.

The latest excursion by the Congress into the operation of pipelines occurred in September, 1978, when Congressman John Seiberling, on the House side, and Senator Edward Kennedy introduced into the "Outer Continental Shelf Lands Act Amendments of 1978,"[1012] an amendment to Section 5 of the OCS Lands Act. Their Amendment provides that, subject to exception by FERC, every permit, license, easement, right-of-way, or other grant of authority for the transportation by pipeline on or across the OCS of oil and gas shall require that the pipeline be operated in accordance with certain "competitive principles." These principles are: "(A) The pipeline must provide open and nondiscriminatory access to both owners and nonowner shippers[; and] (B) Upon the specific request of one or more owner or nonowner shippers able to provide a guaranteed level of throughput, and on the condition that the shipper or shippers requesting such expansion shall be responsible for bearing their proportionate share of the costs and *risks* related thereto, the Federal Energy Regulatory Commission may, upon finding after a full hearing with due notice thereof to the interested parties, that such expansion is within technological limits and economic feasibility, order a subsequent expansion of throughput capacity of any pipeline for which the per-

---

[1012] Pub. L. No. 95-372, § 204, 92 Stat. 639 (1978), amending Section 5 of the Outer Continental Shelf Lands Act, 43 U.S.C. § 1334 (1976).

mit, license, easement, right-of-way, or other grant of authority is approved or issued after the date of enactment of this subparagraph" (emphasis added).[1013]

Because the House Bill did not contain subsection (f)'s "competitive principles," as did the Senate's version, negotiations in the Conference Committee resulted in accepting the Senate's version, but with a significant amendment, *viz.,* "This subparagraph shall not apply to any such grant of authority approved or issued for the Gulf of Mexico or the Santa Barbara Channel." It will be apparent to the reader that this is a "bob-tailed" version of the Justice Department's "Competitive Rules," and the Conference Report notes that in implementing the subsection and applying the "competitive principles," the Secretaries of Energy and FERC are to consult with and give due consideration to the views of the Attorney General, who in turn is to consult with the Federal Trade Commission. Chapter VII, "Proposed Solutions to an Assumed Problem," *infra,* discusses this matter in more detail.

e. Executive Action

The petroleum industry never in recent history has been free from substantial government influence.[1014] In 1954, the President created a Presidential Advisory Committee[1015] on Energy Supplies and Resources Policy to evaluate all factors involving the continuing development of domestic fuel supplies adequate to meet defense needs. The Committee rendered its report on February 26, 1955, to the effect that if the imports of crude and residual oils should exceed significantly the respective ratios that these imports bore to the production of domestic crude in 1954, there would be an inadequate incentive for exploration and development of new domestic sources which could impair the orderly domestic growth which assures the military and civilian supplies and reserves that are necessary to the national defense.[1016]

Congress, which was considering trade agreements legislation that year, enacted as Section 7 of the Trade Agreements Extension Act of 1955,[1017] a provision that whenever the Director of the Office of Defense Mobilization (ODM) had reason to believe that any article was being imported into the

---

[1013] *Id.* at (f)(1).

[1014] DECONTROL REPORT, *supra* note 372, at 1083.

[1015] The Committee was composed of the Director of the Office of Defense Mobilization (ODM), the Secretaries of State, Treasury, Defense, Interior, Commerce and Labor and the Attorney General.

[1016] White House Press Release, Feb. 26, 1955.

[1017] 69 Stat. 166 (1955), 19 U.S.C. § 1342a (b) (1976).

United States in such quantities as to threaten to impair the natural security, he should so advise the President; and, if the President felt that there was reason for such belief, he should cause an immediate investigation to ascertain the facts. If, as a result of such investigation, the President found that the article was being imported into the United States in such quantities as to threaten to impair the national security, he was authorized to take such action as he deemed necessary to adjust such imports to a level which would not threaten to impair national security.[1018] Despite some "jawboning" by the Administration, the price of imported crude was so much lower than domestic crude that imports increased until April 23, 1957, when the Director of ODM advised the President that the quantities of crude imports threatened to impair national security. After receiving a report from his Special Committee, the President directed the Secretary of the Interior and the Director of ODM to take steps as rapidly as possible to effectuate the plan recommended by the Special Committee. Thus the Voluntary Import Program came into being in July, 1957, with the Secretary of the Interior as its Administrator.[1019] After a period of jockeying around for exceptions under the voluntary limitations, the President, by proclamation dated March 10, 1959, instituted the Mandatory Oil Import Program (MOIP) regulating crude and product importation.[1020]

Two effects on pipelines resulted from this action: first, the MOIP had a definite "tilt" in favor of smaller refiners,[1021] which caused certain changes in patterns of movement; second, a direct effect was felt by Laurel Pipe Line which had commenced construction at a time when it appeared that refineries in the Philadelphia area could process low cost imported crude, transport their products by pipeline and deliver products to markets as far west as Cleveland at a price competitive with Midwestern refiners using domestic crude.[1022] The line was just completed when the MOIP was instituted. This action drastically changed the economics of the line, especially that portion from Pittsburgh to Cleveland. As a result, Laurel lost money for years and never has been particularly profitable.[1023]

[1018] Att'y Gen., *Second Report, supra* note 111, at 54.
[1019] *Id.* at 59-60.
[1020] Proclamation No. 3279, 3 C.F.R. § 1959-1963 Comp. at 11 (Mar. 10, 1959). This was subsequently terminated, effective May 1, 1973, by Proclamation No. 4210, 38 Fed. Reg. 9645 (Apr. 19, 1973).
[1021] Att'y Gen., *Fourth Report, supra* note 111, at 21; *cf.* M. CANES, *An Economist's Analysis* in WITNESSES FOR OIL 56 (1976) [hereinafter cited as CANES, *An Economist's Analysis*].
[1022] LIVINGSTON, *supra* note 228, at 22.
[1023] *Id.* at 22-23; *S.1167 Hearings, supra* note 432, at 607 (Statement of Vernon T. Jones); HOWREY & SIMON, *supra* note 3, at 9; PIPELINE PRIMER, *supra* note 331, at 14.

During the 1960's, foreign crude was priced below domestic crude and the right to import a barrel of crude was worth the difference in price, typically about $1.25 per barrel. Once again, there was a tendency to grant more quota rights, known as "tickets," per barrel of capacity to smaller refiners than to larger refiners.[1024] However, as "tickets" could not be "sold" directly but were worked out on an exchange basis, there were certain shifts in pipeline transportation patterns to deliver the domestic crudes to the small inland refiners in exchange for "their" imported crude at seaboard refineries. As the differential between domestic and foreign crudes eroded in the early 1970's, this subsidy disappeared.

Since August 15, 1971, the oil industry has been subject to price control regulations. The effect of freezing the price below the free market price brought its own uncertainties. Price controls exacerbated the imbalance between domestic supply and demand and within a reasonably brief time, independent marketers fell into short supply. This affected pipelines as certain supply relationships were terminated. Then, like the tar-baby story, the shortcomings of the price control strictures were sought to be solved by adding the Emergency Petroleum Allocation Act in November 1973 when the Middle East embargo interrupted the foreign supply situation. The initial allocation regulations required suppliers to sell to their historical customers. This provided partial relief for independent refiners and marketers in aiding their supply situation[1025]—although it again shifted the transportation pattern—but smaller refiners found themselves paying about 18 percent more than the larger companies for their crude supply.[1026] Therefore, yet another layer of regulation, the crude oil cost equalization program (the "entitlements" program) was adopted in November 1974. It was supposed to equalize the effective price that all refiners paid for their crude oil irrespective of its origin as price-controlled domestic oil or foreign crude. However, the entitlements program was carried beyond mere equalization of crude oil prices. It contained an explicit subsidy in favor of small refiners.[1027]

In addition to their impact on certain classes of competitors, the price and allocation controls tend to cause direct distortions away from efficient industry performance as well as to introduce distortions in the structure of the industry.[1028] These programs have an important bearing on the

---

[1024] DECONTROL REPORT, *supra* note 372, at 1038; *Petroleum Industry Hearings, supra* note 192, at 379 (Testimony of Walter R. Peirson).

[1025] *Cf. Consumer Energy Act Hearings, supra* note 154, at 1025 (Statement of Keith Clearwaters).

[1026] DECONTROL REPORT, *supra* note 372, at 1039.

[1027] *Id.; cf.* CANES, *An Economist's Analysis, supra* note 1021, at 56.

[1028] DECONTROL REPORT, *supra* note 372, at 1040.

economics of crude oil supply and distribution. Such external circumstances, which are not safely predictable far into the future, have important effects on pipeline facilities designed to be long-lived investments and, as such, can render uneconomic or obsolete fixed investments considered reasonably safe or less risky in pre-"stabilization control" times.[1029]

Another example of the uncertainties that Executive action can introduce with respect to pipeline operations occurred early in 1977 when the Federal Energy Administration (FEA, now part of DOE) issued a regulation to the effect that the wellhead price of Alaskan North Slope crude oil be controlled at the ceiling price for comparable "upper tier" oil and be treated as "foreign" oil for "entitlements" purposes. The Trans-Alaska Pipeline Authorization Act contemplated favorable treatment for North Slope crude. Just the reconsideration by FEA of this treatment caused great consternation and uncertainty. As Morgan Stanley's Managing Director, David Goodman, testified in the March, 1977, FEA Hearings on Alaskan North Slope Pricing and Entitlement Issues, ". . . the uncertainty of the prospect of these very proceedings here today was enough to cause important lenders not to participate in the most important energy project this country has ever seen."[1030] Although the hearings concluded that the initial beliefs of North Slope Project investors regarding the treatment of North Slope oil was correct, the fact that the hearings were even held affected investors' perceptions of the regulatory-political risk of oil development. It takes no stretch of the imagination to anticipate what would have been the second order effect on TAPS, had the FEA's hearings reached a different conclusion.

Finally, it is self-evident that major changes in energy policy by the Government can alter investment prospects. Current uncertainty about such policies has raised substantially the uncertainties perceived by oil companies when considering pipeline investment.[1031] Vernon Jones, at the time President of Explorer, testified that among the uncertainties facing Explorer in its attempts to break out of its loss position was whether there would be a decline in product consumption brought about by economic or political pressure.[1032] Perhaps the most accurate way to describe the situation is that the unknown impact of the National Energy plan can affect drastically supply and demand over a particular pipeline, or pipelines generally.[1033]

[1029] Gary, *FERC Direct Testimony, supra* note 613, at 16.
[1030] Gary, *Direct TAPS Testimony, supra* note 634, at 37.
[1031] Cooper, *Direct TAPS Testimony, supra* note 718, at 17-18.
[1032] *S.1167 Hearings, supra* note 432, pt.9 at 607 (Statement of Vernon T. Jones).
[1033] *Cf.* MARATHON, *supra* note 202, at 34-35.

2. Acts of Foreign Nations

The effect of acts of foreign nations on the oil industry (and hence the energy consumer) is well known today. Its effect on pipelines is not quite as well recognized, but it can be dramatic. By way of example, the Arab embargo of 1973, which most Americans relate to lines of cars at service stations, cut Explorer's throughput in half between November 1973 and Janaury 1974.[1034]

Another illustration is the Canadian import situation. In 1974, the National Energy Board of Canada adopted a procedure designed to limit exports of Canadian crude any time that 10 years of future Canadian requirements would not be assured from indigenous crude oil and equivalent feedstocks.[1035] In September of 1975, there was a further reduction in allowable export volumes to the United States in light of Canada's growing realization of its own supply deficiencies.[1036] One direct effect on pipelines was the drastic decline in throughput suffered by Platte which had achieved a delivery rate of 231,000 barrels per day in 1974, a substantial part of which was Canadian crude. As a direct result of the curtailment of Canadian crude, movements through Platte were down to 124,957 barrels per day in the first six months of 1977,[1037] and it is reported that they are now down to around 1953 levels.

H. Pipeline Economic Decision-Making

When appraising the economic viability of a proposed pipeline venture, the approach taken is similar to that used by investors in general;[1038] it is what may be termed as required rate of return analysis.[1039] An oil company has widespread operations with numerous investment opportunities bearing different degrees of risk. Because of this, each investment, including pipelines, must be examined individually, and its expected rate of return compared with the opportunity rate of return of other prospective investments with comparable risk characteristics.[1040] In order to simplify procedures, most

---

[1034] *S.1167 Hearings, supra* note 432, pt. 9 at 606 (Statement of Vernon T. Jones); PIPELINE PRIMER, *supra* note 331, at 14; *cf.* Gary, *FERC Direct Testimony, supra* note 613, at 17.

[1035] MARATHON, *supra* note 202, at 34 n.1.

[1036] *Id.* For a schedule of projected Canadian crude oil exports to the United States, see Table 1, NET-III, *supra* note 199, at 6.

[1037] MARATHON, *supra* note 202, at 40.

[1038] Solomon, *TAPS Rebuttal Testimony, supra* note 621, at 5-10; *cf.* Gary, *TAPS Rebuttal Testimony, supra* note 726, at 39.

[1039] Cooper, *Direct TAPS Testimony, supra* note 718, at 5; Ryan, *TAPS Rebuttal Testimony, supra* note 724, at 47.

[1040] Gary, *TAPS Rebuttal Testimony, supra* note 726, at 43; Cooper, *Direct TAPS Testimony, supra* note 718, at 6-10.

Economic Characteristics 157

companies have established "cutoff" or "hurdle" rates for different risk classifications, and projects which do not meet or exceed this "required rate of return," absent some overwhelming peculiar need for the project, will be rejected as an investment.[1041] A proposed pipeline undergoing this screening technique is examined on an "on its own" basis;[1042] it does not get credited with any expected production or marketing profit,[1043] because, as a common carrier, there is no assurance that the owner will be able to utilize a percentage of the line's capacity equal to its ownership interest.[1044] The economic analysis is on a "delevered" basis, *i.e.*, one which takes into account the special tax treatment of interest expense;[1045] the question of how much debt should be incurred by the pipeline is a financing decision,[1046] which is distorted by the *Atlantic Refining Company* Consent Decree.[1047]

The calculation of a proposed pipeline venture is made by the use of a discounted cash flow (DCF) analysis on an after-tax basis, which determines the discount factor (rate) which will equilibrate positive and negative cash flows, utilizing deferred tax accounting, over the life of the project.[1048] This method has been used universally for many years by those corporations which analyze capital budget decision.[1049] The technique employs what operations research types call an "iterative approach." It starts out with a "guesstimated" rate and, depending on whether this is too high or too low, a second rate is chosen and the computations performed again. A second correction in rate is made and the process continues until the rate is found that

---

[1041] AEI, *Oil Pipelines and Public Policy, supra* note 684, Discussion, p. 120 (James Shamas, Chairman of the Board of Texoma Pipeline speaking: "On an ICC rate of return, we earned about 5.8 percent. We sold it to our Getty Oil management on a 14 percent DCF rate of return. We did a three-year follow-up, and we have achieved 13.8 percent DCF rate of return. Today, my management would not accept Texoma pipeline because it only earned 13.8 percent DCF rate of return. *Our new hurdle rate for the corporation is now 14 percent*") [*emphasis added*].

[1042] *Cf. Petroleum Industry Hearings, supra* note 192, at 317 (Statement of Vernon T. Jones).

[1043] Cooper, *Direct TAPS Testimony, supra* note 718, at 13.

[1044] *Id.*; Gary, *Direct TAPS Testimony, supra* note 634, at 15-16.

[1045] Ryan, *TAPS Rebuttal Testimony, supra* note 725, at 34; *cf.* Cooper, *Direct TAPS Testimony, supra* note 718, at 19.

[1046] Gary, *TAPS Rebuttal Testimony, supra* note 726, at 42-43.

[1047] Ryan, *TAPS Rebuttal Testimony, supra* note 725, at 33-34.

[1048] *Id.* at 25; Harrill, *TAPS Rebuttal Testimony, supra* note 774, at 23; Gary, *TAPS Rebuttal Testimony, supra* note 726, at 43; *cf.* Cooper, *Direct TAPS Testimony, supra* note 718, at 22.

[1049] Gary, *TAPS Rebuttal Testimony, supra* note 726, at 45-46.

exactly equalizes the time value of the outgoing and incoming cash flows.[1050] With the proliferation of computers and "desk-top" remote terminals, the calculation is usually done directly. It is simple (in theory) as one seeks the single positive root of an nth degree polynomial. (If the cash flow stream goes from minus to plus or plus to minus more than once, there may be as many solutions as there are such changes in flows and, therefore, no single, unique solution).

[1050] J. CHILDS, PROFIT GOALS AND CAPITAL MANAGEMENT 133-137 (1968). Childs mentions a shortcut method used by some companies which perform these analyses regularly through the construction of a Discounted Cash Flow Template. *Id.* at 136 n. 8.

Chapter IV

Pipeline Industry Structure

A. *Ownership*

1. Presumed Concentration in Hands of "Major" Oil Companies

Critics of the oil industry point to alleged concentration of pipeline ownership by vertically integrated oil companies. The usual thrust of the argument is epitomized by the Kennedy Staff Report which stated: "As the ICC staff has observed, 'pipelines affiliated with major oil companies clearly dominate the industry.'[1051] . . . pipelines affiliated with major oil companies accounted for 74.2 percent of total pipeline mileage and 76.5 percent of total pipeline operating revenue in 1975. Major oil companies' crude pipelines comprised 82.4 percent of total gathering line mileage and 89.3 percent of crude trunkline mileage. These crude trunklines handled 90.3 percent of all crude trunk traffic and 94.5 percent of crude barrel mile movements. Corresponding figures for product pipelines are lower due in part to the inclusion of LPG pipelines and other pipelines carrying products other than refined petroleum products [referring to coal slurry and anhydrous ammonia]. Nevertheless, major oil company affiliated pipelines represented 54.0 percent of total product pipeline mileage in 1975 and handled 72.9 percent of all pipeline product movements, constituting 78.1 percent of product barrel mile traffic."[1052] Assistant Attorney General (Antitrust Division) John Shenefield in his Statement submitted to the Kennedy Subcommittee on June 28, 1978, took a novel new approach: "Another significant pipeline industry characteristic is the level of ownership dominance that vertically integrated petroleum companies have obtained. For example, 1975 data[1053] reveal that there were 104 common carrier pipelines subject to Interstate Commerce Commission (ICC) jurisdiction. The ICC categorized ten of these lines as independents, 59 lines as affiliated with a major oil company, that is, one

---

[1051] Statement of David L. Jones, Acting Chief, Section of Motor, Pipeline and Water Carrier Analysis, Bureau of Economics, ICC, Before I.C.C. Ex Parte No. 308, now F.E.R.C. RM78-2, Valuation of Common Carrier Pipelines, 11 (April 28, 1977).

[1052] KENNEDY STAFF REPORT, *supra* note 184, at 56.

[1053] Drawing upon the same statement by David Jones cited in note 1051, *supra*.

of the top 20 in sales, and 35 as affiliated with non-major, oil-related companies. This listing included 42 pipelines which were joint-venture stock companies, of which 5 were owned by non-major, oil-related companies and the remaining 37 by major oil companies. In addition, the ICC listed separately 27 undivided interest pipeline systems operated as common carriers. In some respects, these figures tend to understate the extent to which petroleum companies (major and non-majors) dominate the pipeline industry. In 1975, the pipelines affiliated with oil companies accounted for 98.6% of all crude barrel miles and 86.9% of all product barrel miles. The corresponding figures for independent pipelines were 1.4% and 13.1%."[1054] Before addressing these comments directly, it should be noted that recent statistical research by Professor Yale Brozen of the University of Chicago, among others, has demonstrated that there is little correlation between concentration ratios and monopoly power,[1055] or monopoly profits across industries.[1056] As Harold Demsetz, Professor of Economics at UCLA, has said: "No serious theoretical basis yet exists for [a doctrine of such correlation] . . ."; thus, concentration does not offer even a presumption of monopoly.[1057] Even so, concentration ratios continue to be used widely as one of the warning indicators, albeit a weak one,[1058] of the possible existence of monopoly power.[1059] Assuming, for the sake of argument, that there is some validity in utilizing concentration ratios as an indicium of market power, the question of quantification immediately presents itself. The academician who pioneered the attempt to quantify concentration ratios is Joe S. Bain, no stranger to researchers in the fields of the oil industry or industrial organization. In 1959,

---

[1054] SHENEFIELD PROPOSED STATEMENT, *supra* note 179, at 4-5.

[1055] Brozen, *Bain's Concentration and Rates of Return Revisited,* 14 J. Law & Econ. 351 (Oct. 1971); *accord,* Ikard, *Competition in the Petroleum Industry: Separating Fact from Myth,* 54 Ore. L. Rev. 583, 597 (1975) [hereinafter cited as Ikard, *Petroleum Industry Competition]; cf. Petroleum Industry Hearings, supra* note 192, at 2342 (Reprint of W. JOHNSON, R. MESSICK, S. VAN VACTOR and F. WYANT, COMPETITION IN THE OIL INDUSTRY, Dec. 1975) [hereinafter cited as JOHNSON et al., *Oil Industry Competition].*

[1056] Statement of Professor Edward J. Mitchell Before the Subcomm. on Antitrust and Monopoly of the Senate Committee on the Judiciary 11 (Aug. 6, 1974) [hereinafter cited as Mitchell, *1974 Senate Testimony];* H. DEMSETZ, THE MARKET CONCENTRATION DOCTRINE (A.E.I. 1973).

[1057] H. DEMSETZ, *Two Systems of Belief About Monopoly* in INDUSTRIAL CONCENTRATION: THE NEW LEARNING 166 (Goldschmid, Mann & Weston eds. 1974); *see Id.* at 177-180, including Tables 8 and 9; *accord,* authorities cited in note 1056, *supra;* Demsetz, *Industry Structure, Market Rivalry, and Public Policy,* 16 J. Law & Econ. 1 (Apr. 1973); *but cf.* L. WEISS, *The Concentration-Profits Relationship and Antitrust* in INDUSTRIAL CONCENTRATION: THE NEW LEARNING 202-277 (Goldschmid, Mann & Weston eds. 1974).

[1058] *Petroleum Industry Hearings, supra* note 192, at 354 (Colloquy between witness W. T. Slick, Jr., and Charles E. Bangert, Subcommittee General Counsel).

[1059] JOHNSON et al., *Oil Industry Competition, supra* note 1055, at 3.

Professor Bain suggested a tabular standard as follows:[1060]

4 Firm Percent of Market	8 Firm Percent of Market	Degree of Concentration
75 Percent or more	90 percent or more	Very high
65 to 75 percent	85 to 90 percent	High
50 to 65 percent	70 to 85 percent	Moderately high
35 to 50 percent	45 to 70 percent	Moderately low
Under 35 percent	Under 45 percent	Low

Later research has modified the standard shown above, and even among those who continue to use concentration ratios as one of many tools to examine an industry for monopoly power, the standard normally used to measure concentration in American manufacturing industries is the share of the market accounted for by the top four firms, weighted by value added.[1061] Professor Neil H. Jacoby, of the Graduate School of Management, UCLA, in his statement before Senator Hart's Subcommittee on Antitrust and Monopoly of the Senate Judiciary Committee on February 18, 1976, submitted a table, reproduced from the U.S. Department of Commerce, Bureau of Census, which was a distribution array of four firm concentration ratios with a weighted average concentration ratio of 40.1 percent.[1062] This was slightly up from the 39 percent average market share of the four largest firms for all manufacturers which was calculated to be 39 percent in 1969.[1063] The data on pipeline concentration ratios over time, measured by barrel-miles of

---

[1060] J. S. BAIN, INDUSTRIAL ORGANIZATION 124-133 (1959); *see also Petroleum Industry Hearings, supra* note 192 at 2230 (Statement of Professor Neil H. Jacoby); Ritchie, *Petroleum Dismemberment,* 29 Vand. L. Rev. 1131, 1142 (1976).

[1061] *Petroleum Industry Hearings, supra* note 192 at 2220, 2230 (Statement by Professor Neil H. Jacoby).

[1062] *Id.* at 2230. The table is as follows:

DISTRIBUTION OF BUREAU OF THE CENSUS FOUR FIRM CONCENTRATION RATIOS FOR MANUFACTURING INDUSTRIES, 1970

4-Firm Concentration Ratio	Share of Value Added	4-Firm Concentration Ratio	Share of Value Added
0—9	4.94	60—69	5.68
10—19	14.71	70—79	7.06
20—29	21.58	80—89	1.67
30—39	14.64	90—100	5.78
40—49	14.46		
50—59	9.48	Weighted average concentration ratio 40.1	

Source: U.S. Dept. Comm., Bureau of Census, "Annual Survey of Manufacturers, 1970, Value of Shipment" Concentration Ratios, M70 (AS)-9 (Wash: U.S. Govt. Printing Office, 1972)

[1063] W. SHEPHERD, MARKET POWER AND ECONOMIC WELFARE 106 (1970), cited in Ikard, *Petroleum Industry Competition, supra* note 1055, at 597-598.

interstate carriers by ownership are as follows:[1064]

Barrel-Miles Handled by Interstate Oil Trunk Lines

Year	4 Firm Percent	8 Firm Percent	15 Firm Percent	20 Firm Percent	30 Firm Percent	40 Firm Percent
1951	44.0	71.1	91.1	96.4		
1962	36.7	61.9	82.3	90.3		
1972	33.7	55.6	77.8	86.6		
1975	32.4	53.6	76.6	85.0		
1976	31.7	52.9	76.0	85.4	90.4	91.7
1977 (preliminary)	30.9	52.3	75.9	84.9	90.1	91.6

The foregoing table not only shows a decreasing concentration over time,[1065] but applying the 1977 figures, which are the latest available, against Professor Bain's standard the four-firm test would indicate a "low" degree of concentration and the eight-firm test would be "moderately low." Using the subsequent four firm, weighted by value-added, modification that Professor Jacoby reported in his testimony before Senator Hart's Subcommittee, pipeline concentration among the Top 4 firms appears to be substantially below the average. Comparing four and eight firm pipeline concentrations with other industries is rather revealing: Locomotives and Parts (98% & 99%); Primary Aluminum (96% & 100%); Flat Glass (94% & 98%); Motor Vehicle (91% & 97%); Chewing Gum (85% & 97%); Cigarettes (81% & 100%); Copper (75% & 98%); Tires and Inner Tubes (72% & 89%); Metal Cans (72% & 83%); Soaps and Detergents (70% & 79%); Radio and Television (48% & 67%); Steel (47% & 65%); Shoes (28% & 36%); Pharmaceutical Preparations (26% & 43%); Newspapers (16% & 24%); Soft Drinks (13% & 20%).[1066]

[1064] MARKET SHARES AND INDIVIDUAL COMPANY DATA FOR U.S. ENERGY MARKETS: 1950-1977 Table 7, p. 14 & Table 72, pp. 127-128 (A.P.I. Discussion Paper #014, Nov. 1978); see also W. SLICK, *A View From a Large Oil Company* in WITNESSES FOR OIL 17 (1976).

[1065] Cf. TEECE, *supra* note 329, at 95 ("National [pipeline] concentration ratios 1951-1971 appear to have declined"); Myers, *supra* note 198, Table 2 between pages 10 and 11 had an 8-firm figure for 1971 which was consistent with this trend. Myers states that "the degree of concentration of revenues by ownership in oil pipelines is about the same as in other segments of the oil industry. The table does not show that the pipeline segment is any less competitive than the others." *Id.* at 9; *but cf.* ERC REPORT, *supra* note 184, at 14 which, while not inconsistent with the trend shown in the chart from 1950 to 1977, indicates a "blip" in 1972-1973, and at iv, which states that crude and product pipeline concentration levels are generally higher than for production, refining and marketing.

[1066] *Petroleum Industry Hearings, supra* 192, at 374-375 (Exhibit E to Statement of W. R. Peirson); *see also Consumer Energy Act Hearings, supra* note 154, at 1043 (Testimony of Senator Dewey F. Bartlett); ERC REPORT, *supra* note 184, at 5 (Table 2).

The concentration ratio of pipelines, like the petroleum industry of which they are a part, when compared with other industries, on the 4, or even 8, firm grouping commonly used by industrial organization economists is very unpersuasive as evidence of monopoly power. Thomas Kauper, when he was Assistant Attorney General, Antitrust Division, testified that the "petroleum industry (hence, similarly, the pipeline segment)[1067] appears to be one of the least concentrated of our nation's major industries."[1068] Perhaps this explains why the authors of the Kennedy Staff Report, quoted above, felt constrained to go to a purported eighteen-firm grouping to achieve percentages that would permit major oil company-affiliated pipelines to meet Bain's eight firm test of "moderately high" and his four firm test of "very high," in order to support their proposed radical measure of divestiture,[1069] despite the fact that 10 out of the 16 industries enumerated above exceeded with 8 firms, and 6 of the 16 exceeded with 4 firms, the pipeline concentration achieved by the Kennedy Staff by aggregating *20* firms.[1070] In the light of this analysis, one can understand the charge made by oil company witnesses at the *Petroleum Industry Hearings* that critics of the industry were "gerrymandering" the numbers in order to achieve sufficiently high concentration ratios to support their case, even if it took 20 company aggregations to do it.[1071] This same technique did not escape the attention of Senator Bartlett of Oklahoma who made a pointed reference in an earlier hearing held by the Subcommittee to the artificiality of comparing 20 firm aggregate concentration ratios to those of four or eight firms in another industry.[1072]

[1067] See note 1065, *supra*.

[1068] S.2387 REPORT, *supra* note 49, pt. 2 at 208 (1976) (Minority Views and Additional Views). *See also* Mitchell, *1974 Senate Testimony, supra* note 1056, at 11-12 ("Most manufacturing industries are more concentrated than the petroleum industry"); *Petroleum Industry Hearings, supra* note 192, at 2228 (Statement of Professor Neil H. Jacoby: "Comparatively unconcentrated"); Ikard, *Petroleum Industry Competition, supra* note 1055, at 598 ("Much less concentrated than most manufacturing industries"); TEECE, *supra* note 329, at 95 ("The concentration of pipeline mileage is quite moderate").

[1069] *Cf.* Ritchie, 29 Vand.L.Rev. 1131, 1140 (1976).

[1070] While the Kennedy Staff Report, at page 30, purported to be using 18 companies for its "major" classification for purposes of the Report, rejecting Occidental and Tenneco (page 30 n.93), when it got to page 56, from which the quotation in this text at note 1052 was taken, the staff drew upon witness David Jones and took his classification of majors, which included "the 20 largest oil companies based on sales." See Jones & Gardner Memo, *supra* note 427.

[1071] *Petroleum Industry Hearings, supra* note 192, at 353 (colloquy between W. T. Slick, Jr., Senior Vice President, Exxon Co., U.S.A., and Charles E. Bangert, Subcommittee General Counsel) ("if you make the sample small enough, you can get the concentration ratio high enough; and if you pick the number of companies high enough, you can get the concentration ratio high enough" — Mr. Bangert: "All right, yes, absolutely, and I agree"); *Id.* at 1821 (Statement of William P. Tavoulareas, President, Mobil Oil Corp.,: "In spite of these facts, there are still some critics determined to make the oil industry into a monopoly, even if they have to count up to 20 companies before they can make the charge stick. What the words '20-company monopoly' mean, I have no idea. Perhaps it is best to leave it as a mystery.").

[1072] *Consumer Energy Act Hearings, supra* note 154, at 1045 (Testimony of Senator Dewey F. Bartlett).

Most industrial organization economists consider 20 firms to be a sufficient number for competitive results.[1073] So much for the Kennedy Staff Report.

Turning, then, to Assistant Attorney General John Shenefield's approach,[1074] it is worth noting that he has extended the aggregation exercise yet another step. He went beyond the device of swelling the ranks of the "major" oil companies by pulling in additional companies such as Occidental and Tenneco[1075] so as to achieve the desired concentration ratio; instead he drew upon protestants' witness David Jones' testimony in the pipeline general rulemaking proceedings (FERC Dkt. RM 78-2, successor to ICC *Ex parte* 308) to develop a "cutting edge" attack. Mr. Shenefield would depart from the classic "major" versus "independent" oil company schism and merge the two industry groups into a new category — "petroleum companies" (all companies in the oil industry which have an ownership interest in an oil pipeline) and produce a new variety of "independent" (a pipeline company which has no affiliation whatsoever with any other person engaged in any other phase of the oil business). In this way, he arrived at concentration ratios of 98.6% of all crude oil barrel-miles and 86.9% of all product barrel-miles for "petroleum companies," leaving only 1.4% and 13.1%, respectively, for the newly classified "independents."

Shenefield's "independent" classification introduces a new definitional problem. For example, Belle Fourche Pipeline Co., Cheyenne Pipeline Co., and Trans Mountain Oil Pipeline Corp. are affiliated with persons in the oil business, a fact obscured by the failure of Shenefield's source,[1076] witness David Jones, to check beyond Standard & Poor's and Moody's to ascertain whether Southern Pacific might have oil interests;[1077] and Williams Pipe Line Company's owner, the Williams Companies, had a subsidiary, Williams Exploration Company, engaged in exploration for, and production of, oil and gas and another subsidiary, Williams Energy Company, which is a substantial "independent" marketer of propane gas.[1078] Suffice it to say that Shenefield's argument

---

[1073] *Petroleum Industry Hearings, supra* note 192, at 1063 (reprint of two chapters of E. ERICKSON and L. WAVERMAN, THE ENERGY QUESTION: AN INTERNATIONAL FAILURE OF POLICY — Volume 1).

[1074] See note 1054, *supra.*

[1075] See note 372 *supra.*

[1076] KENNEDY STAFF REPORT, *supra* note 184, at 45 n.132.

[1077] *RM78-2 Hearings Transcript, supra* note 754, Vol. 3 at 501-504 (Cross Examination by John M. Cleary, representing the State of Alaska, Farmer's Union Central Exchange, Farmland Industries, and other independents and cooperative associations). To set the record straight, Southern Pacific is engaged in oil production. *Petroleum Industry Hearings, supra* note 192, at 308 (Statement of Vernon T. Jones) and Buckeye bought out the owners of Jet Lines.

[1078] *Petroleum Industry Hearings, supra* note 192, at 306 (Statement of Vernon T. Jones, President of Williams Pipe Line Co.).

thereby is strained. W. T. Slick, Jr., pointed out in his testimony before the Hart subcommittee: "All I have to do is to select the sample. You know, right now, the concentration ratio on oil company spokesmen in this room is 100 percent; I've got the microphone."[1079]

Allowing for changes that have taken place since 1975 when David Jones' compilation was made, the corresponding information could be arrayed as follows: As of January 1, 1977, there were 110 regulated common carriers which comprise substantially the oil pipeline network in the United States. Of these 110 companies, 42 are owned exclusively by 18 "major" oil companies [same definition of "major" used by the Kennedy Staff Report, *i.e.*, the PICA, or S.2387, Report pp. 16-17];[1080] 52 are owned exclusively by "non-majors," which includes traditional "independents" and those not related to oil companies; and 16 are owned jointly by both "majors" and "non-majors."[1081] This array appears to be more consonant with the way the average person would look at the structure. Incidentally, bias, if any, is against the "majors" because the small intrastate lines, which are dominated by "independents," are not included in the given universe.

Professor David Teece, who teaches business economics at the Graduate School of Business, Stanford University, summarized the situation succinctly: "Although measures of concentration in pipeline mileage at the national level are very poor measures of competition in pipeline transportation, they nevertheless indicate how the relative involvement of the majors in pipeline ownership has changed over time. National concentration ratios 1951-1971 appear to have declined [reference]. The level of regional concentration is difficult to calculate since pipelines very often connect the various PAD districts. Nevertheless, it is apparent that the concentration of pipeline mileage is quite moderate. Unless this mileage is especially strategically located, it is difficult to see how the alleged monopoly power could be at all pervasive, especially in view of ICC regulation."[1082]

2. Paucity of Pipelines Operated by Non-Oil-Related Owners

As has been demonstrated in earlier sections, oil pipelines are, in most instances, both capital intensive and relatively high risk investments

---

[1079] *Petroleum Industry Hearings, supra* note 192, at 353 (Colloquy between W. T. Slick, Jr., and Charles E. Bangert, Majority General Counsel).

[1080] For the full citation of this report, see note 49, *supra*. The Kennedy Staff Report's adoption of this standard is found in KENNEDY STAFF REPORT, *supra* note 184, at 30.

[1081] PIPELINE PRIMER, *supra* note 331, at 6.

[1082] TEECE, *supra* note 329, at 95.

with limited upside potential due to regulation and no corresponding downside protection.[1083] Because potential investors are well aware of these risks and because most pipelines are of logistical value only to the oil companies which use their facilities,[1084] with few exceptions, petroleum pipelines in the United States have been conceived, financed, and built by the oil companies which they serve.[1085]

The reasons why oil companies have invested in pipelines while others generally do not can be classified roughly into four categories:[1086]

(1) Oil companies are usually the first to recognize that a need for a new pipeline exists. This may be related to a planned refinery expansion or the development of new oil fields. Those involved in the business which will be the source of traffic for a proposed pipeline are usually in the best position to assess the risks involved. They are likely to have more pertinent information concerning the source of supply and the probable growth of demand over the life of the projected pipeline. An outside investor would suffer what the subset of industrial organization economists, known as transactions cost analysts, would term "information impactedness."[1087] As a result, a potential investor in a pipeline who is not in the oil business and had to overcome the foregoing transactional difficulties would not have as strong a belief in the validity of the vital data as would a company which is in the business and has much of the data at hand and a basis for evaluating that which must be obtained externally.[1088]

(2) The oil pipeline business is one of relatively low rates of return on

---

[1083] See Sections III A and B, at notes 606-634, *supra.*

[1084] Gary, *FERC Direct Testimony, supra* note 613, at 5.

[1085] *Id.*; Waidelich, *Rodino Testimony, supra* note 197, at 7; *Consumer Energy Act Hearings, supra* note 154, at 595, 623 (Statement of Jack Vickrey); PIPELINE PRIMER, *supra* note 331, at 5.

[1086] Spahr, *Rodino Testimony, supra* note 255, at 11-12).

[1087] Dr. David J. Teece, in his work, cited in full at note 329, *supra,* defines the term as existing "when pertinent market information is known by one or more parties but cannot be costlessly discerned by, or displayed to others." *Id.* at 125. Teece states that it is a derivative condition which arises mainly on account of uncertainty and "opportunism" [see this text at note 329, *supra*], although "bounded rationality" [see this text at notes 737 and 738 *supra*] is involved as well. *Id.* at 14. Professor Liebeler, of the UCLA Law School, put the matter in a positive way, namely, that vertical integration facilitates the "right" amount of investment in information and, since it reduces the cost of capturing the return on investment therein, it increases the amount of information which is "optimal" in a market system. LIEBELER, *Integration and Competition, supra* note 976, at 11. For a handy Glossary of terms used by transactions analysts, as applied to the oil business, see TEECE, *supra* note 329, at 123-128.

[1088] Spahr, *Rodino Testimony, supra* note 235, at 11; *Petroleum Industry Hearings, supra* note 192, at 1102 (Statement of Charles E. Spahr); *cf.* DE CHAZEAU & KAHN, *supra* note 30, at 337. E. MITCHELL, *Capital Cost Savings of Vertical Integration* in VERTICAL INTEGRATION IN THE OIL INDUSTRY 78 (E. Mitchell ed. 1976) [hereinafter cited as MITCHELL, *Capital Cost Savings*].

total capital.[1089] As witness E. P. Hardin stated in testimony before the Hart Subcommittee: "A 7 percent return on valuation is not a red-hot business deal,"[1090] and there is no guarantee that the line will not be an unmitigated disaster, as has been shown even in the case of integrated ownership lines. Thus, it traditionally has not been attractive to an investor who has no experience in the pipeline business and can readily invest his capital elsewhere for a greater return with less risk.[1091] (Even some members of the industry have indicated that they would rather put their money elsewhere and get a better return.)[1092] Oil companies have made the financial committments for pipelines because they needed the lines and it was the only way to get them.[1093]

(3) Oil companies, which have owned and operated pipelines for many years have a wealth of practical experience upon which to draw. Given an appropriate rate of return, they are the logical ones to build the lines. Not too many non-oil-related companies, if any, could marshal the expertise to design, construct or operate a major new pipeline system.[1094]

(4) The financing of new large pipeline systems is difficult for someone without a substantial reservoir of good credit. Lenders will not lend money to a pipeline without a track record in the absence of full

---

[1089] *Petroleum Industry Hearings, supra* note 192, at 251 (Statement of Thomas diZerega, President, Apco Oil Co. ("not a very high rate of return").).

[1090] *Petroleum Industry Hearings, supra* note 192, at 1102 (Colloquy between E. P. Hardin and Garrett Vaughn, Minority Staff Economist); Judge Keech, the trial judge in the *Arapahoe* proceeding in 1958 asked: "Nobody would contend that in a field which is subjected to hazards as oil transportation that a rate of 7 percent would be unreasonable?" To which question, Attorney Alfred Karsted, of the Antitrust Division, Department of Justice, replied: "No, your Honor." United States v. Atlantic Refining Co., Civil No. 14060, D.D.C., Mar. 24, 1958, Tr. 31-32.

[1091] *Cf.* DE CHAZEAU & KAHN, *supra* note 30, at 336.

[1092] *S.1167 Hearings, supra* note 432, p. 8 at 5929 (Robert E. Yancey, President of Ashland Oil, replying to a question by Senator Tunney: "We would much rather pay the tariff [on Colonial's line] [and] put our money into other areas where we get a much higher rate of return"); *Id.,* pt. 9 at 606 (Statement of Vernon T. Jones); Spahr, *Rodino Testimony, supra* note 235, at 13.

[1093] Spahr, *Rodino Testimony, supra* note 235, at 12; *cf.* Waidelich, *Rodino Testimony, supra* note 197, at 7-8; Gary, *FERC Direct Testimony, supra* note 613, at 5; PIPELINE PRIMER, *supra* note 331, at 5; *S.1167 Hearings, supra* note 432, pt. 9 at 619 (Answer by witness Jones to question by Henry Banta, Staff Counsel); Perhaps the most poignant expression of this point was made by Charles Murphy, Chairman of Murphy Oil, a nonowner shipper on Colonial, at an American Enterprise Institute Round Table Discussion: "Now, on the question of pipelines, we are shippers on Colonial; we're not owners. We built the pipeline [Collins Pipeline] from our refinery to tie into Colonial.

"Now, we needed that pipeline. We would have sat there till doomsday if we had waited for some banking group or some local entrepreneur to come in and build such a pipeline.

"This is why the petroleum industry is vertically integrated. There's an irresistible impulse to vertical integration when a firm has a problem—usually a transportation problem that no one is interested in solving but the firm experiencing the problem." Mitchell, *1974 Senate Testimony, supra* note 1056, at 27.

[1094] Spahr, *Rodino Testimony, supra* note 255, at 12.

access to credit-worthy owners. The lenders must be satisfied that the project is economically viable and can stand on its own feet,[1095] but at the same time they look through the project to the ultimate guarantors, the owners.[1096] In the case where the owner is also a shipper ("shipper-owner") this is done normally by means of a Throughput and Deficiency Agreement under which each shipper-owner agrees to ship, or cause to be shipped, his respective percentage of the total amount of traffic required to meet the pipeline's obligations, and if for *any* reason the pipeline has a cash deficiency when principal or interest on the debt falls due, each shipper-owner will immediately advance its pro rata share (calculated after taking into account his shipments during the accounting period) of the pipeline's cash deficiency as prepaid transportation. This "hell or high water" clause, as it is known, ensures that the debt and interest are serviced.[1097] Thus, the entire risk is shifted to the shipper-owners and the credit rating of the pipeline's obligation tends to rise toward their composite rating, and hence the cost of the borrowing is reduced. Because oil companies, large and small, are the only ones who have the sources of supply and established markets necessary for the pipeline's successful operations and because they have the compelling need for the line, they are the only ones who traditionally have been willing and able to undertake the throughput obligations necessary to obtain the required financing.[1098]

Having laid down the general principles, an examination of the empirical evidence is in order to test the validity, and flesh out the details, of these principles. Early on, there was a promotional interstate crude line built with independent capital to transport crude from the Smackover, Arkansas, field to Vidalia, Louisiana, on the Mississippi River, which failed for lack of throughput and was sold for use as a natural gas line.[1099]

[1095] Remarks of Raymond B. Gary, Partner, Morgan Stanley & Co., to Alaskan Senate and House Committees on Proposed Legislation Concerning Pipeline Regulation, Right-of-Way and State Ownership, Mar. 6, 7 & 8, 1975, reprinted in *Petroleum Industry Hearings, supra* note 192, at 1117 [hereinafter cited as Gary, *Alaskan Bills Testimony*]; Gary, *FERC Direct Testimony, supra* note 613, at 8; Spahr, *Rodino Testimony, supra* note 255, at 11.

[1096] Gary, *Alaskan Bills Testimony, supra* note 1095, at 1117; Gary, *FERC Direct Testimony, supra* note 613, at 8; *S.1167 Hearings, supra* note 432, pt. 9 at 608-609 (Statement of Vernon T. Jones); *cf.* Swenson, *S.1167 Testimony, supra* note 755, at 703.

[1097] Swensen, *S.1167 Testimony, supra* note 755, at 703; Gary, *FERC Direct Testimony, supra* note 613, at 8; Spahr, *Rodino Testimony, supra* note 255, at 12.

[1098] Waidelich, *Rodino Testimony, supra* note 197, at 8; HOWREY & SIMON, *supra* note 3, at 31; PIPELINE PRIMER, *supra* note 331, at 5; *cf. Consumer Energy Act Hearings, supra* note 154, at 595 (Statement of Jack Vickrey). Atwood & Kobrin, Res. Study 005, *supra* note 650, at 4-7 list five reasons for vertical integration: (1) Efficiencies of Integration; (2) Avoidance of Monopoly; (3) Marginal Cost Tariffs Attainable Only Under Integration; (4) No Tariff Exists to Support the Optimum; and (5) Price Regulation.

[1099] WOLBERT, *supra* note 1, at 111-112; *Consumer Energy Act Hearings, supra* note 154, at 623 n.49 (Statement of Jack Vickrey); MILLS, *supra* note 324, at 86-87 (1935).

There have been several other abortive attempts by promoters to put together large diameter projects. In 1948, an outfit called Texas-Western Oil Lines, Inc., proposed a 24-inch line from West Texas to the Los Angeles area. As part of the attempt to secure throughput, the promoters offered ownership interests to producers and refiners (a throwback to the Producers and Refiners Oil Company, Limited, in the Pennsylvania fields in 1890's, except that the initiators of that line had been the producers and refiners themselves).[1100] The $100,000,000 financing requirement proved insuperable, and the existing demands on West Texas crude were too great to justify diversion westward.[1101] In July, 1951, promoters formed the United States Pipe Line Company (no relation to Lewis Emery's company which constructed crude and products lines from western Pennsylvania, to Marcus Hook in the 1890's)[1102] to construct a 963 mile, 16-inch products pipeline from Beaumont to Cincinnati, Ohio,[1103] which, after steel priority was refused, was amended to extend to Newark, New Jersey.[1104] In August, 1953, the American Pipe Line Corporation proposed a 1425 mile 22- to 26-inch products line from Beaumont to Newark, New Jersey. Although each had obtained steel allocations from PAD and had approval for accelerated tax amortization, they failed for lack of financing due to their inability to produce assurance of throughput.[1105] In the other direction, the advent of the Korean emergency revived the question of bringing West Texas crude to the West Coast. The Progress Company of Los Angeles proposed a 24-inch crude line from, and a 10.75-inch products line to, West Texas.[1106] A Dallas group attempted to promote the West Coast Pipe Line Company, a 1,030-mile 22-inch crude line from Midland, Texas, to Norwalk, California. H. Graham Morison,

---

[1100] See text at notes 68-70, *supra*.

[1101] PETROLEUM PIPELINES, *supra* note 45, at 368.

[1102] See text at notes 73-76, *supra*.

[1103] *RM78-2 Hearings Transcript, supra* note 754, Vol. 12 at 1936 (Oral Argument of Jack Vickrey on behalf of Belle Fourche, an independent crude gatherer); HOWREY & SIMON, *supra* note 3, at 29-30 (563 miles).

[1104] *H.Res.5 Subcomm Hearings, supra* note 235, at 141 (Testimony of Fred F. Steingraber); HOWREY & SIMON, *supra* note 3, at 30; *Consumer Energy Act Hearings, supra* 154, at 624 (Statement by Jack Vickrey); PETROLEUM PIPELINES, *supra* note 45, at 375. The promoters offered five-year guaranteed tender contracts to interested shippers and received a "railroad clearance" that the Justice Department would not recommend criminal antitrust prosecution, but no clearance was given on the civil side, nor was any opinion expressed on the possible application of other federal laws, such as the Interstate Commerce Act and the Elkins Act. PETROLEUM PIPELINES, *supra* note 45, at 370.

[1105] *H.Res.5 Subcomm Hearings, supra* note 234, at 141 (Testimony of Fred F. Steingraber); *Consumer Energy Act Hearings, supra* note 154, at 624 (Statement by Jack Vickrey); *RM78-2 Hearings Transcript, supra* note 754, Vol. 12 at 1936 (Oral Argument by Jack Vickrey on behalf of Belle Fourche, an independent crude gatherer); HOWREY & SIMON, *supra* note 3, at 30.

[1106] PETROLEUM PIPELINES, *supra* note 45, at 368.

a former Assistant Attorney General (Antitrust Division), represented the company. Unfortunately, the group owned very little crude and was unable to obtain commitments from either producers or refiners. The resourceful Morison sought to remedy this lack of backup for financing by applying to the ODM for a certificate of essentiality which would generate a government guarantee of the repayment of the loan. There was a substantial question whether the price of West Texas crude plus transportation charges would make the laid down cost competitive with California crudes of comparable quality.[1107] The upshot of the matter was that the ODM found the venture to be economically unsound (which history and the Four Corners' experience have verified), hence the project failed for lack of financing.[1108]

Another insight into the problems faced by non-oil company-affiliated pipelines comes from the experience of the pipelines that were divorced from the Standard Oil group as a result of the 1911 dissolution decree. The Buckeye Pipe Line, Indiana Pipe Line Company, Northern Pipe Line and New York Transit Company were left to make their own way as common carriers, as were the National Transit Company, the Southwest Pennsylvania Pipe Lines, the Eureka Pipe Line Company, and the Southern Pipe Line. The first four lines continued as more or less a common unit, associated in common management, providing a common-carrier crude-oil pipeline network connecting Griffith, Indiana, to Buffalo. For a while they continued to serve their old customers, and although traffic began to erode as the former Standard refining companies found more efficient ways to secure crude oil at more advantageous rates, the lines managed to survive. In 1941, partially due to a disagreement over allocation of joint expenses which culminated in a court decision that New York Transit was bearing more than its proper share,[1109] the four lines merged into Buckeye. The World War II re-creation of the old system, due to the interruption by enemy submarines of tanker movements from the Gulf Coast to East Coast, added some stimulus to the lines but it was obviously limited to the duration of the war. However, under aggressive leadership, Buckeye went into the petroleum products pipeline business, first by purchasing an eight-inch line connecting Ohio's (now Marathon's) refinery at Robinson, Illinois, to

---

[1107] *Id.* at 369; Testimony before the House Armed Services Committee in July, 1955, revealed that West Texas crude delivered in California would be 15 to 20 cents per barrel higher than the cost of California crude of the same gravity. *Consumer Energy Act Hearings, supra* note 154, at 657 (Statement of Jack Vickrey). There also was a problem of quality.

[1108] *Consumer Energy Act Hearings, supra* note 154, at 624, 656-657; PETROLEUM PIPELINES, *supra* note 45, at 368-371.

[1109] PETROLEUM PIPELINES, *supra* note 45, at 365-366.

Indianapolis.[1110] By providing a service to shippers who did not find it economical to build their own products lines, Buckeye sought to build a place for itself in a postwar pipeline world populated by large companies. By 1952, it was providing products service for refineries in Illinois, Michigan, and Ohio to destinations throughout that area, utilizing converted small diameter crude lines where it could and building new ones where it could not. It moved aggressively into the East, transporting products from New York and Philadelphia (via a connection with Harbor Pipeline System at Linden, New Jersey) over a newly constructed 16-, 14-, and 10-inch system through northern Pennsylvania, up to Buffalo and then east to Syracuse, and from New Jersey into the rapidly growing Long Island market. It bought the renovated Tuscarora Pipe Line from Esso Standard, thus linking its Mid-West and Eastern Divisions and provided access by Seaboard refiners to the Cleveland market. It was able to borrow money on its "track record" until it was purchased in the 1960's by an affiliate of the Pennsylvania Railroad.[1111] National Transit was not as strategically located as Buckeye. Its revenue came from its gathering system, serving thousands of "fruit jar" size producing wells in the old Appalachian field; two crude trunk lines, one from the Bradford, Pennsylvania, producing region to Pittsburgh and the other from Kane, Pennsylvania, to Philadelphia; and it owned the National Transit Pump & Machine Company, a substantial manufacturing enterprise. After a World War II induced prosperity, the business began to decline and the Rockefeller Foundation disposed of its substantial holdings in National Transit to a consortium headed by Wertheim & Co., a New York securities firm, which promptly "spun off" the Pump & Machine Company and reduced National Transit's capital to make a substantial cash distribution. This caused so much alarm among the Western Pennsylvania oil refiners who relied on National Transit for transportation to market that in 1948 they acquired a substantial interest in National Transit.[1112] In 1952, Southwest Pennsylvania was merged into National Transit. The Eureka Pipe Line Company was also acquired by Pennsylvania refiners and Southern Pipe Line was bought out by Ashland Oil & Refining Company.[1113]

Before leaving the subject of non-oil-related pipelines, it might be helpful to review witness David Jones' (and by incorporation, Assistant

---

[1110] MCLEAN & HAIGH, *supra* note 35, at 423. Ohio Oil (Marathon) apparently was concerned about the nature and extent of ICC jurisdiction over products lines. Significantly, Ohio Oil was not a signatory party to the *Atlantic Refining Company* Consent Decree although it had been named as a defendant in the "Mother Hubbard" case.

[1111] A. JOHNSON, *Lessons, supra* note 108, at 213; See also text at note 165, *supra*.

[1112] PETROLEUM PIPELINES, *supra* note 45, at 367-368. See text following note 165, *supra*.

[1113] JOHNSON, *Lessons, supra* note 108, at 213. See text at note 166, *supra*.

Attorney General Shenefield's) list of such pipelines. The first pipeline on the list is Air Force Pipeline, Inc., which is owned 100 percent by an affiliate of the Southern Railway. The second is Belle Fourche Pipeline Co., which has already been shown to be related to Dave True and the True Oil Company. The third is Buckeye, discussed above, which comes close to being a new style "independent" but which is 100 percent owned by an affiliate of the Pennsylvania Railroad (now the Penn-Central). Cheyenne Pipeline Company, the fourth on Jones' list, was disclosed to be affiliated with the Nielson interests which are related to Husky Oil. Jet Lines, Inc., a products line operating in the Massachusetts-Connecticut area, has been acquired by Buckeye. Kaneb Pipe Line Company is a genuine non-oil-interest-related independent pipeline company which has, with two exceptions, managed to raise its own financing. In 1952, the Anderson-Pritchard Oil Co., an independent refiner predecessor of Apco, and in 1959 CRA, a subsidiary of Farmland Industries, entered into throughput commitments with Kaneb to support connections which were of sufficient distance to make it unsatisfactory for Kaneb and its lenders to construct without such commitments. They have been satisfied "a long time ago."[1114] Kaneb is the result of the merger of two lines: the Augusta Pipe Line Company, which ran 43 miles from the Anderson-Pritchard refinery at Arkansas City, Kansas, to Augusta, Kansas, (near the Wichita marketing area) and the 246-mile Kaneb line, running north from Augusta, which was directly connected to six other small independent refineries, to Fairmont, Nebraska. The initial capacity of the 10-inch main line was only 25,000 barrels per day and the aggregate funded debt of the two lines was less than $7 million.[1115] This really is not comparable to Colonial or Explorer either in mileage, throughput, or financial requirements. MAPCO, Inc., while shown by Jones' compilation to be held by "10,277 stockholders, the largest owning 5 percent," was promoted and organized by the Katy Railroad[1116] and it is now engaged in the retailing of propane and the exploration for oil and gas.[1117] Southern Pacific Pipelines, Inc., operates over 2,400 miles of refined products lines, consisting of four distinct lines or groups of lines. The largest, over 1,300 miles, runs from connections in Los Angeles east to Phoenix and Tucson, and also west from El Paso, Texas, to Phoenix. Another group radiates from the San Francisco area north to Chico, California, south to San Jose, and east to Reno, Nevada. Shorter lines connect tanker terminals at

---

[1114] *S.1167 Hearings, supra* note 432, p. 8 at 5943 (Testimony of Richard C. Hulbert, President, Kaneb Pipe Line Co.).

[1115] LIVINGSTON, *supra* note 228, at 34-35.

[1116] See text at note 171, *supra;* Burke, *supra* note 9, at 790.

[1117] *Petroleum Industry Hearings, supra* note 192, at 307 (Statement of Vernon T. Jones).

Portland to Eugene, Oregon, and refineries at Bakersfield to Fresno, California. Southern Pacific, a holding company, has total assets of over $4 billion, so financing was not a problem.[1118] It also has the railroad's existing right-of-way,[1119] which was determinative in the construction of Black Mesa, a coal slurry pipeline. It had the motivation of regaining the refined products traffic that it once had enjoyed as a rail carrier but lost to competing forms of transportation. A similar situation is the San Diego Pipeline, jointly owned by the Southern Pacific and the Santa Fe, running from Los Angeles to San Diego. The Santa Fe also operates, through subsidiaries, anhydrous ammonia lines and a LPG line from West Texas to Houston.[1120] Although the point goes more to the definition of "independent" than to the main point of whether related oil operations are necessary to finance the lines, most of the Western Railroads, certainly the Union Pacific (which owns Champlin), the Southern Pacific and the Santa Fe all have oil production, arising out of their "checkerboard" land grants from the government which were given in the late 1800's to support the development of a nationwide rail service.[1121] Trans-Mountain, which is David Jones' next to last "neo-independent" line, clearly is oil related and oil financed.[1122] The last on the list is the Williams Pipe Line Company, owned by the Williams Companies. While Williams Pipe Line Company is one of the largest "neo-independent" pipelines, it is well to remember that Williams purchased the Great Lakes Pipeline System in 1966 at a time when it had a well established operating record commencing from 1931, and that the main portion of the system was originally designed, financed, and constructed by oil company shipper-owners.[1123]

Summing up, the empirical evidence appears to support the opening general principle that, with few exceptions, petroleum pipelines have been conceived, financed, and built by the oil companies who need their services. The leading exceptions had an early history of oil company sponsorship. There has been a recent development where companies of

---

[1118] LIVINGSTON, *supra* note 228, at 32-34.
[1119] *Id.;* Gary, *FERC Direct Testimony, supra* note 613, at 6.
[1120] *Petroleum Industry Hearings, supra* note 192, at 308 (Statement of Vernon T. Jones); *S.1167 Hearings, supra* note 432, pt. 9 at 608 (Statement of Vernon T. Jones).
[1121] *Id.*
[1122] *Cf.* KENNEDY STAFF REPORT, *supra* note 184, at 45 n.133.
[1123] Gary, *FERC Direct Testimony, supra* note 613, at 6; Waidelich, *Rodino Testimony, supra* note 197, at 5; *Consumer Energy Act Hearings, supra* note 154, at 601 (Statement of Jack Vickrey); PIPELINE PRIMER, *supra* note 331, at 5. This point is far more important than the definitional error that Williams is non-oil-related when in fact it is engaged in oil and gas exploration and the marketing of propane gas. *Petroleum Industry Hearings, supra* note 192, at 307 (Statement of Vernon T. Jones, President of Williams Pipe Line Co.).

appropriate financial stature and a motivation and/or a peculiar advantage have entered into the pipeline business. These are typified by the railroads. Finally, there is room for smaller "independent" lines which are able to fulfill a special need — Kaneb being a good example. It would appear that continued shipper ownership of oil pipelines is both necessary and reasonable, and in the words of the President of a non-oil-related "independent" pipeline: "The shippers and the consuming public appear to be best served by a pipeline carrier industry with open entry comprised of independents as well as shipper-owned systems of all shapes and sizes."[1124]

B. Vertical Integration

1. Definition

With the possible exception of market concentration, no antitrust or industrial organization concept has been so widely misunderstood as vertical integration.[1125] The strict industrial organization concept of integration refers to the extent to which a firm carries on the productive process from the extraction of raw materials to the transformation of these materials into a final product.[1126] A variant of this definition is the combining of the owned resources of market transactors (*i.e.*, buyer and seller) in common ownership.[1127] The industrial organization economist is interested more in the functional coordination of one or more units in each of the successive stages of production so that they are operated as a single unified industrial process,[1128] than he is with the particular legal form of the transaction.[1129] Because the industrial organization economist is concerned principally with the relative costs of transacting business via integration as opposed to relying upon the market mechanism,[1130] the goods or services transferred from one stage to another must be able,

---

[1124] *S.1167 Hearings, supra* note 432, pt. 8 at 5941 (Statement of Richard C. Hulbert, President, Kaneb Pipe Line Co.).

[1125] LIEBELER, *Integration and Concentration, supra* note 976, at 5.

[1126] ERC REPORT, *supra* note 184, at 19, citing E. SINGER, ANTITRUST ECNOMICS: SELECTED LEGAL CASES AND ECONOMIC MODELS 206 (1968).

[1127] M. CANES, A THEORY OF THE VERTICAL INTEGRATION OF OIL FIRMS 5 (1976; mimeographed, available through the API) [hereinafter cited as CANES, *Oil Integration Theory*].

[1128] Hale, *Successive Stages, supra* note 104, at 921, citing Frank, *The Significance of Industrial Integration,* 33 J. Pol. Econ. 179 (1925).

[1129] LIEBELER, *Integration and Concentration, supra* note 976, at 7.

[1130] *Cf.* Rostow & Sachs, *Entry Into the Oil Refining Business: Vertical Integration Reexamined,* 61 Yale L.J. 856, 877-878 (1952); CANES, *Oil Integration Theory, supra* note 1127 at 5.

without significant adaptation, to be bought or sold in the market.[1131] From an economist's standpoint, market displacement can take place equally well by contract or by ownership; he is not concerned with the form but by the economic consequences. This is where the antitrust law departs from economic theory and form becomes a factor in and of itself.[1132] However, antitrust law also places great stress on the saleability of intermediate products.[1133] In the petroleum industry, vertical integration refers to the simultaneous operation in more than one of the industry's four principal phases: production, refining, crude and/or product transportation, and marketing.[1134]

2. Extent of Vertical Integration

Preliminarily, one problem in examining the extent of vertical integration in the oil industry is the fact that, given the definitions set forth in the preceding subsection, it is easy, almost inevitable, to start with the particular industry structure *as it happens to exist* and, without reflection, assume it to be a norm; then go on to establish categories in accordance with that structure. As Professor Wesley Liebeler has noted, if shoe manufacturing and distribution had been conducted by separate firms in the past, a merger or other arrangement between a shoe manufacturer and a retailer would be conceived as being vertical integration. Because of the "new look," antitrust enforcement agencies would be prone to examine the situation. But because shoe manufacturing companies have traditionally owned their plants and retail outlets, most people, including antitrust lawyers, would not conceptualize the arrangement in vertical integration terms. Liebeler gives a number of homey examples, such as ownership of houses which could be occupied under rental, do-it-yourself repairs, plumbing, carpentry work, and the like. Yet each of these activities is a supersession of the market mechanism, falling squarely within the definition of vertical integration, but no one harbors a thought about antitrust attack in connection with them.[1135] Moreover, the very concept of different or separable steps in the production process, and the perception of what constitutes a part of it, will vary from time to time,

---

[1131] Adelman, *Integration and the Antitrust Laws,* 63 Harv. L. Rev. 27 (1949); WOLBERT, *supra* note 1, at 91; Hale, *Successive Stages, supra* note 104, at 922, citing BURNS, THE DECLINE OF COMPETITION 421 (1936).

[1132] LIEBELER, *Integration and Concentration, supra* note 976, at 7.

[1133] Hale, *Successive Stages, supra* note 104, at 922; Adelman, *Integration and the Antitrust Laws,* 63 Harv. L. Rev. 27 (1949); BURNS, THE DECLINE OF COMPETITION 421 (1936).

[1134] Ritchie, *Petroleum Dismemberment,* 29 Vand. L. Rev. 1131, 1132 n.2 (1976); CANES, *An Economist's Analysis, supra* note 1021, at 46 n.3; WOLBERT, *supra* note 1, at 3-4 n.3.

[1135] LIEBELER, *Integration and Concentration, supra* note 976, at 5-7.

depending upon technology and the availability and efficiency of alternatives.[1136]

There seems to be a widespread misconception among people outside the petroleum industry that vertical integration is the trademark solely of large companies. Nothing could be further from the truth.[1137] The fact is that vertical integration has not been,[1138] nor is it now, limited to a few large petroleum companies;[1139] most of the smaller ones such as Koch Industries and Crown Central are integrated vertically as well.[1140] Although the *degree* of total integration (*i.e.*, engaged in two to four basic activities) appears to be somewhat higher among the larger refiners than among many smaller refiners,[1141] the differences between "majors" and "independent" refiners with respect to integration are more accurately described as differences in degree than in kind.[1142] If the measuring index

---

[1136] DE CHAZEAU & KAHN, *supra* note 30, at 41.

[1137] *Petroleum Industry Hearings, supra* note 192, at 325 (Statement of W. T. Slick, Jr.); Att'y Gen., *First Report, supra* note 111, at 24, citing MCLEAN & HAIGH, *supra* note 35, at 33-35; *Petroleum Industry Hearings, supra* note 192, at 1856 (Statement of Professor Edward J. Mitchell).

[1138] WOLBERT, *supra* note 1, at 4 n.3; *see also* RUSIN & NEWPORT, VERTICAL INTEGRATION MEASURES FOR A SAMPLE OF OIL FIRMS 81-97 (API Res. Study #009, May 15, 1978).

[1139] LIVINGSTON, *supra* note 228, at 3. See Table 1, *Id.* at 4.

[1140] See note 627, *supra*. In 1974, less than 4 percent of the U.S. refining capacity was operated by companies who were only in the refining end of the business. *Petroleum Industry Hearings, supra* note 192, at 325 (Statement of W. T. Slick, Jr.); *cf.* Ikard, *Petroleum Industry Competition, supra* note 1055, at 600: "A [1950] survey of refiners showed that refiners who were totally nonintegrated accounted for less than 2 percent of domestic refining capacity," citing MCLEAN & HAIGH, *supra* note 35; Mitchell, *1974 Senate Testimony, supra* note 1056, at 18, testified concerning an analysis made in 1960. In that year, the twenty largest domestic refiners had an average ratio of crude production to refinery runs of 49.7 percent. Of the next twenty-five largest refiners, only eighteen offered data adequate to calculate this ratio; their average ratio was 44 percent. Small companies that were completely integrated (all four stages) as of 1973 include Canadian Hydrocarbon, Ltd.; Clark Refining Co.; Coastal States; Crown Central Petroleum Corp.; Crystal Oil Co.; Diamond-Shamrock Corp.; Farmers Union Central Exchange; Farmland Industries; Husky Oil Co.; Koch Industries; Lakeside Refining Co.; Midland Cooperative Inc.; Murphy Oil; Pasco, Inc.; Reserve Oil & Gas Co.; Texas City Refinery Inc.; Total Leonard, Inc.; and Witco Chemical Co. Source: Annual Reports, 10-Ks.

[1141] ERC REPORT, *supra* note 184, at 27-29 (Table I, p. 28 employs an index with 1.0 representing complete integration across all four phases); CANES, *Oil Integration Theory, supra* note 1127, at Table 1, located between pages 21 and 22. Canes ran a regression analysis of the correlation between the integration index (V) and refiner size (R) and found $V = 0.255 + 0.00046R$ $R^2 = 0.44$ F value = 17.8;
(4.2)

This analysis indicates that a 100,000 barrel increase in refiner capacity would be associated with about a 0.05 increase in a company's vertical integration index. *Id.* at 21.

[1142] ERC REPORT, *supra* note 184, at 29. If the index of vertical integration had not been based on the spanning of all four principal stages of operation, virtually all companies in the industry would be integrated in a meaningful sense because they engage in more than one single operation. DE CHAZEAU & KAHN, *supra* note 30, at 20, which takes us back to Liebeler's comments cited in the opening paragraph of this subsection.

of integration had been refiners' integration into pipelines, even the difference in degree would have narrowed sharply. Dr. Michael Canes attributes this to the oil specialization of pipelines and hence the less the diseconomies from nonspecialization of knowledge. He cites as an example that although oil is transported both by barge and pipelines, oil companies have integrated less into ownership of barges than into pipelines, and less into ownership of generalized barges than into barges specially designed for oil transport.[1143] Empirically, pipeline ownership corresponds rather closely with refinery size. Thus, all of the 42 largest refining companies, accounting for 94 percent of total domestic crude-running capacity in January, 1974, owned crude-oil pipelines, and all but three owned refined products lines. Among the smaller refiners, 23 owned crude lines and 11 owned products lines, and the smaller refiners as a group had roughly the same share of pipeline ownership as of refining capacity.[1144] This distribution of pipeline integration should allay concern, expressed in the S.2387 (PICA) Report, that nonintegrated firms in the petroleum industry are disadvantaged.[1145] Professor Edward J. Mitchell testified before the Senate Antitrust and Monopoly Subcommittee in 1974 that the notion that there is a group of "independent" refiners significantly less integrated than "major" refiners apparently was a myth.[1146]

In the Attorney General's First Report to Congress under his obligation to inform that body whether the activities of the member states had conformed to the purpose of Article V of the Interstate Compact to Conserve Oil and Gas, the author(s) inferred that "unlike most other industries," the principal oil companies were fully integrated.[1147] Although it is debatable whether this comparison was accurate even at the time it was made,[1148] the *current* fact is that vertical integration is *not*

[1143] CANES, *Oil Integration Theory, supra* note 1127, at 23. Of course, as will be shown later, there are reasons other than the fact that oil barges require less specialized knowledge than oil pipelines which have caused oil companies to integrate into pipelines more frequently than into barges.

[1144] LIVINGSTON, *supra* note 228, at 3.

[1145] S.2387 REPORT, *supra* note 49, at 19.

[1146] Mitchell, *1974 Senate Testimony, supra* note 1056, at 18; *accord,* Ikard, *Petroleum Industry Competition, supra* note 1055, at 600.

[1147] Att'y Gen., *First Report, supra* note 111, at 24.

[1148] Professor M. A. Adelman's study of vertical integration in the United States indicated that the petroleum industry was less integrated than most other manufacturing industries. M. ADELMAN, *Concept and Statistical Measurement of Vertical Integration* in BUSINESS CONCENTRATION AND PUBLIC POLICY (1955). Adelman's Table, page 302, shows that the petroleum industry was the second least integrated of the nineteen industries arrayed. Professor Edward J. Mitchell made separate calculations from corporate annual reports for the same year on 183 large corporations which confirmed Adelman's conclusion. Of the nineteen manufacturing industries, petroleum was the third least integrated. *Petroleum Industry Hearings, supra* note 192, at 1854 (Statement of Professor Edward J. Mitchell).

unique to the petroleum industry;[1149] To the contrary, as an efficient form of industrial organization designed to serve the consumer better at lower costs, it is a pervasive phenomenon in all advanced industrial economies,[1150] common among U.S. companies[1151] and industries.[1152] In fact, we are surrounded by it.[1153] There are a number of other industries in the United States that are also integrated, such as automobiles, paper, aluminum, food, newspapers, steel, communications, textiles, machinery, wine producers, tobacco, nonferrous metals, chemicals, rubber, beer, and drugs and health products, several of them highly so.[1154] Disinterested observers have stated that the U.S. petroleum industry has relatively little integration compared with U.S. manufacturing in general.[1155]

3. Reasons for Vertical Integration

A key inquiry in examining the structure of the pipeline industry is

---

[1149] *Petroleum Industry Hearings, supra* note 192, at 324 (Statement of W. T. Slick, Jr.).

[1150] *Petroleum Industry Hearings, supra* note 192, at 2228 (Statement of Professor Neil H. Jacoby). For example, the list of foreign privately-owned companies with a significant degree of vertical integration includes: Attock Oil Co. Ltd. (England); Berry Wiggins & Co. Ltd. (England); Canadian Hydrocarbon, Ltd. (Canada); "Delek" The Israel Fuel Corporation Ltd.; Husky Oil Ltd. (Canada); Maruzen Oil Co. Ltd. (Japan); National Refinery Ltd. (Pakistan); Pacific Petroleums Ltd. (Canada); Petrofina S.A. (Belgium); Wintershall Aktiengesellschaft (Germany); and Royal Dutch/Shell Group. Vertical integrated petroleum companies that are entirely or largely owned by foreign governments include: British Petroleum Co. Ltd. (BP); Campaignie Francaise des Petroles (CFP) (France); Enterprise de Recherches et d'Activites Petrolieres (ERAP) (France); Ente Nazionale Idrocarburi (ENI) (Italy); Iraq National Oil Company (INOC) (Iraq); National Iranian Oil Co. (NIOC) (Iran); Neste Oy (Finland); Norske Hydro A.S. (Norway); Oesterreichische Mineralolwervaltung A.G. (OMV) (Austria); Petroleo Brasileiro S.A. (Petrobras) (Brazil); Petroleos Mexicanos (PEMEX) (Mexico); Societe Nationale pour la Recherche, la Production, le Transport, la Transformation, et la Commercialisation des Hydrocarbure (SONATRACH) (Algeria); Venezolana Del Petroleo (CVP) (Venezuela); and Veba A.G. (Germany); *Id.* at 1869 (Statement of Professor Edward J. Mitchell).

[1151] ERC REPORT, *supra* note 184, at 30; CANES, *An Economist's Analysis, supra* note 1021, at 46.

[1152] A. CARD, *A Common Form of Industrial Organization* in WITNESSES FOR OIL 35 (1976) [hereinafter cited as CARD, *Industrial Organization*].

[1153] LIEBELER, *Integration and Competition, supra* note 976, at 18. Senator Bayh, a self-professed hog farmer (*Petroleum Industry Hearings, supra* note 192, at 323) might be surprised to learn that he was vertically integrated. *Petroleum Industry Hearings, supra* note 192, at 2228 (Statement of Professor Neil H. Jacoby).

[1154] ERC REPORT, *supra* note 184, at 30-31; *Petroleum Industry Hearings, supra* note 192, at 325 (Statement of W. T. Slick, Jr.); CARD, *Industrial Organization, supra* note 1152, at 35-36.

[1155] *Petroleum Industry Hearings, supra* note 192, at 2229 (Statement of Professor Neil H. Jacoby); *Id.* at 1850, 1952 (Statement of Professor Edward J. Mitchell); *cf.* M. ADELMAN, *Concept and Statistical Measurement of Vertical Integration* in BUSINESS CONCENTRATION AND PUBLIC POLICY 302 (1955); M. GORT, DIVERSIFICATION AND INTEGRATION IN AMERICAN INDUSTRY 83 (1962).

why firms, and, of special pertinence to this exposition, why oil companies, integrate. It is a black-letter proposition in economics that vertical integration is nothing more or less than the substitution of an internal organization in place of the market.[1156] Stated another way, firms internalize [integrate] their operations when it is more economical to do so than it is to use the marketplace.[1157] Thus, markets and firms can be viewed as alternative vehicles for executing a related set of transactions and the determinative factor of which mode to employ is its efficiency vis-à-vis the other.[1158] Logic would suggest that an inquiry into vertical integration should examine whether, and which, economies can be obtained by bringing certain transactions within the firm.[1159] Unfortunately, however, this line of analysis has been delayed by two errors in the early examinations of the subject. The first was a misconception derived by inductive reasoning from one of the early classic examples of the advantages of the economies of vertical integration, namely the now famous example, first pointed out by Professor Morris Adelman, of the savings inherent in the integration of the steel industry's blast furnace with the open hearth stage. The blast furnace, using coke, produces pig iron plus heat plus combustible gases and by combining with the open hearth stage, much of the pig iron heat is conserved and the coke gases may be utilized, producing obvious savings.[1160]

While the example, as such, cannot be faulted, unfortunately it led to the rather narrow view by some economists, drawing on Professor Joe Bain's early tentative work, that integration savings were associated *only* with purely technical or engineering considerations.[1161] It should be noted

---

[1156] *Petroleum Industry Hearings, supra* note 192, at 1856 (Statement of Professor Edward J. Mitchell); Mitchell, *Capital Cost Savings, supra* note 1088, at 73.

[1157] R. Coase, *The Nature of the Firm* in 4 ECONOMICA (n.s.) 336 (Nov. 1937), reprinted in READINGS IN PRICE THEORY (G. Stigler & K. Boulding eds. 1952) [hereinafter cited as Coase]; Mitchell, *1974 Senate Testimony supra* note 1056, at 12; Ikard, *Petroleum Industry Competition, supra* note 1055, at 598; *Petroleum Industry Hearings, supra* note 192, at 2228 n.3 (Statement of Professor Neil H. Jacoby) ("In theory, vertical integration will occur whenever the costs to the firm of transacting in markets (including premiums for bearing risk) exceed the costs of vertical unification."); F. SCHERER, INDUSTRIAL MARKET STRUCTURE AND ECONOMIC PERFORMANCE 70 (1970) ("The most obvious and pervasive motive for vertical integration is to reduce costs."); Ritchie, *Petroleum Dismemberment*, 29 Vand. L. Rev. 1131, 1132-1133 (1976).

[1158] Oliver Williamson, *supra* note 329, at 1442.

[1159] Mitchell, *1974 Senate Testimony, supra* note 1056, at 12.

[1160] Adelman, *Integration and Antitrust Policy*, 63 Harv. L. Rev. 27, 31-32 (1949); F. SCHERER, INDUSTRIAL MARKET STRUCTURE AND ECONOMIC PERFORMANCE 70 (1970); *cf.* WOLBERT, *supra* note 1, at 92 n.541.

[1161] *Petroleum Industry Hearings, supra* note 192, at 1857 (Statement of Professor Edward J. Mitchell), citing J. BAIN, INDUSTRIAL ORGANIZATION 381 (2d ed 1968). For an example of a statement of the narrow view, see F. SCHERER, INDUSTRIAL MARKET STRUCTURE AND ECONOMIC PERFORMANCE 70 (1970). But even Professor Scherer recognized that the technological interdependency basis was much too narrow. *Id.* at 87. Because nothing was said about ownership, a modern transactions analyst might well view the

that Professor Bain merely said, in effect, that integration having a physical or technical aspect made a *clearer* case for cost savings, a savings *per se* case if you will, and other forms of integration such as the assembly of assorted components of production generally had a less clear case.[1162] Glossing over Bain's self-imposed limitations, his statement[1163] has been seized upon by policy makers who have used it as a limiting parameter for vertical integration measurement.[1164] It took the economic community a substantial period of time to overcome this narrowness of viewpoint and to formulate the broadened and more enlightened modern view. Professor Oliver Williamson has expressed this view as follows: "In more numerous respects than are commonly appreciated the substitution of internal organization for market exchange is attractive less on account of technological economies associated with production but because of what may be referred to broadly as 'transactional failures' in the operations of markets for intermediate goods."[1165] This view is more realistic than the narrow view, and it provides an explanation why integration occurs in some cases and not in others.[1166]

The second delaying theory strand traces its origin to Corwin Edwards who harbored a deep suspicion of the anticompetitive possibilities of misuse of vertical integration.[1167] DeChazeau & Kahn developed a thesis that even though few petroleum firms had complete [crude production to refinery runs] self-sufficiency, it was in their self-interest to hold crude prices up and take their profits in the oil producing stage, so long as they were at least 77 percent crude sufficient because of the then 27½ percent depletion allowance; and this was true even if the firm was only 38.5 percent self-sufficient, provided that 50 percent or more of the crude price increase could be reflected in product price increases. The net result, according to this line of reasoning, was an artificially high crude price and an artificially low refining margin, which

---

problem of supplying hot pig iron to a nearby open hearth as a case of avoidance of a "small numbers bargaining problem."

[1162] J. BAIN, INDUSTRIAL ORGANIZATION 381 (2d ed. 1968).

[1163] Professor Bain candidly acknowledged the tentative nature of his conclusion as being based on "miscellaneous scraps of evidence" and noted the "lack of systematic research endeavor." *Id.*

[1164] *Petroleum Industry Hearings, supra* note 192, at 1857 (Statement of Professor Edward J. Mitchell); Mitchell, *Capital Cost Savings, supra* note 1088, at 8.

[1165] O. WILLIAMSON, *The Vertical Integration of Production: Market Failure Considerations,* 61 Am. Econ. Rev. 112 (May 1971); Williamson's theorem builds on the earlier work of Coase, *supra* note 1157.

[1166] LIEBELER, *Integration and Competition, supra* note 976, at 8.

[1167] C. EDWARDS, MAINTAINING COMPETITION 97-98, 172-174 (1949); see also Edwards, *Vertical Integration and the Monopoly Problem,* 17 J. Mktg. 404-410 (1953).

put the "squeeze" on nonintegrated refiners.[1168] The FTC was so pleased with this "squeezing" thesis that, after adjusting the critical self-sufficiency ratio from 38.5 percent to 40.4 percent because the depletion allowance had been reduced from 27.5 percent to 22.0 percent, it used the argument, first in its 1973 Preliminary Staff Report[1169] and subsequently in the "Big Eight" (*Exxon* et al.) proceeding. The only things wrong with the theory are that the logic is defective, the mathematics are faulty and it is contrary to some commonly known facts about the industry. The logical flaw arises from the fact that although squeezing refining margins would tend to drive smaller nonintegrated refiners out of the market, it would at the time render refining unprofitable for the so-called "squeezing" major company. Given an unprofitable refinery business, a rational integrated firm would cease investing in refineries and invest in "super-profitable" crude production. Capital would continue to flow into production and avoid refining until the rate of return in refining equilibrated with that of production.[1170] Another way of describing the logical flaw is that it ignores the true price determinants, *i.e.*, supply and demand. The theory sidesteps the key question: *how* would the price be raised? If a firm has sufficient market power to raise the price, or conspires to do so, this is the true root of the problem; it is not vertical integration. It can be dealt with directly under the antitrust laws. The mathematical boner, *i.e.*, that with a 22 percent depletion allowance, a firm with a self-sufficiency ratio greater than 40.4 percent would gain from shifting profits from refining to crude production, was pointed out by Professor Richard Mancke of Tufts University, who demonstrated that using the FTC's premises, it would take in excess of 81 percent self-sufficiency for such profit shifting to be beneficial.[1171] Even this figure is too low, because it ignores the effect of severance taxes and royalty payments. When these factors are taken into account, the threshold self-sufficiency ratio for advantageous profit shifting (with a 22 percent depletion allowance) must exceed 93 percent,[1172] a ratio none of the "Big

---

[1168] DE CHAZEAU & KAHN, *supra* note 30, at 221-225 (1959). Interestingly enough, a contemporaneous critic of the industry was arguing simultaneously that integration by the "majors" was being used to depress crude prices. E. ROSTOW, A NATIONAL POLICY FOR THE OIL INDUSTRY 65 (1948).

[1169] Preliminary Federal Trade Commission Staff Report on Its Investigation of the Petroleum Industry, 1973 [hereinafter cited as FTC, *Preliminary Staff Report*].

[1170] Mitchell, *1974 Senate Testimony, supra* note 1056, at 14; Ikard, *Petroleum Industry Competition, supra* note 1055, at 599.

[1171] R. MANCKE, PETROLEUM CONSPIRACY: A COSTLY MYTH (1974).

[1172] R. MANCKE, THE FAILURE OF U.S. ENERGY POLICY (1974), footnote 32 to Chapter 7; Ikard, *Petroleum Industry Competition, supra* note 1055, at 600; JOHNSON et al., *Oil Industry Competition, supra* note 1055, at 2384.

Eight" respondents, nor eight of the nine "second tier" majors, enjoy as evidenced by the FTC staff's own report.[1173] Only Getty, the sixteenth largest, owned enough crude to make shifting pay. The other firms would have lost from three percent to 48 percent on each dollar of shifted profit. The factual deficiencies referred to are that the FTC relied upon "special advantages" allegedly held by the "majors" over their independent competitors in order to make tenable the "squeezing" theory, to wit, import quotas and the depletion allowance. These have passed from the scene,[1174] and as has been discussed previously, large oil companies are not disproportionately integrated as compared to small oil companies.[1175] It would seem that the ghosts of utilization of vertical integration to transfer profits upstream to the producing end of the business and to "squeeze" the refiner's margin have had a respectful burial and should be permitted to rest in peace and not becloud any longer the meaningful investigation of why petroleum companies elect to integrate in some circumstances and not in others.

According to Professor Edward Mitchell, there are two basic conditions where markets are inferior to internal organization. The first of these conditions is where transaction costs are high,[1176] and the second is where information is available more rapidly and/or more cheaply to the integrating firm than to outsiders.[1177] While this discussion will particularize a number of specific motives which industry executives and observers have articulated, Mitchell believes that they fall within one or the other of these conditions.[1178] As Mitchell has observed, businessmen do not conceptualize their motives for vertical integration in Professor Oliver Williamson's transactional analysis terms. However, the importance of reliable sources of supply, assured markets, reduction of risk

---

[1173] FTC, *Preliminary Staff Report, supra* note 1169, at 20.

[1174] Import quotas were abolished in 1973, and the oil depletion allowance was reduced to 22 percent in 1969 and was taken away completely from the larger companies in 1975. MANCKE, *Oil Industry Competition, supra* note 674, at 64; *see also* JOHNSON et al., *supra* note 1055, at 2383-2384. MANCKE, *Id.* at 64 n.38, noted the unprecedented action of the FTC Administrative Law Judge in recommending the charges in the "Big Eight" case be withdrawn in the light of the changed circumstances.

[1175] See text at the notes 1137-1146.

[1176] *Petroleum Industry Hearings, supra* note 192, at 1856 (Statement of Professor Edward J. Mitchell); Mitchell, *Capital Cost Savings, supra* note 1088, at 73; TEECE, *supra* note 329, at 9-12.

[1177] *Petroleum Industry Hearings, supra* note 192, at 1856 (Statement of Professor Edward J. Mitchell); Mitchell, *Capital Cost Savings, supra* note 1088, at 73-74; M. ADELMAN, *Concept and Statistical Measurement of Vertical Integration* in BUSINESS CONCENTRATION AND PUBLIC POLICY (1955); Arrow, *Vertical Integration and Communications,* 6 Bell J. Econ. 173-183 (No. 1, Spring, 1975), TEECE, *supra* note 329, at 12-15.

[1178] *Petroleum Industry Hearings, supra* note 192, at 1856 (Statement of Professor Edward J. Mitchell); Mitchell, *Capital Cost Savings, supra* note 1088, at 74.

and lower financing costs are simply the way a businessman expresses what the economist would describe as the impossibility of drafting an ironclad and complete contract with an upstream supplier that will give the refiner the assurances he needs to operate that refinery as continuously and at as high an operating ratio as he requires. Conversely, no upstream company is possessed of the knowledge of exactly what the refiner needs nor is he likely to acquire such knowledge within a practical time frame. Mitchell translates this back into "economese" by rephrasing the businessman's problem in terms that because of the impacticability of perfectly contracting or the shortcomings of communication, it is cheaper and more timely for the businessman to do it himself.[1179] It also furnishes some protection against market failure, such as the recent cutback in Iranian production.

What, then, are the circumstances that cause members of the industry to choose the internal route rather than rely on the market? One principal factor is that the successive stages of the petroleum industry are each highly capital intensive. The cost of a new "grass-roots" refinery of 250 thousand barrels per day is in excess of $500 million, an installed offshore producing platform and related facilities range in cost from $45 million to over $200 million and a 20 percent share of a pipeline such as TAPS is in excess of $1.5 billion,[1180] closer to $2.0 billion. It is critical, therefore, that companies with investments in assets of this nature operate at a high level with minimum interruption in order to utilize their facilities efficiently and economically.[1181] These facilities are long-term investments, highly dependent on each other,[1182] and hence any arrangement for their source of supply and distribution of their production necessarily must be long-term.[1183] Moreover, once they are in place, they are highly specialized in

---

[1179] *Petroleum Industry Hearings, supra* note 192, at 1860 (Statement of Professor Edward J. Mitchell); Mitchell, *Capital Cost Savings, supra* note 1088, at 80; see also Ritchie, *Petroleum Dismemberment,* 29 Vand. L. Rev. 1131, 1134-1135 n.14 (1976).

[1180] *Petroleum Industry Hearings, supra* note 192, at 332 (Statement of W. T. Slick, Jr.).

[1181] *Id.;* Historically, domestic refiners have operated in the range of 80-85 percent utilization of capacity and have thus managed to balance properly the avoidance of high cost idle capacity on the one hand and danger of too little capacity on the other. *Id.* MCLEAN & HAIGH, *supra* note 35, at 40, found that integration was a far more decisive factor in determining operating rates than was size. See also *Petroleum Industry Hearings, supra* note 192, at 1861 (Statement of Professor Edward J. Mitchell).

[1182] Mitchell, *1974 Senate Testimony, supra* note 1056, at 19; Ikard, *Petroleum Industry Competition, supra* note 1055, at 602; *Petroleum Industry Hearings, supra* note 192, at 331 (Statement of W. T. Slick, Jr.).

[1183] *Oil Daily,* Apr. 18, 1975, p. 6, reported the comment by a Shell Oil Company executive that in order to build a grass roots refinery, a company would have to have at least a twenty year solid source of crude oil; *cf.* Mitchell, *Capital Cost Savings, supra* note 1088, at 76; *Petroleum Industry Hearings, supra* note 192, at 1858 (Statement of Professor Edward J. Mitchell).

purpose, virtually immobile[1184] and of little value for alternative use.[1185] These characteristics subject such investments to the risk that extremely low returns would be earned if flows of crude oil or oil products were interrupted or reduced.[1186] Hence, high on the list of reasons advanced by industry members[1187] and observers of the industry[1188] is security of supply and certainty of outlet. A typical industry statement, cited in note 1187, is that made by witness William P. Tavoulareas, President of Mobil Oil Corporation, in the *Petroleum Industry Hearings*: "I have a tremendous refinery system in the United States, around the world, and knowing I have some supplies in that refinery means I can supply my customers over a long period of time. That has to be efficient, because I am able, then, to enter into contracts on a selling side and contracts on a buying side, where I have security of supply which, as I am sure you understand, can be the greatest thing in order to level out costs . . . Now, let me break down the other part of it. When you build a refinery, you have to know what you are building a refinery for. You could build your refinery for speculation. We have seen people in the tanker business build for speculation. As the market is high, they made a fortune; when the market is low, there is disaster.

[1184] CANES, *An Economist's Analysis, supra* note 1021, at 53; Ritchie, *Petroleum Dismemberment,* 29 Vand. L. Rev. 1131, 1134 (1976); *cf.* Burke, *supra* note 9, at 788.

[1185] Ritchie, 29 Vand. L. Rev. 1131, 1134 (1976); *cf.* Mitchell, *Capital Cost Savings, supra* note 1088, at 76; *Petroleum Industry Hearings, supra* note 192, at 1858 (Statement of Professor Edward J. Mitchell).

[1186] CANES, *An Economist's Analysis, supra* note 1021, at 52; Ritchie, *Petroleum Dismemberment,* 29 Vand. L. Rev. 1131, 1134 (1976).

[1187] *Petroleum Industry Hearings, supra* note 192, at 332 (Statement of W. T. Slick, Jr.); *Id.* at 368 (Statement of Walter R. Peirson); *Id.* at 1834 (Statement of William P. Tavoulareas); Ikard, *Petroleum Industry Competition, supra* note 1055, at 602; Ritchie, *Petroleum Dismemberment,* 29 Vand. L. Rev. 1131, 1134 (1976); Burke, *supra* note 9, at 788. The authors of the S.2387 REPORT, *supra* note 49, took exception to the formulation of this thesis, saying: "The problem with this argument is that it *assumes* intermediate markets cannot work effectively in the petroleum industry. It *assumes* that the inability of the intermediate markets to provide security of supply is an inherent characteristic of the oil business and can only be overcome by vertical integration. There is simply no basis for this *assumption*. The distinguished economist, F. M. Scherer, after a major study of 16 industries concluded that, 'the more prone input markets are to breakdown of price competition, the stronger is a firm's incentive to integrate upstream.' Or to put it another way, the very existence of the *compulsion* to integrate indicates that the supply markets are not adequately competitive." *Id.* at 19. [*Emphasis added*] First, market failure can be caused by factors other than lack of competition. The second Arab-Israeli war led to the closure of the Suez Canal in 1956 and a concomitant crude oil shortage as tankers had to be routed the long way around the Cape of Good Hope. Also, a number of chemical and other manufacturing industries have integrated upstream into crude oil and gas, not because competition was inadequate, but because of market failure. *See* note 1191, *infra*. The FTC Bureau of Economics noted this fact and found it to be competitive in its February, 1979, study. Second, as is demonstrated in a subsequent part of this subsection, see text at notes 1229-1268, especially 1236-1260, *infra*, decisions to integrate are not based on *assumptions* but are made on a case-by-case analysis whether the marketplace or integration is the more economical mode.

"I do not think it is in the interest of the shareholders, [nor that] it is a very smart thing to build a refinery and not know where you are going to put the products, or build a market and not know where you are going to get the products." A similar observation was made by DE Chazeau & Kahn, cited in note 1188, who remarked: "[Integration] advantages are particularly attractive in an industry forced by the nature of its product and technology to place an unusual emphasis on continuity of flow. The fluidity of oil and its products, the enormous volume of their continuing movement, and the essentiality of smooth synchronization of its almost entirely automatic processes make it difficult to conceive of a business system that did not provide assurances of continuous supply of material and acceptance of the product." This need for certainty in the supply system is felt especially keenly when there are distortions in the market such as rationing,[1189] quota schemes,[1190] or price control.[1191] Tying it back to the transactions analysis framework, Professor David Teece has observed that this argument can be translated into a "contractual incompleteness" issue. In other words, the assertion that vertical integration is necessary to assure supply really amounts to a statement that long-and short-term contracts have serious shortcomings in many important circumstances [this will be examined shortly] and therefore, supply reliability is basically a transaction cost problem.[1192]

A second principal reason for integration is economies of centralized management. One such economy arises from coordinated planning.[1193] Because of the highly interdependent nature of the various stages, the success of one stage, say a refinery, is as much, if not more,

---

[1188] *Petroleum Industry Hearings, supra* note 192, at 1860 (Statement of Professor Edward J. Mitchell); Mitchell, *1974 Senate Hearings, supra* note 1056, at 20; Mitchell, *Capital Cost Savings, supra* note 1088, at 80; ERC REPORT, *supra* note 184, at 23; DE CHAZEAU & KAHN, *supra* note 30, at 268; MCLEAN & HAIGH, *supra* note 35, at 665-666.

[1189] ERC REPORT, *supra* note 184, at 23; CANES, *Oil Integration Theory, supra* note 1127, at 15-16.

[1190] Oliver Williamson, *supra* note 329, at 1460. Avoidance of transaction taxes is another similar incentive. *Id.;* Stigler, *The Division of Labor is Limited by the Extent of the Market*, 59 J. Pol. Econ. 190-191 (1951); CANES, *Oil Integration Theory, supra* note 1127, at 15.

[1191] Ritchie, *Petroleum Dismemberment,* 29 Vand. L. Rev. 1131, 1135-1136 (1976); CANES, *Oil Integration Theory, supra* note 1127, at 14-15; Ikard, *Petroleum Industry Competition, supra* note 1055, at 602-603 ("Even companies outside the oil industry, such as chemical, automobile, and steel companies have recognized the reality of price-control induced shortages and have begun to integrate backwards into crude and natural gas production to assure essential supplies"); Mitchell, *1974 Senate Testimony, supra* note 1056, at 20-21 ("The recent controls on crude oil prices have motivated firms as diverse as Bethlehem Steel, Ryder Systems, and Dow Chemical to move into crude oil and natural gas production. The list also includes General Motors, Ford, International Paper, St. Regis Paper, and W. R. Grace").

[1192] TEECE, *supra* note 329, at 12.

[1193] DE CHAZEAU & KAHN, *supra* note 30, at 267; MCLEAN & HAIGH, *supra* note 35, Chapter XI.

attributable to the success in the raw materials, transportation, and products markets as it is to the success of the refining operation itself. Projects being planned today are so intertwined that it is difficult to perceive how they could be done through markets with anywhere near the efficiency that can be achieved by an integrated company.[1194] Start with a simple case, stated by Professor George Harmon,[1195] of the importance of the location and design of a pipeline in facilitating the closing of an uneconomical refinery as part of an overall economy program. Professor Harmon did not spell out the details, but his example fits perfectly the action by Shell Oil in closing down its obsolete East Chicago refinery, expanding and diversifying its Wood River refinery, and converting to products service its old eight inch crude line from Wood River to East Chicago.[1196] Dr. Morris Livingston has described the planning that went into Standard of Indiana's Mandan, North Dakota, refinery, where the refinery, a crude oil feed line and a products pipeline were planned in a coordinated manner.[1197] Then, visualize the planning of a refinery-chemical plant complex with the requisite transportation facilities, raw material supply, and distribution facilities which must come onstream in a closely knit time frame, and it should become obvious that integration has to be more efficient than attempting to do the job completely by the market mechanism. An academician might describe this as achieving a stronger assurance of technical complementarity of successive industrial processes.[1198]

Not only is planning better coordinated, but integration provides closer operational coordination between stages of production,[1199] which is especially important where there is a high degree of interdependency between them,[1200] resulting, among other things, in lower inventory and reduced working capital.[1201] Lower level of inventory means: (a) lower carrying costs of inventories per unit of sale, and (b) smaller fluctuations in inventory value as crude oil and product prices vary and hence smaller

---

[1194] *Cf.* Mitchell, *1974 Senate Hearings, supra* note 1056, at 19.
[1195] Harmon, *Effective Public Policy to Deal with Oil Pipelines,* 4 Am. Bus. L. J. 113, 118 (1966)
[1196] See text at note 858, *supra.*
[1197] LIVINGSTON, *supra* note 228, at 30.
[1198] *Petroleum Industry Hearings, supra* note 192, at 2228 (Statement of Professor Neil H. Jacoby).
[1199] Ritchie, *Petroleum Dismemberment,* 29 Vand. L. Rev. 1131, 1133 (1976).
[1200] *Petroleum Industry Hearings, supra* note 192, at 331 (Statement of W. T. Slick, Jr.); *cf.* DE CHAZEAU & KAHN, *supra* note 30, at 268; MCLEAN & HAIGH, *supra* note 35, at 666.
[1201] *Petroleum Industry hearings, supra* note 192, at 331 (Statement of W. T. Slick, Jr.); *Id.* at 1860 (Statement of Professor Edward J. Mitchell); DE CHAZEAU & KAHN, *supra* note 30, at 267-268; *cf. S.1167 Hearings, supra* note 432, pt. 8 at 6250 (Statement of Charles P. Siess, Jr.); MCLEAN & HAIGH, *supra* note 35, at 667.

variations in unit sales costs.[1202] This operational coordination makes possible a more dependable flow of petroleum products and services of uniform high quality,[1203] and because of its smooth synchronization, a more effective use of available capacity is achieved.[1204] Timing, which is extremely important both from an operational standpoint and the fact that the time value of a more prompt return on invested capital (hence a higher DCF) is enhanced.[1205] For example, witness Walter R. Peirson, President of Amoco Oil Company, testified before the Senate Antitrust & Monopoly Subcommittee on the subject of the value of the timing factor in connection with the Mandan, North Dakota, refinery previously mentioned.[1206] Another operational economy realized by integration is the coordinated use of specialized talents, experience, facilities and resources across successive production stages.[1207]

The third principal advantage of vertical integration in this area is that of risk amelioration and the concomitant reduction in cost of capital.[1208] Risk reduction takes two forms: one is the provision of a tool to deal with uncertainty in the conduct of the business by providing reliability in maintaining orderly and synchronized flow of the production process, high use of capacity and minimization of cost.[1209] The other is that, due to the variation from time to time of the relative profitability of the individual stages of the oil business, integration tends to reduce the variability of an oil firm's profits,[1210] thus an investor, who is the source

---

[1202] *Petroleum Industry Hearings, supra* note 192, at 1860 (Statement of Professor Edward J. Mitchell).

[1203] *Petroleum Industry Hearings, supra* note 192, at 331 (Statement of W. T. Slick, Jr.); *Id.* at 2228 (Statement of Professor Neil H. Jacoby).

[1204] DE CHAZEAU & KAHN, *supra* note 30, at 268: MCLEAN & HAIGH, *supra* note 35, at 40, 312-318; Integration is a far more decisive factor in determining operating rates than is size. *Petroleum Industry Hearings, supra* note 192, at 1861 (Statement of Professor Edward J. Mitchell); Mitchell, *Capital Cost Savings, supra* note 1088, at 82, citing MCLEAN & HAIGH, *supra* note 35, at 40, who had found that in 1950, the typical integrated refining company was able to maintain 81.6 percent of capacity levels as opposed to 54.3 percent by the typical nonintegrated refiner, and the differences in operating levels between companies of different size was significantly less than that; moreover, the differences between successive size groups, were characteristically less than the differences between integrated and nonintegrated refiners within the same size group.

[1205] *Petroleum Industry Hearings, supra* note 192, at 333 (Statement of W. T. Slick, Jr.); *cf.* MCLEAN & HAIGH, *supra* note 35, at 665.

[1206] *Petroleum Industry Hearings, supra* note 192, at 366 (Statement of Walter R. Peirson).

[1207] ERC REPORT, *supra* note 184, at 23-24; DE CHAZEAU & KAHN, *supra* note 30, at 566; *Petroleum Industry Hearings, supra* note 192, at 2229 (Statement of Professor Neil H. Jacoby).

[1208] *Petroleum Industry Hearings, supra* note 192, at 2229 (Statement of Professor Neil H. Jacoby); *Id.* at 331 (Statement of W. T. Slick, Jr.); DE CHAZEAU & KAHN, *supra* note 30, at 566.

[1209] See text at notes 1199-1207, *supra*.

[1210] CANES, *An Economist's Analysis, supra* note 1021, at 53; Ikard, *Petroleum Industry*

of a substantial portion of an oil company's capital, will demand a lesser expected rate of return than he would from a nonintegrated oil company.[1211]

Professor Mitchell's second main category concerned itself with the question of information transmission. In order to schedule and coordinate production efficiently, a firm must possess or gather information on market conditions in order to develop a reasonable set of expectations about the future.[1212] In many instances, the acquisition of market information requires an expenditure or sacrifice of real resources. There is, in effect, a "tooling up" cost. Having once acquired it, however, an integrated firm incurs only the marginal cost of transmitting it to an extra user, whereas a nonintegrated new user would have to expend the entire set-up cost to bring itself abreast of its integrated competitor.[1213]

The existence of these "observational economies," as Professor

---

Competition, supra note 1055, at 602; Mitchell, *1974 Senate Testimony, supra* note 1056, at 19-20 (Over the period 1920 to 1952, McLean and Haigh found a nonintegrated Mid-Continent refiner would have average monthly fluctuations in gross margins four times the size of a fully integrated refiner. Substantial stabilization of earnings were also shown for producers and marketers who integrated). Dr. Canes obtained from the Graduate School of Management, University of Rochester, estimated beta (β) coefficients (which are commonly used in the investment fraternity to measure a firm's "riskiness") and an index of oil firm vertical integration from the Energy Resources Council (see note 1141, *supra*) and ran a regression analysis of β on the vertical integration index (V) for 33 oil firms, yielding

$$\beta = 1.22 - 0.61V \quad R^2 = .25 \quad F \text{ value} = 10.00$$
$$(3.17)$$

Note: The number in parenthesis is the t statistic. By this estimate a 0.1 increase in the firm's vertical integration index (V) reduces its riskiness (β) by about 0.06. CANES, *Oil Integration Theory, supra* note 1127, at 11 n.1.

[1211] Mitchell, *Capital Cost Savings, supra* note 1088, at 90-96; *cf.* CANES, *An Economist's Analysis, supra* note 1021, at 53; Myers, *supra* note 198, at 6; CANES, *Oil Integration Theory, supra* note 1127, at 11; Professor Mitchell analyzed the effect on Standard & Poor's Rating (SPR) of integration into crude (SSF if the index is 50 percent or below and equals 50 percent if the index is 50 percent or greater; SSR if the crude integration index is 50 percent or greater and equals 50 percent if the index is 50 percent or less); of size (TA) of debt-equity ratio (C is shareholder's equity as a percent of capitalization) and of the degree of integration into pipelines (PT). His equation was:

SPR = 6.5 − 0.00018TA − 0.044C − 0.026SSF + 0.011SSR − 0.015PI, R^2 = 0.92, "t" statistics are: (−2.4) (−4.8) (−3.6) (2.3) (−2.0)

Thus: other things constant, each additional billion dollars of assets raises a firm's stock rating by about 0.2 of a risk class (such as A+, A, A−, etc.); an increase of ten percentage points in stockholder's equity/capitalization ratio increases a firm's stock rating by almost half an S&P class; increased integration into crude up to 50 percent self-sufficiency raises the S&P rating about one-quarter of a class for each ten percentage point increase, but beyond 50 percent, each ten percentage point increase in self-sufficiency lowers the rating about a tenth of a class; an increase of ten percentage points in the pipeline index raises the S&P rating by about one-sixth of a class. *Petroleum Industry Hearings, supra* note 192, at 1866 (Statement of Professor Edward J. Mitchell).

[1212] Ritchie, *Petroleum Dismemberment,* 29 Vand. L. Rev. 1131, 1133 (1976).

[1213] TEECE, *supra* note 329, at 12; *cf.* CANES, *Oil Integration Theory, supra* note 1127, at 4, 11.

David Teece has termed them, is enhanced by the common training, experience, linguistics code, and community of interest among the employees of an integrated firm.[1214] Thus the investment in the information tends to be lower and/or the amount of information possessed tends to be greater where the firm is integrated. This fact in and of itself is desirable from an economic point of view.[1215] It also means that the ability to use this information for decision-making in an industry where the raw material and major products are fluids which are relatively costly to store aboveground, and, therefore, possessing the necessary information to maintain continuity from wellhead to service station pump, is an extremely valuable asset.[1216] As DE Chazeau & Kahn expressed it: "Superior knowledge of the final market at many points and managerial control of successive levels of operation certainly make possible significant economies of this kind. The nonintegrated refiner must depend much more on the estimates of final demand reflected back through the changing orders of his wholesaling customers. The integrated firm can act with greater assurance on the basis of its own estimates of the ultimate market situation."[1217]

Even discovering the relevant price signals in the market as outlined by DE Chazeau & Kahn requires an expenditure of resources.[1218] Applying these observations to pipelines, the prospective user or users of pipelines are in the best position to determine their requirements which will form the basis for the projected throughput. This information is a spinoff from their short- and long-run plans for crude oil supplies and their developing marketing strategies. If the company possessing this information were to choose to attempt to persuade someone else to build the line, it would incur the cost of preparing the presentation to such pipeline entrepreneur and the delay inherent in such entrepreneur's independent study and evaluation. Looked at from the other end, the independent pipeline promoter must either rely on the information and projections furnished to him by the prospective users (and thereby incur the risk that he was getting an overoptimistic projection or, more elegantly phrased by the transactional economists, expose himself to the dangers of "opportunism")[1219] or he must duplicate, at a substantial cost, the studies

---

[1214] *Petroleum Industry Hearings, supra* note 92, at 331 (Statement of W. T. Slick, Jr.); *cf. S.1167 Hearings, supra* note 432, pt. 8 at 6250 (Statement of Charles P. Seiss, Jr.).

[1215] LIEBELER, *Integration and Competition, supra* note 976, at 11.

[1216] *Petroleum Industry Hearings, supra* note 192, at 2229 (Statement of Professor Neil H. Jacoby).

[1217] DE CHAZEAU & KAHN, *supra* note 30, at 267.

[1218] Ritchie, *Petroleum Dismemberment,* 29 Vand. L. Rev. 1131, 1133 (1976).

[1219] TEECE, *supra* note 329, at 8, 127.

already made. If the potential numbers on each side of the bargaining table are limited, a situation is encountered which has been described by economists as the "small numbers bargaining" problem.[1220] Vertical integration eliminates that problem and, being simpler for the prospective user or users to build the needed pipeline, this has been the prevailing practice. The pipelines that have been built by interests unrelated to the users have involved numbers of potential shippers that were large relative to the pipeline investment and no one or two shippers predominated. These factors reduced substantially the risk incurred and inhibited the possibility that the users would get together and build their own line.[1221]

At the outset of this discussion, mention was made that integration was not the only way that the industry might operate, and the history and structure of the industry demonstrate that integration has not been the exclusive mode of operation. One of these methods is by contract. But consider the circumstances under which the industry operates. There are substantial variations in supply and demand patterns; for example, the Arab embargo following the "Yom Kippur" war between Israel and Egypt and Syria.[1222] Recently, trouble in Iran has created a supply problem. Prices are subject to the whims of governments, both foreign and domestic. Strikes, natural catastrophes and other emergencies arise, causing logistical dislocations. Governmental policies, emphasizing social concerns, but with seemingly scant consideration of overall energy problems, distort the marketplace. Observe the nature of the investments in the industry: highly capital intensive, long-lived, and special purpose. Now, how does someone less than a divinity, even aided by a computer, conjure up all the permutations and combinations of events that could happen over the life of even a reasonably long-term contract? This is the "bounded rationality" problem that the transactionalists have described.[1223]

Assuming, for the sake of argument, that the oil companies had such a person or persons and the business-machine manufacturers were able to provide a fifth or sixth generation computer so that all conditions could

---

[1220] *Id.* at 8-9, 127-128.

[1221] LIVINGSTON, *supra* note 228, at 68-70. This citation not only refers to the sentence immediately footnoted but applies equally to the text at notes 1219-1220.

[1222] See Tavoulareas' testimony, *Petroleum Industry Hearings, supra* note 192, at 1834-1836. Insofar as appears from the record, Mr. Tavoulareas' question, "if I were not integrated, how would I have done that [diverting non-Arab oil to the United States and Arab oil to non-embargoed countries] job?," was not answered. There is, of course, an answer. In a free market, this could have been accomplished by U.S. refiners bidding up the price of Iranian oil so as to lure it away from historical (non-embargoed country) purchasers who could turn to OPEC sources. *But,* the integration advantage point is not impaired—the price would have risen substantially.

[1223] TEECE, *supra* note 329, at 8,123; Oliver Williamson, *supra* note 329, at 1444.

be spelled out in a document, how could a firm find a supplier willing to undertake such a contract? Again, assume yet another impossibility, *i.e.,* that such a firm could be located, how much would that firm charge as a premium to enter into a contract of this nature? Given our linguistic shortcomings, and the ingenuity and powerful intellect of counsel representing the parties, how long would it take for dispute, requiring at least negotiation and, depending on the money involved, probably litigation, to ensue? What then of the cost of using the marketplace? Suppose instead of this string of improbabilities, the parties just drafted the best contract they could. How long then before the first dispute, the first renegotiation, the first lawsuit? What would be the cost under this arrangement? Moreover, it has become virtually impossible for a large company to enforce contracts when they become unfavorable to sovereign nations or to small companies (see the Sohio example later on in this paragraph). Now, all of this may sound theoretical to some, but consider the record of what actually happened in times far less unpredictable than today. After being divorced from the Standard Oil Group by the 1911 dissolution decree, Standard Oil Company (Ohio) was left as a small refiner and marketer operating in the state of Ohio and without crude production. McLean & Haigh relate how during "the 1920's Sohio lost a large share of the Ohio market" and "by 1928, the company's competitive and economic position had become so precarious that a new management and an entirely new board of directors was placed in charge of the company's affairs."[1224]

In an attempt to solve its crude supply problems, Sohio entered into a long-term crude oil contract with the Carter Oil Company covering its refinery needs, and a joint ownership arrangement with Carter's parent, Standard Oil (New Jersey) and the Pure Oil Company to construct the Ajax pipeline from Glenn Pool, Oklahoma, to Wood River, Illinois, where the crude could be transported to Sohio's refineries via the common carrier Illinois Pipeline Company. The deal with Carter specified posted prices plus a fixed fee for purchasing and gathering. The fee was reasonable in relation to the posted prices when the contract was made but the posted price declined so sharply that the fixed fee became a disproportionate part of the total cost. This was annoying, but the arrangement became economically infeasible when the Illinois field came in and the price of Illinois crude delivered to Ohio refineries made the delivered price of the Carter crude noncompetitive. The situation became so bad that Sohio felt impelled to inform Carter that it could no longer live with the contract. Fortunately for Sohio, Carter let it off the hook but the key point is that Sohio's efforts to obtain security of crude supply by

---

[1224] MCLEAN & HAIGH, *supra* note 35, at 240.

means of the 1930 long-term contract eventually created more problems for Sohio than it solved.[1225]

Looking back at the contract with 20-20 hindsight, Sohio's difficulties arose because it did not anticipate, and provide against, two possibilities: first, it did not conceive of the precipitous drop in crude prices, which caused the fixed "brokerage" charge to become 23 percent of the value of the crude itself; second, it did not visualize the possibility of the Illinois field producing crude that made the laid down cost of Oklahoma crude noncompetitive. Had it owned the Oklahoma crude, it could have shut in the production pending the decline of the Illinois field or it could have sold it locally without the high gathering and purchasing charges. The contract provided for neither contingency nor for any of the options which would have made the occurrence of the contingencies less costly. Sohio learned its lesson and embarked on a program of integrating backward into the crude production stage of the business.[1226] The import is clear: as between long-term contracts between refiners and crude oil suppliers and between refiners and marketers of oil products, and vertical integration of these stages, the long-term contract route will be the more costly because of the negotiation, renegotiation, and enforcement procedures involved in such arrangements. Stochastic (random) variations of crude and product supply and demand imply costly means to adjust within long-term contracts or amendments of these contracts and the larger the economic consequences of such stochastic variations for the contracting parties, the more expensive these adjustments will be.[1227]

The most succinct and explicit summation of the foregoing discussion, although made by a realistic businessman, nevertheless integrates extremely well with current economic language. It is found in the testimony of W. T. Slick, Jr., Senior Vice President of Exxon Co., U.S.A., before the Senate Antitrust and Monopoly Subcommittee. In Slick's words: "Neither long- nor short-term contracts are an adequate substitute for vertical integration in this adaptive, sequential decision-making process. Efficient supply from a petroleum company, as with many companies in other integrated industries, requires large and expensive investment in special purpose long-life equipment. Thus, optimal investment considerations favor the award of long-term contracts to permit the supplier to amortize his investment with confidence.

"Although long-term contracts can result in an industrial structure very similar to that existing under vertical integration, long-term contracts

---

[1225] *Id.* at 246.

[1226] Mitchell, *Capital Cost Savings, supra* note 1088, at 77.

[1227] CANES, *Oil Integration Theory, supra* note 1127, at 20; Ritchie, *Petroleum Dismemberment*, 29 Vand. L. Rev. 1131, 1133 (1976).

are not adequate substitutes for vertical integration. Contingent supply relations must be exhaustively stipulated in long-term contracts to prevent conflict over contract ambiguities. Unfortunately, this is often precluded by the bounds of rationality and the inability to foresee the future. Those stipulations which are specified are often costly in themselves. Suppliers demand a substantial premium when they commit their resources to a particular buyer under specified conditions for a long period of time. Moreover, costs multiply when changing technology and market circumstances force renegotiation. Needed amendments are costly — not only in terms of time and money, but also in the distortions in the decision-making processes they invariably create. Such distortions can include the total cancellation of plans because of the inability of a party to honor a long-term contract. Many industries have experienced severe hardships when suppliers have been unable to deliver vital raw materials.

"The foregoing problems are taken from real world experiences and are the factors which have motivated firms in many businesses to integrate either forward or backward and sometimes both."[1228]

4. Reasons Not To Integrate

One reason not to integrate into another phase of the oil business (including pipelines) is the cost of constructing or acquiring the assets required to engage in a second stage of the productive process. Mention was made previously of the $500 million plus cost of building a "grass-roots" refinery of 250 thousand barrels per day, the $45 million to over $200 million cost of an offshore platform and the $2 billion dollar cost of a 20 percent interest in a line such as TAPS.[1229] These sums caused the "major" integrated companies to "swallow hard"; to a small independent producer or a small independent marketer, they must seem like the Department of Commerce's National Accounts. As IPAA President (and an independent producer) Jack M. Allen testified in the TAPS proceeding: "In many instances, an independent oil producer will have only a small fractional interest in the production from a well. To finance drilling ventures, it is often necessary to divide the working interests into many small pieces of ownership. . . . The owners of small producing interests generally dispose of their production to one of the oil purchasers in the field. . . . Historically, independent producers have had a ready market for their crude production without having to worry about pipeline outlets. . . . Unlike most other industries, the petroleum exploration industry is a high risk in-

---

[1228] *Petroleum Industry Hearings, supra* note 192, at 331-332 (Statement of W. T. Slick, Jr.).

[1229] See text at note 1180, *supra*.

dustry which requires investor capital in hand. Money cannot be borrowed to carry out exploratory drilling programs. I cannot imagine any oil producer now selling oil under a short term division order who would want to enter into long term throughput commitments or guarantees in order to get a few cents tariff reduction for his shipments. The independent producers I know are interested in finding and producing oil—not financing or guaranteeing oil pipelines."[1230]

Thus, the independent producer, with a fractional interest in wells which average approximately 17 barrels per day, who has unlimited access to pipeline transportation at the same tariff rates that the pipeline company's owners pay without having to provide any kind of throughput commitment or financial support to the pipeline project, and facing capital requirements for his own business of exploring for, and producing, oil that already have him straining to stay in the business that he knows, has a strong disincentive to integrate into pipelines.[1231] A similar situation prevails with respect to the small independent marketer. As one independent jobber, Johnson Oil Company, of Morristown, Tennessee, testified before the Senate Antitrust & Monopoly Subcommittee: "The ability of a jobber to associate himself with a vertically integrated [supplier] and the ability to change that association to a similar but competing company is the rock bed of his independence and his competitiveness."[1232]

A second situation where vertical integration appears unable to offer any benefits not already realizable has been described as "geographical advantage."[1233] In the simplest variety of this situation, a small oil field is discovered beneath, or in close proximity to, an existing market, which is located some distance from major sources of crude supply. If the size of the market is less than the producing rate of the field, a refinery equal to the market's consuming capacity will be built to utilize the local crude; the excess crude will find its way to the next closest refinery. The local refinery presumably will be below the optimal scale of refineries and hence only one refinery will be constructed to handle the local crude. Such a refiner would have a "locked in" source of supply and an assured market. He has no incentive to integrate. The next closest refinery which is purchasing some of the crude is not in a viable position to outbid the

---

[1230] Allen, *TAPS Rebuttal Testimony, supra* note 496, at 8-9; see also Allen, *FERC Direct Testimony, supra* note 929, at 7.

[1231] See note 331, *supra*.

[1232] *Petroleum Industry Hearings, supra* note 192, at 2090 (Statement of John R. Johnson).

[1233] The existence, and description, of this situation has been drawn from *Petroleum Industry Hearings, supra* note 192, at 1861-1963 (Statement of Professor Edward J. Mitchell); Mitchell, *Capital Cost Savings, supra* note 1088, at 83-85; and TEECE, *supra* note 329, at 62-68.

local refinery for its crude supply or its customers unless that refiner is so efficient that it can overcome the transportation cost of crude from the local field to it plus the cost of transporting products back to the local market. While a 150,000 to 250,000 barrel per day refinery would be more efficient than the smaller one, the existence of many small crude fields and markets distant from large refineries, coupled with the cost of moving small volumes of oil, favor the probability that it will be common for the local refiner to have a "place differential" advantage.

An example of this type of situation was the Shallow Water Refining Company which owned a 3,000 barrel per day refinery in western Kansas.[1234] The nearest competing refinery was in Wichita, 200 miles away. The round trip transportation advantage was about 58 to 77 cents per barrel. Shallow River's average refining cost per barrel was 54.7 cents, and with crude oil at the Kansas fields in the area being uniform in price, there was no way for outside refiners to undersell Shallow River. Nor is this phenomenon limited to small non-integrated refiners. Ashland had a similar situation in the Kentucky-West Virginia region during the 1920's and 1930's.[1235] Suppose the local crude supply is less than the local market. In this case, the local refinery will be built to handle all the local crude and the remaining part of the market will be supplied by outside refiners. However, the local refiner has the marginal cost advantage and he can always compete in the local market and sell all his output therein. Once again, he has an assured supply and an assured market and hence he has the substance of integration although not the form.

Another form of geographical advantage is really an example of situations where contractual arrangements produce sufficiently satisfactory results that the refiner finds any incremental cost advantage in pipeline integration is more than outweighed by the opportunity value of other investments. Murphy Oil Company owns a 92,500 barrel per day refinery at Meraux, Louisiana, which, because of its location, has convenient and economical access to ocean-going vessels, hence it runs on imported crude brought in by tanker or Gulf Coast crude transported by barge.[1236] Earth Resources Company has a 31,500 barrel per day refinery (Delta Refining Co.) at Memphis. Its crude supply comes up the Mississippi via its wholly-owned Valley Towing Service, which also transports oil for other customers. This really is not an example of *no* integration, but of non-pipeline integration. However, the bulk of Delta's refinery outturn is sold within a range that truck transportation is the most economical means of transportation, which obviates the need to

---

[1234] MCLEAN & HAIGH, *supra* note 35, at 633-639.
[1235] *Id.* at 633-641.
[1236] LIVINGSTON, *supra* note 228, at 9.

integrate into pipelines.[1237] Good Hope Refineries, a subsidiary of Gasland, Inc., (now Good Hope Industries) owns a 29, 450 (now 80,000) barrel per day refinery located at Good Hope, Louisiana, on the Mississippi River near a GATX terminal, which is supplied with either imported crude or Gulf Coast crude by tanker or barge, and ships out its products outturn by tanker to its parent in New England and New York. Because of its location, contractual arrangements with GATX for terminalling and with tankers and barges are cheaper than integrating into pipelines.[1238] Good Hope now uses both a crude oil pipeline (Delta) and a products pipeline (Shell), but in each instance, solely as a shipper and not as a participant in ownership.

In addition to the geographically advantaged refineries, there are a number of refiners whose ready access to someone else's line makes it unnecessary and uneconomic to build their own pipelines. The Murphy Oil Company, which has integrated into pipelines where it was economic to do so, *e.g.*, Butte Pipe Line (20% interest) which carries its crude from its producing fields in Eastern Montana to connections with Platte and Amoco for movement to the St. Louis/Kansas City and Midwestern markets, chose not to integrate into pipelines in connection with its 27,000 barrel per day refinery located at Superior, Wisconsin, which is served by Lakehead Pipe Line (a subsidiary of Interprovincial Pipe Line) which runs close to Superior en route to Sarnia, Ontario, and Chicago. An example comparable to Murphy's Superior refinery is Rock Island Refining Corporation's 29,500 barrel per day Indianapolis refinery. When Rock Island built its refinery, Marathon already had in place a large diameter crude line. All that was needed was a short spur line to connect with the Marathon line and Rock Island's crude transportation problems were solved. Distribution of the refinery's product to the ample nearby markets can be accomplished by truck. If Rock Island should decide to expand and wish to ship to more distant markets, there are several refined products pipelines passing near Indianapolis.[1239] *Ceteris paribus*, integration was not an optimal solution to Rock Island's security of supply and market situation. Subsequently, Rock Island did come in on the Texoma line as an efficient way of bringing imported crude to its refinery.

There are many other instances where a refiner has preferred not to invest in a pipeline venture that would carry its crude, choosing to rely on ICC regulation and the *Atlantic Refining Company* consent decree to ensure reasonable cost transportation, and to invest its money where it could expect a higher rate of return. The extent to which this philosophy

---

[1237] *Id.* at 6-7.
[1238] *Id.* at 7-8.
[1239] *Id.* at 7.

applies was brought out in Section II A 3 c, "Scramble for Traffic," where extensive non-owner usage of pipelines was documented. Going from aggregate numbers to specific situations, Capline Oil, which declined Explorer's invitation to participate in that venture, chose instead to rely upon its common carrier service.[1240]

This is not a new circumstance; way back in the TNEC days before World War II, an independent Houston-based refiner, Louis J. Walsh, testified that his company owned no crude-oil pipelines nor any production but that it was connected to the pipelines of three major companies which gave the refinery access to crude from practically all the fields in Texas and New Mexico, and some fields in Louisiana.[1241] Nor is it limited to small refiners. For over twenty years Colonial's owners shipped over Plantation and Plantation owners (Shell and Exxon) are shipping over Colonial. Recently Standard Oil of California opted not to participate in ownership of TAPS but simply to ship over one of the undivided owner's "lines." Ashland, which was named as a "major" oil company by the S.2387, or PICA, Report,[1242] was quite content to rely upon ICC regulation and the *Atlantic Refining Company* consent decree to see that the tariffs it paid Colonial were reasonable because, in the words of its President, Robert E. Yancey: "We would much rather pay the tariff on that basis [and] put our money into other areas where we get a much higher rate of return."[1243] Yet when Ashland was faced with contractual alternatives more costly than integration in the case of the Capline System, which runs from St. James, Louisiana, to Patoka, Illinois, Ashland came in for a 18.7 percent participation (the largest initial ownership percentage) and even constructed a connecting line from Patoka to its refinery at Cattlettsburg, Kentucky.[1244] Another instance where a major chose the integrated route when "market failure" (in this case, transactions costs and delays) made integration the only feasible answer to timely construction is Amoco's Mandan, North Dakota, refinery, which commenced operations in 1955.[1245] Yet, just one year later, Amoco brought onstream its Yorktown, Virginia, refinery, which was approximately the same size as the Mandan refinery. The Yorktown

---

[1240] *S.1167 Hearings, supra* note 432, p. 9 at 614 (Statement of Vernon T. Jones).

[1241] SPECTRE, *supra* note 104, at 131 n.105, citing *TNEC Hearings, supra* note 6, at 7337.

[1242] S.2387 REPORT, *supra* note 49, at 17.

[1243] *S.1167 Hearings, supra* note 432, pt. 8 at 5929 (Statement of Robert E. Yancey); accord, Spahr, *Rodino Testimony, supra* note 255, at 14 ("Assuming that it could depend on other investors to finance, build and operate the oil pipelines it needs, an oil company would rather invest its money somewhere else, pay the tariff, and let investors own the line and make the limited return").

[1244] LIVINGSTON, *supra* note 228, at 7 (the 20 percent came from Yancey's testimony, cited in note 1243 *supra*).

[1245] *Petroleum Industry Hearings, supra* note 192, at 366 (Statement of Walter R. Peirson).

refinery was built to process imported crudes so there was no crude pipeline integration, and because existing means of transportation provided an adequate alternative to integration, there was no products pipeline integration. As witness Walter Peirson explained the difference in approach: "In each of these instances, Amoco's decisions were made on the basis of the economics of each situation; and the integration of the Mandan project and the non-integration of the Yorktown project both fostered the one basic objective which this proposed bill [S.2387] overlooks — the most efficient and economical production of refined products for consumption by the American people."[1246]

A variation of the use of someone else's line is the situation where the "small numbers bargaining problem" does not exist. One example of such a situation existed in the Bakersfield, California, area. There are 10 small refineries in that area with an aggregate crude running capacity of 169,000 barrels per day. Only a small proportion of their aggregate light products outturn could be expected to find a market in Fresno. The economies of scale dictated that only one line, of 8-inch diameter, would be built. Apparently no one or two of the Bakersfield refiners felt it was worth the risk to construct the line. Given the circumstances, *i.e.,* the large number of potential shippers compared to the relatively small investment required and no one or two refiners that were willing to build a line, Southern Pacific, which had the ability to finance the line, stepped in as an entrepreneur and laid the line.[1247]

An analogous situation was involved in Southern Pacific's Los Angeles to Phoenix and Tucson line. There are 19 refiners in Los Angeles with a spectrum of sizes, having an aggregate capacity of over a million barrels per day but, again, only a very small fraction of their light products moves east into Arizona. Once more, the problem of small numbers was absent and a well-heeled company with an advantageous position respecting right-of-way did not have to rely upon the needs of a few potential users. Closing out this paradigm is the Kaneb Pipe Line, which initially ran from connections with seven small refineries in southern Kansas to a terminal at Fairmont, Nebraska. The number of potential shippers relative to the rather modest investment required,[1248] and the importance to those refiners of the market to which the line provided low cost access, minimized the entrepreneurial risk.[1249]

[1246] *Id.*

[1247] LIVINGSTON, *supra* note 228, at 33. An examination of the right-of-way probably would disclose one of the attributes of integration, *i.e.,* incremental use of assets already held, in this case railroad right-of-way.

[1248] Part of the early risk was ameliorated by Kaneb's securing throughput agreements from two of the refineries which desired connection. *S.1167 Hearings, supra* note 432, p. 8 at 5943 (Statement of Richard C. Hulbert, President, Kaneb Pipe Line Co.).

[1249] LIVINGSTON, *supra* note 228, at 35.

The last example of use of market transactions, rather than integration, is exchange agreements. Unlike open market purchases and sales, there is a strong mutual incentive for the contracting partners to fulfill their end of the bargain. Moreover, exchanges handle two problems at once: the securing of the desired material at the place and time it is needed and the disposition of an unneeded or lesser utility material at the same time. Although exchange agreements have been attacked by critics of the industry as being a device to escape common carrier obligations and anticompetitive in nature as tools with which to discipline price cutters and to thin out the intermediate markets for crude oil and products,[1250] there have been no facts adduced to support these allegations. The Attorney General's Second Report on the Interstate Oil Compact, while mounting an attack against alleged crude price stabilization, was forced to admit that economies in transportation and refinery operation were effected by exchanges.[1251]

Moreover, in the only litigated case reaching an appellate court level that this author knows about, involving an exchange between Frontier Refining Company, which delivered gasoline to the California Company (a subsidiary of Standard Oil Company of California) in Cheyenne, Wyoming, and received an equal amount (5 million gallons per year) from the California Company at Salt Lake, Utah, with a location price differential, the Court of Appeals for the Tenth Circuit held that "The underlying purpose and effect of the exchange agreements are too plain for doubt. In the first place, they were obviously entered into in order to facilitate competition, not to stifle it. They permitted one marketing company to do business at the back door of its competitor's refinery by the exchange of manufactured products. The price differential based on the place of distribution or sale was nothing more than a recognition of the cost of transportation to points of distribution."[1252] In the real world,

---

[1250] KENNEDY STAFF REPORT, *supra* note 184, at 73-74.

[1251] Att'y Gen., *Second Report, supra* note 111, at 77.

[1252] Blue Bell Co. v. Frontier Refining Co., 213 F.2d. 354, 359 (10th Cir. 1954). Two district court cases have touched upon the subject of exhanges. Thomas v. Amerada Hess Corp., 1975-1 TRADE Cas. ¶ 60,126 at p. 65,297 (M.D.Pa. 1975), on a motion for summary judgment ("Exchange agreements are not *per se* illegal as a means of eliminating competition for a substantial market and dividing territories as plaintiffs earnestly suggest. On the contrary, they permit one refiner to do business in the backyard of its competitor's refinery, by the exchange of manufactured products [citing *Blue Bell*]. Without the exchange agreements, oil companies would be confined to selling gasoline in the area where their respective refineries are located and there would be an area of natural monopoly around each oil company's refinery. . . . Companies with distant refineries would be unable to compete in the vicinity of another's [refinery]. Competition would exist only in the fringe areas between refineries. Outlawing the exchange agreement would have a tendency to bring about a needless duplication of refineries with their accompanying drawbacks."); United States v. Standard Oil Co. (Indiana), 1964 TRADE Cas. ¶ 71,215 at p. 79,870 (N.D.Cal. 1964) ("An 'independent' refiner may participate in crude exchanges even though it has no production of its own [referring to oil import ticket transactions].").

as recognized by the Court, exchange agreements are a unique device used by the industry to promote efficiency in the complex logistics of delivering petroleum products throughout the nation. Without such arrangements, the costs to consumers would be substantially higher.[1253] Witness W. T. Slick, Jr., summarized the matter neatly: "In actuality, exchange agreements promote competition by —

"Facilitating entry and expansion in markets where a company does not have its own refining facilities;

"Giving refiners greater flexibility to [secure] crude of the appropriate quality, grade, specific gravity, and sulphur content;

"Reducing inventory and operating costs and supply interruptions caused by temporary shortage/surplus situations at refineries or terminals; and

"Reducing transportation costs."[1254]

Slick then illustrated this exposition by giving two typical examples of exchanges, one a crude oil exchange and the other a products exchange. The crude oil exchange was between Exxon and Lion Oil Company, a small refiner in El Dorado, Union County, Arkansas. Under the arrangement, Lion, which had 2,000 barrels per day of West Texas crude which it had no economical way of getting to its El Dorado refinery, delivered it to Exxon which had a pipeline going to its Baytown, Texas, refinery. In turn, Exxon, delivered to Lion 2,000 barrels per day of East Texas crude which could be transported directly to Lion's refinery by means of the Mid-Valley pipeline. The products exchange involved the receipt by Exxon of 15,000 barrels per month of refined products in Cut Bank and Kevin, Montana, from Thunderbird Petroleums, Inc., a small refiner located in Kevin, and in exchange, Exxon delivered 15,000 barrels per month to Thunderbird at Portland, Oregon, and Spokane, Washington, where Thunderbird marketed in competition with Exxon, but where it could not have otherwise economically accessed from its Montana refinery. Slick's characterization of the transaction is reminiscent of the language of the 10th Circuit Court of Appeals in the *Blue Bell* case. "It is clear that both crude and product exchanges are implemented for a variety of valid economic reasons, and not for the

---

[1253] Ritchie, *Petroleum Dismemberment*, 29 Vand. L. Rev. 1131, 1144-1145 (1976); see also *S.1167 Hearings, supra* note 432, pt. 8 at 5930-5931 (Statement of Robert E. Yancey, President of Ashland Oil, Inc.) ("The exchange function in the oil business has saved the consumer a considerable amount of money, because the exchange always works out that you minimize transportation costs.")

[1254] *Petroleum Industry Hearings, supra* note 192, at 337 (Statement of W.T. Slick, Jr.)

exclusion of any class of competitors, that they do not result in any anticompetitive behavior, and that their abolition would reduce competition, increase consumer prices, and increase the probability of localized product shortages."[1255]

The point being made here simply is that exchanges are forms of market transactions which render integration unnecessary with respect to the particular problems being solved. However, it is pertinent to note that integration already in place does make it possible to avoid the additional integration. Thus it was the presence of Exxon's West Texas to Baytown pipeline and Sun and Sohio's Mid-Valley pipeline that made it feasible for Lion to avoid its own integration into transportation facilities. It is interesting that those who would outlaw exchanges and related transactions thereby would force a greater degree of vertical integration, which they also oppose strongly.

There are some further empirical data available on the integration question, namely examples which appear to support the thesis that decisions whether or not to integrate are economically motivated rather than antitrust inspired. Instances can be cited where changes in circumstances have made integrated operations no longer less costly than marketplace arrangements and, following the economic precepts that companies pursue the less costly alternative mode, voluntary disintegration has taken place. One such instance occurred when the rising costs of company operated service stations, climaxed by the adoption of chain store taxes, made this phase of integration less economic than the market alternative. The response to this was voluntary disintegration of the great majority of service stations under the so-called "Iowa Plan" whereby the stations were leased to independent retailers.[1256]

Perhaps the ultimate example of voluntary disintegration which occurred when acceptable market alternatives had come into being was the sale by the "major" oil company shipper-owners of the Great Lakes Pipe Line Company to the Williams Companies.[1257] Less spectacular, but forming almost a continuous series of marketplace-versus-integration economic decisions, are the constant shifting of arrangements between the more than 25 discrete activities involved in the petroleum industry.[1258] No petroleum company is completely integrated. Even the largest firms in the industry contract out seismological, drilling, refinery and pipeline construction services. All companies buy and sell products in the intermediate

---

[1255] *Id.* at 338.

[1256] WOLBERT, *supra* note 1, at 93 n.548.

[1257] *Cf. Petroleum Industry Hearings, supra* note 192, at 318 (Statement of Vernon T. Jones, President of Williams Pipe Line Co.).

[1258] See *Petroleum Industry Hearings, supra* note 192, at 332 wherein W. T. Slick, Jr. enumerates many of these activities.

markets.[1259] The trend toward using market mechanisms rather than integration in gathering was discussed in Section II A 2, "Gathering Systems." The determination of which way to go in these individual situations does not testify to the inherent advantages of either alternative as an overall proposition; what it does do is to support the contentions of industry spokesmen such as Slick and Peirson that each proposition stands on its own merits and the outcome is determined by strictly economical considerations, and the fact that the decisions reached change from time to time and from circumstance to circumstance as the companies continually adapt their structures to the realities of their surroundings appears to manifest workable competition in the industry.[1260]

While anticompetitive behavior can be a concomitant of vertical integration, it should be recognized that it is not the vertical integration *per se* that dictates the competitive or anticompetitive behavior of a firm. Economists agree generally that vertical integration has a potential for abuse only when excessive market power exists at some stage. Hence, it is not vertical integration that is the culprit but rather excessive market power. Moreover, *potential* for abuse should not be identified as the abuse itself. Appropriate public policy would seem to be to make an evaluation whether abuse has in fact occurred, before mounting an attack on the accompanying structural condition.[1261]

In view of the fact that vertical integration is so universal a phenomenon in both market and centrally planned economies and in both government-owned and privately-owned enterprises, it seems reasonable to conclude that there are weighty advantages to be derived from it.[1262] DeChazeau & Kahn concluded that vertical integration in the petroleum industry had fostered its growth by reducing risks and costs of capital, provided coordinated use of specialized talents, and permitted a better synchronization of the flow of crude and products which, in their words, "are advantages not lightly to be dismissed in the unforseeable contingencies of the future."[1263] Professor Neil Jacoby recently testified that, "It is a reasonable inference that vertical integration in the U.S. petroleum industry — as in other industries — came about as a result of a

---

[1259] *Id.* at 2229 (Statement of Professor Neil H. Jacoby). For a more detailed examination of the subject of intermediate markets, see Section VI C 4 c (3) (G) (i), "The Crude Oil Market," *infra*.

[1260] MCLEAN & HAIGH, *supra* note 35, at 674.

[1261] ERC REPORT, *supra* note 184, at 26-27.

[1262] *Petroleum Industry Hearings, supra* note 192, at 2228 (Statement of Professor Neil H. Jacoby); *Id.* at 244 (Statement of Thomas diZerega) (vertical integration, in and of itself, is not the evil to be addressed).

[1263] DeCHAZEAU & KAHN, *supra* note 30, at 566.

quest for economies of production and limitation of risks, that is, from normal competitive motives."[1264] In the face of this kind of evaluation, the burden of establishing that pipeline integration in the oil industry should be abolished should be upon the proponents of divestiture,[1265] and a clear showing[1266] of necessity for,[1267] and a careful cost/benefit analysis of, any proposed divestiture would appear to be required before any such proposal responsibly could be considered.[1268]

C. *Jointly Owned and Jointly Operated Pipelines*

It has been said that few practices in the oil industry have given rise to

---

[1264] *Petroleum Industry Hearings, supra* note 192, at 2228 (Statement of Professor Neil H. Jacoby); *accord,* TEECE, *supra* note 329, at 121; CANES, *Oil Integration Theory, supra* note 1127, at 26, MCLEAN & HAIGH, *supra* note 35, at 674; SPECTRE, *supra* note 104, at 109; *but cf.* E. ROSTOW, A NATIONAL POLICY FOR THE OIL INDUSTRY 117 (1948).

[1265] *Petroleum Industry Hearings, supra* note 192, at 41 (Opening Statement of Senator Roman L. Hruska) ("the heavy burden, a very responsible burden"); *Id.* at 165 (Senator Bayh agrees with Mr. Ikard that the proponents "have the burden before there is a change"); *Id.* at 1957-1958 (Reply by Peter A. Bator, Davis Polk & Wardell, to question by Senator Hruska) ("a very heavy burden of proof"); *Id.* at 2040 (Statement by Otis Ellis, Private Citizen, Consumer, and Farmer) ("beyond a reasonable doubt").

[1266] SHENEFIELD, PROPOSED STATEMENT, *supra* note 179, at 43 ("convincing case"); *Petroleum Industry Hearings, supra* note 192, at 71 (Statement of Senator Clifford P. Hansen) ("I want to hear the evidence. I don't want vague allegations."); *Id.* at 305 (Statement of Vernon T. Jones) ("absolutely no hard facts have been made available to support a change from the present and historic regulation and structure of the oil pipeline industry."); *Id.* at 319 (Comment by J. Donald Durand) ("We have never seen any hard evidence that pipelines are not operating as common carriers, as they are required to under the law."); *Id.* at 1109 (Statement of Charles E. Spahr) ("substantive evidence" not "vague unsupported assertions"); *Id.* at 2031 (Statement by Otis H. Ellis, Private Citizen, Consumer, and Farmer) ("The language under findings [in S.2387] ranges all the way from that which might be contained in a mother's prayer to that which might be used in describing an economist's nightmare."); CRUDE OIL PIPELINES, *supra* note 158, at 125 ("This demonstration [sufficient to warrant pipeline divorcement] had not yet adequately been made by the proponents of divorcement."); SPECTRE, *supra* note 104, at 110 ("In the absence of *proof* of these assumed facts, the basic argument comes down to the contention that separation of pipelines from major oil companies would [proponents of divorcement say] promote competition, thus there should be a crime *implied* in order to fit or justify the punishment.").

[1267] SHENEFIELD PROPOSED STATEMENT, *supra* note 179, at 43 ("It is up to divestiture *proponents* to make a convincing case that there is a problem and that there is no effective, less drastic solution."); Donald I. Baker, when he was an Assistant Attorney General in charge of the Antitrust Division, addressed the question of showing the necessity for divestiture where other less drastic solutions might be available. In his "series of gates" testimony before the Antitrust and Monopoly Subcommittee he stated: "It is a series of gates. If you decide the case for breaking up isn't strong, you forget about it. If you can't devise a statute that will do the job in a workable way, forget about it. If the cost is excessive, you forget about it." *Hearings on Oversight of Antitrust Enforcement, Before the Subcomm. on Antitrust & Monopoly of the Senate Committee on The Judiciary,* 95th Cong., 1st Sess. 331 (1977); *Petroleum Industry Hearings, supra* note 192, at 240 (Statement of Edwin J. Dryer, General Counsel and Executive Secretary of the Independent Refiners Association of America, "There may be solutions to the problem other than the drastic surgery of divestiture.")

so much criticism and so little research as the widespread phenomenon of joint ownership and joint operation.[1269] These arrangements have roots in antiquity as a common device for the conduct of business, certainly prior to the Christian era, as they were used extensively in the Roman Empire.[1270] So have their critics; Cato the Elder is reported to have expressed qualms about this practice in 160 B.C.[1271] These arrangements take many forms and are widely used in the United States by the construction, railroad, maritime, mining, communications, electric power and air transportation industries as well as the oil industry.[1272]

Most of the published material on jointly owned or jointly conducted operations in the oil industry has been concerned with oil and gas exploration activities, principally offshore leasing.[1273] However, after a

---

[1268] *Petroleum Industry Hearings, supra* note 192, at 41-42 (Opening Statement by Senator Hruska) (not only must the change be shown to be desirable but proponents must show that they have something not only workable and reliable but better than what we have now); *Id.* at 1108-1109 (Statement of Charles E. Spahr); *Id.* at 2016-2017 (Statement of Richard J. Boushka, President of Vickers Energy Corp.); *S.1167 Hearings, supra* note 432, pt.8 at 5916-5917 (Statement of Robert E. Yancey); DE CHAZEAU & KAHN, *supra* note 30, at 566. Deputy Assistant Attorney General (Antitrust Division) Donald Flexner agreed with this at the AEI Conference on Oil Pipelines and Public Policy, replying to a question by the author, as follows: "Before we support a proposal for a flat across-the-board divestiture of existing pipelines we will carefully weigh the benefits against the potential costs. We are sensitive to the potential costs that might attend if we make a mistake. For example, after a close look at Sohio [the proposed Long Beach, California, to Midland, Texas, pipeline] we concluded that there would not be a competitive problem, and there may be other cases like that. We want to approach that part of the issue rather carefully." AEI, *Oil Pipelines and Public Policy, supra* note 684, at 69.

[1269] JOHNSON et al., *Oil Industry Competition, supra* note 1055, at 2411.

[1270] *Consumer Energy Act Hearings, supra* note 154, at 626 (Statement of Jack Vickrey).

[1271] JOHNSON et al., *Oil Industry Competition, supra* note 1055, at 2412, citing F. MACHLUP, THE POLITICAL ECONOMY OF MONOPOLY (1952).

[1272] Broden & Scanlan, *The Legal Status of Joint Venture Corporations,* 11 Vand. L. Rev. 673 (1958). The authors remark that: "Perhaps in no other American industry is there so much use being made at present of the joint venture corporation as in the oil and gas industry. . . . There is a reasonable explanation for this. The considerations which convince businessmen to make use of joint venture corporations probably apply with special force in the petroleum industry. Speculative investments involving a high degree of risk of failure, requiring large and continued expenditures and, very often, a considerable personnel force are frequently the prospects which face those who are engaged in the exploration, development and production of oil and gas in the United States. . . ." *Id.* at 689. It should be noted that joint venture *corporations* are not used in exploration, development and production of oil due to I.T. 3930, 1948-2 C.B. 126 and I.T. 3948, 1949-1 C.B. 161 and the "Kintner" Regulations under Section 7701 of the Internal Revenue Code.

[1273] *E.g.,* Mead, *The Competitive Significane of Joint Ventures,* 12 ANTITRUST BULL. 819 (Fall, 1967) [hereinafter cited as Mead, *Competitive Significance*]; J. MARKHAM, *The Competitive Effects of Joint Bidding by Oil Companies for Offshore Oil Leases* in INDUSTRIAL ORGANIZATION AND ECONOMIC DEVELOPMENT (Markham and Papanek eds. 1950); C. GREMILLION, *Offshore Leases in the Gulf of Mexico — Joint Venture Agreements and Related Matters* in 25TH OIL & GAS INST. (1974); E. ERICKSON & R. SPANN, AN ANALYSIS OF THE COMPETITIVE EFFECTS OF JOINT VENTURES IN THE BIDDING FOR TRACTS IN OCS OFFSHORE LEASE SALES (Feb. 1974) (unpublished paper on file with the Oregon Law Review); T. DUCHESNEAU, COMPETITION IN THE U.S. ENERGY INDUSTRY (1975).

period where criticisms in this area flourished, further examination disclosed that the arrangements were pro-competitive. Dr. Walter Mead, one of the early writers in the field, who had developed a tentative hypothesis that there was a tendency for companies bidding jointly on certain tracts at a lease sale not to bid against each other on the remaining tracts, and to refrain for about two years from bidding against former co-bidders on subsequent sales in the same area,[1274] concluded, after further study, that "joint biddings for outer continental shelf oil and gas leases poses no net threat to competition."[1275] These conclusions, the administrative restrictions on "majors" bidding jointly and exogenous events such as the Arab embargo caused critics to turn their attention away from offshore exploration and toward jointly owned and jointly operated pipelines.

There has been some "carry-over" to the pipeline area of Dr. Mead's formulation of the relevant items to be considered in assessing the impact of joint arrangements which he used in the joint bidding area. Dr. Mead named three possible anticompetitive effects: (1) Competition among horizontally related firms may be restrained because the joint owners may develop a community of interests that discourages arm's length transactions; (2) Where there is a vertical relationship between the joint owners, market foreclosure may result because of preferred treatment toward the jointly-owned or conducted enterprise; and (3) The joint owners may refrain from competing in the same market as the joint enterprise, thus reducing the number of potential competitors in an industry.[1276]

On the other side of the coin, Dr. Mead suggested that the facts should be scrutinized to determine the existence or absence of four basic justifications for these operations: (1) They permit entry into an industry or activity where absolute capital requirements are so high that only a few large firms could otherwise participate; (2) Risks may be so great that only

---

[1274] Mead, *Competitive Significance, supra* note 1273, at 840-843.

[1275] *S.Res.Hearings, supra* note 296, at 1014. (Statement of Dr. Walter J. Mead). It is interesting to note that Dr. Mead, in making his recommendations at the end of his testimony, stated: "Joint ventures (excluding joint bidding) in the United States [petroleum industry] have been largely limited to jointly owned pipeline facilities. Before Congress can legislate in this area, additional study is needed to (1) identify the competitive impact of pipeline joint ventures and (2) examine alternative methods of financing pipeline ventures." *Id.* Recommendation No. 3. For other authorities reaching the conclusion that joint bidding for offshore leases enhances entry into offshore activity by smaller firms, see *Petroleum Industry Hearings, supra* note 192, at 1081 (reprint of E. ERICKSON & R. SPANN, THE ENERGY QUESTION: AN INTERNATIONAL FAILURE OF POLICY — Vol. 2; J. MARKHAM, *The Competitive Effects of Joint Bidding by Oil Companies for Offshore Oil Leases* in INDUSTRIAL ORGANIZATION AND ECONOMIC DEVELOPMENT 116-135 (J. Markham and G. Papanek eds. 1970).

[1276] Mead, *Competitive Significance, supra* note 1273, at 822-823, *see also* JOHNSON et al., *Oil Industry Competition, supra* note 1055, at 2413.

a few, if any, existing firms prudently could, or would be willing to, assume such risk alone; (3) Separate operations by competing firms may be inefficient or economically wasteful; and (4) A large investment may produce external economies that will accrue to all firms in the relevant area from a given endeavor, rather than primarily to the investing firms.[1277]

In order meaningfully to examine the empirical evidence on these cooperative arrangements in the pipeline field, a clear understanding is necessary of the two distinct forms such arrangements have taken in connection with pipelines.[1278] One form is a stock company ("corporate" form) which is formed by more than one company (including sometimes, owners other than oil companies),[1279] and the other is the "undivided interest" line in which the owners hold, as tenants in common, an undivided interest in the pipeline assets themselves. In the corporate form type, participants subscribe to shares of stock, generally in proportion to their expected throughputs, sometimes with a provision for a one-time stock ownership readjustment. The usual procedures are followed for incorporation of the new company and for qualification to conduct business in each state where it is to operate. The new company contracts for construction of the pipeline facilities, publishes tariffs subject to FERC regulations, obtains the required financing (in the beginning almost always on the basis of throughput and deficiency commitments by the owners and usually thereafter, depending on the pipeline's track record and the state of the money market) and performs all the operation, maintenance and record-keeping functions of the business. It establishes its capital cost and operating expense budgets and determines the resulting financing program. The owners receive dividends as they are earned and declared. The participants jointly share the financial risk, rather than assuming individual liability (except as to undertakings under the throughput and deficiency agreements). Colonial, the largest petroleum

---

[1277] Mead, *Competitive Significance, supra* note 1273, at 824-825; *see also* JOHNSON et al., *Oil Industry Competition, supra* note 1055, at 2413.

[1278] The text from this point through note 1280 was derived from the following sources: A.G.'s Deepwater Port Report, *supra* note 446, at 77-78; *Consumer Energy Act Hearings, supra* note 154, at 626-627 (Statement of Jack Vickrey); MARATHON, *supra* note 202 at 43-46; PIPELINE TRANSPORTATION, *supra* note 162, at 43-45; W. J. Williamson, *supra* note 315, at 9-12; PIPELINE PRIMER, *supra* note 331, at 7-9; BLEDSOE, *supra* note 703, at 7-12; KENNEDY STAFF REPORT, *supra* note 184, at 86-87.

[1279] Examples are: Butte Pipeline whose owners are "majors" (Shell Pipe Line and Continental Pipe Line), independents (Murphy Oil and Western Crude) and a railroad (Burlington Northern); San Diego Pipeline (Southern Pacific and Santa Fe Railroads), and Trans Mountain Oil Pipeline Corp. (Trans-Mountain Pipeline Co., Ltd., which in turn has an outstanding public interest in addition to the Canadian oil companies which own a controlling interest).

products pipeline in the world, extending from the Texas Gulf area to northern New Jersey, is an example of a multiple ownership, corporate variety, pipeline, having ten stockholders.

In an undivided interest type of multiple ownership each of the owners holds, as a tenant in common, an undivided interest in the line pipe, rights-of-way, pumping stations and all other property used in connection with the new line. As such, each owner is subject to common carrier regulation, because in concept, the undivided interest line is the equivalent of a series of parallel small-diameter pipelines which are bundled into one large-diameter line, without changing in any way the ownership or legal rights and duties of the pipeline owners to the public or under the regulatory statutes. Each tenant in common posts its own tariffs for movements through its share of the line capacity and acts as a separate common carrier which receives tenders from shippers based on its tariff rates and rules and regulations, which are subject to FERC approval, including minimum tenders, receipt and delivery points, quality of oil received, and the like. Thus, in TAPS, which is the largest example of an undivided multiply-owned system, there are, in effect, eight individual common carriers. A contractor is employed by the tenants in common to construct, maintain and operate the line, in accordance with the terms of the construction and operating contract.[1280] Frequently, one of the participants is chosen to be the contractor, but its role as such is confined to that of a construction, operating and dispatching contractor in accordance with the terms of the contract. The contract operator has nothing to do with the business affairs of the other owners. Under this arrangement, construction and maintenance costs are normally paid by each owner on the basis of his ownership percentage, as are fixed operating costs. Fuel or power and oil losses, which vary with the volume of throughput, normally are borne by owners in proportion to each owner's use of the line for his customers.

From a business standpoint, the corporate form of ownership has the following advantages:

[1280] For an example of a contract for the design and construction of an undivided pipeline system (TAPS), see *Petroleum Industry Hearings, supra* note 192, at 935-967. For an example where one of the participants was chosen as a general contractor by other owners to handle construction, physical operations, and maintenance of the new line, see *Id.* at 828-875 (Ventura Pipeline System). For other jointly owned or jointly operated line agreements, see *Id.* at 786-796 (Cities Service Pipe Line Co.—The Texas Pipe Line Co., East Texas to Sour Lake, Hardin County, Texas—undivided ("braided interest") line); *Id.* at 797-811 Basin Pipe Line System (The Texas Pipe Line Co., Shell Pipe Line Corp., Cities Service Pipe Line Co., and Sinclair Pipe Line Co.—undivided interest); *Id.* at 811-827 Mesa Pipe Line System (Gulf Refining Co., Cities Service Pipe Line Co., Pure Transportation Co., Sun Pipe Line Co., and Sohio Pipe Line Co.—an undivided interest system). Sinclair was awarded the design and construction contract by the prospective owners of Explorer, but it ended up not even being an owner.

(1) Participants share the collective financial risk of the project, rather than assuming individual financial responsibility for an undivided interest in a pipeline project;

(2) Shipper-owner completion and throughput commitments are the basis of the pipeline's economic viability, enabling substantial so-called "project financing," generally through the use of "private placement" loans which can result in up to 90 percent debt. To a certain extent, this presents a "free rider" problem in that a small company with a low credit rating gets a benefit from the rating of the pipeline company, which is a composite of this owners' ratings. An extreme example of this occurred in Texoma where the individual credits varied from AAA to B and one was so bad that the lenders said they would not let it into the corporation. The other nine companies were sufficiently desirous of obtaining diversity of ownership that they "held an umbrella" over this small company.[1281]

(3) The debt incurred for the project, for financial accounting purposes, is considered to be a primary obligation of the pipeline company and is considered a contingent liability, which is not individually quantified, on the owners' financial reports.

(4) No individual participant is responsible for specified throughput volumes so long as the pipeline's revenues are sufficient to satisfy the repayment of principal and interest on the debt. Here again, there is a "free rider" element, which tends to explain why the owners attempt to take an ownership position consistent with their expected usage of the line.

(5) In the case of an extensive new line, a stock company might ease the staffing problem. This is especially advantageous where a participant might not have a pipeline organization in place which could handle the myriad record-keeping and reporting required.

(6) The stock company provides insulation from other parts of the owners' businesses. The pipeline company is a distinct legal entity responsive to its own directors; and

(7) The stock company is less complex. This is especially true if the pipeline is one where owner participation varies widely for geographic or other reasons. (As will be shown below, there may be other advantages that will outweigh the complexity problem.)

The business advantages of the undivided interest form of jointly owning pipelines are as follows:

(1) Because each owner issues his own tariffs, secures his own volume and collects his own revenue, he can compete with other owners for available throughput, and retain all profits on movements through his

---

[1281] AEI, *Oil Pipelines and Public Policy, supra* note 684, at 175 (Discussion statement of James Shamas, Chairman of the Board, Texoma Pipe Line Co.). The nine gave financial guarantees for the smaller company.

portion of the line's capacity;

(2) Costs and expenses arising from undivided interest systems are consolidated into each owner's records. Accounting and reporting requirements are reduced and duplication of records is avoided;

(3) Each owner obtains his financing on his own. Thus *his* credit rating will determine the cost. Loan repayment can be tailored to meet his internally generated cash fund requirements;

(4) Operating agreements may provide that, as an undivided interest line expands, an owner may elect to participate or to refrain, depending on his own need and cash position at the time. [In Section III F 6, "Construction and Technical Problems Now Take the Form of Environmental and Regulatory Difficulties," at note 809, an example of this was shown where Mobil took advantage of the doubling of TAPS's capacity to refrain from participating, thereby reducing its ownership percentage by roughly 50 percent];

(5) Ownership can be divided into line segments to suit the needs of the owners, which would facilitate injection points en route or more than one terminal. The Rancho System provides a good example of each of these advantages;

(6) Under prescribed depreciation policies, which involve the group method and composite rates, earlier recovery of cash through asset depreciation may be possible;

(7) Earnings from an undivided interest system do not involve dividends, hence the small "upstream" or intercorporate dividend tax is not incurred. Also, because capital expenditures are incurred directly by individual owners, a given owner may achieve earlier utilization of investment tax credits; and

(8) Absence of a throughput agreement makes it easier to sell or dispose of an ownership interest.

The first jointly-owned line (corporate form) came about in 1921 when Standard of Indiana purchased from Sinclair Consolidated Oil a one-half interest in Sinclair Pipe Line Company's system from Oklahoma to Chicago. The first jointly-owned line (corporate form) in the sense of planning, formation, financing and construction of a new line was the Texas-Empire Pipe Line, incorporated in Delaware as a common carrier in 1928, and owned 50/50 by The Texas Company and Empire Gas & Fuel Company, a subsidiary of Cities Service. Both Texaco and Cities Service had the same objectives, to bring crude from the flush Mid-Continent fields to market oriented refineries in the Chicago marketing area rather than refine crude in the Mid-Continent and pay the high railroad rate for shipping products to Chicago. In order to achieve the economies of scale, the companies constructed a 12-inch line from Cushing, Oklahoma, to the Missouri-Illinois border, and thence to Chicago via a subsidiary, Texas Empire Pipe Line Company (Illinois). The sharing of the $17,000,000 cost was an obvious benefit.[1282]

In 1930, Ajax Pipe Line Corporation (corporate form), a holding company, and Ajax Pipe Line Company, an operating company, were formed by Jersey Standard (53%), Pure (24%) and Sohio (23%) to build twin 10-inch lines to transport crude from the Glenn Pool, Oklahoma, station of the Oklahoma Pipe Line Company to the Wood River, Illinois, station of the Illinois Pipe Line Company. Jersey advanced the money for construction. The "take out" loan was secured by a forerunner of today's throughput agreement, *i.e.*, agreement by the shippers to purchase, or cause to be purchased, specified amounts of crude oil (Sohio's entire refinery requirements in its case) under a 5-year contract with the Carter Oil Company, a wholly-owned producing subsidiary of Jersey Standard, to be renewed for a subsequent 5-year period upon penalty of forfeiting their stock to the holding company.[1283]

The first jointly-owned products line (corporate form) was the Great Lakes Pipe Line Company, incorporated in 1930, the main stem of which ran from Tulsa, Oklahoma, where feeders connected refineries at Muskogee, Okmulgee, Tulsa, Barnsdall and Ponca City, Oklahoma, to Minneapolis-St. Paul with a branch from Osceola, Iowa to Omaha, and another branch from Des Moines to Chicago, with an interconnection with the Phillips line at Paola, Kansas, (just south of Kansas City) enabling Phillips to move its Borger, Texas-originated products north along the Great Lakes System and the "Tulsa Group" to ship eastward to St. Louis. The owners were Barnsdall, Continental, Skelly, Mid-Continent Petroleum, Phillips and Pure. Later, Texaco and Sinclair joined the ranks of owners, and Cities Service came in as an owner in 1938.[1284]

Just before World War II, Plantation (a corporate form of jointly owned products line with Standard of New Jersey, Standard of Kentucky and Shell as owners) and Southeastern Pipe Line (a corporate products line jointly owned by Gulf and Pure) were formed to penetrate the Southeastern markets. The war and the government's sponsorship of close cooperation in oil supply during the emergency gave a substantial impetus to these so-called "industry" lines.[1285] The first undivided interest system was developed during this period as a means of increasing the ability of Plantation to transport products into the Southeastern part of the country. This system, Bayou, ran from Houston via Beaumont to Baton Rouge where it tied into Plantation. Because of a peculiarity of

---

[1282] See note 139, *supra;* PETROLEUM PIPELINES, *supra* note 45, at 140; KENNEDY STAFF REPORT, *supra* note 184, at 86.

[1283] See note 140, *supra;* PETROLEUM PIPELINES, *supra* note 45, at 140-142; MCLEAN & HAIGH, *supra* note 35, at 243-246.

[1284] See note 138, *supra.*

[1285] *Consumer Energy Act Hearings, supra* note 154, at 1025 (Statement of Keith Clearwaters).

Texas law then in effect, the use of the corporate form was not feasible.

The "Big Inch" and "Little Big Inch" lines, which have been mentioned previously, clearly demonstrated that "big-inch" lines were operationally feasible. The pipe making capacity was there, the operational techniques were proven and the pent-up demand for petroleum created the need. However, as was developed in Section III E, "'Big-Inch' Technology and Jointly Owned and Jointly Operated Pipelines," these pipelines were beyond the practical financial capability of a single company, the risks were too great prudently to concentrate in one project and the throughput volumes required to realize the economies of scale exceeded what a single refiner or, usually, even two refiners could muster, hence jointly owned or operated pipelines, known as "industry lines," became a commonplace.[1286] Since 1950, the majority of the new pipelines constructed have been jointly owned or operated.[1287]

Picking up the thread of criticism of jointly owned or operated pipelines, the influence of Dr. Mead's early work becomes apparent. Industry critics sometimes assert that these arrangements reduce the number of competitors in a given market by combining in the jointly owned or operated line firms which otherwise compete. To the extent that such assertion relies upon an illegality *per se* doctrine, such reliance clearly is misplaced. The courts[1288] and the Justice Department[1289] have stated

---

[1286] See text at notes 702-715, *supra*.

[1287] *Consumer Energy Act Hearings, supra* note 159 at 1026 (Statement of Keith Clearwaters).

[1288] *E.g.*, United States v. Penn-Olin Chemical Co., 378 U.S. 158 (1964); United States v. E. I. duPont de Nemours & Co., 118 F.Supp. 41 (D.Del.1953), *aff'd*, 351 U.S. 377 (1956); United States v. Imperial Chemical Industries, Ltd., 100 F.Supp. 504 (S.D.N.Y. 1951).

[1289] There is a long history of Justice Department's refusals to apply a *per se* approach going back at least to pre-World War II. In 1940, the Justice Department instituted a suit under the Elkins Act against 20 oil companies and 59 pipeline companies to enjoin the payment of dividends by the pipeline owners to their shipper-owners. Among the defendants were ten jointly owned and operated pipelines. No mention was made that the joint ownership created any antitrust problems. The *Atlantic Refining Company* Consent Decree, entered on December 23, 1941, applied to jointly owned and operated lines, both present and future. Implicitly, the Department recognized and expected future jointly owned lines to be built. During World War II, the federal government encouraged and sponsored these industry lines, the leading examples being the "Big Inch" and the "Little Big Inch." Many of the supplementary pipeline projects were jointly owned and operated, and these were cleared with the Justice Department by the Petroleum Administrators for War (PAW) before PAW approved the project. These include extension of the Plantation System from Greensboro, North Carolina, to Richmond, Virginia; the Ohio Emergency Pipe Line, Bayou Pipe Line System, Project V Pipe Line Corporation, Great Lakes Pipe Line extensions, Sun-Susquehanna Pipe Line Company, line loops on the Texas-New Mexico Pipe Line Company and the Texas-Empire System. *Consumer Energy Act Hearings, supra* note 154, at 628-629 (Statement by Jack Vickrey); In 1945, Attorney General (later Supreme Court Justice) Tom C. Clark, in reply to Senator Joseph O'Mahoney's question whether continued joint operation of the War Emergency Pipeline in peacetime would be violative of the antitrust laws, stated that "If the oil companies acquired ownership of the War Emergency Pipelines for operation of the lines as common carriers we would not view such acquisition as a

unequivocally that such arrangements are *not* illegal *per se*. The rationale enunciated by them is very similar to Dr. Mead's economic test, *i.e.*, where the absolute capital requirements and the risks are so great that if the project is to be carried out at all, it may require the combined efforts of two or more firms. Such a venture may very well be procompetitive. Examples of situations requiring the combined efforts of more than one company that come to mind are: The Mid-Valley System, which made Sohio and Sun competitive in the Ohio marketing area and displaced some old smaller lines, thereby reducing the cost of transportation to all shippers moving oil in that direction; Rancho, which has seven owners, including small independents, increased competition in the transportation of West Texas crude to the Houston refinery area, replacing obsolete small diameter lines and lowering the rates for such movements; Colonial, with ten owners, added competition to the only existing pipeline (Plantation) serving the lower part of its route and to the coastwise tankers which previously had the Gulf Coast to New York movement locked up, with the result that rates went down and independents along Colonial's line increased their share of the market; and finally, TAPS, with eight undivided interest owners, which provided a means of transportation that was too heavy financially and too risky for even the State of Alaska to handle; which made North Slope crude available to the refiners in the

---

violation of the antitrust laws in the absence of a conspiracy to eliminate competition among the companies or among possible groups of the companies for acquisition of the lines. Operation as common carriers would require that service be available on equal, nondiscrimiantory and reasonable terms to all those engaged in the indsutry." *Id.* at 651 (reprint of Justice Clark's letter); Mr. (William) John Lamont testified in 1973 that Justice Clark's letter was not commenting on any specific arrangement for the WEP lines, *Id.* at 675 (Statement of W. J. Lamont) but the clearances for the projects listed above were specific, and, moreover, Mr. Lamont was at Deputy Assistant Bruce Wilson's side when Mr. Wilson testified that these arrangements were not illegal per se (see following part of this note), and said nothing; Both the Attorney General's *Second Report* (pp. 73-79) and *Third Report* (p. 45) noted the pervasive extent of jointly owned and undivided interest lines, the former approving their role in providing interconnecting service to shippers and refiners and the latter terming them "favorable developments."; Richard W. McLaren, *Current Antitrust Division Policy on Mergers, Acquisitions and Joint Ventures* 14-15 (address by the Assistant Attorney General; Antitrust Division, in Los Angeles, Calif. May 27, 1969): "A particular project may be beyond the ability of a single firm to undertake alone because, for example, the cost involved is too great. If the project is to be carried out at all, it may require the combined efforts of two or more firms. Such a venture may very well be procompetitive. It may, for example, introduce a new force into a concentrated market and improve its competitive performance."; *H.Res.5 Subcomm. Hearings, supra* note 234, at 206 (Statement of Bruce B. Wilson, Deputy Assistant General, Antitrust Division, Department of Justice); As late as 1976, Attorney General Levi, in his Deepwater Port Report, *supra* note 446, at 47 stated: "The joint venture form of organization of LOOP and SEADOCK is *prima facie* justified because of the economies of scale inherent therein and because of the large economic disincentives to build similar systems in close proximity thereto." This last quotation was made in connection with deepwater ports and cannot be applied *mutatis mutandis* to pipelines but Deputy Assistant Attorney Donald Flexner stated that the Deepwater Port Report analysis was an example of how to analyze [the jointly owned pipeline] problem. Discussion, AEI, *Oil Pipelines and Public Policy, supra* note 684, at 64.

"lower 48" states who were being forced to rely upon imported crude for their refinery intakes.

There have been claims that joint ownership or operation would have an adverse effect upon competing carriers by foreclosing those shipments committed by multiple owners to the joint venture from possible shipment on other carriers even if the other carriers were more efficient.[1290] To the extent that payment under a deficiency agreement plus the cost of the competing carrier's tariff rate make it uneconomic to use the latter line, it is difficult to see how the situation would be any different if pipelines were divested and if, as proponents claim, "majors" would continue to enter into throughput and deficiency agreements. Hence, no antitrust implication should arise from these unilateral decisions based on economic analysis any more than from a decision by such potential shipper to have its traffic carried by a cheaper mode of transportation or to build a more efficient line itself. To the extent that it implies that shipper-owners will be wedded to their own lines when it is cheaper to use other methods of transportation it is flatly contradictory to the empirical evidence. Shell, which has an ownership position in Plantation has consistently used Colonial and, in fact, it was among the very first non-owner shippers on Colonial's line and Exxon, another Plantation owner, now ships over Colonial. Gulf shut down its Northern Pipeline system from Tulsa, Oklahoma, to Dublin, Indiana, and shipped over the Mid-Valley System. Several 8- and 10-inch lines from West Texas were abandoned by their owners to ship over Rancho.

Another issue raised by critics is that jointly owned lines reduce competition among owners by reducing variations in transportation costs among them, thereby promoting parallel pricing.[1291] That charge does not bear up under scrutiny. Ownership and throughput ratios between owners vary and hence so must the cost per barrel. Moreover, the pipeline transportation cost is such a small portion of the delivered price[1292] that any narrowing of costs even if it were factually valid would not be a significant factor compared to the many variables such as crude costs, refinery efficiencies and differentials in handling and distribution expenses, not to mention the competitive effect of alternative routes or different origins of the products with which the transported materials must compete.

[1290] KENNEDY STAFF REPORT, *supra* note 184, at 88. John Lamont, who certainly cannot be accused of being biased in *favor* of major oil companies has suggested in sworn testimony that this is not so. He inferred that some of Explorer's owners left the line to ship over Colonial to realize a better price for their fuel oil. *Consumer Energy Act Hearings, supra* note 154, at 672 (Statement of John Lamont).

[1291] *Id.*

[1292] See text at note 197, *supra*.

It is asserted also that nonowners using the pipeline pay the tariff while owners pay cost, "a sizeable differential up to 20-30 percent."[1293] This, of course, is not limited to jointly owned lines, it applies to any shipper owned line. As an allegation applied to a situation where rates are competitive, it has been termed by an informed and serious observer of pipelines as a "fallacious assertion."[1294] Dr. Stocking, no friend of the oil industry, has discounted the argument as lacking in logic.[1295] Any rate that produced less than a fair market rate of return would result in an unfair and discriminating transfer of income *from* pipeline owners to pipeline shippers,[1296] so the argument at best must rely upon the existence of an alleged excessive rate of return. This question will be discussed in a subsequent section.

Another allegation is that participation in a jointly owned line serves to stabilize market shares among the owners, "since there is no incentive to ship more than the ownership share because of the transportation cost penalty associated with shipments above ownership share."[1297] This suffers from two disabilities. The first is theoretical, being a miniature of the non owner shipper *subsidizing* argument, and *a fortiori* indefensible; the second is that it runs directly contrary to the facts. What the proponents of this allegation have done is distort the normal shipper owner's desire not to be exposed to a greater capital investment and undergo more risk than is necessary to support his shipments, hence he attempts to conform his ownership to his usage. If it were the other way around, as the critics suggest, all the scrambling around for participants would be unnecessary; investors would be clamoring to get in. So far as this author has been able to ascertain, there was no great surge of investors to participate as an owner in TAPS; to the contrary, Mobil took advantage of a capacity expansion to reduce its ownership share to a proportion closer to its expected usage, Standard Oil Company of California "opted out" of ownership as did a number of other North Slope producers, and the State of Alaska did not step up and take its 12.5 percent share.

In the *Petroleum Industry Hearings,* Senator Tunney pressed the point that co-ownership in a jointly owned or undivided interest pipeline

---

[1293] KENNEDY STAFF REPORT, *supra* note 184, at 88.
[1294] CRUDE OIL PIPELINES, *supra* note 158, at 104 (Leslie Cookenboo); *accord,* SPECTRE, *supra* note 104, at 111; Bork, *Vertical Integration — Economic Misconception, supra* note 106, at 200; WOLBERT, *supra* note 1, at 58; Note, 102 U. Pa. L. Rev. 894, 912 (1959); *see also* Adelman, *Effective Competition and the Antitrust Laws,* 61 Harv. L. Rev. 1289 (1948); Adelman, *Integration and Antitrust Policy,* 63 Harv. L. Rev. 27, 42 (1949).
[1295] Stocking, Book Review, 1 Vand. L. Rev. 490, 492 (1948).
[1296] Gary, *Direct Taps Testimony, supra* note 634, at 13.
[1297] KENNEDY STAFF REPORT, *supra* note 184, at 88.

enabled the various companies in an industry to know exactly what were the positions of the other companies in the enterprise.[1298] The serious shortcoming in that line of approach is that the provisions of Section 15(13) of the Interstate Commerce Act[1299] make it unlawful for any common carrier pipeline or any officer, agent or employee thereof knowingly to disclose or permit to be acquired by any person or corporation information as to the nature, kind, quantity, destination, consignee or routing of any other shipper's or consignee's movements without their consent, and Section 15(13) makes it equally unlawful for any other person or corporation to solicit or knowingly receive such information. Moreover, the testimony adduced before the Antitrust and Monopoly Subcommittee which held the hearings was that in *fact,* the pipeline companies have been most diligent in their compliance with that provision.[1300] In the same hearings, Senator Abourezk attempted to lead witness Ikard into acknowledging that executive officers from the various major oil companies "took turns" sitting as pipeline company officers and that Directors of the various major oil companies also sit on the management committees of the jointly owned or undivided interest lines.[1301] Mr. Ikard's oral response to the first part of the question was that the owners' representatives sat on the Boards of the corporate pipelines in which the oil companies had an interest[1302] and his written response to the second part was that major companies which have partial interest in [undivided interest] interstate pipeline [systems] usually are represented on the management committees of those [systems] by representatives who tend to be officers from the major companies' transportation divisions, and in general, these officers are not directors of the parent majors.[1303]

Senator Bayh explored this subject with Charles Waidelich of Cities Service, who had been on several pipeline committees or Boards of Directors, including Colonial's, and was intimately acquainted with the manner in which pipelines preserve the confidentiality of information on

---

[1298] *Petroleum Industry Hearings, supra* note 192, at 146-148.

[1299] 49 U.S.C. § 15(13) (1976). This section was renumbered from 15(11) in 1976. Although the Act was rewritten substantially in 1978 as to other carriers, this section remains applicable to pipelines.

[1300] *Petroleum Industry Hearings, supra* note 192, at 309 (Statement of Vernon T. Jones, President of Williams Pipe Line Co., former President of Explorer Pipeline Co.); *Id.* at 261(Statement of Charles J. Waidelich); Even Counsel for the Association of Oil Pipelines felt constrained to obtain assurance that furnishing such information to a Congressional Committee would not violate the provisions of 49 U.S.C. § 15(13). *Consumer Energy Act Hearings, supra* note 154, at 606 (J. Donald Durand, General Counsel AOPL). Note also the caution expressed in MARATHON, *supra* note 202, at 55 n.1.

[1301] *Petroleum Industry Hearings, supra* note 192, at 148-149.

[1302] *Id.* at 148.

[1303] *Id.* at 175.

individual shipments. Mr. Waidelich assured the Senator that there was, in fact, no exchange of proprietary information and he provided Senator Bayh with a copy of Section 15(13) of the Interstate Commerce Act. More tellingly, he explained to Senator Bayh what the real pipeline world was all about: knowledge after the fact of what were the shipments of another shipper is worthless; no one is about to tell a competitor what someone's future plans are; pertinent market information is readily available from public records such as state gasoline and fuel-tax records; and demographic statistics are used to justify large-sized lines such as Colonial.[1304]

Just as was the case in the arguments against jointly owned and operated pipelines, the case in favor of them largely follows the outline proposed by Dr. Mead. Dr. Mead's first reason favoring the legality of such arrangements was that they permit entry into an activity where absolute capital requirements are so high that only a few large firms could otherwise participate. In the case of the large industry pipelines, the nature and the absolute size of the capital investment required to realize the economies of scale are such that they exceed the practical financial capability of even the largest company in the industry, and the risks are so great that no single company prudently can concentrate the bulk of its financial resources in one project; nor could it muster the volume of traffic necessary to keep the line operating at a percentage of capacity that would achieve the desired economies of scale.[1305]

These circumstances more than meet Dr. Mead's test; in fact they also encompass his second justification as well: "Risks [that] may be so great that few or no existing firms are able or willing to assume such risk alone."[1306] The "fit" between jointly owned or operated pipelines and Dr. Mead's justification criteria was noticed by writers in the field even before Dr. Mead explicitly formulated them. In 1962, Michael Bergman,

---

[1304] *Id.* at 260-262. Mr. Waidelich stated: "In my experience, at least with all of the pipelines that Cities Service has had an interest in, in which I have served on the board of directors or on various committees, I have certainly never found it to be the case where we have exhanged proprietary information." *Id.* at 261. In response to Senator Bayh's remark that "at least one agency of the Government feels that they [the FTC "Big Eight" defendants] do not live up to [presumably, 49 U.S.C. § 15(13), and "lots of rules and regulations"], but that can only be decided in a court case," Mr. Waidelich replied. "Right. I could only say, Mr. Chairman, that in my experience of serving on Colonial and many other pipeline boards, I simply have not run into the situation where a group of men sit around, and if you will, gather in information of a proprietary nature which can be used to hurt someone else." *Id. See also Id.* at 313 (Statement of E. P. Hardin) (the only information a Director has on shipments are his own Company's and the total of all shipments).

[1305] See text at notes 606-655, 702-719, especially 712-712, *supra.*

[1306] See text at notes 720-1036, *supra;* KENNEDY STAFF REPORT, *supra* note 184, at 87; Waidelich, *Rodino Testimony, supra* note 197, 17-18; CARD, *Industrial Organization, supra* note 1152, at 40; *cf.* MARATHON, *supra* note 202, at 51; LIVINGSTON, *supra* note 228 at 67-70.

in an article in the New York University Law Review, stated: ". . . certain forms of horizontal collaboration should be encouraged. This would include the use of the joint venture to undertake a project so large that it would be impossible for any one firm, or, for that matter, any small group of firms, to accomplish it alone. A vast land reclamation project or a nationwide pipeline [summarizing, in a footnote, highlights of the Colonial line] would be examples of such a project. In such cases, the social desirability of the project and the physical necessity of the combination should remove the venture from the operation of the antitrust laws."[1307] Bankers,[1308] Investment Bankers,[1309] industry members,[1310] practicing attorneys,[1311] and even Justice Department officials, with the qualification that participation must be offered to all who are willing and able to join,[1312] have all subscribed to this proposition. The natural evolution of this phenomenon was summed up succinctly by John Winger, Vice President of Chase Manhattan Bank in New York, in the *Consumer Energy Act Hearings:* "In the early days of the petroleum industry there were no major companies — only independent operators. But, in time, the large integrated companies evolved because the capital requirements became larger than the independents could manage.

"Today, the capital and other financial needs are so enormous that even the largest of the major companies are forced to carry on some operations jointly. This is not to suggest that independents are no longer necessary. They continue to play a very important role —but it is completely illogical to think that they could function alone."[1313]

Specifically applying this to modern day pipelines is the testimony of witness Charles J. Waidelich: "Such sums of money [referring to the cost of Colonial and TAPS] are imposing and that is why the Colonial and Trans-Alaska systems, as well as virtually all other large pipelines, are constructed on a joint interest basis. Projects of such scope simply are not feasible for any one oil company.

"Indeed, very few, if any, single corporate entities would be able to

---

[1307] Bergman, *The Corporate Venture Under the Antitrust Laws,* 37 N.Y.U. L. Rev. 712, 732 (1962).

[1308] *Consumer Energy Act Hearings, supra* note 154, at 1308 (Statement of John J. Winger, Vice President, Chase Manhattan Bank).

[1309] Gary, *FERC Direct Testimony, supra* note 613, at 6.

[1310] Waidelich, *Rodino Testimony, supra* note 197, at 4; *Petroleum Industry Hearings, supra* note 192, at 256-257 (Statement of Charles J. Waidelich); CARD, *Industrial Organization, supra* note 1152, at 40; *Consumer Energy Act Hearings, supra* note 154, at 625 (Statement of Jack Vickrey); *cf. S.1167 Hearings, supra* note 432, pt. 8, at 5929 (Statement of Robert E. Yancey); Cooper, *Direct TAPS Testimony, supra* note 718, at 11-12.

[1311] HOWREY & SIMON, *supra* note 3, at 118-120, 169-180.

[1312] See note 1289, *supra.*

[1313] See note 1308, *supra.*

dedicate the larger amounts of cash or debt obligations required to construct a major pipeline system. Joint ownership of pipelines by oil companies is a practice that has been dictated by economic and financial realities that remain relevant. Their purpose has not been to divide markets or inhibit competition, as some people mistakenly have charged."[1314]

There are some critics of the industry, such as Jesse M. Calhoon, President of the National Marine Engineers Beneficial Association (MEBA), who maintain that "the use of joint-venture capital is not required in order to finance and operate new and more efficient pipelines."[1315] Mr. Calhoon's testimony reveals some of the bases of his statement *i.e.*, that "The 10 oil companies which constructed the Colonial Pipeline put up only $36 million cash of the $360 million cost, . . ." and that "There is little financial risk when one considers that the coventurers make long-term commitments [to provide security for the $324 million borrowing] to their own pipeline of crude oil or petroleum products." In light of this obvious display of unfamiliarity with economics and the operations of financial markets; the fact that a large part of MEBA's membership consists of seamen employed on the Gulf Coast to New York coastwise tanker trade which had the petroleum products transport business "locked up" until Colonial introduced it to competition; and the overwhelming evidence that contradicts Mr. Calhoon's statement, it would appear that this line of reasoning deserves slight, if any, credibility.

Dr. Mead's third category of justifications for common ownership of productive facilities was a situation where separate operations by competing firms may be an inefficient or wasteful means of performing a necessary function. Jointly owned or undivided interest pipelines meet this category perfectly. Section III C, "Economies of Scale" and Appendices I and A make it obvious that the large diameter "industry" lines have become a competitive necessity and an attempt to perform their transportation task by multiple small lines or other modes of transportation would be inefficient and wasteful,[1316] and would raise measurably the price of products to consumers.

The fourth justification, *i.e.*, where large investments are expected to produce external economies that will accrue to all firms in the relevant area rather than primarily to the investing group, likewise is present in the case of the "industry lines." Such large scale pipelines reduce transportation costs for all and consequently increase competition in the

---

[1314] *Petroleum Industry Hearings, supra* note 192, at 256 (Statement of Charles J. Waidelich).
[1315] *Id.* at 229 (Statement of Jesse M. Calhoon).
[1316] CRUDE OIL PIPELINES, *supra* note 158, at 128 ("wasteful"); TEECE, *supra* note 329, at 37 ("inefficient").

geographic regions they serve.[1317] In a previous section dealing with the question of whether pipelines come within the definition of a "natural monopoly," the conclusion was reached that the economies of scale inherent in (large diameter) petroleum pipelines were not lastingly internal to the pipeline owners but rather tended to become external to the owners and permanently internal only to the industry.[1318] Partial explanation for this lies in the fact that under the Interstate Commerce Act, pipelines, as common carriers, must make their services available to all shippers on a non-discriminatory basis.[1319] Moreover, the tariff rates which they charge must be "just and reasonable,"[1320] and because of the *Atlantic Refining Company* consent decree, they have tended to be the *lesser* of seven percent of ICC valuation or the competitve rates required to get the business. The service must be provided upon reasonable request to any shipper, and reasonable through or joint rates must be established with all connecting carriers.[1321]

In addition, the pipeline's operating rules, regulations and practices for transporting the traffic tendered to it must be just and reasonable and uniformly applied to all shippers;[1322] there can be no unreasonable preferences or discrimination in furnishing services.[1323] Compliance with these provisions has the effect of passing on to nonowner shippers the economic benefits realized by the economies of scale made possible by the investment made by the owners.[1324] Professor Edward Mitchell testified before the Senate Antitrust and Monopoly Subcommittee in 1974 that he had made an examination of the results of joint ownership of Colonial pipeline by a number of major oil companies to see if other companies were restricted from access to the line and if the economies of scale inherent in the Colonial line were reflected in lower prices in that area. Professor Mitchell reported that the results of his examination were that

---

[1317] CARD, *Industrial Organization, supra* note 1152, at 40.

[1318] See text at notes 672-689, *supra*.

[1319] 49 U.S.C. § 3(1) (1976).

[1320] 49 U.S.C. § 1(4) and (5) (1976).

[1321] 49 U.S.C. § 1(4) (1976).

[1322] 49 U.S.C. § 1(4)-(6) (1976).

[1323] 49 U.S.C. §§ 2 and 3(1) (1976). This is fortified by the Elkins Act, 49 U.S.C. §§ 41-43 (1976).

[1324] Cookenboo in CRUDE OIL PIPELINES, *supra* note 158, at 132 called this an "unjustifiable policy," but his remark was made in the context of "proprietary" (solely owned lines) and his recommendation was a system of forced joint ventures. *Id.* at Chapter IV. Much of what Cookenboo suggested has come about as a natural consequence of the incidence of large diameter "industry lines" and the Justice Department is attempting to develop a theory to accomplish the more difficult and complicated problems of handling variable or "yo-yoing" ownership as an incentive to spur full utilization of scale economies. See L. LEWIS & R. REYNOLDS, APPRAISING ALTERNATIVES TO REGULATION FOR NATURAL MONOPOLIES (Paper presented at American Enterprise Institute Conference on Oil Pipelines and Public Policies, Mar. 1-2, 1979).

in fact prices of gasoline along the Colonial pipeline had fallen significantly since the commencement of Colonial's operations, and at the same time, shares of the market held by Colonial's owners had actually fallen, and done so faster in the Colonial area than in the national gasoline market.[1325] This evidence meshes neatly with the evidence of increasing use of pipelines by nonowner shippers, which was examined in depth in Section II A 3 c, "Scramble for Traffic."[1326]

A "second order," or ripple effect, whereby the scale economies of large jointly owned lines externalize from the shipper-owners to other members of the industry and to the public in general was discussed by witness Charles Waidelich in his testimony of October 31, 1975, also before the Senate Antitrust and Monopoly Subcommittee. Mr. Waidelich, who is the President of Cities Service Company, one of the largest suppliers of gasoline to independent gasoline marketers, described how Cities Service had closed down its East Chicago, Indiana, refinery in late 1972 because of the prohibitive cost of upgrading its operations and environmental protection equipment and how it would have found it extremely difficult to continue marketing in the Midwest without that refinery but for the Explorer System. So Cities Service, whose sole refinery is located in Lake Charles, Louisiana, provides its independent marketing customers in the Midwest and the Southeast with product refined in Lake Charles by means of the Explorer and Colonial Pipelines, respectively. In this manner, these two large diameter "industry pipelines" have extended the benefits of their great economies of scale to many small independent marketers in those areas, thus, making such economies internal to the *industry* in those areas and enhancing competition, with the consumer being the ultimate beneficiary.

Evidence of the external (to the owners) effect is the results of the Lundberg study covering the fifteen state area along the Colonial pipeline route for the years 1963-1974 showing a 59 percent total gasoline demand increase but an increase by "Other than Class I Marketers" (*i.e.,* other than integrated companies marketing under their own brand in twenty or more states) of 205 percent and an almost 100 percent increase in their market shares (*i.e.,* from 14 percent in 1963 to 27 percent in 1974) which was evident in each of the states except Pennsylvania which was influenced by the large Delaware Valley refiners.[1327]

The third way in which competition has been increased by the large industry lines is the creation of opportunities for small companies to get a

---

[1325] *S.1167 Hearings, supra* note 432, pt. 8 at 6044-6045 (Statement of Professor Edward J. Mitchell); *see also* Ritchie, *Petroleum Dismemberment,* 29 Vand. L. Rev. 1131, 1144 (1976).

[1326] See text at notes 410-496, especially 461-496.

[1327] *Petroleum Industry Hearings, supra* note 192, at 779-785.

"piece of the action" and to enjoy the economies of scale which would have been beyond their means without the large lines. As the Kennedy Staff Report put it: "Joint ventures, very importantly, permit vertical integration into large diameter pipelines by small and large companies. Thus, companies that could not afford to build large diameter lines or fully employ them with their own resources can share in the ownership and obtain the significant economies associated with large-diameter pipeline transportation."[1328] Two of the more recent large diameter lines are Texoma, a 472-mile 30-inch crude line from Nederland (near Port Arthur), Texas, to Cushing, Oklahoma, and Seaway, a 510-mile 30-inch crude line from Freeport, Texas, to Cushing. Texoma, which cost initially around $121 million (closer to $135 million now), is owned by Kerr-McGee (10.1 percent); United Refining (7 percent); Western Crude Oil (20 percent); Lion Oil (5 percent); Rock Island Refining (5 percent); Texas Eastern Transmission (5 percent); Vickers Petroleum (2.7 percent); Sun Oil (25 percent); Mobil Oil (10.1 percent); and Skelly (10.1 percent). Thus "independents" have a 54.8 percent interest.[1329] Texoma borrowed $100 million from Chase Manhattan and the bank was reluctant to let one of the smaller companies in because of its extremely low credit rating. The other nine companies had to "carry" (give financial guarantees for) the low credit company, which they did in order to obtain the diversity of ownership.[1330] Seaway, which cost between $170 to 180 million is owned by Apco (2 percent); CRA, Inc. (12 percent); Diamond-Shamrock Oil Company (7 percent); Midland Cooperative, Inc. (9 percent); National Cooperative Refinery Association (12 percent); Continental Oil (15.6 percent); and Phillips Petroleum Company (42.4 percent), a 42 percent ownership by "independents."[1331] These are only the ones who "came in," but by no means all that were invited to participate. Perhaps a few examples will help illustrate the point. Yellowstone, whose main line is a 10-inch products line running 537 miles from Billings, Montana, to Spokane, Washington, invited Husky, Union, Exxon, Conoco, Farmers Union Central Exchange, Phillips, and Texaco (all those who were physically located in such a way as to be potentially interested in such a line) to become owners. Only the first four accepted.

In the case of Capline, Shell Pipe Line Corporation made telephone contact with refiners in the upper Mid-Continent area, and as a result of interest shown by many, held a meeting on April 15, 1966, which

---

[1328] KENNEDY STAFF REPORT, *supra* note 184, at 87. See also PETROLEUM PIPELINES, *supra* note 45, at 386.

[1329] *Petroleum Industry Hearings, supra* note 192, at 316 (Information furnished by J. Donald Durand at the request of Staff Counsel Henry Banta).

[1330] See note 1281, *supra*.

[1331] See note 1329, *supra*.

Texaco, Pan American Petroleum (Amoco), Sinclair, Mobil, Pure, Marathon, Ashland, Clark, Cities Service, Gulf, Sun, Sohio, Humble (Exxon), Chevron (Standard of California), and Continental attended. Shell made a presentation showing the demographics, *i.e.*, decline of supply and increase of demand in PAD II, the increased crude production onshore and offshore south Louisiana and various alternative pipeline plans to solve these problems. After the meeting, additional inquiries were received from Northwestern Refining Company; Delta Refining Company (Earth Resources Company); Madisonville Terminal Corporation; Kerr-McGee; Hess Oil & Chemical (now Amerada-Hess); National Transit Company; Placid Oil Company; and Mid-Valley Pipe Line. Each of these companies was furnished the data discussed at the April meeting and asked if it had an interest, and if so, it would be included in the next meeting. Studies were conducted by Shell Pipe and on April 22, 1966, it sent out a letter to all parties who had attended the April 15th meeting or who had subsequently requested information requesting them to indicate whether they were still interested. Based on the responses, questionnaires were sent out asking for their views and inviting those interested to a meeting at the end of June, 1966. Only Service Pipe Line Company (Amoco), Pure, Ashland, Clark, The Texas Pipe Line Company, Mid-Valley, Marathon, Continental Pipe Line, Humble Pipe Line (Exxon), and Gulf Refining Company were interested; all others indicated they had no further interest in the projects. At the meeting, possible throughputs, capacity, and cost estimates and many details of the project, still in alternative configurations, were discussed and Shell Pipe Line was selected to be the contractor responsible for design, construction, and operation of the system. When commitment time came, Continental, Gulf, and Humble (Exxon) had withdrawn; shortly thereafter, Pure and Clark combined their interest into a new company, Southcap, and the owners became Ashland, Marathon, Mid-Valley, Amoco, Shell Pipe Line, Southcap, and Texaco. Thus, out of the 23 companies "interested," 10 became "very interested" but only 7 (8 if one counts Clark and Pure separately) were "that interested" enough to commit their resources to the project. Finally, Eugene Island Pipeline System, running from the South Marsh Island and Eugene Island areas of the Gulf of Mexico offshore western Louisiana to the mainland invited 56 companies (all companies having production or production interests in the Eugene Island Block 332 area) to participate. Out of this group, only seven accepted ownership.[1332] It is most common to see the names of those who indicated an early interest but who do not become owners, among the ranks of nonowner shippers. When they are not, the place to find them is among

---

[1332] PIPELINE PRIMER, *supra* note 331, at 11.

the owners or shippers of a competing line which has a slightly different route. This was the case in Butte, Explorer, Texoma, and many others. The empirical evidence appears to indicate eagerness to obtain as many participants as possible, and if the invitees, for their own economic reasons, choose not to undertake the risk, to solicit their traffic as shippers. There does not seem to be a viable case to be made for exclusion or forced ownership, rather it appears to be a straightforward economic choice made by the invitees along the lines discussed in reasons to, or not to, integrate.[1333]

D. How Joint Pipeline Projects Are Put Together in the Real World

1. Conception

There are many ways in which the initial inspiration for a new or expanded pipeline can come about. Frequently, a demographer in an oil company's long range planning department believes that population growth in a particular part of the country will require increased volumes of petroleum products for consumers in time frame "x" to "y", or the marketing department may perceive an opportunity through innovative marketing methods to increase the volumes which it can sell in a particular area if the laid down cost were competitive.[1334] Another way the idea for a new line may germinate is where shippers have requested a joint tariff movement and the pipelines concerned discuss the matter in the ordinary course of business and discover that the increased demand would probably be met more economically by the construction of a new-to-industry line rather than "looping" or rearranging the existing lines. Trade publications such as the *Oil and Gas Journal* may run a feature story on a growing need. Perhaps an external event such as the cutback on Canadian crude exports into the United States will cause a reexamination of the supply patterns to the area being deprived of a significant source of material.

Sometimes customers of an existing line will approach the line and ask about additional space on the line or persons who have not theretofore used the line will, because of a dislocation such as the Canadian cutback mentioned above, approach the line about using it to replace their needs formerly served from a different origin. On some of the more established lines such as Colonial or Williams Pipeline, the planning department will conduct systematic long-term (ten year)

[1333] See text at notes 1229-1260.
[1334] Waidelich, *Rodino Testimony, supra* note 197, at 14.

forecasts and contact its shippers to ensure that sufficient capacity will be in place when it is needed.[1335]

2. Preliminary Analysis

Sometimes those considering the necessity for, and the possibility of a new pipeline are fortunate enough to have some hard data, such as Colonial had on Plantations' shipments, as a starting point. In the case of projected products lines, this is either supplemented, or in the absence of such information, the basis is supplied, by public records of what marketing volumes of potential shippers are in a given state.[1336] Refinery information of a rough nature required to make an analysis of potential crude traffic is available from trade magazines or composite statistics published by trade journals and specialized "clipping services," some of which are computerized, or from governmental agencies. Today, in the case of a "new-to-industry" line, the analysis will almost inevitably indicate that no one company would have sufficient new volumes or traffic that it could divert from its current logistical pattern to justify the cost of a modern efficient pipeline with an ability to expand its capacity to meet future needs.[1337]

3. Rough Study of Alternatives

Working with the demographic data mentioned above, supplemented by any information and views that prospective participants or shippers have provided, the party most interested in the project will prepare a rough study of sources of supply (in the case of a crude line, the oil reserves and production in the field or fields which are to be served, and in the case of a products line, projected available outturns of refineries at the point of origin); a market study of possible destinations (in the case of a crude line, the refineries or connecting pipelines which would be potential outlets for the crude and in the case of products lines, the consuming markets to be served); and various alternative transportation plans and economics. Again, there may be input from other companies already interested in the project or it may be done solely by the moving party. In some instances, several companies may be working independently of each other. In any event such study/studies will identify the prospective interested parties.

[1335] MERCHANT, *supra* note 18, at 2.
[1336] *Petroleum Industry Hearings, supra* note 192, at 262 (Statement of Charles J. Waidelich).
[1337] Waidelich, *Rodino Testimony, supra* note 197, at 15; *cf.* Cooper, *Direct TAPS Testimony, suprc* note 718, at 11-12.

4. Approaches to Other Parties Whose Operations Make Them Potential Participants

The next step is to invite all persons who the studies have identified as having a possible interest to a meeting at which the inviting company lays out what the result of its rough study has indicated are the general facts about aggregate supply and demand and the various alternative pipeline plans it has developed as a starting point for discussion. To the extent that some of the parties have already done their own work on the subject, the definition of the project will be sharpened and those at the meeting who have not had an opportunity to consider the problem theretofore are brought abreast of the situation. Sometimes they have information which might correct the assumptions or indicate additional alternatives. These discussions deliberately are not closely held; in fact, the news media frequently are furnished with a story and companies which were not identified as potential participants will contact one of the parties and they are welcomed to participate in future discussions as was illustrated by the Capline example.

5. Formation of a Management Committee

Usually as a result of the meeting and subsequent conversations, a management committee of interested companies is formed. This committee appoints a Feasibility Study Committee which refines the early preliminary study on supply and demand[1338] and makes a detailed economic study running profitability studies (Discounted Cash Flow Analyses)[1339] on the various alternatives. The results of this study are disseminated to the interested group and especially if no single alternative emerges as the clearly superior choice for the great preponderance of the group, a meeting is held to discuss the alternatives. It happens quite frequently that this serves to separate the group into subgroups. In Butte, there were four small capacity lines and three large capacity lines with different origins and destinations. These plans were held in abeyance due to low reserves and the high delivered cost of crude. Subsequently, three alternatives were developed, all three originating from Poplar and Glendive, Montana, but one going east to Clearbrook, Minnesota, to connect with Interprovincial or Minnesota Pipeline for delivery to the Minneapolis/St. Paul area, one to Oregon, Wyoming, for delivery to Platte, and the third to Fort Laramie and Guernsey, Wyoming, for delivery to Service (Amoco), Platte, or Continental. Each route had its proponents and when the greatest number opted for route three, several

---

[1338] For an example of such an analysis, see MARATHON, *supra* note 202, at 26-31.
[1339] See text at notes 1048-1050, *supra*.

of the potentially interested shippers withdrew from the project, which went forward to become the Butte Pipe Line with Shell, Murphy, Placid, and Northwest Improvement Company (an affiliate of Northern Pacific) as the owners.

Explorer had a similar split when one group of potential participants favored a routing westward through Tulsa and the other group known as MATCH (Mobil, Amoco, Texaco, Continental, and Humble [Exxon]) preferred the eastern route (possibly incorporating an existing line owned by Texas Eastern). The route favored by the MATCH group never came to fruition (in effect, this route was put into service when Texas Eastern expanded its line to Seymour and extended to Chicago), and one of the group came over to the western route group which, after a vigorous effort to get any Mid-Continent refiner who might be interested in augmenting its refinery outturn (developing shortages in West Texas crude supplies was threatening to give them problems) or in a pipeline outlet to Midwest markets, and succeeding only in interesting Apco, became the Explorer Pipe Line Company.[1340] Texoma, the idea for which originated with Apco, which foresaw the need to supplement the supply of West Texas crude to its Kansas and Oklahoma refineries by shipments from the Gulf Coast, found agreement among the prospective participants on the destination, Cushing, Oklahoma, but disagreement over the best origin point (the alternatives were Freeport, Houston, and Port Aransas). Later, Sun began studying the feasibility of moving Gulf Coast crude from Nederland, Texas, to Tulsa by reversing its 10-inch line from Longview, building a "bridge" line from Longview to Big Sandy, Texas, where it would tie into an abandoned Gulf line which ran from Big Sandy to Tulsa. After completion of a feasibility study, Sun decided to attempt to develop an "industry line" as a means of moving Gulf Coast crudes to Kansas and Oklahoma refineries. After making contact with a number of the former Texoma group, it became evident that a 10-inch line would not do the job and, moreover, there was a decided preference for Cushing, which permitted many connections to other lines, rather than Tulsa. Sun went back to the drawing board and came up with a Nederland to Cushing 22-inch line, which, as additional participants came in, was changed to a 26-inch line. Continental, Diamond-Shamrock and Phillips withdrew from the project and formed Seaway, which originates at Freeport, Texas, and competes virtually head-to-head with Texoma as it also terminates at Cushing.

If after iterations of the feasibility study, sufficient interest is shown by enough potential participants so that it appears a line is practicable, a right-of-way committee is appointed, which selects a general route by

---

[1340] LIVINGSTON, *supra* note 228, at 23.

utilizing available U.S. Geological Survey maps as well as city and county tax maps. After the route is tentatively selected, a "fly-over" frequently is done to check out questionable spots and to become familiar with the terrain, timber growth, types of soil, marsh areas, amount of rocks, etc., which greatly affect the construction costs.[1341] The Engineering Committee performs preliminary engineering studies, including studies of alternate line sizes, taking into account the possibility of late joiners, allowing for usage by nonowner shippers,[1342] and providing for future growth over the 20-40 year life of the project. Terminals, injection points and tankage are planned, based on indicated needs and those of possible additional shippers.

Generally, at this point another round of solicitations is made to be sure that all those interested in owning or using the line are informed and their views and indications of continued interest are sought. If the project still appears feasible, a Finance Committee is formed to ascertain on an informal basis if funds will be available to construct the line and upon what basis the borrowing is apt to be made. A Legal Committee is formed to prepare the necessary agreements and documents required.[1343]

Another round of solicitation of interested parties takes place and then the Shareholders' Agreements, Completion Agreement, Operating Agreement, and the all-important Throughput and Deficiency Agreements[1344] are executed by those companies that are prepared to undertake the investment and risk of the project.[1345]

In order to round out the picture for those interested in the manner in which these projects proceed, there are "skeletonized" chronologies of two projects, Eugene Island Pipeline System and Texoma, which are attached as appendices M and N, respectively.

[1341] MERCHANT, *supra* note 18, at 3.

[1342] Waidelich, *Rodino Testimony, supra* note 197, at 5-16.

[1343] For the form of Various Agreements, see *Petroleum Industry Hearings, supra* note 192, at 797-811 (Basin Pipe Line System, Revised June 1, 1953); *Id.* at 786-796 (East Texas Main Line System); *Id.* at 811-827 (Mesa Pipeline System); *Id.* at 828-875 (Ventura Pipeline System); *Id.* at 876-967 (TAPS).

[1344] For a form of Throughput and Deficiency Agreement, see *Petroleum Industry Hearings, supra* note 192, at 1110-1114 (Laurel Pipe Line Co.). Although the agreement is headed "Throughput Agreement," examination of Paragraph 6 discloses the undertaking to make good any cash deficiency. The undertakings are several and, theoretically, one who had shipped his portion could stand by and let the deficiency go unpaid. However, as Gary Swenson, Vice President and Director of The First Boston Corporation, a major New York investment banking firm, has pointed out in his testimony before the Senate Antitrust & Monopoly Subcommittee, if the non-defaulting shipper-owners do not cover the deficiency, the lenders may force the pipeline company into bankruptcy and, consequently, there is a strong incentive for the non-defaulting shipper-owners to make arrangements to cover the deficiency. To his knowledge there has never been a bankruptcy of a shipper-owned pipeline. *S.1167 Hearings, supra* note 432, pt. 9 at 703 (Statement of Gary L. Swenson).

[1345] Waidelich, *Rodino Testimony, supra* note 197, at 16.

Chapter V

Financing

A. General Concepts

1. The Required Rate of Return: The Modern Day Cost of Capital

a. Defined

Over the years, the term "cost of capital" has been used in the financial, economic, and regulatory communities in a number of different ways. In modern financial thinking, however, cost of capital is generally treated as being synonymous with the "required rate of return" to those who supply capital.[1346] The required rate of return for a firm may be defined as the minimum rate of return necessary to attract new capital.[1347] Thus, the required rate of return is measured by the "opportunity cost of capital" to the investor, who is the source of this capital. It is the rate of return which his money could be expected to earn if it were employed in its most attractive alternative use[1348] entailing risks comparable to the specific investment being contemplated.[1349] Stated another way, it is the rate of return that the investor must sacrifice or forego in order to invest in the enterprise in question.[1350]

Two important concepts respecting a firm's required rate of return should be noted. First, because the required rate of return is an "opportunity" rate, it must be distinguished from "historical cost," or cost in the actual out-of-pocket cost sense. For example, if a person purchased a house in 1970 for $50,000 and in 1978 prospective purchasers were offering him $100,000 for that house, his "historical" cost would be $50,000, but his "opportunity" cost would be $100,000.[1351] The second important concept is that a firm or industry will not attract new capital

---

[1346] E. SOLOMON & J. PRINGLE, AN INTRODUCTION TO FINANCIAL MANAGEMENT 338-339 (1977) [hereinafter cited as SOLOMON & PRINGLE, *Financial Management*].

[1347] Myers, *supra* note 198, at 3; Ritchie, *Petroleum Dismemberment,* 29 Vand. L. Rev. 1131, 1137 n.23 (1976); *cf.* Gary, *Direct TAPS Testimony, supra* note 634, at 7.

[1348] Ryan, *TAPS Rebuttal Testimony, supra* note 725, at 47; *cf.* J. CHILDS, LONG-TERM FINANCING 314 (4th Print. 1964) [hereinafter cited as CHILDS, *Long-Term Financing*].

[1349] Solomon, *TAPS Rebuttal Testimony, supra* note 621, at 9; Gary, *Direct TAPS Testimony, supra* note 634, at 7; *cf.* Gary, *FERC Direct Testimony, supra* note 613, at 20. See text at notes 1039-1040, *supra; cf.* CHILDS, *Long-Term Financing, supra* note 1348, at 315.

[1350] Cooper, *Direct TAPS Testimony, supra* note 718, at 5; Ryan, *TAPS Rebuttal Testimony, supra* note 725, at 47; Solomon, *FERC Direct Testimony, supra* note 682, at 6.

[1351] Solomon, *TAPS Rebuttal Testimony, supra* note 621, at 8-9.

unless its investment opportunities offer an expected rate of return[1352] at least equal to, or better than, its required rate of return or opportunity cost of capital.[1353] Thus, to use the same homey example, if the going rental rate on houses in the area were 10 percent of value per annum, the homeowner could be expected to reject out of hand an offer to rent his home for $5,000 per year; it would take an offer of at least $10,000, perhaps more, before he would seriously consider it.[1354]

b. Fundamental Factors Determining the Required Rate of Return

The fundamental determinant of the required rate of return is the risk of the assets acquired. Risk results from uncertainty about the future; such uncertainty exposes the investor to the possibility of loss of capital or return.[1355] As discussed previously, those who are intimately acquainted with the pipeline business consider it to be a high risk endeavor,[1356] because of the many risks and uncertainties which were examined in detail in Sections III F & G.[1357] It is quite apparent that investors (sometimes with the aid of investment analysts and rating agencies) are capable of distinguishing in advance between the more risky and less risky investments, and that they demand a premium for, or, expressed another way, assign a higher discount rate to the higher risk investments.[1358]

One has only to compare United States Government Bonds (which, if held to maturity, are considered relatively risk-free) to various long-term industrial bonds, and even as to those, to note the differences in yield levels between bonds in the different risk rating categories. Depending on the state of the bond market, the differential between long-term government bonds and the highest rated industrial bonds (Aaa/AAA)[1359] may be around 12 to 42 basis points.[1360] Similarly a drop in classification from a "triple A" rating to the next highest classification, "Aa/AA," would

---

[1352] Rate of return is calculated by dividing total after-tax return (net income plus estimated after-tax interest expense) by total capital (stockholders' equity plus total debt plus deferred taxes). Harrill, *TAPS Rebuttal Testimony, supra* note 744 at 9.

[1353] *Id.* at 24; Solomon, *TAPS Rebuttal Testimony, supra* note 621, at 9; *cf.* CHILDS, *Long-Term Financing, supra* note 1348, at 327. This is behind the "hurdle" or "cut-off" rate mentioned in the text at notes 1040 and 1041. This should never be below the firm's required rate of return and, absent a shortage of investment opportunities, it should be from 1 to 3 percentage points above such return to allow for the unfortunate gap between expectation and realization. As noted, it must take into account the risk of the type of investment.

[1354] *Cf.* Solomon, *TAPS Rebuttal Testimony, supra* note 621 at 9.

[1355] *Id.* at 25, 29.

[1356] See text at, and notes 622-624, 722, and 735, *supra*.

[1357] See text at notes 720 and 1037, *supra*.

[1358] Mitchell, *Capital Cost Savings, supra* note 1088, at 94; Gary, *FERC Direct Testimony, supra* note 613, at 20.

cost the borrowing company around 10 basis points, and dropping down to a "single A" rating would entail an additional cost to the borrower of 20 to 30 basis points.[1361] "Split" ratings, where one agency might rate the company higher than the other, say an "Aa/AAA" rating, would result in a lesser differential than a full drop in class. Note that the differential increases with the degree of risk, *i.e.*, as the company goes further down the rating ladder. There is a definite discrete loss in borrowing capacity and increase in cost of capital when a company goes from a single "A" rating to a "triple B" (Baa/BBB), due to the existence of laws in many states, and provisions in fiduciary instruments, which limit permissible investments to "A" rated, and above, securities.[1362]

Professor Edward Mitchell found a similar differential in the case of stocks: on average, investors required about one percentage point (actually, 0.95) in return for each reduction in Standard and Poor's stock rating class (say, from A- to B+).[1363] Here again, while the 0.95 coefficient holds true for the broader range (it is right on the money in predicting differentials in yields between the four class drop from A+ to B), there is a lesser drop from A+ to A than from B+ to B, which was same phenomenon observed in the bond comparisons.

As was mentioned previously,[1364] the risk that determines the discount rate used (or premium required) by the investor is the *expected* risk at the time of the investment, not the actual volatility realized in the future.[1365] Thus, the standard for assessing the required rate of return is the rate of return which would be required and accepted by investors for an equivalent risk project in a competitive market at the time the investment is made.[1366]

[1359] For those who might not be familiar with this "shorthand" expression of risk, "Aaa" is Moody's highest rating and "AAA" is Standard and Poor's highest rating, hence a bond top-rated by both rating agencies, Aaa/AAA, is the least risky class of investment rated by the agencies.

[1360] A "basis point" is one hundredth of one percent interest. Thus an investment which yielded a full percent greater interest than another would be said to be returning 100 basis points more than the other investment.

[1361] See Gary, *FERC Direct Testimony, supra* note 613, at 21.

[1362] For an example of bond yield spreads in May 1977, see Gary, *FERC Direct Testimony, supra* note 613, at 21.

[1363] Mitchell, *Capital Cost Savings, supra* note 1088, at 94. See Table 2, p. 92. Mitchell also ran a regression analysis, shown on p. 93, which, using R as the average rate of return (in percent) and C its risk class (using a metric system A+ = 1, A = 2, etc), resulted in an equation $R = 7.2 + 0.95C$,
   (7.6) (3.4), being the "t" statistics

[1364] See note 623, and text at note 724, *supra*.

[1365] Mitchell, *Capital Cost Savings, supra* note 1088, at 93.

[1366] Rebuttal Testimony of Kenneth J. Arrow in Trans Alaska Pipeline System, FERC. Dkt. OR 78-1, 14 [Hereinafter cited as Arrow, *TAPS Rebuttal Testimony*].

Obviously, the required rate of return for any particular asset is largely determined by current conditions in the investment markets. Because the required rate of return is the opportunity cost of capital to investors, it is a current rate of return, recognizing current opportunities for investments and current money market and business conditions. Historically achieved levels of rates of return usually do not reflect current conditions, and indeed may be quite misleading in determining current required returns.[1367] For example, averaging historical levels of interest rates on bond investments reveals nothing about current interest rates. Any analysis of the required rate of return for a bond or any other asset must reflect current market conditions.

Dr. Ezra Solomon, Professor of Finance at the Graduate School of Business at Stanford University and Editor of the Prentice Hall, Inc., series of books entitled *Foundations of Finance,* states that in modern financial theory "[t]he most useful conceptual formulation for the RRR [required rate of return] on long-lived investments (either in financial assets or real assets) is to express it as the sum of two components, $RRR = i + x$, where i is the observable long-term rate of interest on high quality bonds and x is a premium for the differential or extra riskiness, relative to bonds, of the particular asset or class of assets to which investments are committed."[1368] Both components of the formula are market derived, reflecting current conditions.[1369] This conceptual formulation of the required rate of return recognizes the important principle that an asset's RRR is independent of the source of funds used to finance the investment, except for tax effects; it thus avoids problems inherent in a traditional "weighted cost of capital" approach.[1370]

A similar approach is employed by Dr. Kenneth J. Arrow, the 1972 recipient of the Nobel Memorial Prize for Economic Science. Dr. Arrow described the required rate of return for an asset as consisting of two components: the rate of return on riskless investments and the risk premium which will induce the investor to make the investment in the particular asset.[1371] This approach recently was used by the national accounting firm of Ernst & Ernst (now Ernst & Whinney) in preparing a rate of return study for the Department of Energy and by the Federal Energy Regulatory Commission in its *Alaska Gas Pipeline* proceeding.[1372]

[1367] This has been recognized by various economic experts and by the FERC. (*see, e.g.,* United Gas Pipeline Company, FERC Opinion No. 16, Docket Nos. RP75-30, *et al.* (July 3, 1978) at 19.)

[1368] Solomon, *TAPS Rebuttal Testimony, supra* note 621, at 5-6.

[1369] *Id.,* at 12.

[1370] *Id.,* at 13-14.

[1371] Arrow, *TAPS Rebuttal Testimony, supra* note 1366, at 11-13.

[1372] "Costs of Capital and Rates of Return for Industrial Firms and Class A and B Electric Utility Firms," prepared for Economic Regulatory Administration, Department of Energy, By Ernst

Other rate of return experts have introduced a third factor into the required rate of return. This factor is a financing one, *i.e.,* the debt capacity created by the asset acquired.[1373] The difficulty with this third factor is that frequently it is misused by regulatory agencies and financial and economic experts who look not to the debt capacity inherent in the asset, given its risks, but to the actual sources of funds used to finance the asset.[1374] This misuse is dangerous particularly where the level of debt supporting an asset is the product of guarantees and agreements such as those used in financing oil pipelines.[1375] In short, if a financing factor is to be used in determining the required rate of return, it must be employed properly. A careful and proper application of a financing factor is found in the testimony of Raymond B. Gary, a managing director of Morgan Stanley & Co., Inc, in the current oil pipeline proceedings before FERC. Mr. Gary looked to the particular asset involved and its characteristics to determine the true debt capacity of that asset. This debt capacity is weighted into the overall required rate of return at the current cost of long-term debt.[1376] In his testimony in the *Ex Parte No. 308* proceeding before FERC, Mr. Gary determined that a relatively low risk oil pipeline could expect to borrow no more than thirty percent of the value of the asset with the remaining investment financed with equity. The current market rates for debt and equity investment were then weighted by the appropriate percentage to arrive at the overall required rate of return.[1377] This rate of return would be applied regardless of the actual debt or equity used to finance the pipeline.[1378] If the pipeline attempted to issue more debt than the debt capacity created by the pipeline, the savings from the lower cost debt would be "washed out" by the higher rates demanded by equity investors.[1379] Thus, as previously noted, the required rate of return

and Ernst, summarized in Federal Register Notice dated July 31, 1979; FERC Order No. 31, *Order Setting Values for Incentive Rate of Return, Establishing Inflation Adjustment and Change in Scope Procedures, and Determining Applicable Tariff Provisions,* Docket No. RM78-12 (June 8, 1979) at 61.

[1373] Myers, *supra* note 198, at 5; Gary, *FERC Direct Testimony, supra* note 613, at 27.

[1374] Solomon, *TAPS Rebuttal Testimony, supra* note 621, at 12-14. For examples, see direct testimony of protestant witnesses, David Purcell and John W. Wilson in the Trans Alaska Pipeline System, F.E.R.C. Dkt. OR78-1.

[1375] See discussion in text, Section V A 2, *infra.*

[1376] Gary, *FERC Direct Testimony, supra* note 613, at 27.

[1377] *Id.* at 32.

[1378] Direct Testimony of Kenneth J. Arrow before FERC in Williams Pipe Line Case, Docket OR79-1, 10 (June 15, 1979) [hereinafter cited as Arrow, *Williams Direct Testimony*].

[1379] Solomon, *FERC Direct Testimony, supra* note 682, at 8; Swenson, *S. 1167 Testimony, supra* note 755 at 705; *Petroleum Industry Hearings, supra* note 192, at 358 (Statement of W. T. Slick, Jr.). As John Childs points out "As debt is increased it takes only a small increase in the cost of common [stock] to offset the savings resulting from higher debt....there is a basic economic principle to the effect that Cost-of-Capital is determined by the nature of the enterprise and its risk, and not by the way the 'capital structure pie' is divided. In principle, there is no change in the average dollar cost." CHILDS, *Long-Term Financing, supra* note 1348, at 19.

is unaffected by the actual sources of funds used to finance the investment, with the exception of any significant tax effects.[1380]

An important question is how integration or diversification affects the required rates of return for a diversified or integrated company and its respective component parts. Sometimes regulatory agencies and expert witnesses attempt erroneously to determine the required rate of return for a particular asset owned by a diversified company by using the overall required rate of return for the diversified company or investor.[1381] The overall required rate of return of a diversified company or investor reflects an investment in many assets with differing risk characteristics. Obviously, a firm's overall required rate of return will equal that of any of its component assets only by happenstance. Thus, any attempt to measure an asset's required rate of return by reference to its owner's overall rate must be done with great caution, if at all.[1382]

One legitimate measure of risk is the predictability of the future stream of earnings and asset growth.[1383] This predictability typically is increased by diversification.[1384] In a diversified company or portfolio of investments, the earnings from various segments are usually imperfectly correlated at any one point in time.[1385] Diversification thus tends to reduce the variability of a firm's overall earnings;[1386] accordingly, an investor should demand a lesser rate of return than he would from a non-diversified company or security.[1387] Stated otherwise, in diversified companies or portfolios, fluctuations in earnings between investments in individual stocks or individual assets tend to cancel out.[1388] To the investment manager in a diversified company or a diversified portfolio, the required rate of return for each investment must reflect the risks associated with that investment.[1389] The different levels of required rates of return for different levels of risks produce a firm's overall composite rate of return in compensation for the overall composite risk. Clearly, an investment manager should not apply the overall composite required rate

---

[1380] See text at notes 1370-1377, *supra*.
[1381] *See e.g.*, Testimony of J. Rhoads Foster, *TAPS Hearings Transcript, supra* note 785, Vol. 33 at 5674.
[1382] SOLOMON & PRINGLE, *supra* note 1346, at 361.
[1383] Myers, *supra* note 198, at 6.
[1384] *Id.*
[1385] CANES, *An Economist's Analysis, supra* note 1021, at 52.
[1386] See text at note 1210, *supra*.
[1387] *Id;* Gary, *TAPS Hearings Transcript, supra* note 785, Vol. 13 at 2268, explained on cross examination that investors view one function sliced out of the middle of an integrated chain, which is dependent on the two adjacent functions over which they have no control, as requiring a rate of return that is greater than that expected from the integrated whole.
[1388] Myers, *supra* note 198, at 6.
[1389] SOLOMON & PRINGLE, *Financial Management, supra* note 1346, at 338.

of return to each investment. If he did, he would reject low risk investments on grounds they would earn below the composite rate of return, while accepting high risk investments which were earning no more than the composite rate. In that case, the actual rate of return earned on the investments, relative to the risks taken, would be below the competitive level.[1390] A continuation of this process would start a never ending downward spiral of the portfolio's performance.

To a certain extent, vertical integration in the oil business appears to increase the predictability of earnings. Professor Edward Mitchell ran a regression analysis to determine the relationships between the Standard & Poor's stock ratings of 22 diverse oil companies and their size (total assets), equity/capitalization ratios, crude self-sufficiencies, and pipeline integration index (barrel miles of owned pipeline traffic divided by 10,000 times average daily refinery runs)[1391] and found that while each of these variables was a statistically significant determinant of the Standard & Poor's rating (hence the "riskiness" perceived by the investor), size was not nearly as crucial as degree of integration and capital structure.[1392] According to Professor Mitchell's equation, which had a coefficient of correlation of 0.92,[1393] if Standard of Indiana, the largest company in Mitchell's sample, were to exchange its pipeline and production assets for more refinery assets and shift its capital structure to resemble that of Murphy Oil, it would remain just as large but would fall from A+ to B+ in the S&P stock ratings.[1394] In short, because of the apparent increase in predictability of earnings brought about by integration, investors view an integrated company more favorably than they would otherwise, and the cost of capital of the integrated oil company is reduced.[1395] The additional size created by the integration has a "second-order" cost-reducing effect

---

[1390] *Id.* at 361-362; *cf.* Arrow, *TAPS Rebuttal Testimony, supra* note 1366, at 47.

[1391] See note 1211, *supra*.

[1392] *Petroleum Industry Hearings, supra* note 192, at 1867 (Statement of Professor Edward J. Mitchell); Mitchell, *Capital Cost Savings, supra* note 1088, at 96.

[1393] The full name of this coefficient is the Pearson product-moment coefficient of correlation, and it is simply a test of the strength of linear relationships between two variables, or in home spun terms, the "goodness of the fit." 0 is considered very poor and the relationship between the variables is virtually non-existent, whereas 1.0 (or -1.0) is a perfect linear relationship — something not found in the real world. *cf.* J. FREUND & F. WILLIAMS, *Modern Business Statistics* 309 (9th printing, 1964).

[1394] *Petroleum Industry Hearings, supra* note 192, at 1867 (statement of Professor Edward J. Mitchell); Mitchell, *Capital Cost Savings, supra* note 1088, at 96.

[1395] Myers, *supra* note 198, at 6; Mitchell, *1974 Senate Hearings, supra* note 1056, at 23; *S.1167 Hearings, supra* note 432, pt. 9 at 615-616 (Statement of Vernon T. Jones) (the oil pipeline industry, as presently structured, has enjoyed very low debt costs as a result of using the very high credit rating of the stockholder oil companies); *S.1167 Hearings, supra* note 432, pt. 8 at 5945, 5948 (Statement by Richard C. Hulbert) ("There would be a serious question, though, in the case of systems such as Colonial or Kaneb, to whether an independent [non-integrated]

due to economies of scale in large security issues.[1396]

It should be noted that, unlike the case of a typical diversified portfolio, vertical integration in the oil business involves, to a certain extent, investment in interrelated projects. Thus, risks associated with the investment in any one link of the chain can affect an investment in any other link of the chain. As described by Dr. Kenneth J. Arrow, an investment in the Trans Alaska Pipeline System, coupled with an investment in Alaska North Slope production, marine transportation, refinery or marketing can expose a company to increased risks as compared to an investment in only one component of that chain due to the so-called "covariance effect."[1397] This increased risk can be reduced to some extent in a major diversified oil company by securing alternate uses for marine transportation and sources of supply for refinery and marketing operations. Just as was noted with respect to the investment manager of a diversified portfolio, the investment manager of an integrated oil company who demands only the overall required rate of return for an investment in each link of the chain will soon find himself with a diminishing overall rate and a gradually deteriorating company.

Another risk reducing element of the vertical integrated oil company is smooth synchronization brought about by integrated planning and coordinated management.[1398] Such synchronization allows a more effective use of available capacity, lowers the integrated company's inventory and cost of capital requirements, reduces variations in output levels and dampens variations in average unit cost.[1399]

In summary, although integration can reduce the cost of capital to the integrated oil company, the required rate of return for any segment of the integrated company cannot be determined by the overall cost of capital. Moreover, the required rate of return for any single investment is not determined by the costs associated with the sources of capital used to finance the project. With particular reference to oil pipelines, the overall cost of capital to a pipeline's integrated parent company may be irrelevant to the pipeline's required rate of return. Unlike the case of public utilities, where typically all of the activities of a company are regulated and the regulatory commission is concerned with an overall rate of return for a

---

group could get as favorable an interest rate because of the borrowing power of the shipper-owners") ("Again, the interest rate may not be quite as favorable because of the nature of the corporate structure"); *cf.* CARD, *Integrated Organization, supra* note 1152, at 36. The effect has been to reduce financing costs of oil pipelines and encourage investment in oil transportation and distribution facilities. Swenson, *S.1167 Testimony, supra* note 755, at 703.

[1396] Myers, *supra* note 198, at 6.

[1397] Arrow, *TAPS Rebuttal Testimony, supra* note 1366, at 48-50.

[1398] DE CHAZEAU & KAHN, *supra* note 30, at 267-268.

[1399] See text at notes 1193-1204, *supra;* Mitchell, *Capital Cost Savings, supra* note 1088, at 80-82.

company, oil pipelines often have significant non-regulated activities. Thus, the importance of focusing on the particular oil pipeline's investment in question in determining a fair rate of return should be clear to the regulatory agency.

The principles applicable to determining required rates of return discussed in this section are particularly important with respect to an industry such as the oil pipeline industry where the amount of debt in a pipeline company's capital structure is artificial, *i.e.,* unrelated to the debt capacity of its pipeline assets. As discussed in the next Chapter, the capital structures of oil pipelines are severely conditioned by the 1941 Elkins Act Consent Decree.[1400]

2. Capital Structure of Pipelines

a. General Observations

As Professor Stewart Myers remarked in his testimony before the Senate Special Subcommittee on Integrated Oil Operations, financing patterns in the oil pipeline industry are interesting in two respects. First, there is a substantial variation in the debt-to-total capitalization ratio as between the various lines (and ratios of short to long-term debt also display a wide variance). Second, many pipelines have remarkably high debt ratios [especially the more recently constructed "industry lines"]; ratios of debt to total assets of 80 to 90 percent are not uncommon.[1401] Although it is not irrational to infer from this variation in debt ratios that there exists a similarly wide variation in pipelines' rates of return, Myers properly cautions against reaching such a simplistic conclusion. The rate of return depends in part on debt capacity and the fact that a major new pipeline is financed with 90 percent debt does not mean that the pipeline is generating 90 cents of new debt capacity per $1.00 invested. What really has happened is that the pipeline has made a draft upon its shipper-owner's credit through the means of a direct debt guarantee or a through-put and deficiency agreement.

Raymond Gary, in his testimony in the pipeline valuation proceedings (formerly ICC *Ex Parte No. 308,* now FERC RM 78-2), provides the explanation for this phenomenon. Gary examined and displayed a representative sample of industrial companies which had come

---

[1400] See text at notes 1759-1763, *infra.*

[1401] Myers, *supra* note 198, at 10. See bar chart showing distribution of pipeline companies in brackets having an increment of 10% debt to asset ratios, *i.e.* 0-10, 10-20, etc. in his Figure 2, located between pages 10 and 11.

to the long-term debt market during the period 1972-1976, showing the number of issues in the sample, net sales, net income, long-term debt to pro forma[1402] capitalization, pro forma fixed charge coverage before taxes, the cash flow to pro-forma long-term debt coverage and the rating assigned by Moody's and Standard & Poor's.[1403] The sample showed that long-term debt as a percentage of total capitalization for the highest grade (Aaa/AAA) cut off at about 20 percent and the double A's at about 30 percent. The "medium" (Baa/BBB) grade's maximum runs around 43 percent. He then examined and arrayed the ratings of the 29 oil companies included in the Chase Manhattan Bank's Energy Economics Division Survey.[1404] In Gary's judgement, if the group were rated on a composite and weighted basis, the approximate rating would be the second highest, Aa/AA. He noted that the average long-term debt ratios in the years 1971-1975 for the Chase Study group averaged between 21 percent and 24 percent in that period. It is this level of credit rating that permits the shipper-owned pipeline companies to borrow at the rates they do. One might say that they borrow the money from the lenders and the credit from their owners. This is demonstrated by the ratings given the highly leveraged (about 90-10) modern industry lines which tend to reflect the integrated composite credit of their shipper-owners with perhaps a "slippage" of about one credit bracket.[1405] Explorer received a single A rating, Colonial a split A/AA. Dixie was not rated by Moody's but received an AA from Standard and Poor's and Wolverine was a single A.

One immediately asks, how did this structure come about? Prior to the entry of the Consent Decree, the average of oil pipeline debt ratios was below 20 percent.[1406] In December, 1941, the *Atlantic Refining Company* Consent Decree was entered, whereby, effective January 1, 1942, the aggregate of "any earnings, dividends, sums of money or other valuable considerations derived from transportation or other common carrier

---

[1402] The "pro forma" is required by the SEC to give effect to the result of the financing on the issuer's financial statements.

[1403] Gary, *FERC Direct Testimony, supra* note 613, at 11 and Attachment II.

[1404] *Id.* at Attachment III.

[1405] Swenson, *S.1167 Testimony, supra* note 755, at 704. See Swenson's Exhibit A for the individual company ratings.

[1406] Gary, *FERC Direct Testimony, supra* note 613, at Attachment IV. These Debt Ratios were as follows:

Year	Debt Ratio
1935	17.46 percent
1936	19.12 percent
1937	20.21 percent
1938	17.91 percent
1939	21.83 percent

Period average 19.31 percent

services" which was paid by a defendant common carrier to any defendant shipper-owner in any one year could not exceed such defendant shipper-owner's share of seven percentum of the valuation of such common carrier's property (with a provision for additional payment of amounts that could have been paid but were withheld by the pipeline company and a three year "make-up" of the 7 percent "allowable" which was not earned in any year).[1407] Shortly after the decree was entered, the Justice Department challenged the payment, in addition to dividends, by Shell Pipe Line Corporation of debt which it owed to its shipper-owner at the time the decree was entered. Not only was its obligation "pre-decree" debt, but also it was a debt which had arisen because Shell Oil Company made available its credit to its wholly-owned subsidiary, Shell Pipe Line Corporation, by borrowing money, at rates commensurate with its AAA credit rating, and lending that money at the same favorable rate, to Shell Pipe Line to pay off the latter's debt to third parties, thereby saving Shell Pipe Line a tidy differential in interest costs. Notwithstanding these considerations, Assistant Attorney General Wendell Berge took the position that Paragraph III's limitation on payments to shipper-owners prevailed over Paragraph V's permissive language that earnings *in excess* of sums allowed to be paid by Paragraph III could be used "for retiring of any debt outstanding at the time of the entry of this judgement and decree," so long as the debt or refunded debt was originally incurred for the purpose of, and its proceeds expended in, constructing or acquiring common carrier property.[1408] It is reported that a similar position was expressed by Mr. Berge to Ben Harper, Counsel for Pure.[1409] It is now water over the dam, although this author expressed an opinion in 1952, which hasn't changed, that the more specific permission of Paragraph V should prevail over the general prohibition of Paragraph III.[1410]

The Department's position left the oil companies with three alternatives: litigate, continue to advance money to their pipeline subsidiaries and thereby "freeze" the advances in the pipeline company, or have the pipeline companies borrow directly from third parties any money required for additional common carrier facilities. Obviously, the last alternative was the most acceptable, and the debt ratios of pipeline companies began to rise sharply as the statistics mentioned previously reflect. In order to

---

[1407] United States v. Atlantic Refining Co. et al, Civil No. 14060, D.D.C., Dec. 23, 1941, Par. III.

[1408] Letter, dated October 13, 1943, from Wendell Berge, Assistant Attorney General to Cyrus S. Gentry (General Counsel) Shell Oil Company.

[1409] Sun Oil Company's Middlesex Pipe Line Co. also received a similar letter. CELLER REPORT, *supra* note 220, at 182-183, as apparently did Sohio Pipe Line Co. *Celler Hearings, supra* note 162, at 371.

[1410] WOLBERT, *supra* note 1, at 148-149. This view is shared by Professor Johnson. PETROLEUM PIPELINES, *supra* note 45, at 337.

keep the pipeline's cost of capital down, the device of throughput and deficiency agreements was devised so that borrowing on the parent's credit could be achieved without the deleterious legal consequences. About this time, the "off-balance sheet" device was in vogue and the contingent liability of the shipper-owners at first was not even required to be mentioned in its financial statements and later, although it had to be disclosed, it was "bundled up" with a number of other contingencies and not individually quantified. This added to the drive to go the throughput and deficiency route. The clincher came as inflation made the 7 percent return completely unrealistic in terms of modern long-term debt markets, so the only way to cope with the now unrealistic 7 percent limitation was to leverage up the pipeline debt ratios to the maximum amount, thus producing the frequently mentioned 90-10 debt-equity percentage which prevails in today's shipper-owner backed lines.[1411]

b. Artificiality of Pipeline Debt Ratios and their Financial Consequences

As mentioned previously, modern shipper-owned pipelines are financed with debt ratios often as high as 90 percent of their total capitalization (Explorer even went to 100 percent when losses placed it in a negative equity position). The reason lenders accept such high debt ratios is that they recognize the direct debt guarantees and throughput and

---

[1411] Gary, *FERC Direct Testimony, supra* note 613, at 10; Gary, *TAPS Rebuttal Testimony, supra* note 726, at 24-25; *cf. Petroleum Industry Hearings, supra* note 192, at 325 (testimony of E. P. Hardin) ("A 7-percent return on valuation is not a red-hot business deal.") (Comment by Dr. Garrett Vaughn, Minority Economist: "I would agree, and it would seem to me that when you are getting into a risky business, unless you can justify that risk with at least some chance of an outstanding profit, there is really not a whole lot of reason to get into this.") As long ago as 1958, Trial Judge Richmond B. Keech, presiding at the hearing on the Justice Department's motions in the *Arapahoe* case, United States v. Atlantic Refining Co. et al., Civil No. 14060, D.D.C., March 24, 1958, Tr. p. 34 said: "Frankly, I would like the [Justice Department] to deal with the seven percent because I am not shocked in any manner, shape, or form about the seven percent return." And this was when "A" Industrial Bonds were yielding 3.9 percent as compared to May, 1977 yields of 8.20 to 8.60 percent. Gary, *FERC Direct Testimony, supra* rule 613, at 21. Presumably, Judge Keech would be equally unperturbed by an 11.7 percent return in 1977. LeGrange, *FERC Direct Testimony, supra* note 176, in his Attachment, showed that even holding the 1942 risk premium of 5.54 percent [8.0 percent ICC-approved return for "low-risk" pipelines less the 1.46 percent U.S. Government Long-Term Bond yield] constant, the 1976 minimum rate of return for this risk category of pipeline would be 13.40 percent [7.86 percent 1976 Government Bond yield plus the 5.54 percent 1942 risk premium]. Donald Flexner, Deputy Attorney General (Antitrust Division), at the AEI Conference, replied to the author's question, "By definition, does that mean to you that 13 percent is an unreasonable profit?" by saying: "No, I don't think there is any way to pick out a particular number." AEI, *Oil Pipelines and Public Policy, supra* note 684, at 65. For an illustration showing how the Consent Decree forces high debt/equity ratios, *see* Ryan, *TAPS Rebuttal Testimony, supra* note 724, at Exh. JMR-3.

deficiency undertakings to be a full faith and credit obligation of the *oil company or companies* involved. In this sense, the high debt ratios of oil pipeline companies are misleading because the lenders are not taking the risks of failure of the pipeline operation; the throughput and deficiency agreements or guarantees shift the risk from debt-holder to the shipper-owner.[1412] The essence of the transaction is that the shipper-owner oil companies dedicate a valuable asset, their credit or borrowing capacity (which is not unlimited), to the pipeline venture. This asset is just as much an investment in the pipeline as is the cash. The two really cannot be separated economically. In making these commitments, the shipper-owners, from a practical standpoint, are providing the total capitalization of the pipeline, equivalent to a 100 percent equity investment.[1413] The high debt ratios of shipper-owned pipelines must be recognized for what they are, a technique of financing developed to accommodate the now unrealistic 7 percent Consent Decree limitation to the rates of return to the investor that this risk class of investments must earn in today's financial climate. They do not change the intrinsic rate of return determinations.[1414]

B. *How Pipelines Are Financed*

Prior to the 1940's pipeline financing was done either by the shipper-owners lending directly to the pipeline or occasionally pipeline companies floating small issues secured by a mortgage on their pipelines or by the shipper-owner guaranteeing the pipeline's obligations. The Consent Decree brought about a change in financing through the differences in perception which it caused not only on the part of the shipper-owners,

---

[1412] *Consumer Energy Act Hearings, supra* note 154, at 625 (Statement of Jack Vickrey); Gary, *FERC Direct Testimony, supra* note 613, at 8; Swenson, *S.1167 Testimony, supra* note 755, at 704. It should be noted at this point that deficiency payments are treated as advance payments for future transportation, hence a pre-paid expense, but if the line goes bankrupt, the shipper-owner will never realize any benefit and will have to recognize a loss.

[1413] *S. 1167 Hearings, supra* note 432, pt. 9 at 598 (Statement of Fred F. Steingraber); *Id.,* pt 9 at 619-620 (Statement of Vernon T. Jones); Gary, *FERC Direct Testimony, supra* note 613, at 8, 12; Gary, *Direct TAPS Testimony, supra* note 634, at 20; U. LEGRANGE, EFFECTIVENESS OF GOVERNMENT REGULATION 6 (paper presented at the American Enterprise Institute on Oil Pipeline and Public Rolicy, May 1-2, 1979) [hereinafter cited as LEGRANGE, *Government Regulation*]; *cf.* PIPELINE PRIMER, *supra* note 331, at 15; *S.1167 Hearings, supra* note 432, pt. 8 at 5939 (Statement of Richard C. Hulbert).

[1414] *Cf.* Cooper, *Direct Taps Testimony, supra* note 718, at 19. As U. J. LeGrange pointed out in his AEI paper, cited in note 1413, *supra,* at page 4, in considering any investment, a company has two separate and discrete decisions to make: (1) should the investment be made — this decision comes first. It is based on the expected return on the total capital required and generally it is calculated using the discounted cash flow technique. If, and only if, it clears the "hurdle rate," or otherwise passes the investment test, does (2) the question arise of how the financing should be done.

which was discussed above, but also on the part of lenders who viewed the Decree as a factor to be taken into account directly because of its obvious restriction on profitability, and indirectly as a harbinger of even more regulation which could jeopardize the economic viability of pipelines. Lenders perceived the regulatory risks to be such that they demanded that the shipper-owner's credit "backstop" the pipeline's obligation.[1415]

The growing perception of risk on the part of lenders and the movement toward jointly-owned or undivided interest "industry lines" coalesced. Adaptation to the lender's need for "back-up" assurance from a credit worthy source of repayment and the individual oil company's desire to limit its obligation to conform as closely as possible to its anticipated need for transportation led to the development of specialized financing methods for shipper-owned pipelines.

1. Project Financing

Jointly owned pipelines conventionally have been financed on a "project financing" basis utilizing throughput and deficiency agreements whereunder a new pipeline corporation is formed to construct and operate the system and issues debt instruments of 30 to 40 years duration, usually privately placed with a consortium of lenders consisting of banks, pension trusts, savings funds, insurance companies and state agencies. The debt, which typically is about 90 percent of the pipeline system's final cost, is secured by the pledge of Completion Agreements, and Throughput and Deficiency Agreements furnished by the oil company shipper-owners based on their various ownership percentages.[1416]

In project financing, the project itself must be demonstrated to be economic and viable. From the standpoint of the shipper-owner, the project must meet or exceed its cost-of-capital determined "hurdle" or "cut-off" rate of return, utilizing a discounted cash flow analysis. The

---

[1415] MARATHON, *supra* note 202, at 48-49.
[1416] *Id.* at 49-50; *Consumer Energy Act Hearings, supra* note 154, at 624-625 (Statement of Jack Vickrey); Gary, *FERC Direct Testimony, supra* note 613, at 7. Each company is individually liable only to the extent of its undertaking, which almost always corresponds to its percentage of ownership. In rare exceptions, such as was the case in Texoma, the advantage of diversifying ownership is deemed to be of sufficient benefit to justify the "carrying" by the other participants, on a pro rata basis, of the weaker credit of a smaller company whose presence is deemed important. Although the legal position unquestionably is that a shipper-owner who has fulfilled his shipment obligation is not subject to recourse at the hands of the lender, if the non-defaulting shipper-owners do not cover the deficiency, the lenders may force the pipeline company into bankruptcy. Swenson,*S. 1167 Testimony, supra* note 755, at 703. Research has not disclosed any instance of bankruptcy or receivership of a shipper-owned line.*Id.; Consumer Energy Act Hearings, supra* note 154, at 6.

lenders, despite the fact that they will "look through" the project to the credit of the ultimate guarantors (the shipper-owners), will nonetheless require that the project be capable of standing on its own feet.[1417]

A Completion Agreement, which may be a separate agreement or part of the general agreement between the shipper-owners and the pipeline company, specifically makes these undertakings assignable to, or on behalf of, the lenders, and contains an unconditional commitment by the shipper-owners to complete the facilities, to operate them and, if operations are interrupted for any reason, to take the necessary steps to restore its facilities to operation.[1418] It also obliges the shipper-owners to provide the necessary funds to service all of the pipeline company's obligations until the "take-out" provided by the Throughput and Deficiency Agreements.[1419] Although the usual arrangement for financing construction is to sell several issues of serial notes, in one or two instances, notably Explorer, the early construction financing was accomplished by the pipeline company selling commercial notes which were authorized and backed up by the Throughput and Deficiency Agreement. This adaptation of the well established commercial paper technique is reputed to have been initiated by R. C. Thompson, now Financial Vice President of Shell Oil Company.

The Throughput and Deficiency Agreements are instruments whereby each shipper-owner binds itself to ship, or cause to be shipped, through the pipeline is pro rata share of enough oil so that the pipeline will generate sufficient gross cash revenue at government-approved tariff rates to service the interest and principal repayment of the debt and service all operating expenses and other costs of the operation during the entire period of the loan. These obligations are not mere agreements to use the line, although they do require the shipper-owners to commit specific volumes of oil for transportation through the system. They go well beyond that, by virtue of the deficiency agreements, or deficiency paragraphs in the throughput agreements, which contain clauses frequently referred to as "hell or high water" clauses. Under these obligations, if for *any* reason whatsoever, even if the line is inoperable, or the inability to ship is due to causes which under normal commercial dealings would provide a *force majeure* escape, the pipeline does not have sufficient cash on hand to pay the principal and interest on the debt and discharge all its other obligations, the shipper-owners are required to

---

[1417] Gary, *FERC Direct Testimony, supra* note 613, at 8; *Consumer Energy Act Hearings, supra* note 154, at 625 (Statement of Jack Vickrey); Gary, *Alaskan Bills Testimony, supra* note 1095, at 1117.

[1418] *Consumer Energy Act Hearings, supra* note 154, at 613 (Statement of Jack Vickrey); Gary, *FERC Direct Testimony, supra* note 613, at 7; Gary, *Alaskan Bills Testimony, supra* note 1095, at 1117.

[1419] *Cf. Consumer Energy Act Hearings, supra* note 154, at 595 (Statement of Jack Vickrey).

make up the difference by a cash "deficiency payment." This obligation continues as an ever-present possibility for the entire 20-40 year life of the debt.[1420]

Several comments about Throughput and Deficiency Agreements seem in order. The first is that it appears to be the consensus of knowledgeable people that without these agreements or equivalent "stand behind" arrangements that place behind the pipeline's obligations the full faith and credit of companies with the financial resources and credit capacities of the shipper-owners who have built them, modern large diameter pipelines would not have had sufficient funds available to them at the credit terms they have enjoyed.[1421] More than that, several of the larger lines would not even have been financeable on borrowed money without the credit of the oil company owners.[1422]

A second comment is that there appears to be a misconception about throughput and deficiency agreements to the effect that if a shipper-owner owns, say, 20% of the carrier's stock and is "on the hook" for 20 percent of the throughput, that such owner has a guaranteed right to ship over 20 percent of the available capacity. This is erroneous.[1423] Because of the common carrier status of these lines, they must furnish transportation upon reasonable request to any shipper and they are forbidden by law to give any unreasonable preference or discriminate in any way between

---

[1420] *Petroleum Industry Hearings, supra* note 192, at 257 (Statement of Charles J. Waidelich); *Consumer Energy Act Hearings, supra* note 154, at 624-625 (Statement of Jack Vickrey); Swensen, *S.1167 Testimony, supra* note 755, at 703; STEINGRABER, *supra* note 193, at 145-146 (discussing Colonial's situation). Examples of these agreements may be found in *Petroleum Industry Hearings, supra* note 192, at 1110-1115 (Laurel Pipeline Company "Throughput Agreement," the deficiency agreement is contained therein as Paragraph 6, "Cash Payments by Shipping Oil Companies"); *S.1167 Hearings, supra* note 432, pt. 9 at 642-653 (Colonial Throughput Agreement — see Par. 5A for deficiency undertaking); *Id.* at 661 (Explorer Throughput and Deficiency Agreement — see Paragraph 3 for Deficiency Payments By Oil Companies).

[1421] Spahr, *Rodino Testimony, supra* note 255, at 13, Myers, *supra* note 198, at 22; Waidelich, *Rodino Testimony*, note 197, at 9; *S.1167 Hearings, supra* note 415, p. 8 at 5945 (Statement of Richard C. Hulbert); *cf.* Swenson, *S.1167 Testimony, supra* note 755, at 705.

[1422] STEINGRABER, *supra* note 193, at 144 (Colonial); Gary, *Alaskan Bills Testimony, supra* note 1095, at 1119 (TAPS — even the State of Alaska did not have the financial capacity to substitute for them); *cf.* Swenson, *S.1167 Testimony, supra* note 755, at 705 (future pipeline financings). The testimony of witnesses Parcell, Dunn and Wilson in the TAPS hearing, to the effect that TAPS could be financed with 65-70% outside debt without backup by the shipper-owners appears incredible in the light of the experience of the oil pipeline industry over the past 40 years and the advice given to the State of Alaska by Kuhn, Loeb & Co., Salomon Brothers, Bank of America, und Merrill, Lynch, Pierce Fenner & Smith; Gary, *TAPS Rebuttal Testimony, supra* note 726 at 19-21; and Gary's own testimony based on his long experience in pipeline financing with Morgan Stanley. Gary, *Alaskan Bills Testimony, supra* note 1095, at 1119.

[1423] *Petroleum Industry Hearings, supra* note 192, at 257 (Statement of Charles J. Waidelich); Gary, *FERC Direct Testimony, supra* note 613, at 15; Spahr, *Rodino Testimony, supra* note 255, at 14; Waidelich, *Rodino Testimony, supra* note 197, at 12; Gary, *Direct TAPS Testimony, supra* note 634, at 15.

shippers. Hence if the line is full, as has been true from time to time in Colonial, the new shippers' tenders, and on some lines, the increased tenders of nonowners, will "back out" shipper-owner's shipments on a pro rata basis, although the shipper-owners still are obligated to the lenders for any cash shortfall.[1424]

Professor Stewart Myers, in his testimony before the Senate Special Subcommittee on Integrated Operations, characterized the situation this way "Present throughput agreements absorb business risk because they are one-way streets: pipelines are guaranteed revenues to service debt, but they have no obligation to reserve capacity strictly for the firms guaranteeing throughput. If tenders exceed capacity, pipelines' obligations as common carriers prevent them from meeting any firm's capacity demands in full. In other words, the throughput agreements' only function is to support pipeline subsidiaries' debt. This is common in many industries, but I know of no cases in which a firm supports—without special compensation—the debt of another independent firm."[1425] Needless to say, the throughput agreements also are "one-way streets" in the sense that while they obligate fully the shipper-owners to protect the lenders against loss, they most certainly do not guarantee any profits to the investing companies.[1426] No person was willing to guarantee Explorer that it would make money—in fact it lost $46 million in its first five years of its operation and invoked the deficiency agreement to make cash calls on its shipper-owners to the tune of $42 million.[1427]

There have been some suggestions from critics of the industry, some half in jest,[1428] some a little wistful,[1429] and some acrimonious,[1430] to the

---

[1424] See text at notes 416 through 420, *supra*.

[1425] Myers, *supra* note 198, at 21-22.

[1426] *Petroleum Industry Hearings, supra* note 192, at 257 (Statement of Charles J. Waidelich).

[1427] See text at notes 436 & 437, *supra*.

[1428] *Petroleum Industry Hearings, supra* note 192, at 143, Senator Birch Bayh speaking: "It would seem to me that the running of these [Shell Pipe Line, Exxon Pipeline, Chevron and Colonial] pipelines is very profitable and that the reason independents [presumably non-oil related] cannot get in them and cannot get into this, in a truly independent way, is that they cannot get any guarantee to use them from the majors that have the control over most of the supply of oil. "And you cannot get the finances to build even a profitable pipeline, because it is not going to be profitable unless the majors that control the oil run it through those lines.
"I thought this was rather humorous and I thought I might throw it at you in jest [Senator Haskell's remark that if he could get throughput agreements from the companies involved in Alaska, that he would resign from the Senate and go in the pipeline business]."

[1429] *Consumer Energy Act Hearings, supra* note 154, at 1027, Keith Clearwaters, Deputy Assistant Attorney General, Antitrust Division, testifying: "It seems to us that if independent entrepreneurs were given the same kind of throughput commitments from the major prospective users of the line that joint venturers require of themselves when they organize a joint-venture line, the risks would be reduced to acceptable limits and independent operators would be willing and be able to finance even the largest of pipeline projects."

[1430] *H.Res.5 Subcomm Hearings, supra* note 234, at 132, Beverly Moore, appearing on behalf of

effect that pipelines could (and, inferentially, would) be built by non-oil related firms if only the major oil companies would provide the financial backing for such lines by executing throughput and deficiency agreements similar to those used by shipper-owners in connection with the lines in which they had an equity interest. The S.2387, or PICA, Report [at p. 129] cited the statement of Wallace Wilson, Vice-President of the Continental Illinois National Bank, that "We are just as satisfied with a thoroughly satisfactory throughput agreement by a responsible party with whom we have some confidence as we are with an unsecured credit, where a company with substantial credit worthiness would sign the note," for the proposition that an "independent" pipeline company with adequate throughput agreements from reliable shippers would have no difficulty in obtaining financing.

Wilson's statement was made in the context of verifying W. T. Slick's earlier testimony that throughput [*and deficiency*] agreements were virtually the same as debt. Wilson simply verified the essential accuracy of Slick's characterization. If one examines carefully the qualifications "thoroughly satisfactory" [meaning "hell or high water" commitments to repay the debt] and "responsible party" [one whose credit worthiness is commensurate with the amount borrowed] he will find himself in the same position as was outlined in the two preceding paragraphs. The argument then reduces itself to the simple question whether a company with no equity interest would, should or could expend its credit and assume both the business and financial risks of a pipeline project with no opportunity for an equity return. Oil pipeline owners have been willing to invest for the limited return that the Consent Decree and past ICC regulation allowed them because they needed the transportation — even at the risk, under the Interstate Commerce Act, that they might not realize fully the transportation benefit expected because of proration.

In this regard, it is interesting to observe that the Attorney General's Report on the Alaska Natural Gas Transportation Act of 1976[1431] refers to Judge Litt's Initial Decision [at page 426 of the Initial Decision] in which the Judge indicated that investors would *not* be willing to come forward and invest their money in a pipeline venture if others would have an equal ability to use the system without contributing equity. One well-known

---

the Corporate Accountability Research Group, in his prepared statement: "The truth of the matter is not that independent pipeline companies cannot obtain pipeline financing, but that the major oil companies will never tender the necessary throughput commitments as long as monopoly gains lure them to build their own joint venture pipelines."

[1431] Attorney General of the United States, Report Pursuant to Section 19 of the Alaska Natural Gas Transportation Act of 1976, dated July, 1977, p. 65 [hereinafter cited as A.G.'s Alaska Natural Gas Report].

firm of Washington attorneys wrote: "It is unrealistic to expect publicly-held oil companies to lend their credit and bear serious financial risks over a 20-40 year period in the absence of having an equity in the facility which would yield a reasonable return. Indeed, under such an arrangement oil companies would be expected to lend their credit and bear all the risks of ownership without even the 7% return on valuation that they are now permitted under the Elkins Act Consent Decree. The absurdity of that proposition is self-evident."[1432] A Managing Director of a prominent Investment Banking firm, who is widely acknowledged to be an expert in pipeline financing, testified: "No rational businessman would give such an unconditional cash deficiency type of agreement to secure the debt of a third party, since to do so would be utilizing his credit to provide an equity return to that third party. Even to suggest that an oil shipper would consider doing this is so divergent from sound financial policy that it calls into serious question all conclusions...based upon such a premise."[1433]

Industry spokesmen's replies have been lower key, addressing the question with varying degrees of sophistication, ranging from the simple laconic statement that it is not to be expected that an oil company would be willing to undertake the substantial risks involved and obligate itself to ship enough to finance the line over a period of up to forty years without owning an equity interest in the pipeline,[1434] to a statement that to guarantee the debt on someone else's project from which the guarantor received none of the profits would not normally be acting with "financial prudence" because the company would be using up its ability to invest in some other project in which it could earn a return,[1435] to expressions of concern that undertaking such obligations would appear to be inconsistent with the stewardship which the proposed obligor's management owed to its stockholders.[1436]

One industry witness observed that some companies have outstanding indentures which specifically prohibit guarantee of the indebtedness of another entity in amounts other than the guarantor's percentage of ownership in that entity.[1437] While not mentioned by any witness, many of

---

[1432] HOWREY & SIMON, *supra* note 3, Summary at 19-20.

[1433] Gary, *TAPS Rebuttal Testimony, supra* note 726, at 29.

[1434] *H.Res.5 Subcomm Hearings, supra* note 234, at 141 (Statement of Fred F. Steingraber); *Consumer Energy Act Hearings, supra* note 154, at 595, 625 (Statement of Jack Vickrey); *Petroleum Industry Hearings, supra* note 192, at 302 (Statement of E. P. Hardin); Cooper, *Direct TAPS Testimony, supra* note 718, at 19-20; *TAPS Hearings Transcript, supra* note 785, Vol. 12 at 1958-1960 (Cooper reiterated this position on cross examination).

[1435] Spahr, *Rodino Testimony, supra* note 255, at 13.

[1436] *Petroleum Industry Hearings, supra* note 192, at 333 (Statement of W. T. Slick, Jr.); *Id.* at 258 (Statement of Charles J. Waidelich); *Id.* at 315 (Testimony of E. P. Hardin).

[1437] Waidelich, *Rodino Testimony, supra* note 197, at 10.

the shipper-owners are Delaware Corporations, and there is some doubt whether, under Section 123 (formerly Section 78) of the Delaware Corporation Law, such a guarantee would be authorized. Members of the financial community who have testified concerning the proposal uniformly were negative, ranging from an expression that the witness knew of no cases in which a firm supported — without special compensation — the debt of another independent firm,[1438] to the statement that the proposal violated "a cardinal rule of sound finance" for a company which had assumed the cost and risk of building a common carrier system not to be in a position to recover its investment in part through profits derived from the use of the system by others because it otherwise would have used its credit for the benefit of others without compensation.[1439] Moreover, without managerial control, the "guarantor" cannot be assured of efficient pipeline operation. Reluctance to underwrite the risks of some one else's line is not limited to major integrated companies; independent producers want no part of such an arrangement,[1440] nor do independent refiners.[1441]

Finally, a Congressional Research Service Report remarked, in discussing whether a divested pipeline could expect to obtain throughput and deficiency agreements from newly independent producers: "Furthermore, since pipelines are common carriers and therefore required to provide equal access to all shippers at non-discriminatory rates, there would be little incentive for producers to enter into such [agreements] with the pipelines they just divested."[1442]

2. Undivided Interest Systems

The second way of organizing and obtaining financing for multiple ownership pipelines is that of an undivided interest system, or tenancy in

---

[1438] Myers, *supra* note 198, at 22.

[1439] Gary, *Alaskan Bills Testimony, supra* note 1095, at 1118; *cf.* Gary, *TAPS Rebuttal Testimony, supra* note 726, at 27-28.

[1440] Allen, *FERC Direct Testimony, supra* note 929, at 8; Allen, *TAPS Rebuttal Testimony, supra* note 496, at 9-10.

[1441] *H.Res.5 Subcommittee Hearings, supra* note 234, at 22 (Testimony of D. W. Calvert, Executive Vice President, The Williams Companies, recounting his experience with independent refiners whose shipments he was soliciting in order to justify the construction of a pipeline).

[1442] NET-III, *supra* note 199, at 155. In this regard, one notes that Great Lakes' former owners did not provide its purchaser, Williams Brothers Pipe Line Company (now Williams Pipe Line Co.) with such undertakings, although they did take, as part of the purchase price, Subordinated Debentures of Williams Brothers Pipe Line Company in the face amount of $60 million, *H.Res.5 Subcomm. Hearings, supra* note 234, at 31 (Letter to Chairman Neal Smith from D. W. Calvert).

common. Ownership in the pipeline facilities, equipment, communication equipment and the like are vested in the oil company participants, either directly or through their pipeline subsidiaries, each of which operates as a separate and distinct common carrier. Each owner company is responsible for the completion of its own share of the system and it assumes its own distinctive risks. It finances its share of the cost in the same way that it does other items in its corporate budget. These are financed out of all corporate resources including working capital, internal cash flow and the issuance of securities. The effect on the oil company, and the reliance of the lenders is the same as in project financing in the sense that the owner has committed an important portion of its resources and credit to an undertaking which has significant risk and uncertainty, as was discussed in Chapter III. This is as real a use of its resources (including its credit) as any other, and pre-empts its use for other investments.[1443]

Vernon Jones (then President of Explorer, now President of Williams Pipe Line) remarked in a colloquy with Henry Banta during his testimony before the Senate Antitrust and Monopoly Subcommittee, "you can't use credit twice."[1444] One significant distinction from project financing is that each participant is on his own. If a particular carrier cannot obtain sufficient business adequately to utilize its share of the capacity, it still must service the fixed charges and its share of the variable costs, hence it can lose on its investment.[1445] Thus, unlike the corporate form, if several of the other tenants in common are running at full capacity, a non-performer will not get a "free ride" as it might in the project financing. Its only hope is that other participants or other shippers may require space and tender oil to it for transportation. This fact creates a potential for competition among the several carriers. Another significant financing difference is that every undivided interest participant will be borrowing on the credit rating of its shipper-owner. Hence a triple A company, such as Exxon or Mobil, will not suffer the "dilution" of its credit rating that it would in a corporate form project financing when some of the participants will be rated single A or lower and the composite for the project will be the rate to all.

The most striking example of the importance of pipeline companies having the backing of shipper-owners is, of course, TAPS. The TAPS owners were able to maintain high debt ratios solely because of the "backstop" provided by their parent companies. Lenders were willing to commit funds to the debt securities of the TAPS owners despite their

---

[1443] Gary, *Alaskan Bills Testimony, supra* note 1095, at 1117; Gary, *FERC Direct Testimony, supra* note 613, at 20.
[1444] *S.1167 Hearings, supra* note 432, pt. 9 at 620. (Testimony of Vernon T. Jones).
[1445] W. J. Williamson, *supra* note 315, at 11-12.

high degree of leverage *only* because of this support. Every parent company was forced to dedicate a portion of its own available borrowing capacity to support the system. By doing so, the parent companies not only had to forego *pro-tanto* their ability to invest in other projects but they assumed both the business risks and the financial risks of their respective pipeline subsidiaries. This assumption of the full risk of the enterprise, both business and financial, is why the commitment is deemed by the financial community to be equivalent in substance to a direct 100 percent equity investment in the system.[1446]

The crucial importance of the shipper-owner backing to TAPS was evidenced by Raymond Gary's testimony before the Alaskan Senate and House Committees on proposed legislation concerning pipeline regulation, right-of-way and state ownership. Mr. Gary stated that although the tax-exempt market was a large one, the bond issue required if the State of Alaska were to finance the line would be so large that additional reservoirs of capital would have to be tapped. It was Morgan Stanley's opinion that potential purchasers, such as insurance companies and pension funds, who ordinarily were not purchasers of tax-exempt bonds because they received little or no benefit from the tax-exemption arising from the nature of the bonds, would not subscribe for the issue unless it was unconditionally secured by the oil companies involved. In short, the State of Alaska did not have the financial capacity to substitute for the security of these companies.[1447] Other witnesses have testified to the same effect.[1448]

3. "Stand-Alone" Pipeline Ventures

In Section IV A 2, "Paucity of Pipelines Operated by Non-Oil-Related Owners,"[1449] an examination was made of the fortunes (and misfortunes) of pipeline ventures promoted or operated by owners who had no other affiliation with the oil business. There were a number of promotions which never got off the ground, *e.g.*, in 1948, the Trans-Western Oil Lines, Inc. and later (at the time of the Korean conflict), the West Coast Pipe Line Company, which attempted to promote a large

---

[1446] Gary, *Direct TAPS Testimony, supra* note 634, at 19-20.

[1447] Gary, *Alaskan Bills Testimony, supra* note 1095, at 1119. Mr. Gary's opinion was shared by four independent financial consultants to the State of Alaska, who advised the State that TAPS could not be financed by the State on a "stand-alone" basis without the oil company backing. See text at notes 894 and 895, *supra*.

[1448] Tierney, *TAPS Rebuttal Testimony, supra* note 175, at 4-5 (or without sizeable Federal financial assistance [which], based on current programs for the railroads, should be avoided if at all possible); *cf. Petroleum Industry Hearings, supra* note 192, at 144 (Statement of Frank N. Ikard).

[1449] See test at notes 1085-1124, *supra*.

diameter crude oil pipeline from the West Texas producing region to the Los Angeles refining area; in 1951 and 1953 respectively, the United States Pipe Line Company and the American Pipe Line Corporation proposed to build a large diameter products line from Beaumont, Texas, to Newark, New Jersey, all of which failed for lack of financing. A survey was made of the operations of the pipelines severed from the Standard Oil group as a result of the dissolution decree of 1911. Only one line ultimately survived on its own and that was Buckeye, which by virtue of consolidating with three other severed lines had a strategic locational advantage. By the time it expanded into products lines under a progressive and alert management which sought out special situations where it could provide a unique service, it had a "track record" and substantial assets to back its financings, which have been modest in size compared to that required by a modern "grass roots" large diameter long distance line. Another example on a smaller scale of a non-oil related line which found a special need and developed a track record was Kaneb. The other category of lines was those owned by companies which, although not related to the oil business to the extent commonly thought of as integrated oil companies, were of substantial size and borrowing capacity, so that the pipeline venture did not venture into the financial markets on a "stand-alone" basis. Examples were Southern Pacific's pipeline subsidiaries, Santa Fe's pipeline companies and the San Diego Pipeline, which was jointly owned by the two companies. The "bottom line" of these examinations was that pipeline "stand-alones" were possible in small specialized situations especially with a "track record" behind them. Otherwise, an owner of substantial credit capacity would be required to backup the necessary financing.

The effect of the pipeline project's size was illustrated dramatically by the TAPS situation where, according to congressional witnesses, the line could not have been built on a "stand-alone" basis,[1450] in fact, not even on the credit of the State of Alaska.[1451]

In order for the risk category to be satisfactory to lenders, the business risk element must be satisfied by a relatively secure supply and demand situation — witness the Kaneb and Buckeye examples — and the line must exhibit favorable economics and transportation costs which enable it to compete effectively with other modes of transportation. In such cases the ratio of debt to total capitalization which could satisfy lender's criteria for financial risk would be in the range of 30 to 35 per-

---

[1450] Gary, *Direct TAPS Testimony, supra* note 634, at 19.
[1451] See text at notes 1447 and 1448, *supra*.

cent.[1452] Put another way, in order to obtain a single A quality rating, there must be either shipper-owner backing if there is to be high leverage in the pipeline or the leverage must be lowered to the low 30 percent range.[1453]

[1452] Gary, *FERC Direct Testimony, supra* note 613, at 26-27; Gary, *Direct TAPS Testimony, supra* note 634, at 46. This is significantly greater than the average debt ratio for the 29 petroleum companies in the Chase Study, see text at notes 1385 and 1386, *supra*, and that (less than 20%) of oil pipelines prior to the Consent Decree. See text at note 1387, *supra*.

[1453] Swenson, *S. 1167 Testimony, supra* note 755, at 704. Herman G. Roseman, a Justice Department witness in the TAPS proceedings, guessed that a "stand alone" TAPS might be able to raise 10 to 20 percent of its capital in the form of debt. *TAPS Hearings Transcript, supra* note 785, at 9609.

Chapter VI

Current Policy Issues Concerning Pipeline Ownership

A. Introduction

1. Brief History of Public Policy Controversies

The early history of public policy concerning oil pipelines was generated largely by Appalachian producers and refiners who sought first to counter the power of the Standard Oil Group by state legislation such as the "Free Pipe Line Law" in Pennsylvania, followed by similar legislation in New York and eight other states, which granted the power of eminent domain to pipelines, together with a duty to operate as common carriers.[1454] The public did not come into direct contact with the situation, not being a consumer of the crude oil transported by these lines. Hence, the controversy remained largely intra-industry and intrastate until the enactment of the Hepburn Act in 1906. The oil industry, especially the Standard Oil Group, was pulled into the melee because of public resentment against discrimination by railroads and use of that discrimination by Standard Oil to develop a base of power. President Theodore Roosevelt detected a political advantage in supporting the Hepburn Amendment to the Interstate Commerce Act, which would grant to the ICC power over railroad rates. However, the railroads had their supporters, and the proponents of the bill sought to bolster its chance for passage by forming an alliance with members of Congress who were seeking to extend the coverage of the Interstate Commerce Act to pipelines. There was a movement by Kansas producers who, incensed by Prairie Oil & Gas Company's sharp reduction in the posted price of Kansas oil, sought redress by lodging a formal complaint through their Congressman, Philip P. Campbell, with James Garfield, a Commissioner of the U.S. Bureau of Corporations [predecessor to the FTC]. Senator Henry Cabot Lodge of Massachusetts introduced an amendment to the Hepburn bill, which proposed to make oil pipelines common carriers, subject to ICC regulation. Seemingly by coincidence, the "Garfield Report"[1455] [which charged that Standard Oil had profited from secret railroad rebates to the

---

[1454] PETROLEUM PIPELINES, *supra* note 45, at 20-21.
[1455] U.S. BUREAU OF CORPORATIONS, REPORT OF THE COMMISSIONER OF CORPORATIONS ON THE TRANSPORTATION OF PETROLEUM (1906).

detriment of its competitors and that this advantage was solidified by its discriminatory operation of pipelines] was released at the time the Lodge Amendment came before the Senate, and undoubtedly contributed to the 75-0 vote in favor of the amendment's passage.[1456]

Matters became complicated when Senator Stephen Elkins of West Virginia, on behalf of coal producers in his state, introduced yet another amendment to the Hepburn bill, seeking to add a "commodities clause" which would forbid "common carriers" under the Interstate Commerce Act from transporting commodities in which they had an interest. (Elkins' intent was to keep railroads out of the coal business.) However, because of the pivotal description "common carriers," the effect of combining the Lodge and Elkins amendments to the Hepburn Bill would be to preclude the common ownership of pipelines and oil production or refining. Hence, by happenstance, the first pipeline divorcement issue was raised. Many of the Senators who had supported the Lodge Amendment and/or the Elkins Amendment as separate proposals became alarmed that the application of the commodities clause to pipelines would harm the growing petroleum industry, particularly the independent producers who had sought their assistance in supporting the Lodge Ammendment.[1457] Senator Elkins himself was deluged with protests from West Virginia producers who opposed the application of the commodities clause to pipelines. Senator Long of Kansas, from whence had come the original impetus for making pipelines common carriers, was concerned that the divestiture would scuttle the pending effort to connect the Mid-Continent field with the Gulf Coast refinery area. The issue was resolved against divestiture by the Conference Committee, which substituted the word "railroad" for "common carrier" in the designative portion of the Commodities Clause.

Ever since the passage of the Hepburn Act of 1906, there have been sporadic proposals of legislation proscribing oil company ownership of pipelines, but in every instance such proposals have been rejected by the Congress.[1458] Bills seeking, in one way or another, to prohibit oil companies from owning pipelines have been introduced in practically every Congress in the past 48 years.[1459] The earlier bills sought to amend the

---

[1456] PETROLEUM PIPELINES, *supra* note 45, at 24-26. BEARD, *supra* note 39, at 18 cites *In the Matter of Pipe Lines,* 24 I.C.C. 1, 4 (1912) for a timing of a "few moments" between the receipt of President Roosevelt's transmittal to the Senate of the Garfield Report and the introduction of the Lodge Amendment. NET-III, *supra* note 199, at 127 states the vote to have been 74 yeas, 0 nays and 14 not voting, citing 40 Cong. Rec. 6376 (1906).

[1457] PETROLEUM PIPELINES, *supra* note 45, at 28-29.

[1458] HOWREY & SIMON, *supra* note 3, at 67; NET-III, *supra* note 199, at 142-143.

[1459] *Cf. Id.* For a list of bills from 1931 through 1949, see WOLBERT, *supra* note 1, at 4 n.4. For bills introduced in the 83rd Congress through the 95th Congress, see HOWREY & SIMON, *supra* note 3, at 67 n.3. During the 93rd, 94th and the first session of the 95th Congress, approximately 150 bills were introduced calling for an alteration in the shipper-owner

commodities clause so as to read "common carriers," thus extending the clause to pipelines. Subsequent bills have attempted directly to prevent pipeline ownership by persons engaged in one or more other phases of the oil business. Congress consistently has rejected these proposals. The provisions of the National Industrial Recovery Act (NIRA) authorized the President to institute proceedings to divorce from any holding company any pipeline company controlled by it, where such pipeline company by unfair practices or exorbitant rates for transporation tended to create a monopoly.[1460] However, before any action was taken pursuant to this section, the Act was invalidated by the United States Supreme Court as an unconstitutional delegation of legislative power.[1461]

It is not surprising that the commodities clause line of attack has waned over the years. The original enactment of the clause in 1906, which was designed to end discriminatory treatment by the railroads in matters such as rates and services, was adopted at a time when the ICC did not have jurisdiction to halt such disparity in treatment. Since that time, the ICC has been given, and has exercised, authority to remedy every kind of discrimination so that the need for the protection of such a clause no longer exists.[1462] This is true even more today, because the DOE and FERC together have been given powers that even the ICC did not possess.[1463] Moreover, there is a significant difference between the origin and function of railroads and other common carriers on the one hand and oil pipelines on the other. The first category of carriers are engaged in a separate business, transportation of the commodities of many shippers, whereas oil pipelines exist but for a single purpose: to provide a vital link in the productive process of bringing one class of goods, petroleum and its products, from the raw material source (crude oil at the lease tank) to the place of transformation into a useful product and thence to the service station or other point of dispensation to the customer.[1464] In his discussion of the events surrounding the original controversy, Professor Arthur Johnson posed the basic questions of economics and public policy: were oil pipelines basically transportation facilities, and hence properly subject to the same kind of regulatory treatment as railroads, or were they so

---

relationship. NET-III, *supra* note 199, at 143. Sec. *Id.* at 144-146 for description of the key measures. The closest approach to a railroad-oil pipeline analogy is the resemblance between a crude pipeline and a logging spur built into a forest to carry logs to a central saw mill where they are cut into lumber products. Significantly, these "tap lines" were expressly excluded from the commodities clause. Tap Line Cases, 234 U.S. 1 (1914).

[1460] 48 Stat. 200 (1933).

[1461] Schechter Poultry Corp. v. United States, 295 U.S. 495 (1935).

[1462] WOLBERT, *supra* note 1, at 51; HOWREY & SIMON, *supra* note 3, at 68.

[1463] JONES DOE REPORT, *supra* note 372, *passim*.

[1464] HOWREY & SIMON, *supra* note 3, at 69-70; *cf.* Att'y Gen., *Fourth Report, supra* note 111, at 63 ("But the pipeline system is more than a transport agency, it is the crude oil market.").

specialized in their economic relation to other levels of the oil industry as to constitute an integral part of that industry and therefore to call for a different kind of public policy solution? If it were the latter, Professor Johnson questioned whether pipelines even should be declared common carriers.[1465] Congress' action indicates that it did not think in Professor Johnson's "either-or" terms. Rather, it attempted to achieve the "best of both worlds," *i.e.*, regulate oil pipelines as common carriers but give recognition to its peculiar ties with the rest of the industry by refusing to apply the commodities clause to pipelines.[1466] Its decision also recognized the vital distinction between the position of utilities, which are shielded from competition by the grant of exclusive franchises, which gives them the equivalent of a guaranteed profit, and that of oil pipelines, which are subject always to potential competition and usually to intense actual competition (as was discussed in Section III F), and hence, unlike the utilities, have no "safety net" against loss nor enjoy a guaranteed rate of return. Congress' consistent refusal to apply the commodities clause to oil pipelines or otherwise to prohibit their ownership by the oil companies which use them, represents a conscious choice to maintain this competitive element in the oil pipeline industry. Its policy is memorialized in the declaration found in the National Transportation Policy of 1940 which sought to preserve to each type of carrier the inherent advantages it offered to the public.[1467]

The oil industry attempted immediately after the Hepburn enactment to establish the "private carrier" alternative mentioned by Johnson, but the ICC's decision, upheld by the United States Supreme Court in *The Pipe Line Cases*, ended that line of argument in 1914.[1468]

In addition to the virtual continuum of bills seeking pipeline divorcement, there have been several "seventh wave" attacks on pipelines since the 1906 Hepburn Amendment, *e.g.*, the *TNEC Hearings* in 1939-1940; the "Mother Hubbard" and Elkins Act cases just before World War II; the *"Wherry" Committee Hearings* in 1947; an aftermath of the

---

[1465] PETROLEUM PIPELINES, *supra* note 45, at 29.

[1466] Congress' recognition of the inappropriateness of applying railroad regulation concepts without gving due recognition to the function of oil pipelines as a working constituent in a single productive scheme was verified by the famous "Splawn" report in 1933. H.R. REP. No. 2192, 72d Cong., 2d Sess. LXXVIII (1933).

[1467] See text at notes 691-699, 955-959, 962-966, *supra*.

[1468] In the Matter of Pipelines, 24 I.C.C. 1 (1912); Prairie Oil and Gas Co. v. United States, 204 Fed. 798 (Comm. Ct. 1913); The Pipe Line Cases, 234 U.S. 548 (1914). The carriers relied upon the language of the Lodge amendment which applied the ICA to persons "who shall be considered and held to be common carriers within the meaning and purpose of this Act," taking the position that since they purchased all of the oil that moved through their lines, they were *private* carriers and not subject to the Act. Bond, *supra* note 207, at 736. The language was changed to the flat, unambiguous wording of the current statute by the Transportation Act of 1920, 41 Stat. 474 (1920). *Id.* at 737.

Suez Canal closing; and the search for someone to blame following the "Yom Kippur" war and the resultant Arab oil embargo.

Professor Edward Mitchell, in his introduction to the American Enterprise Institute's book, *Vertical Integration in the Oil Industry,* stated: " 'Breaking up the oil companies' has become a popular idea in Washington. It occurs repeatedly in the campaign rhetoric of presidential aspirants and has already given rise to an unsuccessful but close Senate vote on vertical divestiture of the petroleum industry...

"As always, political popularity is derived from public opinion. Polls show that the American people hold a highly unfavorable view of the oil industry. But public opinion is a sound basis for public policy only when that public opinion is informed. And public opinion regarding energy, and the oil industry in particular, is pitifully uninformed. According to polls, the public believes that oil companies make sixty cents of profit on each dollar of sales. In fact, they typically make four to five cents on each dollar of sales.

"The question of oil company profits is a simple one compared to the issue of vertical integration and vertical divestiture. If the public is uninformed or misinformed on oil company profits it cannot have the foggiest idea of what the consequences of vertical divestiture might be. With such a foundation of ignorance it is hardly surprising that the issue should give rise to demagogy."[1469]

2. Sources of Complaints

Reviewing the history of public policy controversy over pipelines, there is a recurring phenomenon which stands out in stark relief, namely that very few complainants, even those appearing before Congressional committees, have been persons engaged in the day-by-day operation of the oil business. There were a few in the *TNEC Hearings* in 1939, but even their testimony was basically directed not at any need for pipeline divorcement but at matters which were often the result of misunderstandings and were capable either of being worked out by negotiation between the parties or by obtaining relief at the hands of state regulatory agencies or the ICC. Louis J. Walsh, Vice President of Eastern States Petroleum Company of New York City, whose company owned and operated a refinery in Houston, complained about buying in a controlled [through proration] production market and selling in a definitely competitive consumer market, and he felt that pipeline rates were too high

---

[1469] E. MITCHELL, *Introduction in* VERTICAL INTEGRATION IN THE OIL INDUSTRY 1 (E. Mitchell ed. 1976).

[twice the cost of operating them according to his testimony]. But, he volunteered the statement that despite his company's lack of production or pipeline ownership, he had connections with three pipelines owned by "major" companies, which made crude available to his company from "practically all the fields in Texas, New Mexico, and some fields in Louisiana."[1470] Another witness in the same hearings remarked that "the wagon haulers seem to have been the earliest advocates of disintegration."[1471] For years, Paul E. Hadlick of the National Oil Marketers Association appeared regularly and complained about "subsidization" by the "majors" of their marketing operations from the profits derived from other phases of the business.

In the *Smith-Conte Hearings,* Hoyt Haddock, Executive Director, AFL-CIO Maritime Committee, appeared and complained bitterly that Colonial Pipeline "would replace 56 T-2 equivalent [each 16,500 dead weight tons] tankers" and mentioned meetings, and a brief filed, with Attorney General Robert Kennedy on April 11, 1963, in opposition to the line, both of which alleged a monopolization of the transportation of oil.[1472] It would seem that 13 years of investigation by the Antitrust Division, Department of Justice, employing voluntary cooperation, visitation and civil investigative demands without finding a basis for suit renders this allegation somewhat suspect.[1473] Alfred Maskin, Executive Director, American Maritime Association, also appeared at the same hearings and complained that Colonial's rate structure discriminated against the tanker trade, alleging that Colonial's tariff from Houston to Collins, Mississippi, cost a shipper .037 cents per barrel-mile, whereas the comparable (overall average) from Houston to Atlanta cost only .029 cents per barrel-mile and for the long haul from Houston to New York the cost was down to .02 cents per barrel-mile.[1474] There was no dispute that Colonial's tariff from Houston to New York was designed to be competitive with tanker competition; it would be unsound economics to spend

---

[1470] *TNEC Hearings, supra* note 6, pt. 14 at 7337 (Statement of Louis J. Walsh).

[1471] *Id.* at 7188 (Statement of J. Howard Pew).

[1472] *H.Res.5 Subcomm. Hearings, supra* note 234, at 36-39 (Statement of Hoyt Haddock). Mr. Haddock "was unable to find" the brief [or the accompanying press release] which was joined in by the Industrial Union Department, AFL-CIO; the National Maritime Union; the Oil, Chemical & Atomic Workers International Union; and the Industrial Union of Marine & Shipbuilding Workers of America, having been prepared by Michael H. Gottesman of Feiler, Bradhoff & Anker, of 1001 Connecticut Avenue, N.W., Washington 6, D.C., so it was not filed as part of the record. However, Haddock did produce, and submit for the record, a similar argument filed by the same parties with Attorney General Nicholas DeB. Katzenbach on April 15, 1965. *Id.* at 40-42.

[1473] See text at notes 421-423, *supra*.

[1474] *H.Res.5 Subcomm. Hearings, supra* note 234, at 47 (Statement of Alfred Maskin). Mr. Maskin's counsel, Joseph A. Klausner, managed to find his press release and brief filed with Attorney General Kennedy and they appear in the record. *Id.* at 59-64.

approximately half a billion dollars on a pipeline between those two points that was not competitive with tankers.[1475] The Ernst & Ernst report, prepared for the Shipbuilders Council of America, and introduced into the *Smith-Conte Hearings* record, showed that the going tanker rate ranged from 25 cents per barrel (for a 1-5 year charter of a 40 to 50 thousand dwt vessel) to 35 cents per barrel for a T-2 tanker; hence the 25 to 35 cents per barrel range was competitive with Colonial, whose rate for the same run was stated by Ernst & Ernst to be 35 cents per barrel.[1476]

There were innuendos that once Colonial had displaced the T-2 tankers, it would raise its tariffs and enjoy the fruits of its "monopoly." Not only is this contrary to economic theory, which would suggest that increased rates would provide an incentive for tankers to come back into the market, but it is contrary to the facts, which are that Colonial *reduced* its Gulf Coast to New York harbor tariffs from 33 cents per barrel to 29 cents during the period 1964 to 1971, whereas tanker rates for average size tankers had increased from 35 to 44 cents per barrel for gasoline and from 39 to 50 cents per barrel for fuel oil over the same period of time.[1477] In contrast, Plantation Pipe Line Company, finding its older, smaller system at a disadvantage with Colonial, expanded and modernized its system, reduced its rates and remained competitive.[1478] The other complaint, namely the alleged long-haul, short-haul "discrimination" (inferred to be forbidden by ICA §4), appears to be misdirected. Arthur J. Cerra, Deputy General Counsel of the ICC, in his testimony immediately following Mr. Maskin's, stated that Section 4 simply prohibits carriers from charging more for a short-haul than for a long-haul. Using witnesses Haddock's and Maskin's example, he noted that the rate from Houston to New York was *not* less than that from Houston to Atlanta, and the ICC consistently had permitted carriers to lower rates to meet competition.[1479] Fred Steingraber, Colonial's President, introduced a chart into the record showing the then current tariff rates for Colonial, Independent Tankers and common carrier trucks, each depicting a decreasing average rate per hundred barrel miles and showing that Colonial's average rates reduced less sharply over distance than did competing tankers and trucks.[1480] The reason that the rate per unit volume mile decreases with the length of the movement

---

[1475] *Id.* at 179 (Statement of Fred F. Steingraber); *Id.* at A73 (Reprint of letter dated Jan. 27, 1972, to Chairman William Proxmire, of the Joint Economic Committee's Subcommittee on Priorities and Economy in Government, from Jack Vickrey).

[1476] *Id.* at 36. According to Mr. Fred. F. Steingraber's Testimony, the actual rate from Houston to Linden, N.J. was 31¢/bbl. *Id.* at 178.

[1477] *Consumer Energy Act Hearings, supra* note 154, at 629 (Statement of Jack Vickrey).

[1478] *Id.* at 629-630; see text at notes 424 to 426, *supra*.

[1479] *H.Res.5 Subcomm. Hearings, supra* note 234, at 90.

[1480] *Id.* at 201-202.

is because each movement involves a front end loading and administrative cost which does not vary with the distance. If, as is the case with pipelines, this converts to a substantial portion of the total cost of the short-haul but becomes a lesser portion of the cost of the total in a long-haul, the conversion of its total rate into a volume-mile basis naturally will produce a decreasing average per mile rate.[1481] Joseph Klausner, Counsel for Mr. Maskin, made sure that the record would contain no suggestion nor inference that Colonial's rate was a non-compensatory rate.[1482]

Beverly C. Moore, Jr., of the Corporate Accountability Research Group, of Washington, D.C., appeared before the *Smith-Conte Hearings* and devoted his main testimony to an attack on the Justice Department's failure to bring an action against Colonial and other "industry lines." He noted the action taken by Assistant Attorney General Richard McLaren against the (Gateway) MATCH group, which had been stirred up by the complaints of certain Mississippi barge operators[1483] with the assistance of Senator Eastland, a friend of one of the barge owners.[1484] He alluded vaguely to "the possibility of political interference from above"; but expressed his belief that the Justice Department's line of distinction was drawn between cases where the new group planned to acquire an existing pipeline — MATCH's proposal included purchase of the "Little Big Inch" from Texas Eastern, and Glacier involved the proposed merger of Exxon's Silvertip Pipeline with Continental's Glacier Pipeline — and the creation of a new-to-industry line. Mr. Moore managed to obtain a "rerun" of his *Proxmire Committee* hearings[1485] in which he had alleged, without any supporting facts, that: (1) pipelines indirectly denied access to new owners by means of their design and operational characteristics [this will be dealt with in a subsequent subsection]; (2) there was a competitive disadvantage to a nonowner-shipper arising out of the difference between the tariff rate and its cost [this was shown to be groundless in Chapter IV, text at notes 1293 to 1296]; (3) jointly-owned pipelines are vast "storage tanks" operated in such a manner as to keep supplies out of the hands of

---

[1481] *Id.* at 91 (Testimony of Ernest R. Olson, Assistant Director, Bureau of Traffic, ICC); *Id.* at 200-201 (Statement of Fred F. Steingraber). If the tariff rates on most pipelines are plotted on a rate in cents per barrel on the ordinate and distance in miles on the abscissa, a least square line will represent an equation rate $(r) = c + a(d)$ where c will be the terminalling and administrative costs and a is the cost per mile (d). A decreasing rate per barrel mile as the length of the haul increases is a recognized principle in rate making. PIPELINE PRIMER, *supra* note 331, at 29; as is the proposition that joint rates usually are lower than the combined local tariff rates of the participating carriers. *Id.*

[1482] H.Res.5 *Subcomm. Hearings, supra* note 234, at 47 (Interposition by Joseph A. Kausner).

[1483] See copy of the Barge Owners' Complaint filed with the Justice Department. *Id.* at A112-137.

[1484] *Id.* at 137 (Statement of Beverly C. Moore, Jr.).

[1485] *Id.* at A58-67.

independent refiners in the case of crude lines and independent terminal operators and branded dealers, in the case of products lines [this is hydraulically inane[1486] and economically unrealistic[1487]]; and (4) that the jointly-owned lines enabled the owners to stabilize their market shares in a regional cartel [Professor Edward Mitchell has discredited this assertion with respect to Colonial].[1488] Mr. Moore closed his "day in court" by interrupting the testimony of the next witness, Mr. Fred Steingraber, and by attempting to cross-examine him.[1489]

In the *Consumer Energy Act Hearings* in 1973 (William) John Lamont, ex-staff member for Senator Guy Gillette, ex-Antitrust Division member, then in private practice of law, in addition to disparaging the ICC and the Justice Department, made two points concerning "industry" lines: the first was an alleged anti-competitive effect of competitors working together on a jointly-owned line and the second was a complaint about pipelines not furnishing input and output facilities for nonowner-shippers [owner-shippers have their own]. He sought to dramatize the latter point by asking the Subcommittee to imagine "a railroad without depots which solicits freight business by saying 'Throw your goods on the flat car as it comes by, or build your own station and maybe we will stop and pick it up.' "[1490] Terminalling will be discussed in a subsequent portion of this Chapter. Suffice it here to say that not only did Mr. Lamont know that the Interstate Commerce Act does not require interstate pipeline companies to furnish storage,[1491] but his analogy to a railroad was an unfortunate one inasmuch as railroads do *not,* as part of their "depot" function, provide tankage for the accumulation or delivery of petroleum but look to their shippers to handle the storage problem precisely in the same manner as do pipelines.[1492] The Supreme Court has even ruled that under the Interstate Commerce Act, the ICC does not have the power to require railroads to furnish tank cars where no

---

[1486] *Id.* at A84 (Letter, dated July 14, 1972, to Chairman Smith from Jack Vickrey).

[1487] See text at notes 1201 to 1205, *supra*. One of the big hurdles to building a strategic crude oil reserve supply so as to ameliorate the impact of embargoes or natural catastrophe supply interruptions was the key question, who was going to finance the inventory?

[1488] Mitchell, *1974 Senate Testimony, supra* note 1056, at 31-40; *see also* text before and after note 1297, *supra*.

[1489] H.Res.5 Subcomm. Hearings, *supra* note 234, at 199.

[1490] *Consumer Energy Act Hearings, supra* note 154, at 670 (Statement of John Lamont).

[1491] WOLBERT, *supra* note 1, at 40, and cases cited therein.

[1492] *Consumer Energy Act Hearings, supra* note 154, at 705 (Letter, dated December 19, 1973, to Senator Magnuson, Chairman of the Senate Commerce Committee, from Jack Vickrey). M. PIETTE, CRUDE OIL AND REFINED PRODUCT PIPELINES IN THE UNITED STATES: AN EXAMINATION OF THE MAJOR ISSUES FOR PUBLIC POLICY 18 (Presentation to the U.S. Dept. of Energy, Jan. 30, 1979 [hereinafter cited as PIETTE, *U.S. Oil Lines-Public Policy*]

discrimination is involved.[1493]

Having lost its skirmish with Colonial, the Marine Engineer's Union (MEBA) appeared again on October 29, 1975, in the *Petroleum Industry Hearings* through its President, Jesse M. Calhoon. After making some vague allusions to collusion between the "majors" and OPEC to raise prices, and an assertion that "the oil companies" decided "how much domestic oil to produce and, therefore, how much oil we must import" [so that] "[t]hey determine how much we pay for oil,"[1494] Mr. Calhoon turned financial expert long enough to opine that "the use of joint venture capital is not required in order to finance and operate new and more efficient pipelines."[1495] Faced with a choice between reliance on Mr. Calhoon's self-serving, non-expert declarations and the testimony cited in the preceding chapter on "Financing,"[1496] it would appear that Mr. Calhoon confused suspicion with fact.

Then came Charles Binsted, Executive Director of The National Congress of Petroleum Retailers, who confined his testimony to divestiture of the retail marketing segment of the business from major oil companies, "which would be financed by the SDA [*sic*, probably the Small Business Administration]." This suggestion troubled Charles Bangert, the Subcommittee's General Counsel, who asked the President of the Independent Terminal Operator's Association, Ronald J. Peterson, whether he had given any thought to the capital expenditures and the need for new capital that might be occasioned by the proposed legislation.[1497] Mr. Peterson's reply was to "spin-off" the companies, and for precedent he relied upon the Pennzoil shareholders' vote to spin off the United Gas Pipeline, a somewhat smaller transaction than divorcing the industry's pipelines, and one involving a franchised business, *i.e.*, the purchase and sale of natural gas together with the sales and purchase contracts which assured it of a profit, as contrasted with a service business, such as an oil pipeline, which is subject to the risks of a competitive marketplace.

The sole industry member who appeared at any of the hearings and testified as to facts was Charles P. Siess, Jr., then President of Apco Oil Corporation. Mr. Siess described an incident wherein Apco, which was a new purchaser of crude in West Texas, outbid Sun Oil by 40 cents a barrel in order to obtain a 22,000 barrel per day contract with General Crude and tendered that same oil to Sun Pipe Line Company for transportation to Gulf's Refinery. General Crude was anxious to have the crude moved

---

[1493] United States v. Pennsylvania R.R., 242 U.S. 208 (1916).
[1494] *Petroleum Industry Hearings*, *supra* note 192, at 226 (Statement of Jesse M. Calhoon).
[1495] *Id.* at 229.
[1496] See text at notes 1447 to 1451, *supra*.
[1497] *Petroleum Industry Hearings*, *supra* note 192, at 189.

within five days, so when Sun Pipe Line demurred on taking the crude [according to Mr. Siess giving an excuse that it did not meet the line's vapor pressure specifications] "push got to shove." Early Friday morning (4 days after the initial tender) Siess called Sun Pipe Line Company's President in Tulsa and told him that unless Sun notified him before Friday noon that its gathering system would move Apco's crude, Siess was going to file suit. By 10:30 a.m. that morning Sun called back and said it would move Apco's crude.[1498] The S.2387, or PICA Report made much of this testimony, citing it as "the most dramatic testimony" on the point that control of pipelines was roughly equal to control of the crude.[1499] The Minority Report and two commentators placed an entirely different construction on the incident, saying that the record showed clearly that the common carrier obligations of the Sun Pipe Line were recognized, that the producer was able to sell its crude to someone other than the pipeline's affiliate, and that the purchaser was able, with some effort, to obtain transportation of the crude.[1500] Significantly, Mr. Siess' successor as President of Apco, Thomas W. diZerega, testifying before the same subcommittee in the next Congress, appeared to concur with the minority's characterization of the incident.[1501]

There has been an abundance of evidence from ICC Commissioners concerning the absence of complaints against pipeline denial of access or discriminatory practices. In 1957, Chairman Owen Clarke testified to that effect before the Celler Subcommittee;[1502] Commissioner Charles Webb stated the same proposition in a 1959 address before the National Petroleum Association;[1503] in 1973, Chairman George Stafford expressed a similar view in his testimony before a Senate Special Subcommittee of

---

[1498] *S.1167 Hearings, supra* note 432, pt. 8 at 6241-6242 (Statement of Charles P. Siess, Jr.)

[1499] S.2387 REPORT, *supra* note 49, pt. 1 at 22-23. The KENNEDY STAFF REPORT, *supra* note 184, at 82-83, cited the incident as evidence of power to exclude nonowner-shippers from the line.

[1500] S.2387 REPORT, *supra* note 49, pt. 2 at 212; *accord,* STEINGRABER, *supra* note 193, at 148 ("According to Mr. Siess' testimony APCO had a dispute with Sun Pipe Line Company, which apparently was satisfactorily worked out before any complaint was filed with the ICC. Of course, disputes occur between pipeline carriers and their shippers. But, in practically all cases, these problems are worked out in the normal course of business to the mutual satisfaction of the parties."); *S.1167 Hearings, supra* note 432, pt. 9 at 636 (Statement of Fred F. Steingraber); JOHNSON et al., *Oil Industry Competition, supra* note 1055, at 2424.

[1501] *Petroleum Industry Hearings, supra* note 192, at 252. This is not the first example of shipper-owner pipelines observing their common carrier obligations where connections have been taken over by a new [independent] purchaser from the carrier's affiliated purchasing company and tendered to the pipeline which ran the oil. WOLBERT, *supra* note 1, at 47 n.261.

[1502] *Celler Hearings, supra* note 162, at 451 (Testimony of Owen Clarke).

[1503] C. WEBB, ICC REGULATION OF THE OIL PIPELINE INDUSTRY 7 (Address before the National Petroleum Association, Sept. 17, 1959), cited in SPECTRE, *supra* note 104, at 115.

the Senate Interior Committee, adding: "Today there are so few complaints and so few problems that I must say it [the oil pipeline group] is one of the best run transportation systems we have..."; [1504] that same year Chairman Stafford sent a letter to Congressman Joe Evins, Chairman of the House Select Committee on Small Business, reiterating the same message. [1505]

There has also been substantial evidence from independent producers. In 1939, Ray M. Johnson, speaking for the Oklahoma Stripper Well Association in his statement in the *TNEC Hearings* in 1939, testified in *favor* of the then current pipeline practices and stated that pending proposals for pipeline divorcement had "no substantial support" among "the thousand of small producers" in his Association. [1506] The Independent Petroleum Association of America (IPAA) consistently has opposed pipeline divestiture bills. In 1973, Tom B. Medders, Jr., as President of IPAA and himself an independent producer from Wichita Falls, Texas, wrote a letter dated July 12, 1973, to every Senator in which he stated: "This is to advise you that we are not aware of any producer having difficulty selling or moving his crude oil, and we do not believe any such discrimination exists."

He continued, in the following paragraph, "Adoption of [Senator Haskell's proposed amendment to the Alaska Pipeline Bill, which would require divestiture of any ownership or interest in common carriers by companies which produce crude oil], therefore, would be of no benefit to independent producers and could be very harmful to some independents who own or have interests in small pipeline systems. Divestiture, we believe, could serve to increase transportation costs for all independents and raise prices for consumers."[1507] On November 6, 1973, L. Dan Jones, General Counsel of IPAA, sent a letter to Senator Adlai Stevenson III, introducer of S.2506, which stated: "When you introduced this bill you said that it is aimed at helping restore competition in the petroleum industry by assuring independent producers access to pipelines owned by major oil companies. This is to advise that we are not aware of any producer who is having difficulty selling or moving his crude oil and we do not believe discrimination exists in this respect. The conclusion that independent crude oil producers may have difficulty securing shipment of their oil, and are subject to discrimination by pipeline companies, is not supported by the experience of independent

---

[1504] *S.Res.45 Hearings, supra* note 296, pt. 3 at 896, 901 (Statement of George Stafford).

[1505] See copy of Stafford's letter reproduced in HOWREY & SIMON, *supra* note 3, Exhibit 9.

[1506] See reprint of Johnson's letter in *Consumer Energy Act Hearings, supra* note 154, at 632-633.

[1507] See copy of Medder's letter reproduced in Allen, *FERC Direct Testimony, supra* note 929, Exhibit "B."

producers."[1508] On October 27, 1978, Jack M. Allen, President of IPAA and the President of Alpar Resources, Inc., an independent exploration and producing company operating principally in the Anadarko Basin, testified in the FERC *Valuation of Common Carrier* proceedings that "The fact that there have been relatively few independent pipelines and a minimum of complaints from independent shippers, indicates that independents have been well served by the existing pipeline industry and have no overwhelming desire to engage in the oil pipeline business .... IPAA has been in existence for 49 years and its records reveal few complaints of any kind against oil pipelines."[1509]

The Antitrust Division of the Justice Department is on record as saying that there is a lack of real evidence of competitive injury by pipelines owned by oil companies.[1510]

What inferences should be drawn from the foregoing? Assistant Attorney General Shenefield in proposed testimony before the Kennedy Subcommittee drew an inference of ineffective regulation when he stated: "Not unexpectedly at hearings such as this, the industry points to the absence of complaints as proof of nondiscriminatory service to the shipping public. In weighing that record, however, Congress should also consider the fact that the nonowner-shipper has in the past been faced with a climate of regulatory indifference in an industry dominated by carrier-affiliated shipments. This is an additional burden to any reluctance that may exist in the nonowner-shipper to disturb a customer/supplier relationship by resort to litigation."[1511] Congressman Neal Smith suggested that people were "a little bit afraid" to complain.[1512] Senator Henry Jackson is quoted by the Kennedy Staff Report

---

[1508] See reprint of Jones' letter in *Consumer Energy Act Hearings, supra* note 154, at 597. W. J. Lamont, former staff member of Guy Gillette and ex-Antitrust Division attorney, asserted in his testimony in the same hearings that the membership in IPAA, while it contained a lot of independent producers, was largely controlled by the same integrated oil companies who own the pipelines, which make up the Association of Oil Pipelines; and when they spoke with respect to pipelines, they spoke "with the same voice as the American Petroleum Institute." *Id.* at 673. Mr. Lamont submitted no statistics or facts in support of this assertion.

[1509] Allen, *FERC Direct Testimony supra* note 929, at 9-10.

[1510] *H.Res.5 Subcomm. Hearings, supra* note 234, at 206-207 (Statement of Bruce B. Wilson, Deputy Assistant Attorney General).

[1511] SHENEFIELD PROPOSED STATEMENT, *supra* note 179, at 14.

[1512] *H.Res.5 Subcommittee Hearings, supra* note 234, at 188 (Interjection by Chairman Neal Smith); In 1977 there were 18 complaints filed with, and considered by, ICC/FERC. Although the number is small, it does appear to demonstrate that present or prospective shippers as well as shippers on competing lines are not afraid, but clearly are ready, to challenge a pipeline's rates and practices when it is in their self-interest to do so. HOWREY & SIMON, *supra* note 3, at 35; PIPELINE PRIMER, *supra* note 331, at 32. LEGRANGE, *Government Regulation, supra* note 1413, at 3 remarked: "When the president of the Independent Petroleum Association of America, representing some 4,000 independent producers, tells the United States Congress that the independents are not aware of anyone having difficulty moving his crude and 'we do not believe discrimination exists,' I have a hard time accepting the argument that he does this only out of fear of the big boys."

as saying, after referring to the lengthy investigation by the Justice Department that "[t]o tell a constituent we are going to get you relief through the Department of Justice is like waiting for the impossible...[people] feel they are being ripped off; they are furious and outraged....when we talk about the length of time involved, it looks to me that there is only one solution and that is to forget the Justice Department....We can only legislate, I think that is the only thing we are left with."[1513]

On the other side, Senator Clifford P. Hansen of Wyoming stated that he didn't want to hear a long list of complaints from those who are really suing for less competition under the guise of promoting competition.[1514] A like note was sounded by Professor Edward Mitchell when he appeared before the Senate Antitrust and Monopoly Subcommittee in 1974: "Looking at these hearings an objective observer might well conclude that the Committee was not concerned about the 'Anticompetitive Impact of Oil Company Ownership of Petroleum Product Pipelines,' as the hearings were entitled, but the *competitive* impact on certain business and labor interests. In this light the charges against the pipelines make more sense. The charge of excess profits may not stem from a desire to reduce rates, but from a desire to discourage pipeline investment. The advocacy of independent ownership may be motivated not by a concern for alleged oil company abuses, but from the knowledge that independent pipelines will either not be built or will involve higher rates and costs of capital."[1515] One witness has noted the historical fact that up until the *Williams Pipe Line* cases, what complaints there have been have originated because pipeline rates were too *low,* not too high.[1516] Another observer testified that there was no evidence that the large vertically integrated oil companies presently are exercising monopoly power in any of the four principal phases of the oil business, including crude oil and products pipelines.[1517]

It has been said that to the extent that pipelines are more economical than alternative modes of transportation, the public benefits because, in

---

[1513] KENNEDY STAFF REPORT, *supra* note 184, at 134.

[1514] *Petroleum Industry Hearings, supra* note 192, at 71 (Statement of Clifford P. Hansen).

[1515] Mitchell, *1974 Senate Testimony, supra* note 1056, at 26-27.

[1516] *Consumer Energy Act Hearings, supra* note 154, at 623 (Statement of Jack Vickrey); text at note 1741, *infra; see e.g.,* Petroleum Rail Shippers' Ass'n v. Alton & Southern R.R., 243 I.C.C. 589, 593 (1941); Pipeline Rates on Propane from Southwest to Midwest, 318 I.C.C. 615 (Div. 2, 1962). For a comparable situation where railroad competitors objected to a pipeline tariff permitting shippers to aggregate tenders so as to enable the smaller shippers to meet the minimum tender requirement, *see* Pipeline Demurrage & Minimum Shipment Rule on Propane, 315 I.C.C. 443, 447-448 (1962).

[1517] *Petroleum Industry Hearings, supra* note 192, at 1894 (Statement of Professor Richard B. Mancke).

the long run, competition forces the integrated companies to pass on to the consuming public a significant portion of such economies.[1518] Thus, it would seem that if a pipeline divorcement bill comes to a vote, public policy makers will be forced to face up to a tough decision. Which interest should receive priority: The well-being of the economy and the consumer or the protection from competition of some specific individuals or companies, even though they are constituents?[1519]

B. Regulation

1. Interstate Commerce Commission — Now Federal Energy Regulatory Commission

The importance of the regulatory influence of the ICC/FERC [these names will be used interchangeably, ICC up to October 1, 1977, and FERC thereafter; "Commission" as to the appropriate one][1520] can be gleaned from the fact that about 80 percent of the nation's oil pipelines are regulated by the Commission.[1521] Interstate oil pipelines were made common carriers, subject to regulation under Part I of the Interstate Commerce Act (ICA),[1522] by the Hepburn Amendment of 1906.[1523] The Hepburn Amendment which, for most practical purposes,[1524] opened up

---

[1518] Emerson, *Salient Characteristics of Petroleum Pipeline Transportation* in 26 LAND. ECON. 27, 39 (No. 1, 1950); SPECTRE, *supra* note 104, at 111.

[1519] *Petroleum Industry Hearings, supra* note 192, at 355 (Testimony of W. T. Slick, Jr.).

[1520] The transfer took place on October 1, 1977; see text at, and note 936, *supra*.

[1521] *Consumer Energy Act Hearings, supra* note 154, at 610 (Statement of Jack Vickrey). The following description of the authority of the Commission has drawn upon *Consumer Energy Act Hearings, supra* note 154, at 594, 611-612, 620-623 (Statement of Jack Vickrey); HOWREY & SIMON, *supra* note 3, at 32-37; D. J. Views and Arguments, *supra* note 656, at 7; JONES DOE REPORT, *supra* note 372, at 16-18, 34-37.

[1522] 49 U.S.C. §§ 1-27 (1976).

[1523] 34 Stat. 584 (1906) (prior to 1920 amendment), 49 U.S.C. § 1 (3) (a) (1976).

[1524] This qualification is necessary in order to recognize the exception, drawn in Justice Holmes' opinion for the Uncle Sam Oil Company which had a pipeline engaged solely in moving its own oil from its own wells in Oklahoma across the state line to its own refinery in Kansas. Holmes dodged the "taking of private property without just compensation" issue by construing Uncle Sam's activities not to be "transportation" within the meaning of ICA §1 (1) (b) (49 U.S.C. § 1 (1) (b)). Chief Justice White, in a concurring opinion, disagreed with Holmes' construction of "transportation," but felt that the application of the statute to the Uncle Sam Oil Company, which was a *private carrier,* would be an unconstitutional "taking," whereas the Standard Company pipelines were actually common carriers in fact. This "private carrier" concept was expanded in the *Champlin II* case to Champlin's 516 mile products line running from its refinery in Enid, Oklahoma, to Hutchinson, Kansas; Superior, Nebraska; and Rock Rapids, Iowa. The right-of-way was purchased without use of eminent domain; Champlin never held itself out as a common carrier, never published tariffs, and had

interstate oil pipelines to public use, was held constitutional by the Supreme Court of the United States in 1914,[1525] In addition, pipeline movements within a state which are part of a continuous flow of oil which has been transported interstate come within FERC's jurisdiction.[1526]

Under the ICA, a common carrier pipeline's rates, rules and regulations must be just and reasonable,[1527] consistent with tariffs required to be filed with FERC prior to commencing the movement,[1528] and be non-discriminatory as between shippers.[1529] Moreover, the line must provide transportation to any shipper upon reasonable request; establish reasonable through (joint) rates with connecting common carrier pipelines and provide equitable division of such joint rates so as not to unduly prefer or prejudice the participating carriers.[1530] Pipelines are expressly barred from giving rebates; or granting any unreasonable preference to, or discriminating in any way between, shippers in the transportation of property, the furnishing of services or the rates that they charge.[1531] The provisions of the ICA requiring compliance with the "long-and-short haul" clause of Section 4, and accounting, reporting and valuation regulations generally are applicable to pipelines.[1532]

Violation of the ICA can be the subject of a proceeding before the Commission, as well as a federal court action, and violators are subject to both fines and imprisonment,[1533] and reparations (damages) to the injured party.[1534] The Commission has broad remedial powers under the ICA: it is empowered to conduct investigations and hearings upon the complaint of

---

never been asked by any other person to use the line. Under these circumstances, the Supreme Court held that Champlin was not a "common carrier" within the meaning of the ICA and did not have to file tariffs. Champlin Refining Company vs. United States, 341 U.S. 290 (1951).

[1525] The Pipe Line Cases, 234 U.S. 548 (1914); Valvoline Oil Company v. United States, 308 U.S. 141 (1939); Champlin Refining Company v. United States, 329 U.S. 29 (1946) (Champlin I — carrier required to file valuation data); *cf.* Schmitt v. War Emergency Pipelines, 175 F.2d. 335 (8th Cir. 1949), *cert. den.*, 338 U.S. 869 (1949).

[1526] Minnelusa Oil Corporation v. Continental Pipe Line Company, 258 I.C.C. 41 (1944 (a 0.56 mile six-inch spur line from Wasatch's refinery in Woods Cross, Utah, to Continental's main interstate line, but Wasatch had participated in joint tariffs and division of rates); Dept. of Defense v. Interstate Storage & Pipeline Company, 353 I.C.C. 397 (Div. 2, 1977). But if the flow of interstate commerce is interrupted, ICC jurisdiction will not attach. Jet Fuel by Pipeline Within the State of Idaho, 311 I.C.C. 439 (Div. 2, 1960).

[1527] 49 U.S.C. §§ 1 (4) and (5) (1976).

[1528] 49 U.S.C. § 6 (1976).

[1529] 49 U.S.C. § 3 (1) (1976).

[1530] 49 U.S.C. § 1 (4) (1976).

[1531] 49 U.S.C. §§ 2, 3 (1), and 41-43 (1976).

[1532] *S.Res.45 Hearings, supra* note 296, pt. 3 at 896 (Statement of Chairman George M. Stafford).

[1533] 49 U.S.C. §§ 6 (10) and 10 (1976).

[1534] 49 U.S.C. §§ 8, 9, 13 and 16 (1976).

a third party[1535] or upon its own motion;[1536] to issue orders approving or invalidating single (called "local") or joint rates;[1537] to determine the division of joint rates among connecting carriers;[1538] and to suspend newly filed rates (including those in initial tariffs)[1539] for up to seven months pending investigation of their lawfulness.[1540]

It is significant that Congress has seen fit *not* to give ICC/FERC certain powers commonly vested in regulatory authorities which deal with public utilities and other franchised monopolies. The most striking example is the exemption by Congress of pipelines from the requirement of obtaining certificates of convenience and necessity prior to the construction of facilities. This has permitted freedom of entry into the business which creates competition, which, in turn, has facilitated the early construction of pipelines in accelerating the development of new fields. It has assisted in the rapid growth of the industry and introduced an element of risk in the oil pipeline which is absent where a business is shielded from competition by the certification process.[1541] Likewise, oil pipelines are not subject to those provisions of Part I of the ICA which deal with the issuance of securities, formation of interlocking directorates, mergers and consolidations, the construction and abandonment of lines or the granting of credit.[1542] Oil pipelines are not subject to the "commodities clause."[1543] These omissions reflect the fact that Congress determined the nature of the industry to be such as to require a less comprehensive range of regulatory devices for the protection of the public interest.[1544] In other words, because of the nature of the oil pipeline industry, Congress chose to rely upon the forces of competition to a greater extent than in any other common carrier industry.[1545]

[1535] 49 U.S.C. § 13 (1) (1976).

[1536] 49 U.S.C. § 13 (2) (1976).

[1537] 49 U.S.C. §§ 15 (1), (3), (6) and (7) (1976).

[1538] 49 U.S.C. § 15 (6) (1976).

[1539] Trans Alaska Pipeline Rate Cases, 436 U.S. 631 (1978).

[1540] 49 U.S.C. § 15 (7) (1976).

[1541] W. J. Williamson, *supra* note 315, at 16.

[1542] *Consumer Energy Act Hearings, supra* note 154, at 1265 (Statement of ICC Chairman George M. Stafford); *S.Res.45 Hearings, supra* note 296, pt. 3 at 896 (Statement of ICC Chairman George M. Stafford); PIPELINE TRANSPORTATION, *supra* note 162, at 48, Bond, *supra* note 207, at 737-738.

[1543] *Id.* See text at notes 1455-1457, *supra* for legislative history of the first consideration by Congress of this question, and 1458-1459, *supra* for subsequent history.

[1544] *S.Res.45 Hearings, supra* note 296, pt. 3 at 896 (Statement of ICC Chairman George M. Stafford).

[1545] Farmers Union Central Exchange v. F.E.R.C., 584 F.2d. 408, 413 (D.C. Cir. 1978), *cert. den.*, 99 S.Ct. 596 (1978); NET-III, *supra* note 199, at 128-129; *cf.* A.G.'s Alaskan Natural Gas Report, *supra* note 1431, at vii.

The differences in the nature of oil pipelines from that of natural gas pipelines, and, even more, from the operation of the "franchised monopoly" utilities, tend to explain the diversity in regulatory approach. Oil pipelines transport for hire, they generally do not own the products they carry, nor do they often buy or sell energy.[1546] Contrast this with the natural gas industry which grew up in a heavily regulated utility atmosphere. At the consumer end, its markets were (and are) served primarily by gas distribution companies operating in franchise-protected markets, usually the only source of a premium fuel in the area. These companies enjoyed, as do the natural gas pipelines today, a constitutionally guaranteed opportunity to earn a fair rate of return on invested capital.[1547] For example, where emergency purchases under the Emergency Natural Gas Act of 1977, which were at prices higher than the area rates, caused additions to the system, the costs thereof were included in the rate base, with the result that customers will pay for the cost of facilities which may never be used after the emergency sale period is over.[1548]

Likewise, pipelines which constructed their lines in optimistic times, anticipating growth, are now charging their non-curtailed customers, as a portion of their natural gas rate, increasing shares of the cost of amortizing the pipeline's rate base—that is, the entire cost of the line is spread over a shrinking volume of gas.[1549] Natural gas pipelines almost always file rates with two components: a demand charge, based on capacity, and designed to recover most of the fixed costs of owning and operating the line (frequently with a minimum bill or a take-or-pay provision), and a commodity charge, which is volume related and is calculated to recover the firm's variable costs. Such a rate structure appears to be precluded by the anti-discrimination provisions of the ICA because an oil pipeline cannot commit its line to a given customer over a long period of time with a minimum "take-or-pay" arrangement, assuring the pipeline that its fixed costs will be covered.

However, it is interesting to note that the Justice Department apparently now believes that a "two-tier" tariff rate structure (whereby those who commit themselves to ship a certain quantity of oil could be charged a lesser rate than non-committed customers) is permissible, at least in the case of the Texas Deepwater Port proposal for state ownership. There is a substantial risk differential between a natural gas pipeline, with a long-term supply contract on one end and a long-term sales contract for delivery at the "city gate" to franchised customers who have a quasi-monopoly and a guaranteed right

---

[1546] *Consumer Energy Act Hearings, supra* note 154, at 610 (Statement of Jack Vickrey); PIPELINE PRIMER, *supra* note 331, at 31.
[1547] W. J. Williamson, *supra* note 315, at 17.
[1548] NET-III, *supra* note 199, at 105-106.
[1549] *Id.* at 120.

to a fair profit at the other end, and an oil pipeline which faces head-to-head, area, and intermodal competition and customers who themselves are engaged in a very competitive business, both on the buying and selling ends.[1550]

The reliance by Congress on competition, ICC "reasonable return on a fair value" regulation and the antitrust laws appear to have been a wise choice. Oil pipeline rate levels, despite persistent inflation, have been quite reasonable;[1551] independent producers' associations have gone on record to state that existing valuation and ratemaking methodology, together with the consent decree dividend limitations, have given the independent producers all of the protection that they need.[1552] Moreover, a vast, interconnected and efficient network of oil pipelines is in existence, giving shippers, owner and nonowner alike, a flexibility and degree of choice unrivaled anywhere in the world.[1553] The ratemaking system has permitted oil pipelines to maintain financial viability;[1554] conduct the safest, most efficient, least expensive, and most environmentally desirable overland method of transporting crude oil and refined petroleum products;[1555] and, in short, provided a climate for oil pipelines to become one of the best run transportation systems in the nation,[1556] and to have done this without government subsidy, in fact, while paying substantial taxes at national, state and local levels.[1557] Moreover, looking back at the past 10 or 30 years, there has been a paucity of challenges to either tariff rates or services.[1558] Quite naturally, to members of the industry[1559] and the ICC,[1560] this indicates that the methods chosen by Congress and their

---

[1550] W. J. Williamson, *supra* note 315, at 18-19.

[1551] Tierney, *TAPS Rebuttal Testimony, supra* note 175, at 5. See Section I B 2, "Anti-Inflationary Rate Trends," notes 188-197, *supra*.

[1552] Allen, *TAPS Rebuttal Testimony, supra* note 496, at 5.

[1553] See Section I B 6, "Flexibility and Alternatives to Shippers," of text at notes 212 to 234, *supra*; Att'y Gen., *Third Report, supra* note 111, at 73-79; LEGRANGE, *Government Regulation, supra* note 1413, at 1.

[1554] Tierney, *TAPS Rebuttal Testimony, supra* note 175, at 20.

[1555] *Petroleum Industry Hearings, supra* note 192, at 256 (Statement of Charles J. Waidelich); see Section 1 B of the text, *supra*.

[1556] *S.Res.45 Hearings, supra* note 296, pt. 3 at 896 (Statement of ICC Chairman George M. Stafford).

[1557] See text at notes 290 and 291, *supra*; LEGRANGE, *Government Regulation, supra* note 1413, at 1.

[1558] *S.Res.45 Hearings, supra* note 296, pt. 3 at 896 (Statement of ICC Chairman George M. Stafford); see text at notes 1502-1510, *supra*; LEGRANGE, *Government Regulation, supra* note 1413, at 3; *cf.* SPECTRE, *supra* note 104, at 114.

[1559] *S.1167 Hearings, supra* note 432, pt. 9 at 606 (Statement of Vernon T. Jones); *Consumer Energy Act Hearings, supra* note 154, at 611 (Statement of Jack Vickrey); STEINGRABER, *supra* note 193, at 136-137.

[1560] *Consumer Energy Act Hearings, supra* note 154, at 1265-1267 (Statement of ICC Chairman George M. Stafford).

implementation by the ICC have proven to be quite adequate. However, critics of the industry have drawn other conclusions. The Kennedy Staff Report stated that "ICC regulation of petroleum pipelines has been woefully inadequate, not simply due to bureaucratic lethargy, but because effective regulation is impossible as long as petroleum pipelines are owned by oil companies."[1561] Later on in the Report, the staff stated "Regulation has failed."[1562] The thrust of the Report, which will be examined in some detail in the latter part of this section, was that the only answer is divorcement legislation. Similar reactions from certain members of Congress were discussed in the preceeding section. Assistant Attorney General Shenefield, characterized ICC Regulation as a "Hear-No-Evil, See-No-Evil" approach; he speculated that the absence of complaints was due to nonowner shippers facing "regulatory indifference" on one hand and the threat of retaliation from the shipper-owners on the other.[1563]

In the meantime, the "pot was stirring" at the Commission. Some 18 cases protesting tariff changes had been filed between January 1, 1969, through February 15, 1974.[1564] Out of these came the Williams (Brothers) Pipe Line Company rate case which, together with the Williams (Brothers) Valuation case and the transfer of jurisdiction from ICC to FERC, have squarely raised the issue of regulatory methodology. Tracing the development, American Petrofina and others protested the Commission's valuation findings in *Williams (Brothers) Pipe Line Company* Valuation Docket No. 1423 (1971 Report) and (1972 Report), and also in 1972, Williams Pipe Line Company filed a new tariff with the Commission which increased its rates 15 percent across the board and initiated joint rates, with Explorer Pipeline, which were 9.5 cents per barrel less than the combined total of the two companies' local rates. Protestants, who had protested the Williams Valuations, also questioned the valuation base in the rate case, urging that "net original cost" be used [meaning the original cost of Great Lakes Pipe Line Company which had sold the line to Williams (Brothers) Pipe Line Company, less accrued depreciation]. The protestants leveled a special barrage at the "Cost of Reproduction New," one of the seven elements considered by the Commission in its determination of pipeline value. Because of the far-reaching consequences of this issue, the Commission, on its own initiative, issued, on August 18,

---

[1561] KENNEDY STAFF REPORT, *supra* note 184, at 4.
[1562] *Id.* at 150. Significantly, the authors of the Report continued: "Antitrust efforts have failed. Congressional oversight has failed." *Id.* One is tempted to ask "failed what;" Failed to find anything wrong, or failed to uncover a problem for which the staff already had a solution?
[1563] SHENEFIELD PROPOSED STATEMENT, *supra* note 179, at 14.
[1564] *S.1167 Hearings, supra* note 432, pt. 9 at 653-654 (Material relating to Testimony of Fred F. Steingraber and Vernon T. Jones).

1974, a Notice of Proposed Rulemaking and Order,[1565] the so-called *Ex Parte No. 308* proceeding, which recited that issues had been raised "...concerning the Commission's valuation process for common carrier pipelines, including the methodology employed; that the issues raised would affect all common carrier pipelines; and that, therefore, for the purpose of permitting interested persons to present their views on whether any modifications are desirable in the rules, regulations and guidelines for pipeline valuation, a proceeding should be instituted. ...."

Specifically, the Commission was interested in views whether there should be any change in its regulations contained in 49 C.F.R. part 1204 entitled "Pipeline Companies List of Instructions and Accounts," in part 1260 of said title 49, entitled "Reporting of Data for Initial Pipeline Valuations," and in part 1261 entitled "Regulations Governing the Reporting of Property Changes; Pipeline Carriers," or "any modification in the methodology of valuation stated in *Ajax Pipe Line Corporation,* 50 Val. Rep. 1."[1566] Notwithstanding this general rulemaking proceeding, the *Williams (Brothers) Pipe Line* rate case continued. The Administrative Law Judge in his initial decision on June 6, 1974, contrary to the protestants' arguments, held "that there is no legal requirement that the same percentage of profit be secured from each type of operation of a transportation company;[1567] that Williams' 5.98 percent rate of return on ICC valuation was not excessive, coming within the 10 percent guideline approved by the Commission in the *Alton Railroad* decision;[1568] [and] that the fair value rate base here meets the end result doctrine of the *Hope Natural Gas Case*." He held further that protestants' contention that if any consideration were to be given to the concept of capital exhaustion, it would have to be based upon the unrecouped balance of the capital originally invested (*i.e. to Great Lakes' original* investment less depreciation), was not sustainable. He noted that the 9.5 cents per barrel (6.5 cents at the time of review by Division 2) disparity between the joint Explorer-Williams tariff and the combination of the local rates of the two pipelines for movement from the Gulf Coast to the Midwest did not support a finding of preference, since the Williams local rate was substantially lower than the joint rates to the same destination. On the basis of these holdings, the Administrative Law Judge rendered his decision in favor of Williams Pipe Line.

Exceptions were filed by protestants, so the case was heard by the Commission's three-Commissioner Division 2. Division 2, in a two-to-one decision,[1569] upheld the Administrative Law Judge on all counts. In reviewing the

---

[1565] 49 C.F.R. Chapter X (1974), service date, September 3, 1974.
[1566] Ajax Pipe Line Corporation, 50 I.C.C. Val. Rep. 1 (1949).
[1567] Relying on No. Pac. Ry v. North Dakota, 236 U.S. 585, 598-599 (1915).
[1568] Petroleum Rail Shippers' Ass'n v. Alton & Southern R.R., 243 I.C.C. 589 (1941).

valuation question, the Division noted that the Commission had permitted Williams to record on its books the price *actually paid* to Great Lakes by Williams for the pipeline and allowed it to accrue depreciation expense based on such purchase price. In fact, the Commission had changed its accounting rules (Uniform System of Accounts) to permit this action. [This privilege was limited to arms-length third party buyers and was denied to purchasers from affiliated companies]. When the Commission did this, said Division 2, it explicitly acknowledged that Williams' purchase price could be used in computing rates of return pursuant to Section 19a of the ICA.

Division 2 went on to note that there were three logical alternative valuation bases: the valuation found by the Commission for Williams in ICC Valuation Docket No. 1423 of $167.6 million; net investment of $287.8 million and net original cost of $101.1 million, which was urged by protestants [derived from *Great Lakes'* original cost, less accrued depreciation of $83.4 million, plus Williams' original cost of $9.7 million of improvements]. The Division said it was obvious that neither Williams nor any other prudent party would invest $293.4 million in property which had a rate base of only $100 million. In fact, protestant's principal witness admitted on the stand that Williams could not service its debt based on a "net original cost" basis. On the other hand, the Division felt that the Docket 1423 valuation rate base produced a reasonable basis upon which to measure the reasonableness of Williams' rates, noting that such valuation fell between the "net original cost" and the "net investment" figures. Noting that there was a "great deal of criticism" directed to the "Cost of Reproduction New" element, which used a 1947 base year and indexed the cost of identical properties to the year in question, the Division remarked that this same indexing technique was used by the Bureau of Labor Statistics in determining the current market value of land and industrial properties, and that the valuation process used by the Commission in deriving Williams' valuation followed the method consistently used by the Commission since the *Ajax Pipe Line* valuation case in 1949, and was recognized by the Commission in its Uniform System of Accounts.

The Division specifically addressed the question whether the *Hope Natural Gas Company* case had rendered obsolete the Commission's "fair value" approach. The Division held that it had not, noting that Section 19a,[1570] the valuation section [which requires that in making valuations, "the Commission *shall* ascertain and report in detail as to each piece of property, other than land, owned or used by said common carrier for its purposes as a common carrier, the original cost to date, *the cost of reproduction new,* the cost of reproduction less depreciation, and an analysis of the methods by

---

[1569] Petroleum Products, Williams Bros. Pipe Line Company, 351 I.C.C. 102 (Div. 2, 1975).
[1570] 49 U.S.C. § 19a (1976).

which these several costs are obtained, and the reason for their differences, if any" (emphasis added)], had not been amended materially since its enactment on March 1, 1913, notwithstanding the *Hope* case.[1571] The Division also commented on the "capital exhaustion" argument of proponents in upholding the Administrative Law Judge, saying "a valuation rate base is a workable method of preventing capital exhaustion. This has become clear in recent years due to the spiralling rate of inflation and can be illustrated by using the Commission's own series of index numbers for pipeline construction," *i.e.* 1966: 179; 1967: 181; 1968: 187; 1969: 191; 1970: 198; 1971: 211; 1972: 222, and 1973: 235. The Division continued, "this means for every $1.79 spent for plant in 1966 it would require $2.35 to buy the same equivalent plant in 1973. In other words, at the end of the eight-year period, $1.79 would have been recouped through depreciation charges and would fall short by $0.56 with respect to the amount required to replace the plant. Since a valuation base reflects to some degree the inflated cost of plant through the reproduction new value, it consequently helps to prevent capital exhaustion and attrition of earnings."[1572]

Having disposed of the instant case, the Division remarked that considerations raised by protestants pertaining to rates of return of pipelines had broad implications affecting the entire pipeline industry, and while it was unnecessary "in this proceeding to resolve these broader issues, the division proposes to recommend to the entire Commission that the Ex Parte No. 308, *Valuation of Common Carrier Pipelines,* be expanded to include the rate of return issue," *i.e.*, whether the 8 percent and 10 percent rate of return on crude petroleum and products, respectively, were still the proper measure of reasonableness.[1573] Thereby, the *Ex Parte No. 308* inquiry was expanded beyond just the question of valuation to include the question of proper rate of return. Division 2's decision, and the Administrative Law Judge's findings, were affirmed, in an opinion filed December 3, 1976, by the full Commission, with one dissenting Commissioner and two not participating.[1574]

In June 1977 seven of the eight TAPS owners filed tariffs with the ICC covering the impending transportation of oil from Prudhoe Bay to Valdez, Alaska. Immediately, the State of Alaska, the Arctic Slope Regional Corporation [one of 13 corporations established pursuant to the Alaska Native Claims Settlement Act],[1575] the Justice Department and the ICC's Bureau of Investigations and Enforcement, all filed formal protests. Because of the im-

---

[1571] Petroleum Products, Williams Bros. Pipe Line Company, 351 I.C.C. 102, 114 (Div. 2, 1975).
[1572] *Id.* at 117.
[1573] *Id.* at 106.
[1574] Petroleum Products, Williams Bros. Pipe Line Co., 355 I.C.C. 479 (1976).

portance of the matter and the obvious need for its expedition, the full Commission,[1576] acting pursuant to Section 15(7) of the ICA,[1577] took the case and by June 28, 1977, issued its decision, finding that it had "reason to believe" the proposed rates were not "just and reasonable;"[1578] and that it had the power to suspend *initial* rates for the full seven month statutory period, based on "probable unlawfulness" and its concern that "maintenance of excessively high rates could act as a deterrent to the use of the line by nonaffiliated oil producers and would also delay the Alaskan interests in obtaining revenues that depend on the well-head price of the oil."[1579]

On the other hand, the Commission found it not to be in the public interest to shut TAPS down for seven months. It made a calculation, based on the carrier's cost data, stated to be its traditional rate of return calculation,[1580] which it said would approximate rates which full investigation would likely reveal to be lawful rates.[1581] In effect, it told the TAPS carriers: "accept this interim rate; file new rates on that basis; and provide for refunds to all affected parties of any amounts in excess over what might finally be found to be reasonable; or shut the line down for seven months." The carriers appealed, on three grounds: (1) that the Commission had no authority to suspend *initial* rates, as opposed to a "change" in rates;[1582] (2) that the Commission could not "set" rates until after a full hearing — a requirement not satisfied by the Commission's summary suspension proceedings; and (3) that the Commission lacked authority to condition a decision not to suspend a tariff on a requirement that the carriers whose tariffs were allowed to go into effect would undertake to make refunds, if such rates were subsequently found to be unlawful. The Court of Appeals for the Fifth Circuit upheld the Com-

---

[1575] 43 U.S.C. §§ 1601-1627 (1976). This one represents the interests of the Inupiat Eskimos who recovered a 2 percent of well-head value interest, up to $500 million, for surrendering their aboriginal land claims in the Prudhoe Bay Area.

[1576] Ordinarily, rate suspension matters, known as I&S cases, go to a staff-suspension board and then to an appellate division of three Commissioners before going to the full Commission. See 49 C.F.R. § 1100.200 (1975).

[1577] 49 U.S.C. § 15 (7) (1976).

[1578] Trans Alaska Pipe Line System, 355 I.C.C. 80, 81 (1977).

[1579] *Id.* at 81-82.

[1580] *Id.* at 85. The Commission used a 10 percent return instead of the traditional 8 percent "in recognition of the extreme risk of the TAPS venture."

[1581] These "interim" rates represented a reduction of between 19 and 26 percent in the filed rates.

[1582] This was a "plain meaning" argument, based on the language of the Mann-Elkins Act of 1910, which added Section 15(7) to the ICA, using the words "any schedule stating a *new* individual or joint rate." (Emphasis added).

mission on all three points, although Judge Roney, in his dissent, said the ICC's action was pure ratemaking and the "carrot technique" could not change that fact.[1583] The Supreme Court granted a stay of the ICC's suspension order, granted certiorari, but on June 6, 1978, affirmed the Court of Appeals on its merits.[1584]

In the meantime, the *Williams Brothers Pipe Line* case was appealed by protestants to the Court of Appeals for the District of Columbia Circuit. A funny thing happened to it on the way to the argument. The respondent ICC was put out of the oil pipeline regulatory business by the Department of Energy Organization Act, which took effect October 1, 1977. Thus, at the time of the argument before the Court of Appeals, ICC was no longer the respondent agency; the respondent party was FERC. At the argument, FERC advised the Court of Appeals that it took no position with respect to the merits and urged the Court to forego adjudication thereon, but instead, to remand the proceedings to FERC so that it could formulate, independently of the ICC, the regulatory principles it found to be suitable to its new responsibilities. The United States, represented by the Department of Justice, which had already revealed its disapproval of the ICC's valuation methods in its Statement of Views and Argument filed in *Ex Parte No. 308,* and in the *TAPS* case, supported FERC's remand request. Thus, Williams Brothers Pipe Line, which had won the battle before the ICC, was left without the usual support furnished by the Government to the party prevailing at the Commission level. The result, described in the text at footnotes 930-939, was a condemnation by the Court of Appeals of the traditional ICC valuation and ratemaking methodology. While one cannot quarrel with the Court's dissatisfaction with the ICC's failure to examine how probative its guideline rates of return, viable in 1940-1948, are in today's climate of continuously spiralling inflation,[1585] there appears to be a serious flaw in the Court's reasoning concerning the basis for the ICC's valuation methodology and the effect that the *Hope Natural Gas* case should have on it. The Court seems mistakenly to have assumed that the ICC's use of "cost of reproduction new" and the other elements used by the ICC in calculating valuation[1586] were based on *Smyth vs. Ames,*[1587] either directly or indirectly because the Valuation Act of 1913[1588] was enacted during a time

---

[1583] Mobil Alaska Pipeline Co. v. United States, 557 F.2d 775 (5th Cir. 1977), *aff'd sub nom.* Trans Alaska Pipeline Rate Cases, 436 U.S. 631 (1978).
[1584] Trans Alaska Pipeline Rate Cases, 436 U.S. 631 (1978).
[1585] See Gary, *FERC Direct Testimony, supra* note 613, at 22 and Attachment V-B where it was shown that in the 1940-1948 period, Long-Term Government Bonds were yielding between 2.1 percent to 2.5 percent per year and triple A Industrials ranged from 2.4 to 2.7 percent, as opposed to 1970-1976 when Governments ended the period at 6.8 percent and the highest rated Industrials at 8.2 percent, more than tripling what they had been when the 8 percent for crude lines and 10 percent for products lines guidelines were laid down by the ICC.

when *Smyth vs. Ames* represented the leading authority on public utility ratemaking.

However, the fact is that the ICC's approach to pipeline valuation is *mandated* by the Valuation Act; *i.e.,* to (1) utilize valuation exclusively as the rate base for pipelines,[1589] and (2) to "ascertain and report in detail as to each piece of property, other than land, owned or used by said common carrier for its purposes as a common carrier, the original cost to date, the cost of reproduction new, the cost of reproduction less depreciation, and an analysis of the methods by which these several costs are obtained, and the reasons for their differences, if any."[1590] The Valuation Act further provides that the final valuation so determined by the Commission shall be *prima facie* evidence of the value of the property in all proceedings under the (ICA).[1591]

The *O'Fallon* case,[1592] cited by the Court of Appeals[1593] is pertinent to the issue because *O'Fallon* held that the evidence of value ascertained by the Commission pursuant to Section 19a is the *exclusive* foundation in carrier rate base consideration, and it held that the ICC had violated its Congressional mandate when it attempted to ignore "Cost of Reproduction New" in the establishing a railroad's rate base because it thought it would result in too high a valuation. The Supreme Court, while sympathetic to the end the Commission was attempting to reach, stated: "No doubt there are some, perhaps many, railroads the ultimate value of which should be placed far below the sum necessary for reproduction. But Congress has directed that values shall be fixed upon a consideration of *present costs* along with all other pertinent facts; *and this mandate must be obeyed.*" (emphasis added)[1594] The Court of Appeals in *Farmers Union*

[1586] For a detailed description of how the ICC arrived at valuation, see Testimony of Jesse C. Oak, formerly with the Valuation Section of the ICC (now with the Department of Energy, FERC) and accompanying Exhibits in FERC Dkt. No. RM 78-2, and his cross examination, *RM 78-2 Hearings Transcript, supra* note 754, Vol. 1 at 13-160. See also Ajax Pipe Line Corporation, 50 I.C.C. Val. Rep. 1 (1949). The clearest shorthand method of expressing the ICC's method is found in SPAVINS, PIPELINE REGULATION, *supra* note 676, Table I at p. 21, which gives the Single Sum Formula: $V = 1.06 [OC (OC/OC + CRN) + CRN (CRN/OC + CRN)] CP + L_1 + L_2 + W$ where V is valuation; OC is Original Cost New; CRN is Reproduction Cost New; CP is "Condition Percent" (a ratio of CRN, less depreciation, to CRN); L_1 is land; L_2 is Right of Way; and W is Working Capital. The 1.06 represents an allowance of 6 percent "going concern" value.

[1587] 169 U.S. 466 (1898).

[1588] 37 Stat 701 (1913), 49 U.S.C. § 19a (1976).

[1589] 49 U.S.C. § 19a (a) (1976).

[1590] 49 U.S.C. § 19a (b) (1976).

[1591] 49 U.S.C. § 19a (i) (1976).

[1592] St. Louis & O'Fallon Ry. v. United States, 279 U.S. 461 (1929).

[1593] Farmers Union Central Exchange v. F.E.R.C., 584 F.2d 408, 414 (1978), *cert. den.*, 99 S.Ct. 596 (1978) [cited on the eighth line, first full paragraph].

[1594] St. Louis & O'Fallon Ry. v. United States, 279 U.S. 461, 487 (1929).

Central Exchange in addition to asserting, on the basis of its erroneous assumption that the ICC valuation methodology was predicated on *Smyth vs. Ames*, that the *Hope Natural Gas* case rendered ICC precedents "products of a bygone era," attempted to bolster its point by adding: "The Commission itself has seen fit to abandon its so-called tradition of valuation computation and ratemaking based thereon in the railroad area, which is equally subject to the Valuation Act."[1595]

Once again, the Court failed to do its homework on the Act's legislative history. The Commission, after *O'Fallon*, requested Congress to amend the valuation process *only* with respect to railroads, because it was having great difficulty in obtaining and processing the required information, and Congress, in response to the ICC's request, in 1933 amended Section 19a(f) under the Emergency Transportation Act of 1933,[1596] giving the ICC discretion in using valuation in the future for existing *railroads* in reviewing their rates.[1597] Moreover, in 1976, Congress reaffirmed its intent *not* to change pipeline valuation and ratemaking methodology when it enacted the Railroad Revitalization and Regulatory Reform Act of 1976.[1598] The House Committee on Interstate and Foreign Commerce stated in its report that: "The Committee has not included oil pipelines in [the new standards of ratemaking], since there does not, at the present time, appear to be a need to do so. Under Part I of the Interstate Commerce Act the tariff rates of interstate commerce carrier oil pipelines are subject to ICC regulation to the same extent as those of railroads. Historically, in determining the reasonableness of the rates of an oil pipeline, the ICC has based such determination upon the annual valuation of the pipeline established by the Commission. Included in the formula used by the Commission in establishing annual valuations of pipeline companies is a factor reflecting the reproduction cost—new of the pipeline system, less depreciation. The use by the ICC of the annual valuation of the pipeline in determining the reasonableness of its rates has permitted the industry to generate common carrier revenues and earnings adequate to engender substantial capital investment in, and needed expansion of,

---

[1595] Farmers Union Central Exchange v. F.E.R.C., 584 F.2d 408, 418 (1978), *cert den.*, 99 S.Ct. 596 (1978).

[1596] 49 Stat. 221 (1933), 49 U.S.C. § 19a (f) (1976).

[1597] See the use of the word "may" in connection with *railroad properties,* and Ex Parte No. 271, Net Investment-Railroad Rate Base and Rate of Return, 345 I.C.C. 1492, 1517 (1976) which quoted the permissive language of the amendment for the grant of discretion to the ICC in *railroad cases.* See also the testimony of Commission witness Jesse C. Oak on cross-examination in the FERC Valuation of Common Carrier Pipelines proceedings, *RM 78-2 Hearings Transcript, supra* note 754, Vol. 1 at 93-98, that the ICC had interpreted the 1933 amendment to Section 19a as making no change whatsoever in valuation procedures with respect to pipelines.

[1598] P.L. 94-210, 90 Stat 31. (1976).

the nation's oil pipeline network."[1599]

In transferring jurisdiction over pipelines from ICC to FERC, the Congressional intent not to change ICC methodology can reasonably be inferred from the Ribicoff-Hansen dialogue, which was placed in the Congressional Record as a memorial of understanding before Senator Hansen would go along with the DOE Organization Act.[1600] Summing up, it appears that the Court of Appeals' great familiarity with FPC type regulation[1601] led it to overlook the marked differences between oil pipelines and natural gas pipelines; to assume that ICC pipeline ratemaking authority rested on *Smyth vs. Ames* instead of Section 19a of the ICA; and, therefore, mistakenly to conclude that the *Hope Natural Gas* controlled the issue.[1602]

Because the *Williams Pipe Line* case is pending, it would not be appropriate to debate the merits in this text. However, it would be remiss not to refer to an earlier section of the text on the subject of "Economic Characteristics" of pipelines which discussed precedents in this area and recorded the unfavorable reaction of independent producers, the financial community and others to the prospects of application of Natural Gas Act methodology to Interstate Commerce Act carriers[1603] and to note the dismal result which has been produced by such methodology.[1604] Departing from ICC methodology in the case of railroads, under the authority of Congressional change in the Valuation Section because of regulatory cost, and the accompanying regulation of the railroads,

---

[1599] H.R. Rep No. 93-1381, 93d Congress, 2d Sess. 26 n.1 (1974).

[1600] See text at note 960 and note 936, *supra*.

[1601] Protestants in Natural Gas Act cases where the FPC decision is in favor of producers or natural gas carriers are permitted by the Natural Gas Act to lodge their appeals only with the Court of Appeals for the District of Columbia Circuit; hence, the Court hears a substantial number of such cases.

[1602] It should be noted that the *Hope Natural Gas* case did *not* overrule *O'Fallon*. The holding in *Hope* simply was that a valuation rate base was required neither by the Constitution nor by the Natural Gas Act; it did not confer upon the ICC any discretion to disregard the mandate of ICA Section 19a. Nor has this analysis on the text of Congressional intent overlooked Congress' action in repealing Section 15(a)(4) in its enactment of the "Emergency Railroad Transportation Act, 1933." The very title of the Act reveals its tenor. Section 15(a)(4), rather than "freeing the ICC from the requirements of *O'Fallon*," simply deleted what was a section made surplus by its specific reference to "the value of railway property." H.R. Rep. No. 193, 73d Cong., 1st Sess. 30 (1933) cannot properly be cited for dispensation of *O'Fallon*. The true purpose of the enactment addressed Sections (f) and (g) of 19a to relieve the ICC of the requirement to make reports to Congress, at each regular session thereof, of the corrected valuations of *railroad properties;* but simply to keep itself informed as to new construction, extensions, improvements and other changes of such properties. It was a cost-saving measure, requested by the ICC.

[1603] See text at notes 921-927 and 945-954, *supra*.

[1604] *See e.g., S.1167 Hearings, supra* note 432, pt. 9 at 609 (Testimony of Vernon T. Jones); A.G.'s Alaska Natural Gas Report, *supra* note 1431, at 16; Allen, *TAPS Rebuttal Testimony, supra* note 496, at 6, 14-15.

disabled them entirely from competing, with well-known results.

Regulation under the Natural Gas Act using the methodology which has been proposed for oil pipelines, together with the accompanying regulation of virtually every action of the transmission companies and the independent producers made subject to this regulation by the *Phillips Petroleum case*,[1605] produced another fiasco. Common sense and logic suggest that application of FPC-type regulation to even as robust an industry as oil pipelines will produce an enervating influence, which one witness analogized to the security of a prison or the warmth of a bear's hug,[1606] and, inevitably, stagnation will set in. To repeat the same mistake three times in a row in the light of the evidence adduced in the *RM78-2* and *TAPS* proceedings would seem unwisely to ignore the lessons of history.

2. State Regulation

State regulation of pipelines goes back to the early days of the industry when railroads sought to preserve their pre-eminent position by refusing to grant rights-of-way across railroad lands.[1607] In order to overcome this obstacle, pipelines sought the power of eminent domain, which was granted by New York in 1878, followed by Pennsylvania in 1883 and by 1906 ten states[1608] had conferred the right of eminent domain [power to condemn] on pipelines.[1609] Most of these laws, and those which followed, imposed a concomitant duty to act as common carriers.[1610] By 1931, 21 states had laws on the subject of common carrier responsibilities.[1611] Courts and legislatures were rather ingenious in finding a basis for fastening the mantle of common carriage on pipelines. Any voluntary dedication to public use, even the carriage of oil for others as a favor; public grants, such as disposition of public lands, leasing of public lands for exploration and production, or permits to cross public property; and general business legislation, such as domestication of foreign corporations have all been used to find or impose common carrier respon-

---

[1605] Phillips Petroleum Co. v. Wisconsin, 347 U.S. 672 (1954), *reh. den.*, 348 U.S. 851 (1954).
[1606] Ryan, *TAPS Rebuttal Testimony, supra* note 724, at 60-61.
[1607] See text at note 73, *supra*; KENNEDY STAFF REPORT, *supra* note 184, at 113.
[1608] California, Colorado, Indiana, Kentucky, New York, Ohio, Pennsylvania, Texas, West Virginia and Wyoming.
[1609] PETROLEUM PIPELINES, *supra* note 45, at 21; KENNEDY STAFF REPORT, *supra*, note 184, at 113.
[1610] PETROLEUM PIPELINES, *supra* note 45, at 21; WOLBERT, *supra* note 1, at 115-120; *cf.* BEARD, *supra* note 45, at 39-40.
[1611] KENNEDY STAFF REPORT, *supra* note 184, at 114.

sibilities on pipelines.[1612]

"Common purchaser" laws abound in many oil producing states. The thrust of these laws is to require purchasers of crude oil to purchase ratably from among all producers and not to discriminate in favor of one producer as against another.[1613] While once extremely important, the decline in production and the increase in demand have so changed the producer's position vis-a-vis prospective purchasers that the regulators are beginning to turn this idea around, or at least the thrust of the entitlements program would give that appearance.

The Kennedy Staff Report makes light of state regulation stating, "Although State regulation is intended to assist the independent producer and shipper, the effectiveness of such regulation is open to doubt. With little or no rate regulation, and little or no enforcement or [sic] existing laws, intrastate pipelines appear to be free to operate as they see fit with little active oversight by state regulatory agencies."[1614] This statement does not appear to be supported by empirical evidence. For one thing, many pipelines transporting oil from one point in a state to another point in the same state may well be within the jurisdiction of FERC.[1615] The states almost invariably will not permit higher intrastate rates between the same points covered by an interstate rate.[1616] Frequently, the Rules and Regulations established by the pertinent state regulatory agency, [in Texas, the Texas Railroad Commission] are even more stringent than are FERC's.

For example, in Texas if the oil tendered for shipment does not differ materially in character from that usually produced in the field and being transported therefrom by the pipeline, the minimum tender is 500 barrels or one tank carload, as opposed to FERC's 10,000 barrel minimum.[1617] Where segregated shipments are desired by the shipper or required by the difference in quality of his shipment from the stream going through the pipeline, a 500 barrel minimum is impracticable. This is usually the case in

---

[1612] BEARD, *supra* note 45, at 30-45; WOLBERT, *supra* note 1, at 114-117; KENNEDY STAFF REPORT, *supra* note 184, at 114.

[1613] See text at notes 352-359, *supra*.

[1614] KENNEDY STAFF REPORT, *supra* note 184, at 115.

[1615] See text at note 1526, *supra*. For example, Exxon Pipeline Comapny, Local Tariff, F.E.R.C. No. 130, effective March 1, 1978, covers crude oil gathering and trunk line rates from fields in Ector, Crane, Gaines, Andrews, Reagan, Crockett, Winkler and Pecos counties in West Texas to Baytown, Texas.

[1616] Identical intrastate and interstate rates are the norm. Comparison of the rates in Exxon Pipeline Company Crude Petroleum Local Tariff, Texas R.R.C. No. 148, effective February 1, 1978, with the rates in its F.E.R.C. No. 130, effective March 1, 1978, between the same origins and destinations shows them to be identical. If the states did permit differences in rates which affected interstate commerce, FERC could step in. Houston, East & West Texas Ry. and Houston & Shreveport R.R. v. United States, 234 U.S. 342 (1914) ("Shreveport Rate Cases").

interstate shipments even though the origin and destinations may correspond to intrastate points in certain instances.[1618]

Sometimes this problem is handled by providing that orders for individual shipments must be of sufficient size or sufficient buffers furnished (with the non-conforming shipper accepting the interface contamination) as will protect the quality of other shippers' products.[1619] Usually, some administratively convenient minimum, patterned after the FERC's rule of 10,000 barrels, is used, pending a more explicit showing that a greater or lesser quantity can be handled without undue contamination problems.[1620] There are other differences; for example, in the Texas intrastate tariffs, there is a provision for pipeline storage at destination for five days.[1621] While this discussion by no means purports to be exhaustive, it should indicate that to a greater or lesser degree, depending on the importance of oil to the State, the length of time the State has been an important oil producing state and upon its political constituencies, there appears to be a significant degree of state regulation of oil pipelines. By way of example, New Mexico and Louisiana recently

---

[1617] *See e.g.,* Exxon Pipeline Company Crude Local and Joint Tariff (with Gulf Refining, Shell Pipe Line, Sun Pipe Line, Texas-New Mexico, The Texas Pipe Line and West Texas Gulf Pipe Line, as participating carriers) Texas R.R.C. No. 147, effective February 1, 1978) (Item 7); Shell Pipe Line Corporation, Texas Local Tariff No. 678, effective September 1, 1977 (Item 7); Crown-Rancho Pipe Line Corporation, Texas Joint Tariff No. 43 (in connection with Shell Pipe Line Corporation) effective August 1, 1978 (Item No. 7); Gulf Refining Company, Joint Tariff R.C.T. No. 451 (with Chevron Pipe Line Co. and West Texas Gulf Pipe Line Co.), effective August 1, 1978 (Item No. 7); Marathon Pipe Line Co., Proportional Local Tariff, Texas R.R.C. No. 16, effective December 29, 1975 (Item 7). See text at notes 499-501, *supra* for the promulgation of the FERC 10,000 barrel minimum tender.

[1618] Compare Exxon's Texas R.R.C. 147 which has a 500 barrel minimum with its interstate tariff which has similar origin and destinations, F.E.R.C. No. 129, effective March 1, 1978 (Item 20: 10,000 barrels). *But see* Exxon Pipeline Co., Local and Proportional Tariff, F.E.R.C. No. 143, effective November 10, 1978, which has no minimum because it is operated as a common stream line.

[1619] Marathon Pipe Line Company, Local Tariff, Ind. Rate Sheet No. 11, effective April 1, 1975 (Item 3).

[1620] *See e.g.,* Exxon Pipeline Co., Local Tariff, Montana, P.S.C. No. 3, effective November 1, 1974 (Item 4: 10,000 barrels); Marathon Pipe Line Co., Joint Tariff, Wyoming P.S.C. No. 75, effective July 1, 1978 (Item 2: 10,000 barrels); Shell Pipe Line Corporation, Local Proportional Tariff, Michigan Rate Sheet No. 31, effective October 1, 1977 (Item 30: 10,000 barrels); Shell Pipe Line Corporation, Local Tariff, O.C.C. No. 16, effective October 1, 1977 (Item 30: 10,000 barrels); *but see* Gulf Refining Co., Local Tariff, L.P.S.C. No. 77, effective September 15, 1975 (Item 2: no minimum tender where crude of same quality and characteristics are being moved through the line. If crude of different characteristics is desired to be shipped, then the transportation will be under such terms as the shipper and the pipeline may agree.)

[1621] Marathon Pipe Line Co., Proportional Local Tariff, Texas R.R.C. No. 16, effective December 29, 1975 (Item 5); Exxon Pipeline Co., Local and Joint Tariff, Texas R.R.C. No. 147, effective February 1, 1978 (Item 5); Shell Pipe Line Corporation, Texas Local Tariff No. 678, effective September 1, 1977 (Item 5); Gulf Refining Co., Joint Tariff, R.C.T. No. 541, effective August 1, 1978 (Item 5); Crown-Rancho Pipe Line Corporation, Texas Joint Tariff No. 43, effective August 1, 1978 (Item 5).

have adopted the practice of requiring a hearing on any pipeline tariff change.

3. Department of Justice

Section III G 1 b of this text, entitled "In-Place Enforcement Agency Changes in Antitrust Theory," contains a description of Justice Department action involving pipelines in the *API* or *"Mother Hubbard"* case; the three Elkins Act cases which culminated in the *Atlantic Refining Company* Consent Decree; and the abortive attempt to rewrite the consent decree in the *Arapahoe* case.[1622] After World War II, the *"Mother Hubbard"* case, which antitrust experts doubted ever was a "triable case,"[1623] had become so obscured by the many changes in industry ownership and conditions during its wartime suspension that the Antitrust Division abandoned the suit and it was dismissed in June 1951. In its place, the Justice Department filed a series of "segment suits," designed to deal, on a manageable basis, with a number of the issues raised in the *Mother Hubbard* case. The most important of these, and the only one which involved pipelines, was the so-called *West Coast* case,[1624] which charged Socal and six other major Pacific Coast oil companies with conspiracy to monopolize and to suppress competition at all levels of operation. The relief requested was an injunction against continuation of all the specifically alleged wrongdoings and divorcement of marketing facilities. At a pretrial hearing, Judge Carter announced that he would not order divestiture even if the Government proved a Sherman Act violation. As a result, the Government commenced negotiations with the defendants. In June 1959, a settlement was reached with all of the defendants except Texaco, and a consent decree, which had a 15-year term, prohibited a variety of behavior which the Government alleged to be illegal and the defendants claimed they weren't doing anyway. Relevant to this discussion is the fact that the decree did *not* create a common carrier pipeline system in California (where the ICA is not applicable). The suit against Texaco was dropped.

The Department has been investigating jointly owned and jointly operated pipeline operations since 1963, when the Federal Trade Commission transferred its investigation of Colonial, begun in 1962, to the

---

[1622] See text at notes 967-996, *supra*.

[1623] *H.Res.5 Subcommittee Hearings, supra* note 234, at 69 (Testimony of Robert L. Wright, Antitrust Lawyer for 40 years, 15 of which were with the Antitrust division); *cf.* Att'y Gen., *Third Report, supra* note 111, at 67.

[1624] United States v. Standard Oil Co., of California, Civil No 11584-C, S.D.Calif., May 12, 1950.

Department.[1625] The Antitrust Division, using compulsory process (Civil Investigative Demands, generally called "CID's") under the Antitrust Civil Process Act, began an extensive investigation of the formation of Colonial. Substantial further investigation was undertaken in 1966[1626] to update the earlier reports in the context of actual operation of the pipeline.[1627] According to the Kennedy Staff Report, the matter bounced around the Division, went up to the Attorney General, was farmed out to outside consultants, was massaged again within the Department where the emphasis was directed to the question of whether nonowner shippers were being granted access on reasonable terms, and, after an updating of its information on this question, the Antitrust Division found that there were no meaningful denials of access, so the Department closed its investigation in 1976, 13 years after it had commenced its examination.[1628] At one time or another, investigations, usually employing CID's, were conducted by the Justice Department into the proposed formation or the operation of Olympic, Gateway-MATCH, Explorer, Glacier, and TAPS. In the cases of Glacier and MATCH, which involved proposals by a combination of companies to acquire pipelines already in existence, the Department took a firm position against their legality and the proposals were abandoned. In none of the other situations were cases filed, although the testimony indicates that there were some staff members who strongly favored bringing a suit. One situation even got as far as to the Deputy Assistant Attorney General level before it was turned down.[1629]

The Justice Department has been building on its knowledge about

---

[1625] KENNEDY STAFF REPORT, *supra* note 184, at 125-126; *H.Res.5 Subcomm. Hearings, supra* note 234, at 204-205 (Statement of Bruce B. Wilson, Deputy Assistant Attorney General, Antitrust Division); Perhaps some incentive might have been added by the American Maritime Association's letter and brief sent to Attorney General Robert Kennedy on April 10, 1962 (reprinted together with Press Release in *H.Res.5 Subcomm. Hearings, supra* note 234, at 59-64); a similar communication was sent to Mr. Kennedy on April 11, 1963 by the AFL-CIO Industrial Union Department, joined by the NMU, OCAW, AFL-CIO Maritime Committee and the Industrial Union of Marine and Shipbuilding Workers of America.

[1626] Shortly before the "second round" of investigation, the Justice Department was the recipient of another broadside, this time directed to Attorney General Nicholas DeB. Katzenbach, from the same five union complainants which had sent the earlier missive. See reprint of same in *H. Res.5 Subcomm. Hearings, supra* note 234, at 40-42.

[1627] *Id.* at 205 (Statement of Bruce B. Wilson).

[1628] KENNEDY STAFF REPORT, *supra* note 184, at 125-128. This apparently embittered (William) John Lamont, who had worked on the investigation for many years before leaving the Division in 1972, and presumably felt that a complaint should issue. He attributed the failure not to bring suit to "other reasons [which may appear] in this year of Watergate." *Consumer Energy Act Hearings, supra* note 154, at 675-676.

[1629] *H.Res.5 Subcomm. Hearings, supra* note 234, at 127-140 (Statement of Beverly Moore, Corporate Accountability Research Group); *Id.* at 203-209 (Statement of Bruce B. Wilson, Deputy Assistant Attorney General); *Consumer Energy Act Hearings, supra* note 154, at 1029-1031 (Statement of Keith Clearwaters, Deputy Assistant Attorney General); KENNEDY STAFF REPORT, *supra* note 184, at 125-132.

pipelines through its advisory duties in connection with other phases of the oil industry and other energy industries.

In its Deepwater Port Report, the Department had an opportunity to observe the process of how jointly owned projects are put together. While members of the industry may feel, with some justification, that the Department's recommendations respecting its LOOP and SEADOCK licenses were too abstract, took too little account of the economic realities of the situation, and that the proposed "competitive rules" were ill-fitted to the real world, cognizance must be taken of the difficulty under which the Department operates — it is always an "outsider" and despite its CID powers, it never has the opportunity to become acquainted with the facts and to recognize their significance to the business in a manner comparable to that of the members of the industry. Hence, the Department is virtually compelled by necessity to theorize and to rely upon abstract reasoning in formulating its approach, leaving it up to industry to step up and tell them why its proposals won't work so that it can reconsider its rationale with the benefit of additional insight.

Both the Department and the industry can benefit from the type of dialogue that took place at the American Enterprise Institute's Conference on Oil Pipelines and Public Policy in Washington on March 12, 1979. Moreover, the lessons learned by the Department from its investigations which produced the *Alaska Natural Gas Report,* the so-called "SOHIO Report,"[1630] and even to a limited extent, the "Coal Report,"[1631] plus the repercussions caused by the submersion of SEADOCK, have sensitized the Department to the complexity of relationships in the oil pipeline industry and, while still a long way from producing workable solutions, the Department's more recent pronouncements disclose a definite increase in awareness of the complexity of the problems and a willingness to approach their solution with a much more pragmatic attitude than they were able to do at the time of the Deepwater Port examination.

4. Federal Trade Commission

On July 17, 1973, the Federal Trade Commission issued a complaint against the eight largest integrated companies in the industry.[1632] Because

---

[1630] D.J., SOHIO REPORT, *supra* note 633.

[1631] REPORT OF THE U.S. DEPARTMENT OF JUSTICE PURSUANT TO SECTION 8 OF THE FEDERAL COAL LEASING AMENDMENTS ACT OF 1975, May 1978 [hereinafter cited as DJ COAL REPORT].

[1632] In the Matter of Exxon Corporation, et al., F.T.C. Dkt. No. 8934 (July 1973).

respondents were listed in descending order of size, Exxon was the first named company, hence the case is frequently referred to as the *Exxon* case. The Complaint was preceded by a formal investigation, initiated in 1971 and apparently not fully completed before the case was instituted. The Complaint, issued under Section 5 of the Federal Trade Commission Act,[1633] alleged that the respondent "eight majors" had engaged in unlawful concert of action and individual activities in restraint of trade, all of which were monopolistic in purpose (particularly as to refining) and destructive as to effect on the "independents." The Complaint alleged that such acts were made possible by what it contended was a "structural defect" in the industry, namely, vertical integration of the majors; excessive cooperation (interdependence) among the majors, and too high a concentration of the industry's facilities and reserves in the hands of the majors. Although the Complaint did not request any particular relief,[1634] FTC Counsel Supporting the Complaint stated that their suggested remedies included divestiture of 40-60 percent of Respondents' refining capacity with the formation of 10 to 13 new firms; divestiture of "some" crude and products pipelines; a ban on future acquisitions of refineries by Respondents, and the prohibition of "some" joint ventures, processing arrangements, and crude and product exchanges. As the Kennedy staff report described the status of the case, "Five years following the filing of the initial complaint, the proceedings are still mired down in procedural and discovery problems. Respondents have attacked the Commission's authority to consider pipeline issues in their proceedings."[1635]

A substantial portion of the problem arose because of the premature filing of the complaint. While this action served to alleviate the political pressure on the Commission, it left Counsel supporting the Complaint in the unenviable position of not having enough information (which usually is available as a result of a full pre-filing investigation, which has the broadest kind of information-requiring and access powers)[1636] to proceed in conformity with the FTC's own rules of practice governing litigated proceedings. After being rebuffed in an attempt to launch a massive, omnibus discovery without defining the issues so that Respondents would know how to deal with the discovery request, Staff Counsel filed a motion with the Commission asserting that the FTC's Rules of Practice had crippled their efforts. They requested a change in the FTC's own rules — to reflect the Federal Manual for Complex Litigation [which deals with

---

[1633] 15 U.S.C. § 45 (1976) which provides that "Unfair methods of competition in commerce and unfair or deceptive acts or practices in commerce, are declared unlawful."

[1634] It is reported that the tie-breaking vote of the Commission required to issue the complaint was secured by an undertaking *not* to ask for divorcement.

[1635] KENNEDY STAFF REPORT, *supra* note 184, at 133.

[1636] 49 U.S.C. §§ 46 (b) and 49 (1976).

situations where there has been no comparable ability to obtain pre-trial information] and they sought the appointment of a different Administrative Law Judge.[1637]

It has been extremely difficult for Respondents to come to grips with exactly what is being alleged. Their first attempt to ascertain just exactly what they were being charged with was answered by a document that appeared to be a listing of all the conclusionary statements issued in Congressional Committee reports dating back to the *TNEC Hearings*. There was clearly a charge of "squeezing," adapting the DE Chazeau & Kahn thesis, which was discussed in Section IV B 3 of this text entitled, "Reasons for Vertical Integration."[1638]

Without questioning the validity or spurious nature of the charges in the original framework from which they emerged, it does appear difficult to see their relevance in today's situation. As Professor Richard Mancke noted in his article *Competition in the Oil Industry*,[1639] "In an unprecedented move, the FTC judge hearing the [*Exxon*, et al.] case issued a brief (October 1975) arguing that, because of changed circumstances since the charge was brought (especially OPEC's success at raising world oil prices and the abolition of both oil import quotas and the oil depletion allowance), the charge should be withdrawn [and further investigation could take place]. The full Commission ruled against this suggestion."

It would be extremely helpful for disinterested observers to have available to them, in forming their own opinions on the matter, the full report of a panel of distinguished economists the Commission retained to examine the complaint to see whether, as alleged, it was supportable from an economic standpoint, and, if not, whether there was any economic rationale which would support a modified complaint. At any rate, the case record consisted of more than 60,000 pages of proceedings with another 10,000 pages of non-substantive depositions and interviews even before any substantive discovery started. Complaint Counsel's schedule for its proposed discovery and pretrial preparation would appear to indicate that trial is not expected to start before 1985 or to be completed earlier than 1987. The Kennedy Staff Report contains an estimate that trial at the FTC level and ensuing appeals will not conclude before the early 1990's.[1640]

In January 1979, Senator Edward M. Kennedy filed with the FTC a petition requesting the Commission to initiate a rulemaking proceeding for the purpose of issuing "a trade regulation rule declaring it to be an unfair trade practice for an oil company to have any ownership interest in

---

[1637] Ikard, *Petroleum Industry Competition, supra* note 1055, at 583.
[1638] See text at notes 1167-1175, *supra*.
[1639] MANCKE, *Oil Industry Competition, supra* note 674, at 64 n.38.
[1640] KENNEDY STAFF REPORT, *supra* note 184, at 132-133.

petroleum pipelines of certain classes." The Petition recited that under Section 6 (g) of the FTC Act, the Commission has the authority to issue rules defining unfair methods of competition, infractions of which would constitute violations of Section 5 of the FTC Act,[1641] and that rulemaking "...is especially appropriate where an elaborate factual record need not be developed," because "[t]o a large extent, the issues raised by this petition do not require an evidentiary hearing" but "...are more matters of law and public policy than questions of fact," and hence "...rulemaking rather than adjudication is the most efficient way to address this question."[1642]

At the outset, there are substantial questions concerning the jurisdiction of the FTC to order massive divestiture of an industry as large and important as the pipeline industry on the asserted basis of public policy,[1643] and the constitutionality of an irrebutable presumption of illegality from the mere existence of vertical integration. These issues are being addressed by the comments of affected parties. However, suffice it here to say that basing a plea for a sweeping divestiture order on the authority of the *National Petroleum Refiners' Association* case, which, incidentally, did not reach the question of whether Section 6 (g) authorized the promulgation of substantive rules, particularly in the antitrust area,[1644] is somewhat akin to attempting to transport the Prudhoe Bay production to Valdez through the use of ¼-inch copper tubing. Senator Kennedy's petition in its "Conclusion" would substitute for the painstaking effort and evidentiary factual hearings that must underpin a divestiture order, the *ipse dixit* declaration that, "Since 1906, there have been numerous studies, investigations and reports on the question of pipeline ownership . . . . virtually all have called for divorcing ownership of pipelines from the shipper function." [Citing the Kennedy Staff Report]. These reports were directed to Congress which has resolved repeatedly the public policy issue against divorcement of pipelines. The FTC presently is engaged, in

---

[1641] United States of America, Before Federal Trade Commission, Petition for the Initiation of a Rulemaking Proceeding Prohibiting Ownership of Petroleum Pipelines by Petroleum Companies, filed January 4, 1979, p. 2 [hereinafter cited as Kennedy FTC Rulemaking Petition], citing National Petroleum Refiners' Ass'n v. FTC, 482 F.2d 672 (D.C. Cir. 1973), *cert. den.,* 415 U.S. 951 (1974) ("Octane Ratings Case").

[1642] Kennedy FTC Rulemaking Petition, *supra* note 1641, at 2, 10.

[1643] FTC v. Eastman Kodak Co., 274 U.S. 619, 623, 625 (1927).

[1644] Senator Kennedy's attempt to sweep divestiture for antitrust purposes within the ambit of Section 6(g) rests on a citation of FCC v. National Citizens Committee for Broadcasting, 436 U.S. 775 (1978). This reliance ignores completely the difference between the FTC's limited mandate to protect the consumer from fraud and deception, and the pervasive regulatory authority of the FCC to allocate the limited number of radio and television freqencies available through the specifically granted power to grant and renew licenses under a legislative standard contained in the Federal Communications Act.

the *Exxon* case, in attempting to develop the very kind of information that is the minimum requirement before divestiture could be ordered, and then questionably, and only against the respondents before it. The other string to Senator Kennedy's bow was the statement that "Reinforcing this mass of accumulated expertise [note, not "evidence"] is the overwhelmingly persuasive analysis done by the Department of Justice." After reading sections B 6 and C 3 and 4 of this chapter, the reader can determine for himself the validity of the Senator's "Conclusions" in the petition.

The FTC has gone on record clearly disclaiming its rulemaking authority in the antitrust area. For example, during the debate on the Magnuson-Moss Federal Trade Commission Improvement Act, FTC Chairman Lewis Engman was asked point-blank by Congressman James T. Broyhill of North Carolina (who was a Conferee on the bill), "Do you feel that rulemaking authority is an appropriate method of dealing with so-called antitrust matters?" Chairman Engman replied: 'I think that it becomes very difficult to attempt to promulgate a rule which would apply in restraint of trade cases. The Commission in the past has generally felt that this could pose a problem, as opposed to the consumer protection area, where deceptive practices, whether in advertising ["F&M" cases] or in other areas, are more susceptible of being dealt with on a broader basis." Congressman Broyhill pressed harder: "If your rulemaking authority is upheld by the Court [in the *Refiners' Association* case], would you then contemplate using this authority in the antitrust area?" To which, Chairman Engman replied: "I would not, no."[1645]

Not only has the FTC not promulgated any substantive antitrust rules in the six years since the *Petroleum Refiners' Association* case, but as late as October 1978, Alfred Dougherty, Director, Bureau of Competition, FTC, was quoted as saying that he believed the possibilities for antitrust rulemaking to be quite limited and that there was enough case-by-case enforcement to be done that the FTC's resources should be spent bringing cases.[1646] Surely, its resources must be strained already by the *Exxon* case without taking on this additional burden.

[1645] *Hearings on H.R. 20 & H.R. 5201 before the Subcomm. on Commerce and Finance of the House Comm. on Interstate and Foreign Commerce*, 93d Cong., 1st Sess 86 (1973); *See* 120 Cong. Rec. S. 21,978 (daily ed. Dec. 18, 1974) (Sen. Hart, another Conferee, speaking: "Because S.356 primarily concerns consumer protection matters, the procedural requirements in title 2 respecting FTC rulemaking are limited to unfair or deceptive acts or practices rules") and 120 Cong. Rec. H 12,348 (daily ed. Dec. 19, 1974) (Congressman James Broyhill speaking: "We have made clear that the new bill does not deal with the antitrust laws.")

[1646] 884 ANTITRUST & TRADE REG. REP. (BNA) A-15 (Oct. 12, 1978).

5. States' Attorneys-General MDL-150 Cases

Shortly after the FTC brought the *Exxon et al* case, the Attorneys-General of Florida, Connecticut, Kansas, California, Arizona, Oregon and Washington filed actions against the leading major oil companies usually doing business in their respective states which challenged existing business practices in the petroleum industry and alleged that the companies conspired to restrict supplies and to control prices of oil and oil products in violation of state and federal antitrust laws. These cases seek damages and injunctive relief on behalf of the states and their local subdivisions and residents or consumers affected by the alleged restrictive practices.[1647] After jousting with Attorney-General Shevlin of Florida concerning his authority under the Florida statutes to bring such action and some dismissals of parties defendant in certain of the states wherein they did not do any business, the cases were consolidated in the United States District Court for the Central District of California for pre-trial discovery under the multi-district panel procedures of the Federal Rules. The defendants brought into the consolidation the *City of Long Beach* suit which has the interesting aspect of the State of California, which has a pecuniary interest in Long Beach oil production, urging a position diametrically opposite to that taken by it in its principal suit brought by its Attorney-General. The cases, under the management of Judge Gray in Los Angeles, are collectively referred to as Multi-District Litigation-Docket No. 150 (or by its shorthand nomenclature of MDL-150).

During February 1977, all plaintiffs except Long Beach filed amended complaints conforming generally with that of the State of California's. All state plaintiffs allege that defendants have: (1) combined unreasonably to restrain trade and to monopolize trade by their vertically integrated structure and their many horizontal combinations and agreements; (2) conspired to create an artificial scarcity of crude oil and refined products; and, (3) violated price controls and crude oil "entitlements" under the FEA regulations. Some state plaintiffs allege also that defendants have conspired to raise or stabilize prices for refined products and have fraudulently concealed such conduct. Virtually every

---

[1647] State of Florida v. Exxon, et al., Eq. No. 730380, D.C.Fla., July 9, 1973; State of Connecticut v. Amerada Hess, et al., Eq. No. 730420, D.C.Conn., July 25, 1973; State of Kansas v. Exxon, et al., Eq. No. 740535, D. Kan., Oct. 8, 1974; State of California v. Standard Oil Co. of California, et al., Eq. No. 750372, D.C.Calif. June 25, 1975; State of Arizona v. Standard Oil Co. of California, et al., Eq. No. 760422, D.C.Calif., July 22, 1976; State of Oregon v. Standard Oil Co. of California, et al., Eq. No. 770115, D.C.Calif., Feb. 22, 1977; State of Washington v. Standard Oil Co. of Calif., Eq. No. 720487, D.C.Wash., Aug. 15, 1977; City of Long Beach, et al. v. Standard Oil Co. of California, et al., Eq. No. 750378, D.C.Calif., June 27, 1975.

aspect of operation in a vertically integrated oil company is attacked as a violation of the antitrust laws.

Each state asks for treble damages and injunctive relief: (a) in behalf of the state in its proprietary capacity; (b) as *parens patriae* for its citizens; and (c) as class representative for a class consisting of its political subdivisions and public entities, as well as a class of private consumers resident in the state. Only Florida, Connecticut and Kansas specifically ask for divestiture of exploration and production functions. Connecticut and Kansas also seek pipeline divorcement.

Hearings on motions so far have: (1) resulted in denial of defendants' motion challenging the constitutional application of *parens patriae* legislation [which was enacted to overturn the holding of *Hawaii vs. Standard Oil Company*];[1648] (2) required plaintiff states to detail their allegations of fraudulent concealment or suffer the allegation to be stricken from the complaints; (3) granted defendant's motions to strike plaintiffs' claims that defendants conspired to control access to petroleum pipelines; (4) denied, for the time being, plaintiffs' motions to bifurcate the trial of liability and damage issues; and (5) taken under advisement defense motions based on the "indirect purchaser" doctrine[1649] attacking plaintiffs' standing to sue. Discovery, except as to the question of whether the cases can properly be maintained as class actions and a look at essential corporate organizational and record keeping data, has been curtailed by Judge Gray until the states detail individually what their specific legal theories are; the information presently known to plaintiffs which support such theories; and exactly what discovery the plaintiffs seek to supplement the evidence they now are relying upon. *Long Beach,* which is basically a suit charging, contrary to the other cases, that defendants conspired to keep the prices of crude at unreasonably *low* levels, has been proceeding somewhat more rapidly than the other cases. It does not appear to have material pipeline ramifications.

6. Staff Report of the Antitrust and Monopoly Subcommittee of the Senate Judiciary Committee *(Kennedy Staff Report)*

In June 1978, a Committee Print entitled "Oil Company Ownership of Pipelines," a Staff Report of the Subcommittee on Antitrust and Monopoly of the Senate Judiciary Committee, 95th Cong., 2nd Sess.

[1648] 405 U.S. 251 (1972). See also California v. Frito-Lay, Inc. 474 F.2d 774 (9th Cir. 1973), *cert. den.*, 412 U.S. 908 (1973).
[1649] Illinois Brick Co. v. Illinois, 431 U.S. 720 (1977).

(1978), was introduced into the literature on the subject of pipelines.[1650] Although there was some grumbling on the part of minority members of the (Kennedy) Subcommittee concerning the issuance of the Report before they had a chance to consider it, and that it might be misconstrued as being an official report of the Subcommittee, these procedural aberrations fade into insignificance when the merits, or more accurately, the lack of merit of the Report itself is considered. This examination is required, not only to warn against the use of the Report without the most careful examination of the source material purported to be supportive of the assertions contained in the Report, but also because it is the linchpin used by Senator Kennedy as the basis of his "factual background" in his FTC Rulemaking Petition.[1651]

There are a number of useful tidbits of non-controversial factual (general operating) information scattered throughout the Report. However, the more a reader goes to the cited source material to verify the information or to look for "leads" to more exhaustive treatment of subjects of interest, the more difficult one finds its use as a building block. Perhaps the compilers of the report were operating against an unrealistic deadline and were asked for too much, too soon. First, discounting typographical errors and occasional miscitations to the wrong page, which are almost impossible to avoid in the preparation of a document of this length, especially if, indeed, the time pressure was great, there are some mistakes which can mislead the reader, are inaccurate, or are inconsistent with other sections of the Report. For example, Tables A & B on page 44, while correctly showing the proper ownership percentages, are titled in such manner that a casual reader can well be misled into believing that "majors" control all or part of the 20 largest crude oil lines. Also, while purporting to exclude Tenneco and Occidental as majors in its analysis, the staff report used the FTC's concentration data for its concentration statistics (see page 31 of the Report), which pulled the two companies back into the group, although the staff did specify that the data related to the *20* largest companies.

Second, there are a number of inaccurate statements in the Report. In some instances, this is accomplished by innuendo. For example, on two separate occasions,[1652] we are told that in 1914 the Attorney General of Oklahoma took a strong stand for pipeline divorcement. Mr. West, in

---

[1650] KENNEDY STAFF REPORT, *supra* note 184. The Report appears to be a "scissors and paste" "lift" from a "Draft Report on Oil Pipelines," written by Leonard L. Coburn, Department of Justice, Antitrust Division, Energy Section, dated February, 1978, with the addition of the Justice Department's "undersizing" theory.

[1651] Kennedy FTC Rulemaking Petition, *supra* note 1641, at 4 n.*.

[1652] KENNEDY STAFF REPORT, *supra* note 184, at 11 and 92.

fact, was the originator of the term "divorcement."[1653] The Report does not, however, acknowledge a substantial number of differing views of much more recent vintage, among them four such expressions from ICC Commissioners, three from presidents of independent oil-producers associations, one from IPAA's General Counsel, and testimony in 1976 by a learned observer of the industry.[1654] With such an abundance of more current information readily available, *much of which has been submitted to the staff in its official capacity*, the repeated recourse to a 1914 statement as if it established an immutable truth for all time and circumstances, raises doubts concerning the accuracy and objectivity of the Report.

Another reason why there should be a degree of caution in relying on the Kennedy Staff Report is its incompleteness. For example, the Report, at page 77, alleges that Explorer "set a level [of rates] to earn the maximum allowed under the [*United States vs. Atlantic Refining Company*] consent decree but it also set a level no lower than the alternative barge shipments to similar destinations." There is no specification of which route(s) of Explorer or what barge competition is involved. Procedural due process and a sense of fair play would seem to dictate that before a company is divested of its pipelines, it should be entitled to see and test the facts which are supposed to substantiate the allegations made against it. For another example of this type of "non-evidence" see this text at note 448.

There are a number of illogical statements in the Report. For example, still discussing Explorer, the Report states: "Even though it was possible to charge lower rates and still earn an acceptable return on investment, it chose not to." (Report, at 77). This statement is economically unrealistic. Intelligent businessmen seek to earn the highest return possible, and would not lower rates voluntarily to earn rates which are merely "acceptable." This would be true whether the pipeline were owned by oil companies or "independent" firms. Moreover, it has a hidden "hooker" — "an acceptable return on investment" is disingenuously used in the sentence. "Investment" is silently equated to *equity,* which in Explorer's case is somewhat less than 10 percent of valuation. Explorer started out 90 percent "leveraged," *i.e.,* 90 percent debt to 10 percent equity,[1655] and being a new line, its total investment approximated closely its valuation. Also, rather than earning an "acceptable return," because of the competition it faced, even at the rates the Staff was castigating, Explorer lost $46 million during the first five years of operation, requiring

---

[1653] See note 105, *supra*.

[1654] See text at, and notes 1592-1509 and 1517, *supra*.

[1655] See text at notes 1401-1441, *supra,* for reasons why this leveraging came into vogue.

the shipper-owners to meet cash calls of some $42 million.[1656]

The Report did not wait until page 77 before using this type of reasoning. On page 7 of the Report, at the end of the first paragraph on the subject of prorationing a full line, the staff remarked: "Moreover, pipeline owners can overwhelm both other historical shippers and prospective new shippers with the amounts of crude oil or petroleum products which they can nominate and tender for shipment." This is contrary to the requirements of Section 3 of the ICA,[1657] which mandates non-discriminatory service and this obligation is reflected in the tariffs of most pipeline companies.[1658] Under these provisions, everyone gets a fair proportionate share of the available capacity. As to those lines which prorate on a "historical basis," the new entrant can get even a better deal. The reason for this is that the initial shippers are almost always the shipper-owners; they financed and built the line because they needed the transportation. Then, when others come into the line and volume expands by growth, these self-same shipper-owners *are* the historical shippers. Colonial's owners were in just such a position.[1659] During the periods of Colonial prorationing, the shipper-owners, far from "overwhelming" prospective shippers, actually had their shipments *reduced* to make room for new shippers, with the result that nonowner-shippers' traffic rose to 15 times what it was before prorationing commenced and the shipper-owners' movements decreased 4.5 percent, or stated another way, the new nonowner shippers received the *entire* increase in line capacity.[1660] The result was that the gasoline volumes sold by independent marketers in the 15 state area served by Colonial increased during 1963-1974 by 205 percent whereas overall gasoline volume increased only 59 percent, and more meaningfully, the market share of these independents climbed from 14 percent in 1963 to 27 percent in 1974 while the market shares of the "majors," including Colonial's owners, declined steadily.[1661]

The Report, at page 82, alleges that product specifications may be used to deny non-owner shippers access to pipelines. In support of this proposition, the Report cited the Siess-Sun Pipe Line incident described in Section VI A 2, "Sources of Complaints,"[1662] without hinting that there

---

[1656] See text at notes 432-437, *supra*.

[1657] 49 U.S.C. § 3 (1976).

[1658] *See e.g.,* Item 16 in the Shell Pipe Line Corporation Tariff, Exhibit B; Exxon F.E.R.C. Tariff No. 129, eff. 3/1/78 (Item 30); Colonial ICC 29, eff. 2/10/78 (Item 90); Crown-Rancho Pipe Line Corp. Texas Joint Tariff No. 43, eff. 8/1/78 (Rule 16); Gulf Refining Co. R.C.T. No. 541, eff. 8/1/78 (Item 16); Marathon Pipe Line Co., L.P.S.C. No. 4, eff. 10/5/78 (Rule 11).

[1659] See text at notes 418 through 421, *supra*.

[1660] See text at note 421, *supra*.

[1661] See Lundberg Survey Inc. Study reprinted in *Petroleum Industry Hearings, supra* note 192, at 779. Dixie Pipeline had a similar experience.

were any differing opinions, especially in Part 2 of the S.2387 or PICA Report, upon which the Kennedy Staff Report drew heavily.[1663]

The most serious flaw in the Kennedy Staff Report is the misstatement of cited testimony and authorities. In its least reprehensible form, this consists of what might be called a "nobody home" technique: when the citation is examined, there is nothing there which even faintly resembles the proposition for which it is cited. For example, the staff sets forth (page 86 of the Report) a proposition that the return earned by oil company owners of pipelines gives them an unfair competitive advantage over nonowner shippers. The Report states: "Only cost based rates or independent ownership could effectively equalize this competitive advantage." In support of the quoted conclusion, it cites Dr. Walter Mead's testimony in the *S.Res.45 Hearings.*[1664] Dr. Mead's testimony never even discussed the point. The only reference to pipelines in his testimony was to jointly-owned lines, and as to that he made the common sense observation that before Congress can legislate in the area, additional study was needed to identify the competitive impact of pipeline joint ventures and to examine alternative methods of financing large-scale pipeline ventures.[1665] Conceivably, this type of thing could in some instances be tracked down and be nothing more than another example of editorial slippage. But no such excuse can be found for the following examples. At pages 4-5, the Report states, "For example, independent pipelines *generally* provide storage tankage for all prospective shippers" (emphasis added). On page 79 (third full paragraph, last sentence), the Report attempts to drive the point home by stating: "Shippers on pipelines have emphasized the importance of provision of storage facilities to increased access to pipelines." Thus the picture drawn is that the "independent" pipelines [this means lines owned by non-oil interests] *all* provide storage facilities and the "majors" do not, to the end of denying access to "independent" [other than a "major" oil company] shippers.

A neat package on the surface, but what really was said, and by whom? The witnesses cited for the first passage were D. W. Calvert, Executive Vice President of the Williams Companies, owner of Williams (Brothers) Pipe Line Company, and Richard C. Hulbert of Kaneb Pipeline Company, and for the second, Mr. Hulbert. There were no shippers. Mr. Calvert testified that Williams (Brothers) Pipe Line had 31 company-owned storage facilities which are provided as part of its overall

---

[1662] See text at notes 1498-1501, *supra*.
[1663] S.2387 REPORT, *supra* note 49, pt. 2 at 212 (Minority Views and Additional Individual Views).
[1664] *S.Res.45 Hearings, supra* note 296, pt. 3 at 1001 (Statement of Walter J. Mead).
[1665] *Id.* at 1005.

transportation service,[1666] and he stated in a letter dated July 14, 1972, to Congressman Neal Smith, Subcommittee Chairman, in answer to the latter's question, that the following pipeline companies provide common carrier terminalling services: Kaneb Pipe Line Company, Texas Eastern Transmission Corporation, Dixie Pipeline Company, MAPCO, Inc., Pioneer Pipeline Company, Continental Pipe Line Company, and Calnev Pipe Line Company.[1667]

The very same letter shows, what is common knowledge, that Williams Brothers purchased the system from the Great Lakes Pipe Line Company. The original owners of Great Lakes were Continental, Barnsdall, Skelly, Mid-Continent Petroleum, Phillips and Pure, subsequently joined by Texaco and Sinclair and finally Cities Service.[1668] Each of these has been classified by critics of the industry as "majors."[1669] They were the ones, because of the peculiarities of the Great Lakes line, *e.g.,* operating in an area of comparatively sparse markets, with relatively low volume for the individual marketer (as distinguished from Colonial, which serves relatively concentrated metropolitan markets where 28 existing terminals of 13 non-owners, located in 15 cities, were connected to the system at the beginning of operations) which instituted the common terminalling service which Williams is continuing and expanding.[1670] Dixie is owned by 13 companies, 11 of which are "majors," being operated by Exxon Pipeline Company, hardly an "independent" pipeline; Continental Pipeline Company is 100 percent owned by Continental Oil Company; Pioneer is owned 80 percent by Continental; Calnev is presently owned 100 percent by Champlin Refining Company, [it was originally promoted by an individual named J.B. Harshman, with the bulk of the capital supplied by the Union Pacific Railroad].[1671]

So up to this point, the cited references really stand for the proposition that *some* "independent" pipelines and *some* "major" pipelines furnish this service. Mr. Hulbert's testimony adds something, but not that for which it is cited by the Kennedy Staff Report. What Mr. Hulbert said, when pressed by Senator Tunney to agree that oil company-owned pipelines discriminate against shippers by not providing storage, was actually a denial of the charge: "I don't really feel there is any form of discrimination or specialized treatment in those cases." He explained the

---

[1666] *H.Res.5 Subcomm. Hearings, supra* note 234, at 13 (Statement of D. W. Calvert).
[1667] *Id.* at 31.
[1668] See note 138, *supra*.
[1669] See note 372, *supra*.
[1670] *S.1167 Hearings, supra* note 432, pt. 9 at 609-610 (Statement of Vernon T. Jones).
[1671] LIVINGSTON, *supra* note 228, at 34; Jones & Gardner Memo., *supra* note 427, app. 1 at 3 (Continental, Dixie), 5 (Pioneer), and 8 (Calnev).

true distinction to be that where pipelines are handling common specification materials as Kaneb does [and is the case with Williams Brothers, and its predecessor Great Lakes], it is more apt to provide terminal facilities than where there are individual movements of segregated batches of materials moving to specific shipper-owner terminals.[1672] So the real distinction appears clearly to be operationally oriented and not based on whether the line is owned by an "independent" or by a "major" oil company.

Another example appears at pages 80-81 of the Report. After citing a 1924 La Follette Report and the *TNEC Hearings* in the late 1930's, the staff attempted to establish a current pipeline routing malpractice by citing ICC Chairman Stafford for the proposition that he "voiced" the "same problem," *viz,* that pipeline location and independent refinery location diverge, creating substantial problems for independent refiners; and therefore the route design benefits the owners to the detriment of nonowners.

In addition to citing the wrong page (900), the Report misstates Chairman Stafford's testimony on page 901. Actually, what Chairman Stafford said was: "Advocates of divorcing pipelines from the oil industry believe that only in this way can real competitive conditions be restored in the oil industry. Others, however, question whether separation could at this late date in the development of the industry be of any substantial benefit to independent refiners who have mostly established themselves at locations near the oil fields and who have little occasion to make use of crude oil lines now owned by the integrated oil companies."[1673] The reader might wish to glance at the quotation of Chairman Stafford cited from two paragraphs later on the same page 901.[1674] The disparity between what the staff cited his testimony for and what he plainly said should tell the story.

Another instance of misuse of a cited source is found at page 86 of the Report, wherein an assertion is made that the difference between cost [to the shipper-owner] and the tariff may constitute an illegal rebate, and this "rebate," whether legal or illegal, provides a subsidy to the pipeline owners which may be used to subsidize other segments of the vertically

---

[1672] *S.1167 Hearings, supra* note 432, pt. 8 at 5947 (Statement of Richard C. Hulbert).

[1673] *S.Res.45 Hearings, supra* note 296, at 901 (Statement of ICC Chairman George M. Stafford).

[1674] *Id.:* "In conclusion, it would appear that except for certain impediments brought about because of environmental considerations, pipelines have been constructed on as-needed basis and generally provide good service. It has been our experience that pipeline rates are just and reasonable. . . . We have received no complaints in recent years involving allegations relative to the size of tender, the failure to publish through routes and joint rates, or to provide service to independents."

integrated company that are operating at a loss. The quoted portion cited Wolbert, *American Pipe Lines*, at 57. What Wolbert actually said at the cited page was that there was an *allegation* to that effect; but the sentences immediately following that statement are, in fact, simply supportive of the efficiencies of integration and, if anything, tend to discredit the "undersizing" argument.

More examples could be given but they would be simply duplicative evidence. By now, the reader should have a rather good idea of the methods used in compiling the Report.

C. *Specific Policy Questions*

1. So-Called "Excessive Rates" of Return

Before one can begin to inquire into the question of "excessive" rates of return, there has to be a search for a "norm," against which a rate can be compared to see if it is "excessive." This requires going back to the basic questions: (1) What is the goal of regulation, and (2) How is this goal translated into the setting of a norm? A frequently encountered expression of the regulatory goal is "to replicate results which would occur in a competitive market economy,"[1675] or, stated another way, regulation should act as a "surrogate for competition."[1676] The criteria used most commonly to quantify this goal into a desirable rate of return are that such rate should: (1) be commensurate with returns on investment in other enterprises having comparable risks; (2) assure confidence in the financial integrity of the enterprise and permit it to maintain its credit standing; and (3) enable the enterprise to attract new capital.[1677] The protection of the consumer has long been recognized, but as the FPC natural gas regulation proved so clearly, protection of the consumer is not realized by hammering prices down to achieve the lowest possible short-run cost to consumer, but rather to achieve the lowest rate consistent with maintaining a long-run adequate supply at that price.[1678]

[1675] Solomon, *TAPS Rebuttal Testimony, supra* note 621, at 4.

[1676] Gary, *TAPS Rebuttal Testimony, supra* note 726, at 50.

[1677] *Id.* at 49-50. In order to do this, the allowed rate of return must exceed the cost of capital. *TAPS Hearings Transcript, supra* note 785, at 9727 (Testimony of Justice Department Witness Herman G. Roseman on cross-examination); *cf. Id.* at 10,068 (Testimony of Dr. Charles F. Phillips on cross-examination).

[1678] Ryan, *TAPS Rebuttal Testimony, supra* note 724, at 48.

If the regulated price is set so low that it produces a rate of return less than the opportunity cost of capital, there will be a flight of investors' funds away from the industry. As the Statement of the Department of Justice in the *Valuation of Common Carrier Pipelines* proceeding aptly observed: "All of this is not to say that the Commission's responsibilities end with ensuring that pipeline rates are not excessive. It is equally important to our national interests and, a large part of our purpose here, to assure that pipeline rates are sufficiently renumerative to attract investment to new pipeline projects, for inadequate pipeline investment might have consequences as serious for national energy and transportation goals as excessive pipeline rates."[1679] This thought was expressed pragmatically by Ulyesse LeGrange when he said, "One significant aspect of regulation is the end result — that is, in what shape is the industry after a number of years of regulation?"[1680]

In applying the foregoing three-factor test to rates of return, special care must be exercised with respect to the first one: "Comparable earnings." If the regulator looks *exclusively to other utilities,* there is a danger that circularity will be introduced into the picture because the second regulated industry's rate of return could well have been based on the first industry's rates in an earlier period.

Moreover, the returns used in such comparisons invariably are *realized* rates of return, which do not necessarily reflect the decisions of the relevant regulatory commissions concerning the appropriately *allowed* returns.[1681] Few public utilities have been able to earn the returns allowed to them by their regulatory commissions and to set the next period's allowable return on the basis of the prior period's experienced returns will have a ratchet effect which will ensure a continued financial deterioration of the industry. In addition, comparability is lost if a regulator resorts to stable, mature concerns against which to compare a new grass-roots venture. Such a comparison ignores the fact that the investor, from whom

---

[1679] Kaplan, Rm 78-2-Statement, *supra* note 641, at 7. It is significant that the Department of Energy, whose job it is to see that the nation's energy resources are managed most efficiently, appeared in the TAPS hearing by its Assistant General Counsel, Cameron Graham, who stated: "It is the Department's position that the objective of the Commission [FERC] in setting rates for the oil pipelines should be rates that are sufficient to achieve the investments necessary for the industry, basically cost of capital standard. . . . it is extremely important that [the rate level] approximate [sic] the cost of investment opportunities. It insures the lowest customer prices, which is an interest of the Department of Energy. On the other hand, it allows the necessary investments to insure adequate energy supplies when they are needed. Underinvestments due to insufficient revenues is simply a misallocation of resources and would be a constraint on existing markets and transportation systems, whereas overinvestments due to excessive revenue levels would provide similar defects on the other side." *RM 78-2 Hearings Transcript, supra* note 754, Vol. 11 at 1656 (Oral argument of Cameron Graham). *See also* SHENEFIELD, PIPELINE POLICY, *supra* note 640, at 11.

[1680] LEGRANGE, *Government Regulation, supra* note 1413, at 107.

[1681] Gary, *TAPS Rebuttal Testimony, supra* note 726, at 51; Ryan, *TAPS Rebuttal Testimony, supra* note 725, at 46.

the necessary capital must come, is going to expect a higher rate of return from the new venture than from the going concern because of the differences in the risk he perceives, with the result that all three guidelines are violated.

The foregoing principles found substantial agreement among the parties involved in, and observers of, the recent FERC valuation and rate proceedings; differences in opinion arise in their application. Positions taken prior to the proceedings had been rather far apart. Professor Stewart Myers, testifying on February 20, 1974, before the Special Subcommittee on Integrated Oil Operations of the Senate Committee on Interior and Insular Affairs, reported that the mean rate of return,[1682] using ICC reports (based on "flow-through," rather than "normalized" accounting)[1683] for 97 out of the 98 pipelines reporting in 1971[1684] was 9.4 percent and the median was 8.4 percent,[1685] with a wide degree of variation among the companies, ranging from a negative of 10 to over 22.5 percent.[1686] The corresponding mean and median figures, using normalized accounting, were 7.8 and 7.4 percent, respectively. The dispersion about the mean was somewhat less than was the case in the flow-through method, but there was still a substantial spread between successful lines, from a negative 2.5 percent to over 17.5 percent.[1687] Among other observations that can be drawn from these data, is the indication of risk that arises from such dispersion.[1688] Public utilities, in contrast, tend to cluster rather closely around the mean.[1689] Professor

---

[1682] Myers' calculation of rate of return was based on the ratio of after-tax operating income over net total assets. He corrected the income by adding back interest less the tax effect of such interest divided by net total assets. Myers, *supra* note 198, at 11.

[1683] "Flow-through" accounting takes into income the tax benefits of accelerated depreciation and investment tax credit in the year in which they are taken on their tax return, whereas "normalized" accounting operates on the basis that these benefits should be apportioned over the life of the assets involved, records them as a deferred charge and parcels them out proportionally over the respective years of their life.

[1684] One of the pipelines had to be excluded from the universe because of missing data. *Id.* at 12.

[1685] The difference between mean and median was due largely to Crown-Rancho, whose rate of return was nearly 70 percent. *Id.*

[1686] See Myers' histogram, Figure 3, located between pages 12 and 13. The model class was 7.5-12.5 *Id.* HOWREY & SIMON, *supra* note 3, at 84, cites the same figures from Myers' testimony in the *Energy Industry Investigation, Hearings on Energy Before the Subcommittee on Monopolies and Commercial Law of the House Committee on the Judiciary,* 94th Cong., 1st Sess. 725 (1975).

[1687] Myers, *supra* note 198, Figure 4 between pages 12 and 13. Again, Crown Rancho's 35 percent produced the mean-median spread.

[1688] *S.1167 Hearings, supra* note 432, pt. 9 at 608 (Statement of Vernon T. Jones).

[1689] See, for example, Statement of George A. Hay, Director of Economic Policy, Antitrust Division, Department of Justice, Before the Interstate Commerce Commission, Valuation of Common Carrier Pipelines, Dkt. Ex Parte No. 308, Table II, p. A-7 (May 26, 1977) [hereinafter cited as Hay, *FERC Direct Statement*].

Myers made two other comments relevant to this subject. The first was based on the results of weighting the averages of the above rates of return, which he did by calculating the ratios of total industry operating income, after tax, to total net assets for the industry getting the following results:

	Weighted Average Rate of Return	Unweighted Average Rate of Return
Flow-through accounting	8.7%	9.4%
Normalized accounting	7.7%	7.8%

Myers' observation from these figures was as follows: "Since the weighted averages are below the unweighted one, we can reject the view that large pipelines are more profitable, on a book basis, than small ones. If anything, the evidence supports the opposite conclusion."[1690] The second comment by Professor Myers, made after deriving A T & T's cost of capital and calculating a 7.8 percent rate of return on a normalized basis which, as can be seen from the chart above, was exactly equal to the oil pipeline industry's unweighted average rate of return and 0.1 percent *above* the weighted average rate of return of oil pipelines on a comparable accounting basis, was that "...it is hard to see how the average [oil] pipeline return could be viewed as unfairly high."[1691] Professor Myers also examined the rates of return *allowed* by the FPC to 18 natural gas transmission companies (using an imputed cost of capital calculated by using an 8 percent interest rate on debt and preferred stock and a 50 percent marginal tax rate).[1692] The average allowed rate was 6.66 percent. Speculating what this would equate to if gas pipelines were financed at 40 percent debt and 60 percent equity [which is slightly more leveraged than the "stand-alone" pipeline debt of 30-35 percent debt discussed in Section V B 3, "Financing; How Pipelines Are Financed; 'Stand-Alone' Pipeline Ventures."],[1693] the 6.66 percent allowed at the 64.6 percent debt level shown in Myers' Table 4 [cited in footnote 1692] would equate to 7.9 percent at the 40 percent level; hence FPC decisions would imply a cost of

---

[1690] Myers, *supra* note 198, at 13.

[1691] *Id.* at 15. See also HOWREY & SIMON, *supra* note 3, at 84; *S.1167 Hearings, supra* note 432, pt. 9 at 606 (Statement of Vernon T. Jones); *cf.* Tierney, *TAPS Rebuttal Testimony, supra* note 175, at 5 ("... our information clearly indicates to us that oil pipelines rate levels have been quite reasonable under ICC regulation.").

[1692] Myers, *supra* note 198, at Table 4 on page 16.

[1693] See text at notes 1432-1433, *supra*.

capital for oil pipelines of approximately 7.9 percent.[1694] This must be viewed with some caution because it involves a number of assumptions and it compares returns *allowed* to natural gas pipelines with *actual* returns on oil pipelines without a "risk premium" being assigned to the difference in the risks of oil pipelines over those associated with natural gas pipelines.

On the other end of the spectrum, Assistant Attorney General John Shenefield in his Proposed Statement before the Kennedy Subcommittee wrote, "...in an evaluation of pipeline profitability conducted by our Economic Policy Office last year [1977], our economists concluded that the rates of return for many oil pipelines are significantly above what is encountered in the *most profitable* electric utility and natural gas pipeline companies."[1695]

Because the Kennedy Subcommittee never held the hearings before which Mr. Shenefield was scheduled to testify, he thus was not accorded an opportunity to answer questions. One such question might well have been: Which lines are you talking about, and why didn't you use the *average* rate of return since your statement advocates[1696] divorcement of "pipelines?" A second question could have been: How do you reconcile your thesis that pipeline rates are so high with your theory, also expounded in your statement,[1697] that vertically integrated pipelines "circumvent" rate regulation by undersizing their lines so as to reap "monopoly profits" *downstream* [and upstream] by forcing other suppliers to use "more expensive" transportation modes? It is traditional economic theory that vertically integrated monopolies can take only one monopoly profit.[1698] Common sense, as well as the practical impossibilities of "fine tuning" line sizing explained in Section III F 5, "Fluctuation in Throughput,"[1699] would suggest that if oil pipelines did have monopoly power, as Mr. Shenefield implies, and the regulation is as inept as he asserts, they would simply take their profits the easiest way, *i.e.*, in the pipeline phase.[1700] Because Mr. Shenefield did not have an opportunity

---

[1694] Myers, *supra* note 198, at 18.

[1695] SHENEFIELD PROPOSED STATEMENT, *supra* note 179, at 19-20, citing Hay, *FERC Direct Statement, supra* note 1689, at 6-7.

[1696] *Id.* at 37.

[1697] *Id.* at 21-23.

[1698] Bork, *Vertical Integration-Economic Misconception, supra* note 106, at 196. For a more detailed exposition of this concept, see Section VI C 4 a, "Fatal Flaws in the Economic Theory Involved in the Undersizing Argument", text at notes 2023-2041, *infra*.

[1699] See text at notes 786 to, and immediately following, 799, *supra*.

[1700] See comment to this same effect by Professor Edmund W. Kitch, of the University of Chicago Law School, AEI, *Oil Pipelines and Public Policy, supra* note 684, at 127.

to answer these questions by the untimely cancellation of the Kennedy Subcommittee hearings, these questions will merely be posed and attention will be directed to certain constructive developments that have taken place since the public release of his proposed statement.

At the American Enterprise Institute's Conference on Oil Pipelines and Public Policy, March 12, 1979, papers which dealt with rates of return were delivered by Thomas C. Spavins,[1701] an economist in the Economic Policy Office of the Antitrust Division, and Ulyesse J. LeGrange, Controller, Exxon Corporation.[1702] The methods used by each were marked more by their similarities than by their differences. First, both agreed that a return standard should be in terms of the *total capital invested,* not just the equity capital. Second, both went beyond the 1978 Forbes magazine study of 1,005 public companies to the S & P Compustat Services data, which cover nearly 2,500 public companies, to acquire the broadest available data base so as to make the comparison between industries as reliable as possible. Spavins excluded ARCO, B.P., Exxon and Sohio from the sample because of the distorting effect of the construction work in progress on TAPS. Both papers calculated the return on *total capital.* Spavins' weighted average return for the years 1975 through 1977 for oil pipeline companies was 13.0 percent, which made it the seventh highest among the 30 industries in his sample.[1703] LeGrange's median return for oil pipeline companies was 10 percent, as opposed to 9.6 percent for *all* industries [which would place it 15th among 31 industries]; his average return for oil pipeline companies was 9.5 percent, compared to 9.9 percent for all industry; and the mean weighted average for oil pipeline companies was 7.6 percent, as against about 10 percent for general industry.[1704]

LeGrange went on to compare oil pipeline returns to 171 gas, electric and telephone utility companies. He found the mean, median, and mean weighted average returns of oil pipeline companies of 9.5 percent, 10 percent, and 7.6 percent somewhat higher than the comparable returns of

---

[1701] SPAVINS, PIPELINE REGULATION, *supra* note 676. It should be noted that the views expressed in Spavins' paper were not necessarily those of the Antitrust Division, and no inferences should be drawn that the Justice Department sponsors them. However, Spavins has participated in virtually all of the Division's recent work on pipelines.

[1702] LEGRANGE, *Government Regulations, supra* note 1413. LeGrange has served as President of Exxon Pipeline Company and is a veteran of the SEADOCK debacle. For a detailed backup paper describing the methodology used, see *An Analysis of the Rates of Return on Petroleum Pipeline Investments* in AEI, *Oil Pipelines and Public Policy supra* note 684, at 261-318. *See also* Harrill, *TAPS Rebuttal Testimony, supra* note 744, *passim.*

[1703] SPAVINS, PIPELINE REGULATION, *supra* note 676, at 35 (Table VIII), and 34 (Table VII).

[1704] LEGRANGE, *Government Regulation, supra* note 1413, at 7. For a more detailed description of the methodology see Harrill, *TAPS Rebuttal Testimony, supra* note 744, at 16-21.

6.9 percent, 6.6 percent, and 6.7 percent for the utilities. However, he sought to account for this difference by differences in risk. As he pointed out, oil pipelines do not have a market monopoly — no one is forced to ship over a pipeline. Other alternatives are available and are used. This should produce greater volatility in pipeline returns. Second, pipeline companies are unable to sign long-term contracts guaranteeing capacity because they must take on all comers under their common carrier obligations. This means shippers are not "locked in" and can ship or not as their economics direct. Third, pipelines have a ceiling on earnings but, like railroads, no guaranty of earnings, whereas most utilities generally will be protected by the regulator against less than a minimum return.

Each of these should cause a volatility difference between the industries. LeGrange then examined the data to see if the empirical results verified the expected result, which they did. Ten of the 79 oil pipeline companies had returns of less than 4 percent whereas none of the 171 utilities returned less than 4 percent. Moreover, the dispersion of utility returns was not great — they clustered around the 6-8 percent level, with the lowest about 4 percent and the highest at 16 percent. Oil pipeline dispersion of rates, on the other hand was wide, from a negative 8 percent to a positive 18 percent. From these data, LeGrange concluded that his intuitive belief had been confirmed, *i.e.,* that oil pipelines indeed were riskier than utilities, and as Chapter V demonstrated, this differential in volatility causes investors to demand a higher rate of return. Because of the commonality in approach, it is easy to discern where the difference lies. LeGrange computed his rates of return by using total capital invested as the denominator. He got his numerator by adding back to net income the interest paid on borrowed capital, but he adjusted the interest figure by taking into account the tax effect of such interest payments. In short, he used net income plus after-tax interest divided by total capital invested as the measure of return.[1705] As Professor Shyam Sunder of the University of Chicago Graduate School of Business, who was one of the designated discussants of the two papers, observed: "This rate has taken out the effect of leverage and reduced all the data of all firms within the pipeline industry and across the industries to a comparable basis after adjusting it for differences in leverage."[1706] Tom Spavins did not make the tax effect adjustment in his paper, so his rate of return computation was made by dividing the sum of after-tax profits plus the interest payments by the sum

---

[1705] Professor Myers, in his 1974 Testimony, also adjusted the interest add-back by the "tax shield provided by [the] interest." Myers, *supra* note 198, at 11. Ryan, *TAPS Rebuttal Testimony, supra* note 725, at 36 explained the procedure as reconstructing the income stream as it would have been if the project had been financed totally by equity.

[1706] Comments of Discussant Shyam Sunder, AEI, *Oil Pipelines and Public Policy, supra* note 684, at 114.

of the stockholders' equity, long term debt, notes payable, and the current liability portion of long term debt.[1707] This produced the higher rate of return number.

Rather than quibble about which method is "right," it seems more appropriate simply to note that Messrs. LeGrange and Spavins have contributed a significant step to the orderly process of fact finding. There is still another step that has to be taken before any really meaningful conclusions can be reached about oil pipeline rates of return, namely, to derive a risk premium that will permit across-industry comparisons. As was mentioned in Section V A 1 b, "Fundamental Factors Determining the Required Rate of Return,"[1708] the risk involved is the primary determinant of rate of return. With Spavins' calculation, one must consider the total risk differential (both business and leverage) between the industries being compared. As Professor David J. Teece, the other discussant, noted, Spavins, unlike Donald Flexner, Deputy Assistant Attorney General, did not draw a monopoly power explanation from the numbers, adding '...nor do I think he would want to make one without first seeing if risk could explain the difference that exists.''[1709] Spavins elucidated his approach in response to the discussants' comments by remarking that profit rates do not mean anything [in regard to monopoly power] *by themselves,* but they were very useful in the context of a number of other factors such as evidence on the technology and the structure of the industry being studied. He referred to the Division's monograph on the coal industry (mostly the bituminous coal industry) where despite the fact that it was "the one industry in the energy, transportation, and utility [sectors] which had significantly higher rates of return than the oil pipeline industry had," the Division had concluded strongly that it was workably competitive,[1710] whereas, prior to deregulation, the airline industry was not competitive in some ways and it had the lowest rate of return. Spavins' second point was that in conducting an inter-industry study, the different effects that the federal

---

[1707] SPAVINS, PIPELINE REGUALTION, *supra* note 682, at 32. Spavins ought not to be held to the Department's prior position, but it is interesting to note that the Department, in the Deepwater Port Report, while preferring a return on equity position, suggest as an alternate the method followed by LeGrange. See A. G.'s Deepwater Port Report, *supra* note 446, at 56: "The second [method]. . . is net profits plus 50% of interest payments as a percent of the sum of equity and debt investments." See also *Id.* at 62:" The difficulty with this formula [used by ICC and the Consent Decree in computing allowable returns] is that in computing the net income figure, the ICC does *not* include interest payments *(net of tax benefits)* that constitute the return to the debt-supported assets of the rate base" [emphasis added to the parenthetical modifier].

[1708] See text at notes 1355-1370, *supra.*

[1709] Comments of Discussant David J. Teece, AEI, *Oil Pipelines and Public Policy, supra* note 684, at 116.

[1710] D. J. COAL REPORT, *supra* note 1631, at 130.

corporate income tax has on different income streams must be borne in mind, hence proper adjustment of net income was not just a financial risk question but also a tax incidence question, which presumably he felt was not addressed adequately by LeGrange's method.

It appears that there is a respectable intellectual difference of opinion here, but as Professor Sunder pointed out: the analysis of the relevant risk elements should be capable of being developed; historical accounting rates of return are misleading; and only after the necessary analysis is done to achieve comparability can the matter be settled in a more objective manner.

Before leaving the subject of rates of return as such, there is one other point that should be mentioned. In Spavins' paper, he noted that the oil pipeline industry, under ICC regulation, has been operating under broad guidelines for overall returns, and it has evolved over time in response to them. In that connection, he observed that much of the investment in oil pipelines has been made by diversified oil pipeline companies which are subject to overall average rate of return guidelines, not specific project guidelines. Mr. Spavins commented: "This point is of interest because the averaging process enables a pipeline company to insure itself against projects that fail to yield the maximum allowed return if there are other parts of the pipeline system that charge less than the unregulated monopoly tariff. The pipeline system can increase the tariff on those other parts of the system to compensate for these losses."[1711] This raises the question of segment or individual rates. Professor George Harmon reads the *Minnelusa Oil Corporation* case[1712] as holding that individual crude rates in excess of the Commission's [8%] guideline would be judged unreasonable.[1713] Professor William K. Jones, in his DOE Report, stated that the ICC did not examine the reasonableness of Williams (Brothers) Pipe Line's rates on particular movements, on the ground that its rate structure had not been challenged.[1714] The language of the Administrative Law Judge was rather broad, *i.e.,* "there is no legal requirement that the same percentage of profit be secured from each type of operation of a transportation company."[1715] It would, of course, be appropriate to the challenged joint rate which definitely was at issue.

[1711] SPAVINS, PIPELINE REGULATION, *supra* note 676, at 15n.33. In a subsequent part of his paper, he noted, but was careful not to suggest that it existed, the possibility that if there were parts of a system that transport essentially only the oil of its shipper-owner and that were charging tariffs low relative to costs, the effective earnings on the part of the system that was carrying for others may be much higher than the system-wide average. *Id.* at 47-48.

[1712] Minnelusa Oil Corp. vs. Continental Pipeline Co., 258 I.C.C. 41 (1974).

[1713] Harmon, *Effective Public Policy to Deal with Oil Pipelines,* 4 Am. Bus. L. J. 113, 123 (1966).

[1714] JONES DOE REPORT, *supra* note 372, at 43.

[1715] See text at note 1567, *supra*.

There was considerable criticism of Colonial's rates in the *Smith-Conte Hearings* by the union witnesses, led by Congressman Conte's questions: "I understand that the rate from Houston to Atlanta, inland, is about 22.5 cents a barrel. Is that right? And from Atlanta to New York is only about 8 cents more. What is the reason for this?"[1716] However, not only did the Deputy General Counsel of the ICC point out to the Subcommittee that this definitely was not a violation of the long haul-short haul section of the ICA [Section 4],[1717] but Colonial's President, Fred F. Steingraber, introduced into the record a chart showing how common it was for oil transportation modes to have a decreasing average rate per hundred barrel miles.[1718] Moreover, Professor Edward Mitchell in his 1974 Statement, submitted to the Senate Antitrust and Monopoly Subcommittee, referred to Hoyt Haddock's testimony embodied in Congressman Conte's question and gave the economic justification for Colonial's rate structure: "When the Colonial pipeline was being planned it might have run just from Houston to Atlanta. However, adding a segment from Atlanta to New Jersey permits the use of a larger pipeline and lower costs per barrel mile *all along the line.* If one were to spread the costs of the Houston-New Jersey pipeline evenly over the whole line it is likely that the Houston-New Jersey run could not have competed with tanker shipment at the time the line was planned. If it were not possible to allocate the costs in such a way as to make the Atlanta-New Jersey segment cheaper per mile than the Houston-Atlanta segment it is possible that the segment beyond Atlanta might not have reached all the way to New Jersey, and therefore that average per barrel costs would have been higher at all points along the line, including Atlanta. Thus, by charging a tariff closer to the incremental cost of the Atlanta-New Jersey segment it was possible *to benefit all users* (emphasis supplied).''[1719]

If, indeed, the goal of regulation is to replicate the results which would occur in a competitive market economy, this is exactly the result which one would expect. This explains Spavins' care in making his comment. Notwithstanding that, his point is valid: averaging over different systems could give rise to favoritism of shipper-owner movements. There appears to be several safeguards against this. The antidiscrimination provisions of the ICA would seem to bear on this; the divergent interests of the multiple owners of a jointly owned system tend to have an inherent "check and balance" effect, and the feeling among

---

[1716] *H.Res.5 Subcomm. Hearings, supra* note 234, at 43 (Question by Congressman Silvio Conte, of Mass.)

[1717] See text at note 1479, *supra.*

[1718] See text at note 1480, *supra.*

[1719] Mitchell, *1974 Senate Testimony, supra* note 1056, at 25.

pipeliners is that significant projects which are separate —self-contained, if you will — projects, such as TAPS, in the future will be judged on their own figures in meeting a rate of return test.[1720]

2. Regulation of Rates

a. By the Interstate Commerce Commission (Now FERC)

It must be recognized that before the ICC could begin to discharge its rate-making duties with respect to pipelines, it had to accomplish three prerequisite tasks: (1) establish its jurisdiction over pipelines; (2) devise a system of accounts for pipelines that would permit uniformity of its action; and (3) develop a valuation method pursuant to the mandate of the Valuation Act (Section 19 of the ICA). Immediately after the enactment of the Hepburn Act in 1906, the ICC circulated letters among the pipeline companies, seeking information and recommendations relative to the establishment of a common system of accounts for pipelines.[1721] It first established a classification of operating expenses, then revenues and a form for general balance sheet statements.[1722] It then issued a form of special report which was to be rendered by oil pipelines effective January 1, 1911.[1723] Not unexpectedly with a new effort as broadly based as this was, there was a wide divergence of interpretations by the respondent companies and the desired degree of uniformity was not achieved by this first attempt.[1724]

At that point, further efforts to improve the report were held in abeyance until the question of the ICC's jurisdiction over pipelines was settled. The Commission had started the process in motion by its 1911 hearing on the issue, culminating in a decision entitled *In the Matter of Pipelines,*[1725] which ruled that all interstate oil pipelines were included within the scope of the Hepburn Act and hence were subject to ICC jurisdiction. After the carriers obtained from the Commerce Court an

---

[1720] See LeGrange's remarks in the general discussion held on the second day of AEI, *Oil Pipelines and Public Policy, supra* note 684, at 184.
[1721] 20 ICC ANN. REP. 62 (1906).
[1722] 23 ICC ANN. REP. 57-58 (1909).
[1723] 24 ICC ANN. REP. 31 (1910).
[1724] FTC, REPORT ON PIPELINE TRANSPORTATION OF PETROLEUM 20 (1916).
[1725] 24 I.C.C. 1 (1912).

injunction against the enforcement by the ICC of its order directing pipelines to file tariffs,[1726] the United States Supreme Court approved the inclusion within the ICA of all pipelines engaged in the interstate transportation of oil.[1727]

With the way thus cleared, the ICC resumed its task of improving the comparability of pipeline carriers' reports. It issued an order in 1915 which included carrier investments,[1728] and on June 23, 1919, promulgated a form of Annual Report for Pipeline Companies, commencing with the calendar year 1918.[1729] Shortly thereafter, the Transportation Act of 1920[1730] directed the ICC to prescribe, "as soon as practicable," a system of depreciation accounting, which, after a substantial amount of controversy from the carriers, who objected to the application of railroad depreciation to the dissimilar operations of pipelines, was resolved by the Commission's decision in *Depreciation Charges of Carriers by Pipe Lines*.[1731]

The ICC'S valuation efforts had an equally difficult time, commencing in 1934, when it requested the American Petroleum Institute (API) to form a committee to expedite comments and suggestions by the pipeline companies,[1732] and involving a test case to determine its jurisdiction to require valuation information from interstate private carriers.[1733] This time, however, the Commission, spurred on by the obvious need for valuations in order to proceed in its rate investigation, did not wait for the final outcome but commenced the field work of verifying physical inventory, land valuation, audit and compilation of reports; priced the inventories; and ground out tentative valuations. Eight

---

[1726] Prairie Oil and Gas Co. vs. United States, 204 Fed. 798, 821 (Comm. Ct. 1913). The Court's basis was that Congress' action in opening up all private interstate oil pipelines for public use was an unconstitutional taking of private property for public use without just compensation. *Id.* at 817.

[1727] The Pipe Line Cases, 234 U.S. 548 (1914).

[1728] BEARD, *supra* note 45, at 59; MILLS, *supra* note 324, at 48.

[1729] 33 ICC ANN. REP. 37 (1919).

[1730] 41 Stat. 493-494, as amended, 49 U.S.C. § 20 (4) (1976).

[1731] 205 I.C.C. 33 (1934).

[1732] The Commission had begun work promptly after the enactment of the Valuation Act to attempt to establish valuations for railroads but encountered substantial resistance, including litigation which continued until 1933. It found out the hard way, by its experience with the railroads, that cooperation with the regulatees expedited agreement on valuation facts and principles; so when Petroleum Administrator Ickes urged the ICC in 1933 to order valuations as a necessary prelude to rate determination, the ICC decided not to repeat the same mistake with pipelines but to commence such work on a cooperative basis. PETROLEUM PIPELINES, *supra* note 45, at 240.

[1733] Valvoline Oil Company Petition in Valuation of Pipe Lines, 47 I.C.C. Val. Rep. 534 (1937); Valvoline Oil CO. vs. United States, 25 F.Supp. 460 (W.D.Pa. 1938) (upholding ICC's order); *aff'd,* 308 U.S. 141 (1939).

tentative valuations were served on carriers during 1937,[1734] 25 in 1938[1735] and 13 in 1939.[1736] It held protest hearings and the number of final valuations began to rise, reaching 35 by the end of 1939.[1737] The basic program of pipeline valuation was completed in 1940.[1738]

While this preparatory process was taking place, the Commission held a "rate" case in 1920, arising out of an objection by a shipper to a proposed cancellation of a joint rate posted by the Prairie Pipe Line Company and the National Transit Company covering the transportation of crude oil from the Midcontinent producing field to Warren, Pennsylvania. The ICC suspended the cancellation and, at the hearing, the carriers agreed to continue the joint rate provided certain rate increases that had been made after the suspension would be allowed to continue. This arrangement satisfied the parties and the joint tariffs remained in force.[1739] In 1922, the Commission upheld the rates from Kansas and Oklahoma origins to Franklin and Lacy Stations, Pennsylvania, but ordered the pipeline carriers to reduce their minimum tender requirements to 10,000 barrels.[1740]

In 1934, the ICC commenced its first investigation of pipeline rates and gathering charges. Ironically, the underlying factor for the investigation was the proposed *reduction* in rates by the Shell, Texas, Stanolind (Amoco) and Texas-Empire pipelines connecting the Midcontinent fields to the Chicago refining area. John E. Shatford, representing the Louisiana-Arkansas Refiners' Association, protested the reductions on the ground that their field-oriented refineries in Arkansas and Louisiana would be placed at a competitive disadvantage in such markets because they had to use railroad transportation at rates which made their laid-down price in Chicago non-competitive. Division 2 of the ICC denied the request for suspension of the protested rate reductions on June 20, 1934, but the ICC on its own motion, ordered a general investigation not only of the rates in question but of all rates which had been reduced since the rate reductions in question. The resultant investigation thus was broadened to include 37 out of the 49 carriers reporting to ICC in 1933.[1741] Questionnaires were sent to each respondent pipeline company seeking data on physical facilities, shippers, history of rates and pertinent

---

[1734] 51 ICC ANN. REP 102 (1937).
[1735] 52 ICC ANN. REP. 118 (1938).
[1736] 53 ICC ANN. REP. 135 (1939).
[1737] BEARD, *supra* note 45, at 62.
[1738] 54 ICC ANN. REP. 133 (1940). Under this program, the properties of 52 operating companies and 1 lessor were valued as of December 31, 1934. WOLBERT, *supra* note 1, at 140.
[1739] Crude Petroleum Oil from Kansas and Oklahoma to Lacy Station, Pa., 59 I.C.C. 483 (1920).
[1740] Brundred Brothers vs. Prairie Pipe Line Company, 68 I.C.C. 458 (Div. 2, 1922).

financial data as of the close of 1933. On July 23, 1935, a hearing was held before Commissioner Clyde B. Aitchison which only 25 persons attended. Just one witness, John Shatford, testified and his testimony was directed principally to an alleged damage to the field refiner and to the railroads if the reduced rates were permitted to continue. The responses to the questionnaire disclosed a rate history of gradual reduction since 1931, a wide diversity of minimum tender requirements, and very substantial earnings in relation to the *equity* capitalization of the lines (whose owners had rather universally followed the practice of "thinly capitalizing" the lines to avoid franchise taxes and to permit repayment of advances as debt instead of dividends).

Because of the desultory showing, Commissioner Aitchison attempted to establish a framework for further investigation by posing six questions: (1) Were rate schedules or individual rates excessive in terms of aggregate earnings, or in comparison with other rates, costs of service, or effect on movement of traffic or value of service; (2) Should the rates be considered with respect to the respondents individually and should their effect on other pipelines or rail or water carriers be considered; (3) Were the disparities and differences in rates justified and what accounted for them; (4) Were varying minimum tenders justified and was any violation of the *Brundred Brothers* decision involved; (5) If earnings of carriers were found to be unduly high, what rate or other action was required and (6) Was it desirable or necessary to have greater uniformity in the rules governing shipments?[1742] Commissioner Aitchison invited the carriers and Shatford to submit briefs on the questions and he adjourned the hearing. In response to Aitchison's invitation, only Shell Pipe Line Corporation filed a brief. In April 1936, Hearing Examiner J. Paul Kelley issued a proposed report. Relying principally on the answers to the questionnaires, Kelley recommended that the Commission take three actions: set a general guideline of 10,000 barrel minimum tender; order additional information updating the questionnaire to cover 1934 and 1935; and order respondents to show cause why their rates should not be found unreasonable to the extent they exceeded 65 percent of those in existence on December 31, 1933.

This action stung the parties into action and the illusion of unity across the industry was shattered when Sohio and National Refining Company intervened in support of Examiner Kelley's findings. Many of the companies had no basic objection to the lower rates, but the rationale

---

[1741] PETROLEUM PIPELINES, *supra* note 45, at 242. Professor Johnson reports that within the industry, the ICC's action was attributed to urging by Secretary Harold Ickes, in his role as Administrator of its Code of Fair Competition for the petroleum industry. *Id.* at 527 n.22.

[1742] *Id.* at 243.

of reducing rates because past pipeline earnings appeared to be too high, without any consideration of the effect of the 1934 or 1935 reductions on earnings, or the relationship between current earnings and a fair return on the value of the properties, was bothersome to some, and others were concerned about the failure to distinguish between the different positions of the lines.

In November 1938, another hearing was held and the updated information was introduced into the record. Several companies, especially Oklahoma Pipe Line Company and Ajax Pipe Line Company, recounted the difficulties they were having arising from decline in throughput. Presentations were made in an effort to ensure that Examiner Kelley was thoroughly familiar with the intricacies of pipeline operations and ratemaking. Examiner Kelley apparently saw the problem in a new light because in his second report, issued in February 1939, he subscribed to the view of pipelines as plant facilities and found an absence of foundation upon which to declare that pipeline rates were unreasonable or that the tariff rules and regulations were unlawful or discriminatory. Accordingly, he recommended that the proceedings be terminated.[1743]

However, when the case went up to the full Commission, a majority of the Commissioners decided that action by the Commission was required, and it issued an order to that effect on December 23, 1940.[1744] In this order, the ICC adopted the *Brundred Brothers* holding that minimum tenders in excess of 10,000 barrels were unreasonable and it gave respondents 60 days to show cause why such a finding should not be entered[1745] and why rates on crude oil pipeline should not be reduced so as to limit returns on valuation to an 8 percent maximum.[1746]

The Commission addressed the question of rates of return on products pipelines the following year.[1747] Certain independent Mid-Continent refiners asserted that they were being crowded out of the Midwestern and North Central gasoline markets because of their disadvantaged position in shipping over railroads which had excessively high rates from the Mid-Continent, while short-haul rates from various Midwestern refining points had been reduced to meet the products pipeline competition. Although the Commission found that complainants' decrease in sales was due substantially to a 63 percent increase in refinery outturn in the marketing area,[1748] it ordered pipeline rates not

---

[1743] *Id.* at 246-248.
[1744] Reduced Pipe Line Rates and Gathering Charges, 243 I.C.C., 115 (1940).
[1745] *Id.* at 136-137.
[1746] WOLBERT, *supra* note 1, at 138; PETROLEUM PIPELINES, *supra* note 45, at 249-250.
[1747] Petroleum Rail Shippers' Ass'n vs. Alton & Southern R.R., 243 ICC 589 (1941).
[1748] *Id.* at 593.

to exceed 10 percent on ICC valuation and minimum tenders not to exceed 5,000 barrels of a single product from one shipper to one consignee, subject to delay until the carrier had accumulated 25,000 barrels of the same specification product for movement.[1749]

In 1944, the ICC handed down its decision in the *Minnelusa Oil Corporation vs. Continental Pipeline Company*,[1750] which refused to depart from its prior guideline of 8 percent return on crude lines which it had established in its 1940 *Reduced Pipe Line Rates and Gathering Charges* decision.

In 1948, the Commission issued its final decision in *Reduced Pipe Line Rates and Gathering Charges*,[1751] which found that as a result of a series of rate reductions by crude oil pipelines, rates had been reduced to a level of less than an 8 percent return.[1752]

The ICC had an occasion to consider a rate case again in 1962, when certain railroads protested a rate reduction by a propane line running from the Southwest to the Midwest. The railroads claimed that they were being damaged by this "act of destructive competition." However, the Commission refused to suspend the reduced rate.[1753]

The regulatory actions taken by the ICC/FERC since that time were detailed in Section VI B 1, "Interstate Commerce Commission — Now Federal Energy Regulatory Commission."[1754]

b. Indirect, But Powerful, Effects of the Elkins Act Consent Decree

There seems to be little question that the *Atlantic Refining Company* consent decree has exerted a downward pressure on pipeline rates.[1755]

A graphic example of this was provided by Colonial which had set its initial Houston-New Jersey rate at 35 cents to compete with T-2 tanker rates. The tanker rates subsequently rose to above a dollar a barrel. The Colonial line was full, so it increased the size of its line and the incremental barrels reduced its average operating cost per barrel. It reduced its rates to 28 cents per barrel when it had no competition from tankers,

---

[1749] *Id.* at 663-665.
[1750] 258 I.C.C. 41 (1944).
[1751] 272 I.C.C. 375 (1948).
[1752] SPECTRE, *supra* note 104, at 113.
[1753] Pipeline Rates on Propane from Southwest to Midwest, 318 I.C.C. 615 (Div. 2, 1962).
[1754] See text at notes 1564-1581, *supra*.
[1755] W. J. Williamson, *supra* note 315, at 14-15; *Consumer Energy Act Hearings*, *supra* note 154, at 594 (Statement of Jack Vickrey); *S.1167 Hearings*, *supra* note 432, pt. 9 at 604 (Statement of Vernon T. Jones); *cf.* A.G.'s Deepwater Port Report, *supra* note 446, at 61.

and it did so because of the Consent Decree constraint.[1756]

The reason for this is essentially the questionable usefulness of any earnings in excess of the allowable seven percent, coupled with the fact that ICC has a crude rate limitation of 8 percent.[1757] In testifying before the House Small Business Committee in 1972 (*Smith-Conte Hearings*) Arthur J. Cerra, Deputy General Counsel of the ICC, stated: "We think that the consent decree has presently been a very effective means of limiting the oil companies in obtaining profits."[1758]

The Consent Decree has had another effect on pipeline financial thinking, namely, that the seven percent limitation on annual dividends permitted by the Decree has been an important contributing factor to the increase in pipeline capitalization leverage;[1759] in the case of some companies it has been the *primary* factor leading to the current degree of pipeline borrowing.[1760] As a leading investment banker put it: "[the Consent Decree limitation] invariably dictates that oil companies maximize the use of debt, the objective being to match as closely as possible permitted ICC depreciation with repayment of debt."[1761] If they didn't do so, they would be limiting their return in a highly risky venture for the possibility of earning a 7 percent return, which is a "palpably inadequate rate of return under today's conditions."[1762] The result of this has been to increase the debt to total capitalization of all pipelines reporting to the ICC from about 35 percent in 1948 to almost 52 percent in 1958 and to 57 percent in 1971.[1763] Because today's lines are usually financed on a 90 percent basis, this figure must be even higher now. Of course, this fact has not gone unnoticed in government circles. Assistant Attorney General John Shenefield in his Proposed Statement before the Kennedy Subcommittee stated that "[i]n response to this situation, debt financing in the industry has risen sharply since 1941. Prior to 1941 pipelines were funded almost entirely from equity funds provided by their shipper-owners, and investments in oil pipelines from outside sources were rare. Oil pipelines had capital stock outstanding of over $264 million in 1940, while total debt of pipeline companies reporting to the ICC that year was under $21 million, and that amount was attributed to only eight

---

[1756] *RM 78-2 Hearings Transcript, supra* note 754, Vol. 12 at 1939 (Oral Argument of Jack Vickrey).
[1757] PIPELINE TRANSPORTATION, *supra* note 162, at 45.
[1758] *H.Res.5 Subcomm. Hearings, supra* note 234, at 95.
[1759] PIPELINE PRIMER, *supra* note 331, at 16.
[1760] Ryan, *TAPS Rebuttal Testimony, supra* 724, at 30.
[1761] Gary, *Alaskan Bills Testimony, supra* note 1095, at 1117.
[1762] Gary, *TAPS Rebuttal Testimony, supra* 726, at 25.
[1763] PIPELINE TRANSPORTATION, *supra* note 162, at 45.

companies. Today, heavy debt financing with minimal equity contribution is commonplace, with debt-equity ratios of 90:10 or higher [Explorer's went above 100 percent when its equity was wiped out by losses during the first five years of operation].[1764] Professor William K. Jones, in his DOE Report, observed: "The nature of the formula is such that it encourages the maximum amount of debt financing and imposes only the most haphazard (and extremely liberal) limit on the return to equity interests."[1765]

The first attempt to challenge the effect of this leverage which critics have claimed permits the carriers to charge "excessive rates" was the ill-fated attempt to rewrite the Consent Decree by the device of the so-called "Motion(s) for Order for Carrying Out Final Judgment" in the *Arapahoe* case, discussed in Section III G 1 b, "In-Place Enforcement Agency Changes in Antitrust Theory."[1766] The outcome of that effort has been summarized by Professor William K. Jones in his report prepared for the Department of Energy: "In a subsequent dispute as to the interpretation of the decree, the Supreme Court held that the seven percent in allowable dividends was to be computed by reference to the entire valuation of the common carrier pipeline, not the portion represented by the equity investment of the owner-shipper, *United States vs. Atlantic Refining Co.*, 360 U.S. 19 (1959)."[1767]

After the *Arapahoe* decision, industry critics have used two lines of argument in an attempt to combat the *prima facie* reasonableness of industry rates. One strand, which flies squarely in the face of *Arapahoe*, appears to be merely a recidivism to the "good old days" of the *TNEC Hearings* and Senator Joseph O'Mahoney. Perhaps the lure of sensationalism is just too strong for an office-seeker to resist. Thus, Senator Birch Bayh "testified" during Frank Ikard's testimony. "If you look at the 1973 returns on *equity* for just a few [of the oil pipelines], you see that the Shell Pipeline Co. [*sic*] made 18.7 percent; that Exxon made 31.7 percent; Chevron 61.6 percent; and Colonial Pipeline 65.3 percent." (Emphasis added).[1768] At the next hearing session, the Senator led witness Thomas D. Jenkins, President of Permian Corp., through the following line of questioning—Senator Bayh: "Let us say, this last year [1975], what was the rate of return for Permian, relative to assets?" Mr. Jenkins: "Relative to total assets, approximately 12 to 14 percent." Senator Bayh:

[1764] SHENEFIELD PROPOSED STATEMENT, *supra* note 179, at 11. See also Kaplan, RM 78-2 Statement, *supra* note 641 at 16, which parrots Shenefield.
[1765] JONES DOE REPORT, *supra* note 372, at 46.
[1766] See text at notes 990-996, *supra*.
[1767] JONES DOE REPORT, *supra* note 372, at 45.
[1768] *Petroleum Industry Hearings*, *supra* note 192, at 143.

"What about equity?" Mr. Jenkins: "Equity, approximately—this year, equity, it would be approximately 40 to 42 percent." Senator Bayh: "What about last year on equity?" Mr. Jenkins: "Last year exceeded this year." Senator Bayh: "By how much?" Mr. Jenkins: "Oh, approximately, on total assets—approximately 15 percent. Equity was approximately 48 percent, to the best of my recollection." Senator Bayh: "What about the year before that?" Mr. Jenkins: "The year before that was not as good—1973 was not as good as 1974. I do not have the exact figures in front of me."[1769]

Senator Howard Metzenbaum, who was scheduled to give a luncheon address at the AEI's Conference on Oil Pipelines and Public Policy, but who was unable to attend, submitted prepared remarks in which he stated: "The futility of government attempts to control the petroleum pipeline monopolies is reflected in pipeline profits. The rates of return on shareholders [sic] *equity* for all pipelines have increased rapidly from 15.7 percent in 1974 to an astounding 26.2 percent in 1976.

"The rates of return of some extremely important lines are even more startling. The largest product line — Colonial — which is owned entirely by vertically integrated oil companies, had a 121.4 percent rate of return on *equity* in 1976." (Emphasis added).[1770]

As might be expected from the accusatory tenor of the Kennedy Staff Report, it indulged in the return on equity sleight-of-hand to equate the Consent Decree's 7 percent limitation to a 70 percent return on equity [based on an assumed 90-10 debt to total capitalization ratio].[1771] Even the normally professional Justice Department has succumbed to the temptation to raise this discredited issue yet another time.[1772]

Industry spokesmen are quick to point out that this method distorts economic reality by completely ignoring the fact that the shipper-owner is completely obligated, come "hell or high water" to repay the debt.[1773] William P. Tavoulareas, President of Mobil Oil Corporation, made a "homey" analogy when he testified: "You know, it is like saying, 'You own your house, and you put 10 percent in your house, you have got a mortgage [on] the other 90, so you do not really have the 100 percent in your house.' You surely have the 100 percent in your house, because you

---

[1769] *Id.* at 216.
[1770] Metzenbaum, Remarks in Absentia, *supra* note 658, at 2.
[1771] KENNEDY STAFF REPORT, *supra* note 184, at 8-9.
[1772] D.J. Views and Arguments, *supra* 656, at 25; see also Hay, *FERC Direct Statement, supra* note 1689, at 7. Hay cited *Forbes Magazine's* list of 1000 [sic] industrial firms but *Forbes*, in the same article, clearly warns against use of this method. *S.1167 Hearings*, pt. 9 *supra* note 432, at 598 (Statement of Fred F. Steingraber).
[1773] *Petroleum Industry Hearings, supra* note 192, at 303 (Statement of E. P. Hardin); *Id.* at 1836-1837 (William P. Tavoulareas); LEGRANGE, *Government Regulation, supra* note 1413, at 110.

are responsible for it."[1774] L. D. Wooddy, who was President of Exxon Pipeline Company at the time of his testimony in the *TAPS Hearing,* illustrated the meaningless nature of a return-on-equity analysis by referring to protestant's (State of Alaska's) witness Dr. Michael J. Ileo's testimony, principally his Exhibit 563 (MJI-1.14), in which Dr. Ileo developed what he considered to be a reasonable level of tariff rates. But when Dr. Ileo converted those very same tariff rates to returns on equity, he produced numbers ranging from 3.4 percent to 158.2 percent for the various TAPS owners. As Wooddy testified: "[Dr. Ileo's] own calculations thus illustrate the error of attaching any importance to returns on oil pipeline equity."[1775]

Dr. Ezra Solomon, Dean Witter Professor of Finance, Stanford University Graduate School of Business, in his testimony in the Commission's *Valuation of Common Carrier Pipelines* proceedings [then *ICC Ex Parte No. 308*], illustrated the fallacy of return on equity from a different point of view. He illustrated how the rate of return must be dependent solely on the perceived riskiness of the investment *itself* and that such rate was independent of the particular source from which funds are derived to make that investment. His example was an investor who had three $10,000 bank accounts — one given to him by his grandfather, one consisting of money he found unclaimed in the streets, and one which he had accumulated after years of extremely hard work and representing difficult savings. As Dr. Solomon pointed out, the rate of return from a minimum risk investment from *each* of these three sources of investible funds was 8 percent. It would be absurd, testified Dr. Solomon, to expect a zero return on the first two sources because both were "free" in the sense of cost and it would be equally absurd to expect a 20 percent per annum yield from the third fund just because it has been so difficult to accumulate.[1776]

The second strand of "fall-back" attack by critics after losing the *Arapahoe* case is an asserted unfairness or "excess profits," derived by pipeline companies in calculating their rates by including interest expense in the computation of net income and then adding a return element to such net income sufficient to provide them a 7 percent return on total investment (equity plus debt). Professor William K. Jones objected decorously to this practice in his report to the Department of Energy: "Nor is [the allowable rate limitation of the Consent Decree] a measure of overall earning capacity, since the numerator excludes interest on debt

---

[1774] *Petroleum Industry Hearings, supra* note 192, at 1837. (Statement of William P. Tavoulareas).

[1775] Wooddy, *TAPS Rebuttal Testimony, supra* note 725, at 31.

[1776] Solomon, *FERC Direct Testimony, supra* note 682, at 5-6.

(which is included in a calculation of a conventional rate of return on rate base of a regulated utility)."[1777] Advocates, such as the attorneys representing the Alaska Intervenors in the initial *TAPS* proceedings in the Supreme Court employed the perjorative expression "double-dipping" to decry the method.[1778] Assistant Attorney General John Shenefield used the term "double-counted."[1779]

The practice of permitting interest which is paid to lending agencies or financial institutions not in any way connected with or related to a pipeline company's parents or stockholders to be deducted before computing the allowable dividend harks back to the so-called "Bergson letter" of September 14, 1950,[1780] and, according to Shenefield's Proposed Statement, was construed by the Supreme Court in the *Arapahoe* case to be permitted by the Consent Decree.[1781]

The Consent Decree, and by implication, the ICC method, has its supporters. In his oral argument in the *Valuation of Common Carrier Pipeline* proceedings, Jack Vickrey, representing Belle Fourche, an independent pipeline company, and speaking also in behalf of the IPAA and the Transportation Association of America, defended it on a pragmatic basis saying:[1782]

"The decree has several things in its favor.

"In the first place, consent decree earnings are based on valuation, which takes into account the cost of reproduction new, and partly recognizes inflation.

"In the second place, interest is treated as an operating expense. This is not a common practice in public utility regulation, but this treatment has provided flexibility by giving recognition to the wide variance in interest costs occurring during the planning, execution and operation of a particular project.

"In contrast, public utility regulators, who treat interest as a cost of capital component, have been unable to cope effectively with the problem of inflation."

The problem of appropriate compensation for inflation has brought forth a second "double counting" or "double dipping" charge against the use of ICC valuation methodology. The critics claim that the use of a rate base such as ICC valuation, which is adjusted for past inflation, in

---

[1777] JONES DOE REPORT, *supra* 372, at 46.

[1778] Trans Alaska Pipeline Rate Cases, 436 U.S. 631 (1978), Brief of Alaska Respondents 13.

[1779] SHENEFIELD PROPOSED STATEMENT, *supra* note 179, at 12.

[1780] Letter dated September 14, 1950 from Herbert A. Bergson, Assistant Attorney General, to J. L. Burke, Esq., President of Service Pipe Line Co., reprinted in full in *Celler Hearings, supra* note 162, at 212-213.

[1781] SHENEFIELD PROPOSED STATEMENT, *supra* note 179, at 10-11.

[1782] *RM 78-2 Hearings Transcript, supra* note 754, Vol. 12 at 1938.

conjunction with a market-derived rate of return, which contains a component for anticipated inflation, results in double compensation for inflation.

The critics argue that inflation could be compensated for by granting a rate of return which would be "inflation adjusted" and applying that return to a net original cost rate base. In effect, this was the approach taken by witness Herman G. Roseman in the *TAPS Proceedings.* His argument was essentially that "fair value" was a "will-o'-the-wisp" and that it was easier to use historical cost and to derive an adequate rate of return by adjusting the intended *real* rate[1783] (which he assumed for purposes of illustration to be 5 percent), for anticipated inflation (which he assumed again for illustration to be 7 percent), producing a total nominal rate of 12.35 percent which is applied to a net original cost rate base.[1784] Application of "inflation adjusted rate of return" of 12.35 percent to a rate basis which is itself adjusted in whole or in part for the 7 percent inflation rate, said Mr. Roseman, would provide a rate of return which exceeded the intended real rate of return of 5 percent, thereby resulting in a "double counting" for inflation.[1785]

Mr. Roseman's arithmetic cannot be faulted. The difficulty lies with his unrealistic assumptions underlying his proposed regulation scheme; in particular, Mr. Roseman's model assumes that at all points of time the nominal rate of return will fully and exactly reflect the future rate of inflation over the relevant life of the pipeline, which can be 30-40 years, and that the regulatory process will function with perfect efficiency in adjusting rates of return to current market conditions. Professor Ezra Solomon, in his TAPS rebuttal testimony, demonstrated the frailty of both these basic assumptions. He examined the experience of the long-term rate of interest on high-grade corporate bonds over the past 35 years. Dr. Solomon found that such rate *consistently* had underestimated the long-run rates of inflation. Hence the assumption that the current observable long-term rate of interest would reflect properly the inflation rate that this country is going to experience over the next quarter of a century is not a realistic one. He illustrated how this was so by showing the actual experience over the past 25 years: the yield on new issues of long-term high-grade corporate bonds averaged 3 percent over the five years preceding 1954, and in that year the yield was just under 3 percent. Since bond investors expect a real rate of return of around 3 percent, the observable *nominal* long-term rate of 3 percent per annum to maturity as of

---

[1783] The "real" rate of return is one that would be expected by investors in the absence of *any* inflation.

[1784] *TAPS Hearings Transcript, supra* note 754, at 9279-9280 (Direct testimony of Herman G. Roseman).

[1785] *Id.* at 9292.

1954 must have reflected an expected rate of inflation close to *zero* over the life of the bonds, say, 20–25 years. In fact, measured by the broadest available yardstick of general inflation (the Gross National Product implicit price-deflator), the actual rate of inflation experienced over the succeeding 24 years (1954–1978) was 4 percent compounded. Obviously, the observable long-term nominal rate in 1954 seriously under-anticipated the actual course of future prices. Dr. Solomon confirmed this experience by showing the same result during the 25-year period following the *Reduced Pipe Line Rates and Gathering Charges* decision in 1940.[1786]

Even if the nominal rate to be applied against the original cost rate base is changed promptly each year by the regulatory agency for increases in the observed nominal rates during each preceding year, and the company is able to charge and receive tariffs incorporating its full rate of return [clearly an unrealistically optimistic assumption] the method proposed by Mr. Roseman would not allow the regulated company to receive its full, intended rate of return.[1787] Not resting on the negative, Dr. Solomon applied the same empirical test to the existing ICC method which showed that it produced results much closer to the appropriate level, although even it, because the valuation base is always weighted to a certain extent by an element of original cost, less depreciation, was *below* what full adjustment for actual inflation would have produced.[1788] Dr. Solomon attributed the ICC method's more correct result (compared to that reached by use of the original cost method) to the ICC method's greater flexibility and adaptability to whatever rates of inflation occur, *as the change takes place*, thus producing results more consistent with what occurs in the unregulated competitive market. The present method also accords recognition to the fact that pipelines, unlike franchised monopolies which enjoy virtually exclusive access to a captive end-market, must depend upon end-markets which are not only geographically dispersed but which have access also to alternative modes and routes for their supply.[1789]

Dr. John M. Ryan, whose experience combines an impressive academic background with extensive experience in the industry,[1790] also provided in his TAPS hearing testimony a scholarly rebuttal to the "double counting"/"double dip" argument. He explained how the advocates of

---

[1786] Solomon, *TAPS Rebuttal Testimony, supra* note 621, at 40-42 and Exhibit ES-3.

[1787] *Id.* at 46-50 and Exhibits ES-4 through ES-6.

[1788] *Id.* at 51-57 and Exhibits ES-7 and ES-8.

[1789] *Id.* at 61-63.

[1790] See Ryan's curriculum vitae in Ryan, *TAPS Rebuttal Testimony, supra* note 725, at Exhibit JMR-1, which couples a Ph.D. in economics with minors in mathematics and statistics from the University of North Carolina with 20 years experience in the oil industry.

that argument, even Herman Roseman, who advanced the most logical rationale in its support, had ignored a very important economic fact of life, namely that inflation creates what economists call "economic rents." Dr. Ryan illustrated this by posing a situation where an investor built a widget factory today for $100x dollars and the price of widgets was such that the investor earned a fair rate of return after adjusting for inflation in the manner that Justice Department witness Roseman had outlined. Dr. Ryan then described the situation, which is familiar to everyone today, where, a short period later, the cost of building an identical widget factory cost $200x. If the demand for widgets is growing and a new factory is required to meet that demand, the factory will not be built until the price of widgets has risen (or production costs have fallen) to the point that the person building the new plant can realize a rate of return on the inflated cost of the new plant equal to the opportunity cost of capital. When, and only when, that happens will the new factory be built.

Now, if the economic incentive to build that new plant was created by a rise in the price of widgets, the first investor will be making widgets for about one-half the cost that the second investor is incurring, but he will be selling them at the same price. Therefore, the first investor will be making a considerably higher profit than the second investor. The differential is what economists call an "economic rent." The first investor can realize this "rent" either by remaining in the business of selling widgets at a higher profit margin than his competitor or by selling his plant at its inflated or then current fair value. Simply by calling this a "double dip" or "double counting" does not change the basic economic fact that in a free market the integrity of the original investment is generally protected and the investor is simply enabled to earn a market-determined rate of return on that fair value rate base.[1791]

Witness Roseman, while ignoring the question of economic rent in his explicit argument, implicitly was suggesting that the first investor does not require this economic rent to induce him to invest; all he needs to receive is his inflation-adjusted real rate of return on his original cost because this is all that was required to induce him to invest in the first instance. While there is some logic behind the argument, it fails to take into account the alternatives available to the investor. Only a portion of the economy is subject to original cost rate regulation. If economic rents are to be pre-empted for the investor in such regulated sector, but not in others, then in an inflationary environment, the investor in that sector is being discriminated against, and he will attempt to shift his capital to the higher return, unregulated sector. While the investor who already has his investment in place may be "stuck" with his investment and have no

---

[1791] Ryan, *TAPS Rebuttal Testimony, supra* note 724, at 53-56.

recourse but to continue producing widgets in an effort to get as much of his original investment back as he can, it is unreasonable to expect him — and others who have perceived what had happened — to make similar investments. The result will be a flight of capital from industries suffering from this type of regulation.

The real life proof of this is evident from a comparison of the Standard & Poor's Security Price Index for all Industrial Stocks with that for the utility companies. Since 1965, when the high inflation rates really commenced, the S & P index for industrials has managed to rise while that of the utility companies is well below its 1965 level. Investors dumped utility stocks to buy assets with greater growth potential, and suffered capital losses in the process.[1792] As Dr. Ryan described it: "The combination of original cost rate making and the inflationary environment simply leads to confiscation of capital and thereby precipitates a flight of capital."[1793] Nor does the consumer benefit in the long-run; he either does not get the service he wants at a price he is willing to pay [what better example than natural gas?] or he gets shoddy or less reliable service due to the regulated firm's depressed situation.

In the AEI's Conference on Oil Pipelines and Public Policy, Professor Edward W. Kitch, of the University of Chicago Law School, expressed much the same idea concerning economic rents when he commented:[1794] "An issue that runs through the whole discussion is whether these organizer-owners have, in the beginning, a special advantage in organizing pipelines which could come from know-how, or from the fact that they have an important position in the field that is being exploited.

"I am implicitly suggesting that some of the returns could be characterized as rents. And to make a general point about price theory, there are times at which the Department of Justice's position seems to me to fall into the fallacy — or to teeter on the brink of the fallacy — that the existence of a low-cost firm in a market makes that firm a monopoly. In fact, low-cost and high-cost firms can be in the same market. The competitive price will be the marginal cost of the high-cost firm. The low-cost

---

[1792] *Id.* at 56-59. See also Ryan's Exhibit JMR-4, an excerpted portion of which is as follows: Monthly averages of daily indexes (1941-1943) = 10).

Year	Industrials	Utilities
1965	93.48	76.08
1969	107.2	62.64
1973	120.5	53.47
1977	108.4	54.23

[1793] *Id.* at 59.

[1794] AEI, *Oil Pipelines and Public Policy, supra* note 684, at 122.

firm will be very profitable, and it would not be correct to characterize that as a monopoly return.

"Now, the other part of this is, there is also implicit, it seems to me, in the Justice Department's position an assumption that entry by non-petroleum entities, nonpresent participants in the oil industry, is an equally attractive substitute, which would solve some of the complexities of vertical integration. This assumes that the oil companies do not possess specialized know-how, planning capability, technology, or position that is of value to the pipeline company. If they do have such a special position, then the insistence on the use of outside organizers will raise the cost of the organized pipelines to consumers."

Mr. Roseman objected also to the use of ICC methodology on the basis that original cost regulation is easier to apply. Industry supporters, in response to this objection, point to the fact that the array of original cost experts who testified in the TAPS proceedings could not agree among themselves how it should be done. In an effort to create a proxy for a competitive rate, each of the "original cost" witnesses sought, in his own way, to create a hypothetical pipeline company with an imaginary capital structure that could "stand alone" without owner guarantees. The only thing they did agree upon was that the existing highly leveraged line would not fit into the FPC mold. Instead of producing a more simple method of regulation, their suggestions only made the Consent Decree/ICC regulatory method appear more simple and straight-forward. Nor was there any explanation offered as to how the pipeline companies subject to the Consent Decree would be able to live with two forms of regulation. It is difficult to see how anyone other than the old-line FPC staff, who would have to develop expertise in a new (to them) method, would benefit from a change in oil pipeline rulemaking. The testimony indicates that the position of oil pipeline companies would be substantially worse to the point where it is questionable whether they could attract capital; the independent producers have gone on record as opposing it; the Justice Department supports it but admittedly hasn't thought through all the ramifications. As ICC Commissioner Webb said in a similar situation, "[i]f it is not necessary to change, it is necessary not to change."[1795]

Having discussed the contentions of both sides of the argument, perhaps the examination can be brought into sharper focus. First, the *Hope Natural Gas* case clearly did *not* hold that the ICC methodology was invalid. Dean Eugene V. Rostow, hardly a friend of the oil pipeline industry, in his provocative monograph, "A National Policy for the Oil Industry," after analyzing the *Hope Natural Gas* case, correctly stated that the Court was not much interested in the rate-making theory the FPC

[1795] Tierney, *TAPS Rebuttal Testimony, supra* note 175, at 20.

used but in the result achieved,[1796] and further remarked: "The *Hope* case confirms the probable legality of the Interstate Commerce Commission's opinion in its pipe-line investigation."[1797] This "result achieved" is the product of a joint determination of the way in which the property rate base is valued and the rate of return allowed on such base. When price levels are stable, the choice of original cost or reproduction cost or some mixture really is not an issue — each would produce results which would satisfy the statutory test of a "just and reasonable rate." However, when price levels change, the original cost of an asset commences to lose economic significance as an indicator of value. This effect becomes even more pronounced as the level of inflation rises, as it has done in this country, to the point where investment analysts, the Financial Accounting Standards Board (FASB) and the Securities and Exchange Commission (SEC) are requiring adjustment in financial statements to reflect the changing purchase price of the dollar. In effect, inflation-adjusted asset valuation, which the ICC has used for decades in oil pipeline regulation is coming into more widespread use.

The adoption of this system by the ICC in accordance with its mandate under the Valuation Act of 1913 (which has not been changed materially as to oil pipelines)[1798] is especially apposite to oil pipelines because use of an original cost basis for valuation of their assets leads to an even larger error than it does in the case of most other enterprises whose assets are placed in service over a continuing period of time, whereas the great bulk of investment in an oil pipeline is made at the time of its original construction and the size of additional plant, equipment and inventory are relatively small compared to the original investment.

Thus, as inflation grows, the differential between the original cost of total assets of an oil pipeline increases far more rapidly than it does in a firm whose continuous investments tend to narrow the gap between original cost and true value, due to the fact that the additional assets reflect at least *pro tanto* the effect of inflation up to the point of time they are added to the base. Thus to use the original cost of a pipeline built, say, 15 years ago, as a base upon which to apply the currently required rate of return would result in *"under-counting" or "half-dipping,"* and raise questions of expropriation of the pipeline owner's assets for public use without just compensation.[1799]

[1796] Federal Power Commission v. Hope Natural Gas Co., 320 U.S. 591, 602 (1944) ("Under the statutory standard of 'just and reasonable' it is the result reached not the method employed which is controlling. . . . If the total effect . . . cannot be said to be unjust and unreasonable, judicial inquiry is at an end.")

[1797] ROSTOW, A NATIONAL POLICY FOR THE OIL INDUSTRY 62 (1948).

[1798] See text at notes 1586-1602, *supra*.

[1799] Solomon, *FERC Direct Testimony, supra* note 682, at 14-16.

c. Where Did the "Monopoly Profits" Go — Why Isn't Everyone Clamoring to Get In?

Many economists consider profits as a hallmark of monopoly power. Professor Edward Mitchell has cautioned against the automatic equation of high profits with monopoly. For example, they could be the result merely of an abnormal risk situation, or an "economic rent" where one firm has technological superiority. Nonetheless, the persistence of abnormally high profits over long periods of time in a particular industry make it more likely the industry is monopolistic than competitive. Professor Mitchell has stressed the qualified nature of this relationship, *i.e.* it is probabilistic; the time span of high profits must persist for long periods before the probability can be seriously considered; and the abnormally high profits must extend across the industry, not just a few firms or there will be a danger of confusing economic rents with monopoly profits.

With this background in mind, what kinds of allegations are currently being made by pipeline industry critics? The Kennedy Staff Report mentions pipeline profits in two places, *i.e.,* page 10 wherein it states: "The natural monopoly characteristics of pipelines and their absolute cost advantages relative to other modes of petroleum transportation virtually immunize pipelines from competition. Oil company owners of petroleum pipelines have succeeded, despite common carrier regulation, in using their pipelines to secure anticompetitive advantages or monopoly profits at all levels of the petroleum industry." On page 144 of the Report, referring to the Justice Department, the authors assert: "The data presented by the Department in a companion statement [filed by Justice in the FERC valuation hearings, *i.e.,* the George Hay Statement cited in footnote 1695] demonstrates [sic] that *many* oil pipelines have profit levels one would expect of an unregulated natural monopolist; indicating that ICC rate regulation has failed to curtail monopoly profits. For example, 20 pipelines earned returns to *equity* of greater than 40 percent, while 11 had returns to *equity* of 60 percent or better" (emphasis supplied). The reader is urged to examine the Hay Statement, which is part of the FERC RM 78-2 *Valuation of Common Carrier Pipelines* record, and is attached to the Statement of Views and Arguments filed by the Department of Justice in the proceeding on May 27, 1977. However, the Kennedy Staff Report's statement provides an obvious indication of the nature of this argument, *i.e.*, a bald conclusionary statement combined with the vacuous rate of return on equity argument which has made political headlines since the TNEC days but which has been thoroughly discredited by reputable economists, investment bankers and financial experts.[1800] Significantly, Thomas Spavins, an economist for the Department of Justice, studiously avoided this technique and focused on the question of the appropriate numerator to be used in the calculation of the rate of return.[1801]

Assistant Attorney General Schenefield, in his Proposed Statement before the Kennedy Subcommittee, at page 19-20, made this statement: "Our study of pipeline returns in connection with our rate reform efforts has provided what we consider to be strong empirical evidence that many [no quantification] pipelines have substantial market power — power which rate regulation has failed to curb. For example, in an evaluation of pipeline profitability conducted by our Economic Policy Office last year [1977], our economists concluded that the rates of return for many oil pipelines are significantly above what is encountered in the *most profitable* electric utility and natural gas pipeline companies [citing the Hay Statement]. This bears out our historical impression that pipelines have exhibited monopoly power all through the years of ineffective regulation."

One notices that Mr. Shenefield's statement meets none of the probability criteria and qualifications used by economists in their careful use of profits as an indicator of monopoly power. First he cites "many," without saying how many, and who they are, and definitely not the *industry*. As a matter of fact, Hay's Table 1 shows that, listed in order of size of return to capital, only 26 out of 98 oil pipelines exceeded the rate of return to capital enjoyed by the highest natural gas line, and 9 of the 26 (including the 1st, 2nd, 4th, and 5th highest) were "independents," and the time span is not over a long period of time, but simply two years. No mention is made of the fact that 7 of the lines had *negative* rates of return and 15 made less than 5 percent, well below the 12.1 percent lowest rate of the five natural gas lines shown. Nor was any account taken of the difference in risk in the industries. The dispersion in returns of the natural gas lines was only from 13.7 percent to 12.1 percent, whereas the return rates on oil pipelines varied from 305.6 percent to −10.2 percent. While the difference may be exaggerated by the small sample of natural gas lines to that of the oil pipelines, it certainly is an indicator of a differential in risk between the industries being compared.

Rates of return of oil pipelines were computed and comparisons to other industries were made on a more professional manner in the works

---

[1800] *E.g.* Arrow, *Williams Direct Testimony, supra* note 1378, at 10 (Economist); Gary, *Direct TAPS Testimony, supra* note 634 at 43 (Investment Banker); Solomon, *FERC Direct Testimony, supra* note 682, at 5-6 (Financial Expert); *see also* text at note 1775, *supra*.

[1801] See text at notes 1701-1704, *supra*.

cited in Section VI C 1 "So-called 'Excessive Rates' of Return."[1802] As the reader will no doubt recall, Professor Stewart Myers, using 1971 figures, derived a weighted average rate of return, using ICC "Form P" Reports, adjusted for "normalized" accounting, of 7.7 percent which he compared with AT&T's 7.8 percent. Ulyesse LeGrange presented a paper based on 1977-1978 figures which showed a mean weighted average rate of return for oil pipelines of 7.6 percent, as against 10 percent for general industry and 6.7 percent for 171 gas, electric and telephone utility companies. Thomas Spavins of the Antitrust Division, whose method differed from LeGrange's in that he did not adjust the interest "add-back" to net income to reflect the tax effect, found a 13.0 percent return for oil pipelines. LeGrange, Spavins and Professor Shyam Sunder, who discussed their papers, were in agreement that the absolute numbers could not be taken as an indicator of monopoly power until a rational analysis of the relevant elements of risk differential between industries could be developed.

John Shenefield himself raised the second part of the question: if the pipeline industry is so profitable, why haven't more firms entered the market to earn some of these monopoly profits? His answer to the question was that to the extent rate regulation is at present ineffective, potential shippers would not want to pay the excessive tariffs that "independent" (non-oil related) pipeline companies would charge. Thus, he alleges the current throughput and deficiency agreement form of financing is an effective barrier to entry by independents. This has a bit of a hollow ring to it in the light of the entry of MAPCO, Kaneb, Southern Pacific, Santa Fe and others into the business. What seems really to have happened is that the Consent Decree and competition have held pipeline rates down to levels that are unattractive to investors, both oil and non-oil related, and hence in the absence of some special need for the line or a particular situation where in-place investments or talents can be used incrementally in building a line there is no incentive to enter into it. As E. P. Hardin, President of Mobil Pipe Line Company, expressed it pithily, "A 7 percent return on valuation is not a red-hot business deal."[1803] Robert E. Yancey, President of Ashland, testified in 1974 that "We would much rather pay the [Colonial] tariff on [the Consent Decree] basis [and] put our money into other areas where we get a much higher rate of return."[1804] Charles E. Spahr, Chairman of the Board of Sohio, in his statement submitted to the Senate Antitrust and Monopoly subcommittee in 1975, remarked: "In the past and today, if an investor were to show an

---

[1802] See text at notes 1682-1709, *supra.* See also Exxon Pipeline Company/Exxon Company, U.S.A., *An analysis of the Rates of Return on Petroleum Pipeline Investments* in AEI, *Oil Pipelines and Public Policy, supra* note 684, at 261-315.

[1803] *Petroleum Industry Hearings, supra* note 192, at 315 (Statement of E. P. Hardin).

[1804] *S.1167 Hearings, supra* note 432, pt. 8 at 5929 (Statement of Robert E. Yancey).

interest in owning a part of the Trans-Alaska Pipeline, and if he were willing to participate on the basis of his own credit, he would be welcomed to the project....At present we are actively looking for others to join us in the building of the Trans-U.S. Pipeline [from Long Beach, California, to Midland, Texas]. They need not be oil companies. We ask only two things. They must be interested in owning and operating a pipeline, and they must be able to provide for their own capital resources without relying on Sohio's credit."[1805] It is a matter of common knowledge that no one stepped up to take advantage of Mr. Spahr's offer. In fact, Standard Oil Company of California opted not to take any interest in TAPS, as did the state of Alaska, and no one has been beating on Sohio's door to get in on the Trans-U.S. Pipeline. The empirical evidence seems to indicate that investment in oil pipelines just is not a "red-hot deal" and the fact that investors have not been clamoring to get in but have been shying away from it appears to be more consistent with that explanation than with Mr. Shenefield's.[1806]

3. Asserted Denial of Access to Pipelines

a. Claimed Direct Denial of Access

Because of the importance of oil pipelines to the rest of the business, access to pipelines traditionally had been regarded as a competitive necessity.[1807] Despite the forthright testimony of Deputy Assistant Attorney General Bruce B. Wilson before the Smith-Conte Subcommittee of the House Small Business Committee on June 15, 1972, that "...[pipeline owners] assert that access to the pipeline is open and available to all without regard to ownership interest. This is a contention difficult to refute when we have few, if any, who will testify that they have been excluded,"[1808] there have been allusions to the subject of access, which critics have used to claim that such denial of access does exist. One example sometimes referred to is the testimony of Robert E. Yancey, President of Ashland, on August 6, 1974, before the Senate Antitrust and Monopoly Subcommittee, wherein the questions and answers were as follows:[1809]

[1805] *Petroleum Industry Hearings, supra* note 192, at 1098-1099 (Statement of Charles E. Spahr).
[1806] *Cf.* Ryan, *TAPS Rebuttal Testimony, supra* note 725, at 72-73.
[1807] Harmon, *Effective Public Policy to Deal with Oil Pipelines,* 4 Am. Bus. L. J. 113 (1966).
[1808] *H.Res.5 Subcomm. Hearings, supra* note 234, at 206 (Statement of Bruce B. Wilson).
[1809] *S.1167 Hearings, supra* note 432,, pt. 8 at 5928-5929 (Statement of Robert E. Yancey).

"Senator Tunney. Do you have any problem getting your product into pipelines that are common carriers, joint ventures of the major companies?

"Mr. Yancey. We have only had one instance where we had a problem of getting products into a pipeline, and that was in moving from the gulf coast to the east coast through Colonial. We tendered some oil to that system, and they had to roll over and give us some space because of the ICC regulations.

"Senator Tunney. They had to give you space?

"Mr. Yancey. Yes sir. Well, we didn't get the space we asked for; we did get it on a pro rata basis.

"Senator Tunney. What was your scheduling? Were you able to put it in the pipeline on the schedule that you wanted?

"Mr. Yancey. Yes, sir.

"Senator Tunney. Were there any product standards applied that were discriminatory?

"Mr. Yancey. As I said, we had to go through the motions of tendering the product to the pipeline, and after that within a month or 6 weeks that we were able to get the product through the line.

"Senator Tunney. But you couldn't get all that you wanted?

"Mr. Yancey. No sir, only because we hadn't been in there originally, and the way Colonial works is that you take it on a pro-rata share. I think we tendered 15,000 barrels a day and ended up getting less than half that space.

"Then we had a historical position, so the next year we tendered a little more and got a little more space. So now I think we have some 10,000 barrels a day space in that pipeline and we pay the tariff."

It seems difficult to read this exchange as representing anything other than that Ashland tendered to a full line and had its tender prorated on a non-discriminatory basis with everyone else. In fact, it may have received even better treatment than the owner-shippers, who had their historical shipments prorated while Ashland had its new tender prorated.[1810] This interpretation is buttressed by Mr. Yancey's statement just five paragraphs later in the cited testimony when he said: "Those companies in Colonial are under a consent decree, and they are only permitted to

---

[1810] This happened because the shipper-owners were shipping at about their refinery outturns and since under the "historical" method of prorationing, the new shippers were prorated at the ratio that their tenders bore to total tenders, whereas the historical shippers' tenders, after being reduced as a group by this "first round," then were further prorationed on the basis that their historical shipments bore to reduced "historical shippers" group allocation. The nonowner historical shippers, who were still increasing their tenders, picked up space while the shipper-owners' movements actually decreased. See text at notes 420, 1657-1660, *supra*.

earn at the rate of 9 percent [actually, only 7 percent] on their invested capital on the pipeline. They have to set a tariff that will not give them any higher return. We would much rather pay the tariff on that basis [and] put our money into other areas where we get a much higher rate of return."[1811] Moreover, Mr. Fred G. Steingraber, President and Chief Executive Officer of Colonial, clarified the incident. According to Mr. Steingraber, Ashland was invited to, and attended, a prospective shippers' meeting called by Colonial on May 7, 1963, attended by 23 prospective shippers (including nine Colonial owners). This meeting was held *prior* to the beginning of Colonial's operations and while the system was still under construction. Ashland's first request to become a shipper on Colonial was made by a letter, dated September 4, 1970, which was answered by Colonial on September 9th, at which time Colonial was advised by Ashland what their initial requirements would be December 1970. Colonial urged Ashland, in a letter dated October 21, 1970, to meet with Colonial's scheduling people as soon as possible in order to work their allocation into Colonial's schedule in such manner as to meet Ashland's timetable. Ashland actually commenced shipping on January 1, 1971. Space allocated to Ashland was, in accordance with its forecast, reduced by the percentage of proration in effect at that time on the Colonial line (which had been on proration since July 1967).[1812] The clinching items of evidence on the proper construction of Mr. Yancey's statement are his recent forthright statement on the subject of access (reproduced verbatim in this text at note 1819) and the fact that the Justice Department, after investigating Colonial from 1963 to 1976, found there were *no* meaningful denials of access.[1813]

There was some criticism[1814] of the Department of Interior's exemption of "operating lines," as "plant facilities" or small "Uncle Sam Oil Companies" if you will, from the provisions of Section 5 (c) of the Outer Continental Shelf Lands Act. This exemption extended to pipelines used solely for purposes such as moving production to a central point for gathering, treating, storing or measuring; delivery of production to a point of sale; delivery of production to a pipeline operated by a transportation company; or moving fluids in connection with lease operations, such as injection purposes. The point urged was that as private lines, the

---

[1811] *S.1167 Hearings, supra* note 432, pt. 8 at 5929 (Statement of Robert E. Yancey).

[1812] STEINGRABER, *supra* note 193, at 140-141.

[1813] KENNEDY STAFF REPORT, *supra* note 184, at 128. See also *S.1167 Hearings, supra* note 432, pt. 8 at 6044-6045 (Statement of Professor Edward J. Mitchell).

[1814] *Petroleum Industry Hearings, supra* note 192, at 59-61 (Statement of Walter S. Measday, Chief Economist of the Senate Antitrust and Monopoly Subcommittee); *Id.* at 311 (Colloquy between Henry Banta, Assistant Staff Counsel and J. R. Kinzer, Counsel for Mobil Pipe Line Co.).

owner could exclude altogether, or admit on whatever terms he saw fit, other producers who wanted to use such lines for the transportation of their offshore production. Dr. Walter Measday, Chief Staff Economist, felt that some of the pipelines through which the oil from offshore wells was flowing to onshore terminals were rather large to be mere "plant facilities." Dr. Measday cited, for examples of his point, Gulf's No. 2 South Timbalier line which is about 28 miles long and, before going onshore, crosses a number of leases owned by other companies. Dr. Measday testified that No. 2 carries only Gulf's oil, and throughput in 1974 was only 24,313 barrels a day although it is an 18-inch line, which, on the OCS, should have a maximum capacity much higher than that, perhaps as much as 200,000 barrels per day. He also cited Gulf's No. 43 line in West Delta No. 2 as another example, stating that it "winds around," covering more than 35 miles, "connecting wildly [sic] scattered leases" and crossing leases of other companies. It is a 16-inch line, which "probably had a maximum capacity in the range of 145,000 barrels a day, but which only averaged a throughput of 9,600 barrels a day in 1964."

Gulf provided an explanation for these two lines later on in the same hearings. No. 2 South Timbalier line is a multiphase line (oil, water, and gas) in which operationally it was more efficient to bring the contents onshore to treat, separate products and, in the case of the gas, to compress at its onshore terminal. Because of hurricanes and other external damage to the line, the inlet pressure had to be reduced to a maximum of 150 p.s.i. The capacity of the line, under these conditions, was 51,000 barrels per day liquids and 500,000 cubic feet of gas. Actual throughput was 34,000 barrels per day of liquids and 500,000 cubic feet of gas. No. 43 line, running from West Delta block 117 over to Gulf's West Delta Block 41 field and hence to the onshore terminal for processing, also is a multiphase line. It is a telescoped line, 10 inches from West Delta Block 117 to West Delta Block 41, and 16 inches from West Delta Block 41 to the onshore terminal. Both segments were operating at capacity. The line configuration, rather than "winding around," actually saved 17 miles over that which would have been required by two separate lines. These lines met fully the "operating line" qualifications (2) & (3) and Section 5 (c).[1815]

Considerable attention was directed to the so-called MCN line (an acronym for Mobil, Continental and Newmont) which was an undivided interest line, the ranks of whose owners had increased to 11 companies at the time of the testimony. The line was originally installed by two majors and an independent. Gulf became the fourth owner and since that time, nine other partipants have joined — of which seven are "independents."

[1815] *Id.* at 1026-1027 (Gulf Oil Response to Dr. Measday's Testimony).

Gulf's records do not indicate that any application for ownership has been turned down. Contrary to Dr. Measday's statement that the operator (Mobil) may require each shipper to provide "extensive treatment for its oil," the agreement in fact simply placed a limitation of 5 percent on the amount of water that can be transported with the oil. This is a level normally achievable by simply separating the free water (pure mechanical gravity separation) which applies to *all* shippers alike, including Mobil. Dr. Measday's statement that "Mobil had the right to shut in the wells of any party who exceeded his portion of the storage facility at the Burns Terminal" (on East Cote Blanche Bay) is a distortion of a contractual policy precisely similar to that which enforcement agencies would have required of a common carrier, *i.e.*, non-discriminatory prorationing of a common facility when it is at full capacity, thus ensuring that all parties, majors and independents alike, receive equitable treatment.[1816]

In the Attorney General's Deepwater Port Report there were extracts from some internal documents obtained by the Justice Department from the files of prospective owners which have been referred to by the Department and others as indicative of exclusionary practices. The first is an excerpt from the Gulf files where the memorandum, or letter, stated: "It was learned from SEADOCK participants that the joint pipeline from the terminal near Freeport will probably be an undivided interest system. If so, and if the volumes are an indication of Gulf's future requirements, Gulf would benefit by becoming an owner rather than a shipper, as a shipper would have to deal with several of the owners to obtain space and could not be assured of obtaining the required capacity."[1817] One could infer from these words that there might be a problem of access. However, read in the light of the fact that this is an internal memorandum in which the writer's job was to anticipate all possible difficulties and to ensure that the Company would be advised of the most secure and administratively efficient method of ensuring orderly flow of its supply stream, it appears more reasonable to construe the memorandum as saying that because Gulf's requirements would be such that no one of the undivided interest carriers would *by itself* have enough spare capacity to accommodate, without prorationing, both its shipment and *all* of Gulf's, Gulf would have to tender to two or more of the undivided interest carriers in order to get its transportation requirements satisfied. Moreover, if Gulf's volumes *in the future* were expected to expand considerably, it would be unwise from a business standpoint to expect the current owners then to construct and finance unused capacity awaiting the time when, to them, Gulf *might*

[1816] *Id.* at 1028-1029.

[1817] A. G.'s Deepwater Port Report, *supra* note 446, at 79. For a parroting reference, see DOUGHERTY, REVISED PROPOSED STATEMENT, *supra* note 656, at 10.

possibly expand its shipments, without seeking some type of financial commitment from Gulf. This being so, it would make sense for Gulf, which had the best information on its future requirements, to provide for such requirements by the ownership route.

Since making this surmise, this author has received a communication from the writer of the cited document. He states unequivocally that the Department's interpretation is *not* correct and he was kind enough to set forth the true meaning of the statement contained in his memorandum, to wit: "A pipeline carrier must have an indication of volumes which will move through its line in order to decide on the capacity of the line. Unless shippers, by either investing in the line, or providing throughput agreements give such indications, the carrier has scant information on which to size the line. This is true whether the line in question be in undivided or stock ownership, but the problem of a non-owner utilizing capacity is somewhat more complex in the undivided ownership case, as he may find it necessary to contact a number of owners rather than the one entity he would contact in the stock company case. It is only sound business for a carrier to construct capacity for volumes which are committed to the line at the time of construction, as to oversize the line may prove wasteful and if fortuitous volumes do not develop, futile. Therefore, a prospective shipper who foresees future increases in his capacity requirements in a particular carrier would ordinarily find it prudent to commit capital or through-put to assure future capacity expansion.

"There is nothing in the statement which I made in my memorandum [of February 9, 1973] which can be taken to imply that the owners of pipeline carriers whether in a stock company or an undivided interest system would attempt to refuse to handle a non-owner's traffic on a non-discriminatory and where necessary, pro rata basis as required by law."[1818]

There were several excerpts from Ashland documents which can be taken fairly to indicate at least their author's view that shipper-owner lines, especially undivided interest lines, discriminate against nonowner-shippers, and one of these documents named Colonial and Explorer as examples of such lines. While we have seen that Colonial was given a "clean bill of health" by the Justice Department after 13 years of investigation and it sounds somewhat ludicrous for Explorer, which was losing $46 million during the period these memoranda were written, to be excluding *anybody,* the most direct refutation of this evidence comes from Robert E. Yancey, President of Ashland, who was disturbed by the

---

[1818] Letter, dated June 21, 1979, to this author from Hugh L. Scott, author of the cited memorandum.

use of the memoranda, which as he states, reflected only an impression by an Ashland Pipeline Company employee many years ago. In an effort to settle once and for all Ashland's position in the matter, Mr. Yancey authorized the use of the following official position:[1819]

> *"Statement of Position of Ashland Oil, Inc.*
> *With Regard to Petroleum Pipeline Ownership*
>
> As evidenced by the Department of Justice's Report and Recommendations on the Deepwater Ports Act Licenses in 1976, the belief has developed that several memoranda, written by an Ashland Pipeline Company employee in 1972, expresses the position of Ashland Oil, Inc. in 1972 and describes facts supposedly existing then and now to the effect that non-owner shippers are 'locked out' from shipment on petroleum pipelines.
>
> Ashland Oil Inc.'s management did not share that writer's view of pipeline structure or behavior in 1972. We are satisfied that pipelines do not deny access to any shipper, owner or non-owner, either by behavior or discretionary structure. Ashland Oil Inc.'s experience as a non-owner shipper through many pipelines, and as an owner and operator in others, does not support that belief today, and did not support it in 1972.
>
> With refineries in seven locations and marketing in twenty-seven states, pipeline access is essential to Ashland Oil, Inc.'s survival. We have no reason to believe that Ashland will be denied access to any pipeline in which it does not share ownership. We base our decisions to build or participate in a pipeline on the economics of the pipeline itself, our need for pipeline transportation, and our conclusion as to whether or not the line would be built at all without our participation. We do not base our decisions on an imagined fear of being "locked out" of common carriers in which Ashland lacks equity interests.
>
> Simply stated, Ashland would not be able to provide the competition it does to shipper-owners were there any truth to the accusations being made."

There was another item of "evidence" in the Deepwater Report, found in the footnote on page 90, which says "When [Mobil] was a

---

[1819] Statement of Position of Ashland Oil, Inc. With Regard to Petroleum Pipeline Ownership, attached to letter, dated May 2, 1978 from Robert E. Yancey, President Ashland Oil, Inc. to James B. Atkin, Committee on Industrial Organization, American Petroleum Institute.

member of LOOP, it too was concerned about space availability in Capline. One way of acquiring more space was to buy into Capline. But the major obstacle was getting someone to sell. As [Mobil] stated: 'Other than money and the availability of substantial volume that would average down everybodies' [sic] cost, [Mobil] has no particular leverage in obtaining ownership. It may be possible that one or more companies such as Union would be willing to exchange support for [Mobil] ownership in Capline for [Mobil] support in their obtaining ownership in such other pipelines as Olympic.'" The Antitrust Division suggests that this supports its contention that Capline owners were not willing to sell ownership interests and that an ownership position was necessary to obtain shipping space in Capline. This is a misuse of a preliminary *draft* internal memorandum from one Mobil staff member to another and the passage was not included in the final position paper which was submitted to Mobil management.

More significant, however, is the fact that the extract from the memorandum was *not discussing access or lack of access,* to pipeline space, since Mobil recognized that it could at any time ship on Capline at tariff, but was referring to a possible basis for purchasing an interest in Capline at a price as low as possible. The quoted extract did *not* reflect Mobil management's belief that access to Capline would be denied if Mobil did not have an ownership interest. Mobil's true position with regard to shipper access to Capline is that Mobil, as a nonowner, just as any other nonowner, *could* ship on Capline consistent with the several Capline owners' common carrier obligations. Mobil's position is directly contrary to the Attorney General's inference and is confirmed clearly by its action in November 1976 when Mobil began shipping on Capline as a nonowner, and since that date continuously has shipped substantial volumes through Capline, paying the published tariffs of the carrier.[1820]

There is ample support for the proposition that there is uninhibited access to oil pipelines by nonowner-shippers. This evidence ranges from the broadest, overall sweep in the form of statistical compilations documenting the large, and growing, use by nonowner-shippers of oil pipelines, which was discussed in Section II A 3 c, "Scramble for Traffic,"[1821] to a number of public statements or testimony by ICC Commissioners,[1822] and official spokesmen for independent producers'

[1820] Communication dated March 29, 1979, from Charles S. Lindberg, Esq., Mobil Oil Corporation, N.Y., N.Y.

[1821] See text at notes 410-496, especially 461-496, *supra*.

[1822] See text at notes 1502-1505, *supra*. There was one instance mentioned by ICC Chairman George M. Stafford where a dispute concerning access was considered by the Commission but it became moot when the carrier provided the requested service. *Consumer Energy Act Hearings, supra* note 154, at 1266 (Statement of ICC Chairman George M. Stafford).

associations[1823] to the effect that they had few, if any, complaints about pipeline service — to the contrary, the pipelines appear to be serving independent shippers well. In addition, there is also testimony concerning individual situations which support the proposition that major oil company-owned pipelines are readily accessible to nonowners including other majors,[1824] independent crude gatherers,[1825] independent producers,[1826] independent refiners,[1827] and independent marketers alike.[1828]

b. Alleged Techniques to Deny Access Indirectly

Rebuffed in their attempts to procure respectable evidence to support a charge of major oil company denial of access to pipelines, critics have resorted to vague allusions about "indirect denials of access." In the Attorney General's *Alaska Natural Gas* Report, at page 39 (footnote) it is stated: "For example, there have been *allegations* that vertically integrated oil pipelines have employed discriminatory devices to avoid honoring their common carrier obligations under ICC regulation. Such devices include (1) tailored routing and sizing of the pipeline for the sole convenience of the owners; (2) denying input connections; (3) refusing to provide sufficient ancillary facilities to accommodate nonowner shippers; (4) refusing to carry small shipments; (5) granting only irregular shipping

---

[1823] See text at notes 1506-1509, *supra*.

[1824] *Petroleum Industry Hearings, supra* note 192, at 1105 (Statement of Charles E. Spahr, Chairman of the Board and Chief Executive Officer, Sohio) ("In the past few years, Sohio has shipped, as a non-owner, through many lines, including Buckeye, Cherokee, Interprovincial, Lakehead, Marathon, Tecumseh, Texas and Texas-New Mexico. Our records do not indicate that we have ever been denied access to a common carrier pipeline. Those of us in Sohio who have been associated with this activity for many years have no knowledge of any such refusal.")

[1825] *Id.* at 219 (Statement of Thomas D. Jenkins, President, Permian Corporation) (now a subsidiary of Occidental) ("I am not familiar with the various policies of the various major oil companies with respect to their pipelines, but I do know that they are, and have been available to us, and I presume others, for transporting crude.").

[1826] Allen, *TAPS Rebuttal Testimony, supra* note 496, at 8-9; *cf. Petroleum Industry Hearings, supra* note 192, at 221 (Statement of Thomas D. Jenkins).

[1827] *Cf. Petroleum Industry Hearings, supra* note 192, at 2018 (Statement of Richard J. Boushka, President of Vickers Energy Corporation) ("We feel the pipeline subject has been overplayed and a simple business made mysterious. Alternatives are too easy to come by for anyone to be injured over a long period of time.")

[1828] *H.Res.5 Subcomm. Hearings, supra* note 234, at 112 (Statement of James W. Emison, Vice President of Oskey Gasoline and Oil Co., Inc.) (" . . . Explorer Pipeline provides its function adequately and fairly. . . . In these days of an energy shortage and dislocated supplies, the importance of the Explorer line cannot be over emphasized. I might add that it required great vision and foresight together with an enormous amount of money to build this new system.") (For Mr. Emison's minor complaints, see text on minimum tenders and terminalling, *infra*).

dates; and (6) imposing unreasonable commodity quality specifications" (emphasis added). A more direct, but still speculative, statement appears on page 60-61 of the Report: "In our experience with common carrier oil pipelines, conditions relating to product specification, product cycles, batch size, tankage ownership and the like *may* have acted to preclude use of the line to some shippers even with common carrier obligations imposed on the system" (emphasis added). Examination of such broad, nonspecific allegations is best done topic by topic.

(1) Routing

The Kennedy Staff Report claimed that "The location of a pipeline is a more subtle form of access denial."; that "Pipelines are configured by their owners to best serve their needs."; and that "The location or needs of nonowners are given little thought."[1829] No hard evidence, no specific instance, not even a single potential shipper is named as having been denied by virtue of a pipeline's location. Instead, there is a reference to a *1924* La Follette Committee Report and the *1939 TNEC Hearings* where some unnamed person was alleged to have said "they [the pipelines] don't go where they need to go to reach the small man's product." The Staff Report made a "bootstrap" attempt to bring this up to date by citing ICC Chairman Stafford's 1974 testimony before the Senate Special Subcommittee on Integrated Oil Operations as voicing the same problem. This miscitation of Chairman Stafford, who in fact testified squarely to the contrary, was discussed in Section VI B 6, "Staff Report of the Antitrust and Monopoly Subcommittee of the Senate Judiciary Committee (Kennedy Staff Report)."[1830]

Witnesses who have testified concerning the layout of routing universally have testified that the origin and destinations are selected to serve the greatest number of prospective owners and users within the constraints of engineering, environmental considerations, and right of way restrictions. Fred F. Steingraber testified in 1972 concerning the selection of the Colonial route, *i.e.,* from the Gulf Coast and Baton Rouge, Louisiana refinery centers to the chief market areas in the southeast, such as Birmingham, Atlanta, Spartanburg, Charlotte, Greensboro, thence to Washington, Baltimore and the New York metropolitan market. Because many of the southeastern terminals were already in place as a result of Plantation's lines, it was economically

---

[1829] KENNEDY STAFF REPORT, *supra* note 184, at 80.
[1830] See text at notes 1650-1674, especially 1674, *supra*.

efficient to tie these into the system as it was built. There was a GATX public terminal in Houston which provided a service for many nonowners, so it was connected to the line as well.[1831]

Vernon T. Jones, then President of Explorer, described how the routing of the Explorer project was selected. As he stated, the line has to follow the traffic that is indicated at the time the project is being financed and it will be that which is the most economical to serve both the present and foreseeable business in the area.[1832] The TAPS line origin was fixed by the Prudhoe oil field and the terminals had to be located at a deepwater year-round ice-free port which had suitable terrain for constructing tankage and terminal facilities. The most feasible location for such a terminal was Valdez, which had deep water, was the most northerly ice-free port and had suitable (and available) terrain for terminal facilities. The route of the pipeline was influenced more by environmental considerations than any previous line, but it was selected by soil testing to maximize the amount of line that could be buried. The final routing had to be approved by both United States and Alaskan authorities.

If one were to summarize the considerations that go into routing selections, it would probably be along these lines: pipeline routing must follow the traffic indicated at the time of planning. Assured shipments by owners of the line and nonowners, in addition to potential shipments by others, must all be taken into consideration. It is a financial necessity to locate pipelines along routes which will maximize the regular, high-volume use of the line's facilities. The final decision is a resultant of assured and potential demand, the origin and destination potentials and the influence of right-of-way difficulties, environmental contraints and terrain factors. Failure to observe these principles means that a pipeline cannot be operated economically or efficiently. Inefficient operation means higher tariffs and costs to the consumer and lower returns to the operator. Obviously, the routing necessarily will be more convenient to some and less convenient to others. As was discussed in Section IV D, "How Joint Pipeline Projects Are Put Together in the Real World,"[1833] study groups of all those who might be interested in participating as owners and/or shippers are formed and many iterations of plans are made in an effort to find the routing which will be satisfactory to the greatest number. This exercise is not unlike a "least square" analysis of the divergent points on a graph. It frequently happens that the original study group splits up into two or more subgroups such as was the case in the

---

[1831] *H.Res.5 Subcomm. Hearings, supra* note 234, at 150-153 (Statement of Fred F. Steingraber).
[1832] *Id.* at 115 (Statement of Vernon T. Jones).
[1833] See text at notes 1334-1345, *supra*.

Gateway (western route)/MATCH (eastern route) split, out of which came the Explorer line, or the Gulf Coast to Tulsa study, which ended up in two lines, Texoma and Seaway. *Regardless of ownership,* pipelines have been, and will continue to be, routed between the same points of origin and destination. No one can change, by legislation, regulation, or litigation, the location of crude reserves, major refining centers or markets. Within each major urban market area, terminals have been built in concentrated groupings. Local ordinances, zoning laws, availability of major highway access, access to water transportation and other factors tend to fix terminal locations within a rather small area. Any pipeline routed to provide the most economical service, and situated to acquire a base load, will of necessity connect these same general terminaling areas.[1834]

(2) Design Characterisitics

From time to time, there have been comments by critics that in designing the lines, owners have had their own interests primarily in mind and thus owners frequently are better served than are nonowners.[1835] Assistant Attorney John Shenefield in his proposed statement, remarked,"In the deepwater port situation, for example, where a critical question was the initial sizing of the facilities, our study of LOOP and Seadock clearly revealed that their initial sizing decisions were originally based almost exclusively on the needs of the owners, conservatively stated, rather than the shipping public at large."[1836] There are two aspects to the sizing allegation; one, that, inadvertently, the line subsequently turns out to be too small for the use which develops — this was discussed rather thoroughly in Sections II A 3 c, "Scramble for Traffic"[1837] and III F 1, "Risks Involved in Pipeline Ventures — Incorrect Forecast of Traffic,"[1838] where it was demonstrated that conservatism was virtually an economic necessity; and second, the recently-evolved theory of the Justice Department on "deliberate undersizing" to secure the fruits of a regulated pipeline in adjacent stages of the productive process (which will be discussed in the subsequent major subsections). With this in mind,

---

[1834] *Consumer Energy Act Hearings, supra* note 154, at 1035 (Supplemental submission by Vernon T. Jones).

[1835] *H.Res.5 Subcomm. Hearings, supra* note 234, at 213 (Supplementary submission by Bruce B. Wilson, Deputy Assistant Attorney General).

[1836] SHENEFIELD PROPOSED STATEMENT, *supra* note 179, at 28, citing the A.G.'s Deepwater Port Report, *supra* note 446, at 19-21, 70-73, and 114.

[1837] See text following note 496, *supra*.

[1838] See text at notes 727-746, *supra*.

examination of the cited pages of the Deepwater Port Report indicates at most a certain degree of conservatism because of the completely new and untried technical operation being attempted; the "yo-yoing" of indications of shipper interest in using the projected facilities, and the uncertainties of usage—for example, Mexican imports would bypass the ports; timing (which can drastically change the economics) could be skewed by actions of foreign countries or the United States government — witness the Iranian situation — and quota and rationing plans informal and formal, being bandied about. Despite all these factors, there was still a spare capacity factor of 15 percent above the latest nominations (from both owners and definitely interested shippers) built in LOOP's initial design.[1839]

(3) Connections, Input, Interchange and Delivery Points

Although the amorphous omnibus allegations of indirect denial of access have mentioned difficulties in obtaining connections, input points, interchange and delivery points, there never has been much in the way of specifics. The question of interchange or joint tariffs was settled rather early in the history of pipelines in 1920 when the ICC suspended a proposed cancellation of a joint tariff rate covering movements over the Prairie Pipeline Company and the National Transit company.[1840]

The question of alleged refusals to "grant gathering connections to independent leases," thereby forcing "independent producers to resort to higher cost truck transportation," occupied a page and one-quarter of the Kennedy Staff Report [1841] which, in turn, rested on the Attorney General's Second Report of 1957.[1842] This subject was a cause celebre for a brief period at the time of the Suez Canal crisis. Senator Joseph O'Mahoney had a great time with it.[1843] However, as was noted in Section II A 1, "Pregathering Procedures-Trucking," [1844] the case was examined thoroughly by the Texas Railroad Commission, which found that the

---

[1839] A.G.'s Deepwater Port Report, *supra* note 446, at 21. Citation to the Deepwater Port Report should not be taken as an indication that this author considers pipelines and deepwater ports are the same "creatures"; he has gone on record that they are not. See AEI, *Oil Pipelines and Public Policy, supra* note 684, at 60; and see Donald Flexner's disclaimer that the Deepwater Port Report was intended to apply to every pipeline — it was merely an example of how to analyze the problem. *Id.* at 64.

[1840] Crude Petroleum Oil from Kansas & Oklahoma to Lacy Station, Pa., 59 I.C.C. 483 (1920); *see* text at note 1739, *supra*; BEARD, *supra* note 45, at 91; WOLBERT, *supra* note 1, at 133.

[1841] KENNEDY STAFF REPORT, *supra* note 184, at 84-85.

[1842] Att'y Gen., *Second Report, supra* note 111, at 94-101.

[1843] O'MAHONEY REPORT, *supra* note 313, at 28-31.

[1844] See text at notes 312 to 314, *supra*.

individual complaining witnesses were really looking for a market rather than for pipeline capacity. The Railroad Commission established a procedure and criteria for adjudicating future specific unconnected well cases, of which only one was filed, and that was withdrawn. As was indicated by the testimony of Thomas D. Jenkins, President of Permian Corporation, before the Senate Antitrust and Monopoly Subcommittee in 1975, this situation has changed completely and Permian and its fellow independent buyers/gatherers and the major oil companies are desirous of purchasing any production that becomes available.[1845] Even the Kennedy Staff Report concluded: "It is unlikely that this problem will in the future plague newly discovered wells. With current high levels of demand by integrated company purchasers, their interests appear to lie in connecting every well practicable."[1846] However, the authors couldn't resist adding: "But the problem is symptomatic of the industry generally: the recurrent attempts by major integrated oil companies to maintain control over price and supply at the crude production level."[1847] The crude oil market will be addressed in a subsequent subsection in connection with the now well-publicized "undersizing" argument.

The *Smith-Conte Hearings* in 1972 produced two types of complaints about connections. The first was the Meyer Kopolow allegation that Explorer Pipeline required him to execute a "ship-or-pay" contract before it would build a connecting spur to his terminal, whereas "none of the other shippers (owners or nonowners, including J. D. Street, an independent) were asked to do so." This allegation and its answer, and the statements of other parties concerned, were discussed at length in Section II A 3 c, "Scramble for Traffic."[1848] The reader will recall that one of the parties whose name was volunteered by Mr. Kopolow to verify his version, Triangle Refineries, stated in a letter answering Chairman Smith's follow-up inquiry, that although Triangle had several discussions with Explorer Pipeline with regard to tying its terminal into the Explorer line, no formal request had been made by it to Explorer for a tie-in because Triangle's requirements for a terminal at St. Louis (in the same general area as Mr. Kopolow's terminal) were not sufficiently definite to pursue the matter but *"[i]t was our understanding we would have to commit a definite thru-put over a period of time...."* In answer to Chairman Smith's question, "Can you tell us of any other companies which have had difficulties in seeking to ship on Explorer or any other joint venture pipeline?" [Chairman Smith offered to treat the source as

---

[1845] *Petroleum Industry Hearings, supra* note 192, at 222-223 (Statement of Thomas D. Jenkins).
[1846] KENNEDY STAFF REPORT, *supra* note 184, at 85.
[1847] *Id.*
[1848] See text at notes 450-460, *supra.*

confidential, if requested], Triangle's J. H. Barksdale replied, "I do not know of any other companies that have experienced difficulties seeking to ship on Explorer or any other joint venture line."[1849] A second referral, Apex Oil Co., answered by telephone that Apex was interested in connecting to and shipping over Explorer's lines but that nothing had come of the discussions, not because of a lack of good faith on Explorer's part, but because of Apex's inability to secure supplies.[1850] Apparently, Apex subsequently was able to obtain supplies because it appears as a shipper over Explorer's line in the list of nonowner-shippers at note 441. The third of Mr. Kopolow's references was Martin Oil Company, whose response indicated a belief, based on "conversation with others [unnamed, but perhaps Kopolow] who have tried to get on the Explorer Pipeline" that "they [Explorer] are not interested in business other than their own." However Martin's spokesman, Hugh E. Watson, admitted that he had never shipped products on Explorer, nor had he made a formal request to Explorer for a connection. Moreover, as his letter stated, neither had he used (Williams Pipe Line) or any other "independent" pipeline, "...as our terminals are on the Ohio and Mississippi Rivers...."[1851]

The other type of complaint voiced at the *Smith-Conte Hearings* came from Beverly Moore, of the Corporate Accountability Research Group, Washington, D.C. Mr. Moore did not allege that nonowner-shippers were denied the right to connect to Colonial's line, but he raised the question whether there was a differential cost to such a shipper compared to what it cost "the owners." Mr. Moore was not content just to "price out" the marginal cost of a new tie-in connection after the line already was in existence; he wanted a weighted average cost per barrel of *all* nonowners compared to the weighted average cost to *all* owners.[1852]

In addition to the fact that Colonial had no way of obtaining such figures, this was a "loaded question" because the cost of all new connections, which predominately would be those of nonowner-shippers, was bound to be higher because the later in time, the greater would be the inflation factor and the cost effect of accessing the line over terrain which had become more difficult and costly as housing, commercial and other development ran up the cost in the interim. The essence of Moore's question really was the same as the routing issue discussed in the preceeding subsection. Mr. Steingraber, whose testimony Mr. Moore

---

[1849] *H.Res.5 Subcomm. Hearings, supra* note 234, at A56 (Letter, dated July 20, 1972, to Chairman Smith from J. H. Barksdale, Triangle Refineries, Inc.).

[1850] *Id.* at A57.

[1851] *Id.* at A56.

[1852] *Id.* at 195. Even Chairman Smith felt constrained to object to Mr. Moore's unorthodox behavior by interjecting: "Any new matter raised you want to —" *Id.,* but Mr. Steingraber thought it best to address the question on the spot.

interrupted as if he were a Congressman or a member of the Subcommittee staff, answered the question forthrightly insofar as Colonial had information, *i.e.,* that Ashland had incurred no expense to get into the system — they used existing connections; Crown Central had to lay a small line and install a pump at its Pasadena, Texas, refinery, which adjoined Colonial's receiving station in Houston; Coastal States used an existing line from Corpus Christi to the Houston GATX public terminal to which Colonial had laid a line; Lion-Monsanto had existing connections; Marathon came in through existing connections, as did Metropolitan Petroleum, Charter Oil, and Texas City Refining. Only two companies, Murphy and Tenneco, whose refineries were a substantial distance from their input point to Colonial, had to build a line — the Collins pipeline, which they built jointly. The other aspect of the "loading" in the question is that it ignored completely the fact that the nonowner-shippers do not have to carry the business and financial risk of the line, *i.e.,* bear the many risks of volume diversion and loss discussed in Sections II A 3 C, "Scramble for Traffic"[1853] and III F and G, "Risks" and "Uncertainties," respectively,[1854] or the financial risk in repayment of the 40-year debt obligation.

As was stated in the preceding subsection on routing, new pipelines are planned to follow routes best able to meet demand requirements, and the basic routing of the main line and lateral lines must be determined in the early planning stages with a view toward accommodating all those who were *then* either willing to participate as owners, or as nonowner-shippers whose projected use was firm enough to justify economically the capital expenditures and debt obligations involved. After the line is built, connections with individual terminals of shippers who determine subsequently that they desire to ship through the line, depending on their location with respect to the line, may require a commitment from the user to ship a quantity of oil through the line sufficient to amortize its incremental investment, or, at *such shipper's option,* to build its own connecting line. These agreements are not unique to major shipper-owned lines[1855] or to the oil industry. For many years, American railroads have required similar agreements in connection with construction of side tracks, and natural gas transmission lines routinely require "take or pay" contracts of their industrial customers, as frequently do electric utilities. Provision for such commitments is made in most tariffs, which gives the

---

[1853] See Table I and closing paragraph of the cited Subsection.
[1854] See text at notes 720-917, and at notes 918-1037, *supra.*
[1855] *H.Res.5 Subcomm. Hearings, supra* note 234, at 21 (Statement of D. W. Calvert, Exec. Vice President, The Williams Cos.).

regulatory body an opportunity to judge their reasonableness.[1856]

Governmental agencies have addressed the subject; in the specific instance of Beverly Moore's complaint about Colonial, the reader will recall the Justice Department's conclusion, after 13 years of investigation, that there was no meaningful denial of access[1857] and the overall "clean bill of health" which was given by ICC Chairman George M. Stafford, with whom such tariffs were filed, in his 1973 testimony before the Senate Special Subcommittee on Integrated Oil Operations.[1858]

(4) Expansion Practices

The question of expansion is of fairly recent vintage, probably a "spin-off" from the Attorney General's Deepwater Port examination. The Kennedy Staff Report seemed to "throw it in" with an omnibus declaration that: "The convenience and objectives of the owner-shippers dictate the routing, extension, expansion and sizing of their pipelines and the positioning of input and overtake points."[1859] Consistent with other passages of the Report, this was a bold assertion, clothed with no supporting facts.

The Antitrust Division dug itself a neat little hole in the Deepwater Port Report by first expressing an antipathy to undivided interest lines based on its aversion to the "averaging effect" made possible by the addition of the higher income from new large-diameter trunk lines to offset the less lucrative small-diameter trunk lines and gathering facilities.[1860] This was consistent with the Department's long-standing "market stabilizing" aversion with respect to undivided interest products lines which forced Colonial to become a joint stock company rather than an undivided interest line, as most of the participants desired.[1861] But later, when the idea surfaced concerning the desirability of frequent

---

[1856] PIPELINE PRIMER, *supra* note 331, at 25-26.
[1857] See text at note 1813, *supra*.
[1858] *S.Res.45 Hearings, supra* note 296, at 901 (Statement of I.C.C. Chairman George M. Stafford).
[1859] KENNEDY STAFF REPORT, *supra* note 184, at 5.
[1860] A.G.'s Deepwater Port Report, *supra* note 446, at 82-83.
[1861] See comments by Mr. James Shamas, Director of Transportation, Getty Oil Company in AEI, *Oil Pipelines and Public Policy, supra* note 684, at 41. This was the position taken by Robert A. Bicks, former Assistant Attorney General. However, at the AEI Conference, Donald I. Baker, himself a former Assistant Attorney General, advised that merely because the Department "frowned" on undivided interest lines in Colonial's day, and presumably even when he was in office, should not determine the issue because "the decision-makers change, and the amount of information available to the department also changes." *Id.* at 182-183.

ownership changes, especially as they might be occasioned by a nonowner desiring an expansion, the Antitrust Division apparently changed its mind, due to the fact that undivided interest lines are far more amenable than are "corporate " lines to ownership change.[1862] If, as was the case with TAPS, the rates are defended on a discrete pipeline system basis, the last vestige of objection to undivided interest lines by the Division will probably be subordinated to the more favorable ability to accommodate expansions. Both Rancho and Capline provide for unilateral expansion on a basis analogous to the "sole benefit" clauses in joint operating agreements in the production area. If the Division is willing to approach the problem realistically, such as giving appropriate recognition to the fact that the company which initiated the expansion must also bear the consequences on the overall operation of the line,[1863] there would seem to be a basis for some give-and-take mutual resolution of the problem agreeable to both the industry and the Division. Such an arrangement would have to be among the owning parties because the lenders are not apt to permit the credits upon which they have invested their money to be replaced by less credit-worthy companies so as to jeopardize their security.

(5) Storage and Terminalling

Oil pipelines provide operating or "working" tankage along the line to accommodate the efficient movement of shipments from the main line to smaller lateral lines and to maintain quality control.[1864] Generally, however, interstate lines do not offer tankage at points of origin and/or

---

[1862] One important problem that is obviated by this form of organization is that the original owners' liability to the lenders will not be changed by the incremental investment in the expansion.

[1863] See the down-to-earth approach to this problem expressed by Scctt Harvey, Economist, Bureau of Economics, Federal Trade Commission, in the AEI discussion. AEI, *Oil Pipelines and Public Policy, supra* note 684, at 181: "I would expect the company that initiated the expansion to be made responsible. It puts up the money to finance the expansion and takes the consequences [including the loss to the rest of the System occasioned by the drop in throughput] if the expansion turns out to be unwarranted."

[1864] The need for "breakout" tankage can occur where the velocities of two line segments vary and tankage must be utilized to maintain stable flow and preserve product integrity; *cf., Consumer Energy Act Hearings, supra* note 154, at 605 (Statement of Jack Vickery). Working tankage frequently is required at points where traffic is received from other carriers unless the movements are synchronized sufficiently to "tightline" its transfer. It is also used to accumulate "batches" large enough so that the amount of contamination will be small percentage-wise. Note, *Legal & Practical Aspects of Petroleum Products Pipe Line Rates and Terminal Tankage*, 102 U. Pa. L. Rev. 894, 921 (1954) [hereinafter cited as C. I. Thompson, Jr.]

destination,[1865] and the tariffs of most such carriers so specify.[1866] This fact has been the subject of substantial criticism. For example, the Kennedy Staff Report twice mentions the subject: (1) "Pipelines controlled by major oil companies, unlike independent pipelines, generally do not provide common carriage storage and terminal facilities";[1867] and (2) "As early as 1914, Oklahoma's Attorney General West, testifying before Congress stated that common carrier pipeline transportation was of no advantage to independent operators unless there was storage."[1868] After noting that "others [had] commented on the lack of storage facilities for use by independent shippers as well as the large capital investment required to furnish sufficient tankage at input and delivery points," the Staff Report, at page 79, cited Williams [Pipe Line], an "independent," as "a prime example of a pipeline with common tankage available for all shippers" and contrasted Williams' operation with "the experience [at an unnamed point of time] of an [unnamed] independent marketer desiring to use Explorer pipeline for shipment and delivery to the Dallas area. This independent marketer was able to arrange for a contract for gasoline with [an unnamed] refiner-owner connected to the Explorer system. The marketer did not have a terminal connected to Explorer for deliveries in the Dallas area, nor did he have a terminal close enough to the pipeline route to make a connection economically attractive. The marketer contracted several [unnamed] companies with terminals connected to Explorer but was unable to secure any space." Apparently neither the truth of this allegation nor the circumstances under which it allegedly took place were ever tested in any formal proceeding, for no authority is cited. Certainly, due process, fairness and good sense would seem to require a more explicit exposition before such an incident would be reported in a Senate document as a fact. What objective evidence is known casts considerable doubt on the statement. First, the wording would lead a casual reader to believe that Williams Brothers [now Williams] Pipe Line provides this service as a matter of course. It does not. In the first place, Williams does not provide terminalling at origin points,[1869] nor is there an

---

[1865] WOLBERT, *supra* note 1, at 40; PIPELINE PRIMER, *supra* note 331, at 26; KENNEDY STAFF REPORT, *supra* note 184, at 79; *cf.* MARATHON, *supra* note 202, at 23.

[1866] *See e.g.,* Exxon Pipeline Co., F.E.R.C. No. 129, eff. 3/1/78 (Item 45-no destination facilities); Colonial Pipeline Co. ICC 29, eff. 2/10/78 (Item 35-no origin or destination tankage); C.I. Thompson, Jr., *supra* note 1864, at 920, and tariffs cited in footnote 163 thereof.

[1867] KENNEDY STAFF REPORT, *supra* note 184, at 5-6. The report cited WOLBERT, *supra* note 1, at 40, but Wolbert drew no distinction between independent and major lines in the cited reference.

[1868] *Id.* at 79.

[1869] See Williams Pipe Line Co., I.C.C. No. 10, effective 12/1/76, Item No. 60A "The carrier will not provide storage facilities at points of origin."

express undertaking to provide storage facilities at all destination points.[1870] As a matter of fact, it provides such service for only slightly more than one-half its destinations.[1871] Second, not only does it strain credulity to believe that Explorer, which was losing $46 million and making cash calls on its owners of $42 million during this period,[1872] would rebuff any potential business, but also there is specific testimony by Vernon T. Jones, then president of Explorer, now President of Williams Pipe Line Company, which stated categorically that "Explorer has never been approached by any independent and asked to provide terminal facilities."[1873]

The S.2387, or PICA Report, at pages 55-56, citing W. John Lamont and his partner, Martin Lobel, made a reference to failure to provide input tankage as an indirect device to preclude nonowner shippers access to pipelines by utilizing the colorful imagery: "Imagine a railroad without depots, which 'solicits' freight business by saying, 'throw your goods on the flatcar as it comes by or build your own station and maybe we'll stop' [and pick it up]."[1874] Again, there was an inference that independents had white hats and offered uniformly this service because they were interested only in transporting oil whereas the "majors" all had black hats and were more intent on excluding outsiders than in making money. Earlier in this chapter, Mr. Lamont's metaphoric method of expression was admired but his analogical analysis was found to be seriously flawed.[1875]

Ironically, the very witnesses cited by the Kennedy Staff Report as supporting its independent-major behavioral dichotomy have provided the true insight into what is the real life situation. Mr. D. W. Calvert's testimony,[1876] cited by the Staff Report at page 5, footnote 16, showed that four of the seven pipelines providing common terminalling were "majors." Plantation Pipe Line, which would also be termed a "major," provides origin tankage at its origin point in Baton Rouge, tanker dock space which Plantation furnishes under contract with Exxon, and storage

---

[1870] *Id.* "Petroleum Products will be accepted for transportation only when consignor has provided equipment and facilities satisfactory to the carrier and when consignor or consignee has ascertained from the carrier or has furnished evidence satisfactory to the carrier that there are adequate facilities at destination which are available for receipt of the shipment as it arrives without delay."

[1871] *H.Res.5 Subcomm. Hearings, supra* note 234, at 14 (Statement of D. W. Calvert, Exec. Vice President, The Williams Co.) ("Of the delivery terminals, 31 are owned by Williams Bros. Pipe Line Co., and 29 are owned by shippers or other persons.").

[1872] See text at notes 432-437 and 1656, *supra*.

[1873] *S.1167 Hearings, supra* note 432, pt. 9 at 609 (Statement of Vernon T. Jones).

[1874] Mr. Lamont got a "rerun" on that one in the *Consumer Energy Act Hearings, supra* note 154, at 670 (Statement of W. John Lamont).

[1875] See text at notes 1490-1493, *supra*.

[1876] *H.Res.5 Subcomm. Hearings, supra* note 234, at 31 (Letter to Chairman Smith by D. W. Calvert, Executive Vice President, The Williams Cos., dated July 14, 1972).

at Spartanburg, South Carolina.[1877] One also notices that Williams Pipe Line was singled out by the Staff Report as a "prime example of a common pipeline with common tankage available to all shippers,"[1878] with the innuendo that this is the case because it was independently owned,[1879] whereas, in fact the Williams system, which does provide some terminals, was originally built by a consortium of oil companies, and it was while the system was under the (major) company ownership that it was put into the terminalling business,[1880] building almost two-thirds of the present terminals.[1881]

The second witness cited by the Staff Report, Mr. Richard C. Hulbert, President of Kaneb Pipe Line Company, an "independent" pipeline, provided the first breakthrough on *why* some pipelines provide the service and some do not. Kaneb, which handles "common specification" materials, is far more apt to find the economics of providing common storage favorable than would lines which handle segregated (by shippers) movements, especially where the individual marketers already own their terminals.[1882] Mr. Hulbert saw no form of discrimination involved; to him it was simply a matter of the nature of the types of services performed.[1883]

In addition to the common specification versus segregated shipment distinction, there is also the nature of the market served. In areas with a number of comparatively sparse markets wherein individual companies shipped relatively small volumes, as was true when Great Lakes (Williams' predecessor) constructed its line in 1931, and when Kaneb built its line, carrier-owned terminals shared by all shippers made economic sense. When Colonial built its line, the extent of the market dictated large volume shipments and there were 28 existing terminals of 13 different nonowner shippers located in 15 cities in seven states ready to be connected *in addition* to those of the shipper-owners.[1884] Under those conditions, it would have been economically wasteful to build carrier storage.

[1877] Plantation Pipe Line Co., FERC 52, eff. 8/1/78, Items 115, 130, and 140.

[1878] KENNEDY STAFF REPORT, *supra* note 184, at 80.

[1879] COMMENTS OF THE AMERICAN PETROLEUM INSTITUTE ON THE PETITION FOR THE INITIATION OF A RULEMAKING PROCEEDING PROHIBITING OWNERSHIP OF PETROLEUM PIPELINES BY PETROLEUM COMPANIES FILED BY SENATOR EDWARD M. KENNEDY 37 (Mar. 27, 1979) [hereinafter cited as API's Comments on Kennedy FTC Petition].

[1880] *S.1167 Hearings, supra* note 432, pt. 9 at 609 (Statement of Vernon T. Jones).

[1881] *H.Res.5 Subcomm. Hearings, supra* note 234, at 13 (now have 31), *Id.* at 19 (added 11) (Statement of D. W. Calvert, Exec. Vice President of The Williams Cos.) (31 minus 11 added equals 20 taken over from Great Lakes, 20/31 is close to two-thirds).

[1882] *S.1167 Hearings, supra* note 432, pt. 8 at 5947 (Statement of Richard C. Hulbert).

[1883] *Id.*

[1884] *Id.* pt. 9 at 610 (Statement of Vernon T. Jones).

The legal position of the carriers is quite sound. The business of common carriage is transportation, not storage.[1885] Moreover, the Commission does not have jurisdiction over storage facilities under its general power to enforce the duty to provide transportation.[1886] Storage of the goods being transported is a transportation service only to the extent that it is "necessarily incidental" to the transportation and a carrier's duty is coextensive with that obligation.[1887] However, this is a fact question and where a carrier, such as Williams or Kaneb, has found that there was a necessity for the storage which arose from their particular operating circumstances and the use of such tankage is an integral part of the operation, the tankage probably will be found to be "necessarily incidental" to the transportation and properly will be included in the rate base by the regulatory agency.[1888]

In the case of a Colonial or Explorer type operation, where shippers are moving large volumes of segregated products —for example, Shell's unleaded gasoline which has the highest octane rating of all the unleaded gasolines with the possible exception of some of Amoco's—common storage would be an anathema and it would be difficult to find common storage to be a "necessary incident." To superimpose the expense of common tankage for a service that most shippers do not want would only increase the cost, which would be prohibitive[1889] to the tariff if provided by the carrier.[1890] Terminals would be no bonanza to the carrier because the rate of return on terminal facilities experienced by those carriers who do provide the service "has not been highly attractive."[1891]

Charles I. Thompson, Jr., in a 1954 University of Pennsylvania Law

---

[1885] WOLBERT, *supra* note 1, at 40, and cases cited in note 217 thereof.

[1886] C. I. Thompson, Jr., *supra* note 1879, at 920, citing United States v. Pennsylvania R.R., 242 U.S. 208, 222 (1916) where the I.C.C., sought, under Section 1(4), to compel defendant to furnish petroleum tank cars, relying on Section 12 giving it broad powers to enforce the ICA and Section 15(1) giving it powers to regulate practices. See also Jacksonville Port Terminal Operators Ass'n v. Alabama, Tennessee & Northern R.R., 263 I.C.C., 111, 116 (1945); KENNEDY STAFF REPORT, *supra* note 184, at 6. *See Consumer Energy Act Hearings, supra* note 154, at 1266 (Statement of I.C.C. Chairman George M. Stafford).

[1887] WOLBERT, *supra* note 1, at 40; C. I. Thompson, Jr., *supra* note 1879, at 921; *see* Propriety of Operating Practices - New York Warehousing, 216 I.C.C. 291, 349 (1936); Guaranty Claim of the Central Elevator & Warehouse Co., 72 I.C.C. 169, 176 (1922); Reconsignment and Storage of Lumber and Shingles, 27 I.C.C. 451 (1913); *see also* Atlantic Pipe Line Co., 47 I.C.C. Val. Rep. 541, 546 (1937).

[1888] *See, e.g.* The Texas Pipe Line Co., 48 I.C.C. Val. Rep. 249, 250-252 (1938); Atlantic Pipe Line Co., 47 I.C.C. Val. Rep. 541, 543-549 (1937); C. I. Thompson, Jr., *supra* note 1879, at 921.

[1889] C. I. Thompson, Jr. *supra* note 1879, at 931.

[1890] *Cf.* PIETTE, *U.S. Oil Lines - Public Policy, supra* note 1492, at 18, noted that private companies, unconnected with the oil industry, do not presently offer this service: railroads do not provide grain elevators for free; and trucking firms do not provide warehouses without charge.

Review note carefully reviewed the practical obstacles to the forced requirement of tankage for small, spot and "in-and-out" shippers. He reached a conclusion that, absent special circumstances, such as pertain in the Kaneb and Williams situations, the request by a shipper or a group of shippers for service on anything other than a long-term basis would probably not be deemed to meet the "upon reasonable request" requirement of Section 1 (4) of the ICA.[1892] The most sensible approach appears to be that followed by Colonial and Explorer which are connected to a public GATX terminal in Houston, which by virtue of these connections has sufficient commercial potential to justify the operation.

In closing this subsection, it seems that once again the "majors" are faced with inconsistent charges. On the one hand, it has been claimed that their pipelines earn rates far in excess of the cost of capital while, simultaneously, they are criticized for allegedly keeping nonowner shippers off their lines by not constructing terminal facilities. The "Averch-Johnson" theorem that utilities whose rates of return are comfortably in excess of the cost of capital will "gold-plate" their systems, *i.e.*, overinvest in capital equipment, is now black-letter textbook economics.[1893] Both economic theory and the empirical evidence suggest that the absence of common terminal facilities on the large "industry" lines support the pipelines' attribution of this phenomenon to the low, virtual sub-marginal, rate of return on terminals rather than the critics' vague allegations that the explanation is an exclusionary motive.[1894]

(6) Specifications

Under tariff rules and regulations, both owner and nonowner shippers alike must tender product meeting designated specifications such as color, gravity, temperature, water and sediment, and vapor pressure. The purpose of these specifications is to facilitate efficient operation of the system and to preserve the integrity of each shipper's product. Pipelines basically utilize two methods of handling shipments tendered to them for delivery. One is the so-called "common-stream" or "common specification" method wherein a specification for a particular grade of oil

---

[1891] *S.1167 Hearings, supra* note 432, pt. 8 at 5953 (Statement of Richard C. Hulbert, President of Kaneb). Moreover, the Commission sees that it doesn't get too attractive. In. I. & S. Dkt. No. 8823 — Propane via Dixie Pipe Line Co., the Commission ordered cancellation of schedules which sought to establish an unlawful terminal service charge. *Consumer Energy Act Hearings, supra* note 154, at 1265 (Statement of I.C.C. Chairman George M. Stafford).

[1892] C. I. Thompson, Jr., *supra* note 1879, at 924-928.

[1893] Averch & Johnson, *Behavior of the Firm Under Regulatory Constraint,* 52 Am. Econ. Rev. 1052-1069 (1962).

[1894] *Cf.* Mitchell, *1974 Senate Testimony supra* note 1056, at 29.

is established, which must be met by all shippers. All shippers' movements of that particular grade will be commingled in order to secure the advantages of large scale movement and avoid the costs of segregation. The other method segregates each shipper's traffic from the shipments of other shippers even though the products of the other shippers may be roughly similar in quality. Some pipelines, such as Williams Pipe Line, operate solely on the common specification basis, others are principally segregated shipment operations.[1895] Colonial is principally a segregated shipment system, but in order to accommodate smaller shippers, it runs a common specification "batch," externally segregated from other shipments, but internally commingled to permit the aggregation of small individual shipments into a single batch size which will meet the minimum tender required by the operational characteristics of the line.[1896]

The Kennedy Staff Report referred to the subject, saying (at page 6), "Product quality specifications regarding, for example, specific gravity and permissible amounts of water and sediment, ostensibly imposed for the purposes of avoiding contamination of other shipments on the pipeline, *may* be so narrowly drawn as to foreclose use of the line by particular shippers" (emphasis added). On page 81 of the Staff Report the following comment appears: "Mention has been made that other service requirements *may* be devices to deny access to pipelines. Thus, product specifications and shipping schedules *may* be used to delay access to pipelines and lead to virtual denials of access" (emphasis added). Once again, one is appalled by the concept that operational provisions with admittedly proper purposes but which *may* be misused are made cornerstones upon which to build a case for divorcement of as important and progressive an industry as oil pipelines. Two items were suggested as supportive of the staff's assertion. The first item was the Attorney General's Deepwater Port Report, which dealt with a subject matter analogous to oil pipelines solely in the sense that both offshore ports and pipelines employ pipe to transport oil. Significantly, however, and unmentioned by the Staff Report, was the Attorney General's basic decision to recommend *against prohibition* of port ownership by integrated oil companies, despite all the ghosts that the fertile and suspicious minds in the Antitrust Division were able to conjure.[1897] The second was the dubious Siess testimony[1898] which was stressed in the S.2387, or PICA Report,[1899] but which most observers, including the Minority Report and Mr. Siess' successor, believe stands for

---

[1895] PIPELINE PRIMER, *supra* note 331, at 27-28.

[1896] *Consumer Energy Act Hearings*, *supra* note 154, at 607 (Statement of Jack Vickrey).

[1897] A. G.'s Deepwater Port Report, *supra* note 446, at 104-105.

[1898] *S.1167 Hearings*, *supra* note 432, pt. 8 at 6240-6242 (Statement of Charles P. Siess, Jr.).

[1899] S.2387 REPORT, *supra* note 49, at 23-25.

exactly the opposite proposition, namely that when the chips are down, all pipelines observe their common carrier obligations, even to the extent of carrying crude which was tendered by someone who had "stolen" a connection from the carrier's affiliate.[1900]

Lest the impression be left that a prospective shipper must resort to the courts for relief when he believes that specifications are unfairly drawn or administered, the reader's attention is directed to the case of *Denver Oil Company vs. Platte Pipe Line Company*[1901] where a shipper felt that the carrier had violated Sections 1 (6) and 3 of the ICA by refusing to accept the shipper's crude as part of its "asphalt sour stream." Platte's refusal was sustained only because the tendered oil did not meet the quality standards for the asphalt sour stream. However, it *did* meet the requirements for movement in the pipeline's "general sour stream," and Platte was required to spell out standards in its tariff for its different batches or "streams."[1902]

(7) Minimum Tenders and Demurrage

The efficient operation of a typical oil pipeline and the preservation of each shipper's right to receive the product he shipped, undegraded by the shipments of his fellow shippers, gives rise to service rules such as the quality specifications and terminalling facilities which were discussed in the previous subsections. In addition, most pipelines have established, using criteria discussed in Section II B 3, "Technical Aspects," a minimum quantity of oil which will be accepted at a single point of origin from one shipper as a segregated "batch," frequently referred to as a "minimum tender." In order to accommodate the needs of small shippers, the tariffs of many pipelines provide that smaller quantities, or tenders, will be accepted from a shipper or shippers at multiple origins, subject to delay until the quantities become part of a joint batch, equal to or exceeding the lines' minimum tender before entering the main line.[1903] A further accommodation is made by many lines by permitting two or more shippers to combine their products of similar specification (in what might be characterized as "intra-batch" common specification operation) so as to enable the smaller shippers to meet the pipeline's minimum tender

---

[1900] See text at, and notes 1498-1501, *supra*.
[1901] 316 I.C.C. 599 (Div. 2, 1962), 319 I.C.C. 725 (Div. 2, 1963).
[1902] JONES DOE REPORT, *supra* note 372, at 53.
[1903] MARATHON, *supra* note 202, at 82-83; HOWREY & SIMON, *supra* note 3, at 43-44.

quantity for the batch shipment entering at a single point of origin.[1904] In this connection, the unfavorable comparison frequently made by critics between Colonial and Williams Pipe Line, when placed on a *truly comparable basis,* evaporates like pentane on a hot day; that is to say, Williams, which operates on a common specification basis, has a 25,000 barrel minimum tender.[1905] Colonial, with a substantially larger line, for which physics and economics would suggest a larger minimum tender,[1906] in its "fungible batch" provision, has *exactly the same* 25,000 barrel minimum quantity requirement.[1907] Moreover, where a crude line is run on a wholly "common stream" basis, such as Exxon's line from offshore Louisiana to onshore points, there is no minimum tender at all since the oil is not batched but is commingled.[1908]

Notwithstanding the above, the Kennedy Staff Report commented adversely on minimum tenders. At page 78 of the Report, the staff devoted four paragraphs to a history of unreasonable minimum tenders which had preceded the corrective action taken by the ICC in the 1930's. Although acknowledging that the ICC's decisions had lowered "the egregiously high minimum tenders," the report continued: "...however, the problem has not disappeared. New, modern large diameter pipelines have imposed a high minimum tender and minimum offtake requirements and have

---

[1904] Pipeline Demurrage & Minimum Shipment Rule on Propane, 315 I.C.C. 443, 447 (1962); *H. Res.5 Subcomm. Hearings, supra* note 234, at 899 (Statement of I.C.C. Chairman George M. Stafford); C. I. Thompson, Jr., *supra* note 1864, at 919; *cf.* Brundred Brothers v. Prairie Pipeline Co., 68 I.C.C. 458, 463 (1922).

[1905] Williams Pipe Line Co., I.C.C. No. 10, eff. 12/1/76, Item 80 Section A. "A shipment of 25,000 barrels or more of Petroleum Products of the same required specifications shall be accepted for transportation at one point of origin or one transit point from one consignor."

[1906] *Consumer Energy Act Hearings, supra* note 154, at 607 (Jack Vickrey testifying: "Senator [Stevenson], minimum tenders are based upon engineering requirements of the size of the line, . . . In the Colonial line, since it has such a high volume, moving 1,400,000 barrels a day [substantially greater today], and being a 36-inch line, the size of the minimum tenders has to be large in order to prevent contamination. When you have 27 front end shippers moving a range of 6 or 7 different products each in 10-day cycles, and you will have a batch of kerosene following gasoline and home heating oils [No. 2 distillate] and other things, and you are delivering this out to the various shippers, you obviously want to keep the amount of contamination that interfaces down as small as possible.") *Id.* at 607-608 (John E. Green, President of Shell Pipe Line Corp., taking up the discussion: "I don't know what the actual figures are for Colonial Pipeline, but I would estimate that it would be something like this, that when two batches of product are put in the pipeline one following the other, in Houston, Tex., the beginning point of Colonial, and these two batches are moved continuously to New York, some 1900 miles, the contamination zone between these two products will be in the neighborhood of 1,000 barrels of intermixture.").

[1907] Colonial Pipeline Co. I.C.C. 29, eff. 2/10/78, Item 15(c) "The minimum quantity of petroleum product which will be accepted, at one point of origin by the carrier from one shipper, for participation in a joint or fungible batch shall be 25,000 barrels, and will be accepted only when such petroleum product can be joined during one cycle with other petroleum products in other shipments at the same or other origin points to form a joint or fungible batch of not less than 75,000 barrels."

[1908] Exxon Pipeline Co., F.E.R.C. No. 143, eff. 11/20/78.

justified these on the need to maintain product integrity in the operation of the pipeline."[1909] This appears to be a willfull disregard of the increased technological contamination problems, discussed in detail in Section II B 3 "Technical Aspects,"[1910] and the ICC's actions taken not only as early as 1922 in the *Brundred Brothers*[1911] case but continuously thereafter; including the 1940 *Reduced Pipe Line Rates and Gathering Charges*;[1912] the 1941 *Petroleum Rail Shippers Association* decision;[1913] and the 1962 case of *Pipeline Demurrage and Minimum Shipment*.[1914]

It would appear that minimum tender provisions are determined by the technological, engineering and operating requirements of the particular line and cannot accurately be described as an indirect discrimination "so subtle as to be scarcely detectable." They are uniformly applicable to all shippers, and have a necessary purpose of enabling a pipeline to accept, transport and deliver all grades of petroleum on the same terms for all shippers. They are openly announced and filed with the regulatory commission, now FERC, to whose expertise Congress has committed their regulation. Neither Congressional committees nor antitrust agencies are qualified to pass judgment on them. No showing has been made that the agency's remedies are not sufficient to protect the interest of all concerned.

Demurrage is an analogous situation. Here again, the Commission has acted. In 1956, in the case of *Farmers Union Central Exchange, Inc. vs. Great Lakes Pipe Line Company*,[1915] the Commission determined which of several pipeline tariffs was applicable. In the *Pipeline Demurrage and Minimum Shipment* case mentioned in the preceding paragraph, the ICC approved a tariff provision that required the shipper to remove its product within five days after notification of deliverability, or be subject to a demurrage charge. The rule was occasioned by the

---

[1909] KENNEDY STAFF REPORT, *supra* note 184, at 78.

[1910] See text at notes 534-580, *supra*.

[1911] Brundred Brothers v. Prairie Pipe Line Co., 68 I.C.C. 458 (1922) (ordered Prairie Pipe Line Company to reduce its minimum tenders on crude movements to 10,000 barrels).

[1912] 243 I.C.C. 115 (1940) (extended the *Brundred Brothers* 10,000 barrels minimum tender to all crude lines under I.C.C. jurisdiction).

[1913] Petroleum Rail Shippers' Ass'n v. Alton & Southern R.R., 243 I.C.C. 589, 665 (1941) (ordered product lines' tenders reduced to 5000 barrels of the same specifications from one shipper to one consignee, subject to delay until the carrier had accumulated 25,000 barrels of the same specifications).

[1914] 315 I.C.C. 443, 447, 448 (Div. 2, 1962) wherein the Commission rejected protests by certain railroads to a revision in the demurrage and minimum tender regulations of a Liquified Petroleum Gas (LPG) line. Under the proposed reduction, the shippers were permitted to tender 10,000 barrels to the pipeline *within one week,* and there were no restrictions as to number of consignors, consignees, origins or destinations. The Commission based its decision, among other things, on the ground that the regulation would aid small shippers.

[1915] 297 I.C.C. 645 (Div. 2, 1956).

hydraulics of pipeline operation. Orderly movement out of the line is required to enable the extremely complex and highly automated pipeline systems to function at reasonable levels of efficiency, and the appropriateness of the regulations adopted by the carrier to achieve that end is a matter for the expertise of a designated regulatory agency.

(8) Alleged Discriminatory Prorationing

As a common carrier, oil pipelines are prohibited from discriminating in favor of their shipper-owners or giving any unreasonable preference in any way in the furnishing of services to any shipper or group of shippers.[1916]

Compliance with these requirements means that when the volume of crude or products tendered for shipment exceeds the capacity of the line, the capacity must be ratably apportioned among the shippers desiring to ship.[1917] In keeping with the mandate, tariffs filed with the FERC so provide.

Because there have been very few occasions when a pipeline has been under sustained prorationing[1918] and the fact that the few lines which have been faced with the problem have complied with their obligations, there have been no complaints alleging discriminatory prorationing filed with the ICC.[1919]

However, the Kennedy Staff Report asserted: "The manner in which space on the major oil companies' pipelines is prorationed among shippers also works to the disadvantage of nonowners. The major oil companies' pipelines are not prorationed automatically when full capacity is reached. In order for limited space on these pipelines to be prorationed, a shipper, at the risk of damaging its relationship with companies upon which it may rely for its supply of crude petroleum products, must insist that the line proration space. The pipeline then requires each prospective shipper to nominate the amount it wishes to ship over a given future period of time. Among historical shippers, space is allocated on the basis of the amount each has shipped during some prior period. New shippers

---

[1916] 49 U.S.C. § 3(1) (1976).

[1917] W. J. Williamson, *supra* note 315, at 15; MARATHON, *supra* note 202, at 53; *Consumer Energy Act Hearings, supra* note 154, at 611 (Statement of Jack Vickrey); HOWREY & SIMON, *supra* note 3, at 37; PIPELINE PRIMER, *supra* note 331, at 30.

[1918] See text at Section II A 3 c, "Scramble for Traffic," immediately preceeding Section II A 3 d, "Tenders," *supra*.

[1919] MARATHON, *supra* note 202, at 53-54; *cf.* PIPELINE PRIMER, *supra* note 331, at 30; HOWREY & SIMON, *supra* note 3, at 37-38.

are allocated a percentage of their individual nominations based on the ratio of pipeline capacity to total nominations of all shippers. Since owner-shippers generally ship much higher volumes than nonowners from the very start of pipeline operations, they have a substantial advantage over other historical shippers. Moreover, pipeline owners can overwhelm both other historical shippers and prospective new shippers with the amounts of crude oil or petroleum products which they can nominate and tender for shipment."[1920] It is somewhat difficult to see the foregoing recitation as anything other than an attack on *size*.

Certainly if two small, unrelated shippers were shipping equal amounts over an unrelated line and the first shipper, due to superior efficiencies, prospered in the marketplace and over a period of years steadily increased its shipments until it was shipping twice the amount that the second shipper put through the line, and subsequently the line reached full capacity, would the Kennedy Staff have the carrier *automatically* notify each shipper that henceforth each could ship one-half of the line's capacity? One hardly would assume that it would, but that is exactly what the quoted language implies. The carrier has to have a basis for equitable distribution of space and obtaining nominations is the only way the carrier can initially discover if there is really a problem. As a matter of day-by-day operation for scheduling purposes and efficient operation, the carrier has to receive nominations even when the line is nowhere near capacity so that it can coordinate receipts and deliveries to accommodate the shipper's requirements, and where "batching" is involved, to plan its cycles, and the positioning within cycles, so as to treat all shippers on an equitable basis. If it turns out that shipper X, for some reason such as a refinery shutdown or a sudden need for its oil elsewhere, is not going to ship its usual amount next month, if the carrier had automatically reduced every shipper's movements, say 20 percent, only to find that they could have shipped their 100 percent requirements, it would not only have a bunch of angry shippers on its hands but it would have caused needlessly the line to operate at less than capacity and thereby decreased its revenue.

The Staff's description would lead one to believe that the only method of proration is the so-called "historical" movements method. That simply is not true. Many pipelines' allocation procedures are based on an allocation among *all* shippers in proportion to the amounts *tendered by each*. Marathon's allocation procedure, for example, set forth in its rules and regulations filed with the ICC/FERC, states that its capacity will be thus apportioned. Shippers submit tenders by the 12th of the month preceding shipment and must have the tendered volumes readily

---

[1920] KENNEDY STAFF REPORT, *supra* note 184, at 6-7.

accessible. If a shipper is allocated space and fails to use it, its volumes for the second succeeding month may be reduced by the amount of its underutilized allocation.

Marathon has never had to invoke that policy, although there have been times when shippers were informed that the line was at capacity. What happened in such instances was that tenders submitted on the 12th were below capacity, and later that month, or during the actual shipping month, shippers desired to increase their volumes above those nominated. The fact that proration was unnecessary reflected the fact that shippers were able to find acceptable alternative means of pipeline transportation. Marathon operates on a "first come, first served" basis up to the capacity remaining after the tenders of the 12th are secured, reasoning that it would be unjust to penalize someone who had submitted his tender on time by prorating him out of some space in order to accommodate someone who had not submitted his tender on a timely basis. The same rationale is applicable to a shipper who requested additional space on the 20th of the preceding month vis-à-vis another shipper who proffered shipments on the 10th of the actual business month. It is impossible for a pipeline to anticipate the exact needs of each shipper, so Marathon simply makes available any space remaining after servicing the timely tenders to shippers as they request it until line capacity is reached.[1921]

The Staff Report, in describing the "historical experience" method, gives it a sinister twist as if it "loaded" the allocation in favor of shipper-owners. Insofar as the new shipper is concerned, he gets the same proportionate share as he would under an "allocation across the board" method because the first "cut" prorates his tender on the ratio that the line capacity bears to the total of all tenders. The only difference is that the space allocated to the "historical" shippers, as a group, by the first "cut" is then suballocated between the "regular" or "historical" shippers on the basis of the ratio that each such shipper's movements during the historical period, say the 12-month period ending one month prior to the allocation month, bears to the total movements of all such shippers during that period.[1922] After a "new" shipper has shipped during the qualifying period[1923] so as to verify the reliability of his tenders, he becomes a historical shipper himself. The working of this method in practice was attested to by the testimony of Robert E. Yancey,

---

[1921] MARATHON, *supra* note 202, at 80-81. Marathon's products lines operate in much the same manner except that the tenders are made on the 10th of the preceeding month, instead of the 12th, *Id.* at 82.

[1922] *H.Res.5 Subcomm. Hearings, supra* note 234, at A243; *S.1167 Hearings, supra* note 432, pt. 9 at 640 (both reprints of Colonial's Proration Policy).

[1923] In Colonial's case, this period is at the end of 13 months from the beginning of the first month in which the new shipper received deliveries. *Id.*

President of Ashland Oil, Inc., in his 1974 testimony before the Senate Antitrust and Monopoly Subcommittee wherein he described how Ashland started out, during a period when Colonial was on proration, [forcing Colonial "to roll over and give us some space because of the ICC regulations"] tendering 15,000 barrels per day, prorated back, on a proportional basis, to something "less than half that space," but building up to 10,000 barrels per day. As Mr. Yancey remarked, "we would much rather pay the tariff on [the consent decree] basis [and] put our money into other areas where we get a much higher rate of return."[1924]

The charge of restricting current shippers on an "as is" basis falls of its own weight when one examines the actualities of the situation as demonstrated by the Colonial example. As described earlier in this chapter,[1925] far from "overwhelming" other historical shippers and prospective new shippers, Colonial's owner-shippers had their shipments reduced by 4.5 percent and the new nonowner-shippers increased their traffic some 15 fold, with the result that the gasoline market share of independent marketers in the 15 state area served by Colonial climbed from 14 percent in 1963 to 27 percent in 1974 while the share of Colonial's owners declined steadily. There is another facet to the "historical shipper" method which the staff neglected to mention. Certain lines are subject to fluctuations caused by seasonal changes in demand. Dixie, for example, has a number of faithful shippers who move their products throughout the year, either to year-round customers or to storage at the point of usage. Then, when winter comes and the "snow birds" flock in to use the line, an interesting question is raised. Should these shippers who provide nothing in the way of support for the line during nine months of the year, be given a barrel for barrel allocation with those shipper-owners and nonowners alike who have been using the line regularly all year? Granted that the "three months a year" types should not be frozen out of the line, it does seem more equitable to allocate space on a 12-month average basis than to give them as much weight for their tenders as those whose use of the line has been continuous throughout the year. Dixie expanded and obviated this problem, but fundamental fairness would seem to suggest a historical proration has a definite role in this type of situation.

There presently is an interesting proceeding before FERC[1926] in which Powder River Pipe Line Company and its affiliate, The Crude Company, have alleged that Amoco Pipe Line Company discriminated against them

---

[1924] *S.1167 Hearings, supra* note 432, pt. 8 at 5928-5929 (Statement of Robert E. Yancey).

[1925] See text at notes 1657-1661, *supra*.

[1926] Powder River Pipeline Corp. and The Crude Company v. Amoco Pipeline Co., F.E.R.C. Dkt. No. OR78-6. (Jan. 4, 1978).

in favor of Western Oil Transportation in Amoco's allocation of its pipeline capacity serving the Hartzog and Heldt Draw crude fields in Wyoming. Powder River's complaint, filed on January 4, 1978, claims that Amoco gave preferential treatment, in violation of Sections 1 (4) and 3 (1) of the ICA by (1) allocating capacity on the basis of historical shipments rather than by current tenders and (2) allocating capacity at its Reno terminal among Powder River, Western and Shell without taking into account the fact that Western had extra capacity available in *its* own pipeline which connected to Amoco's Sussex-Salt Creek line at Sussex station. In a motion to dismiss, filed on April 13, 1978, Amoco answered the first count by averring that it had allocated capacity on the basis of current tenders since August 1977, when proration first became necessary. Its answer to the second charge was that the lines serving Reno station and Sussex station were separate. (The Sussex station-Salt Creek line is a 13.85 mile 8-inch line built in 1949, and the Reno station-Salt Creek line is a 31.87 mile, 10-inch line built in 1965; there is a third line, not involved in the case, which is a 29.5 mile, 12-inch line built in 1958 from North Fork station to Salt Creek Station, each of which has a separate tariff and receives separate tenders.) While Amoco recognized its statutory duty under §1 (4) to provide transportation on either of its pipelines to all persons making reasonable requests therefor, it did not have an obligation to require Western to route its crude oil by way of the Sussex station rather than the Reno station in order to accommodate Powder River. Finally, Amoco argued that the issues raised in the complaint were moot because there was presently spare capacity at Reno station. The case is still pending at this time. It would appear that Powder River's real complaint, if any, should be directed at Western for allegedly "juggling" its movements and not at Amoco Pipeline, which followed an allocation method not only approved, but directed, by the Federal District Court of Colorado.

(9) Scheduling

The Kennedy Staff Report contained the same nebulous quality of complaint with respect to scheduling that was evident in its other regulation and service claims. On page 6 of the Report, the staff asserts that "Other pipeline operational features, such as product quality standards and *shipping schedules, may* also serve to disadvantage prospective outside shippers" (emphasis added). On page 81, after repeating the foregoing allegation, the Staff amplified the statement as follows: "Shipping schedules normally are developed through computerized control and advance knowledge of the number and types of shipments. Such schedules are developed far enough in advance to accommodate the needs of all shippers. Variations in the shipping schedule *could* affect the ability of any shipper to meet the schedule and therefore create additional problems in using the pipeline" (emphasis added). The Report then attempts to obtain some support from the Attorney General's Deepwater Port Report with the assertion that: "Pipeline operations are essentially similar to the deepwater port situation and the same principles apply. Thus, pipeline owners have turned to more subtle, sophisticated methods to deny access to potential nonowner shippers."[1927] It is difficult intelligently to answer this type of argument other than to point out the inaptness of the analogy and the fact that, after analysis, the last sentence hints much but says nothing. The reader is urged to refresh his recollection of scheduling as discussed in Section II B 3 b, "Scheduling and Dispatching,"[1928] and determine for himself whether or not the alleged subtle discrimination really is not just a gossamer suspicion.

c. Empirical Evidence on the Access Issue

(1) Nonowner Usage

The growth of nonowner use of pipelines has been a long-term observable trend; one which has accelerated substantially since the advent of the large diameter "industry" lines. The reader will recall that in the early 1930's, over 60 percent of the large integrated companies carried no oil for nonowners in their gathering lines and about one-half of them did not transport nonowners' oil in their trunk lines.[1929] The reason for this exclusivity was attributed to the fact that pipeline technology at that time limited the economic capacity of the lines in that area to throughput volumes which barely were able to serve the refineries of the companies that had built the pipelines and there was an economic disincentive to seek out non-proprietary traffic.[1930] By the *TNEC* days of 1939-1940, just under 10 percent of the crude oil and 20 percent of the gasoline transported by pipelines belonged to non-affiliated shippers. During the fuel "shortage" in 1947, the Wherry Committee investigated the matter and the testimony tended to show an increase in nonowner use of pipelines.[1931]

[1927] KENNEDY STAFF REPORT, *supra* note 184, at 82.
[1928] See text at notes 581-588, *supra*.
[1929] H.R. REP. No. 2192, 72d Cong., 2d Sess. LXIII (1933).
[1930] See text at note 112, *supra*.
[1931] WOLBERT, *supra* note 1, at 44-45.

During the late 1940's and in the 1950's a surge of large diameter "industry" lines were constructed which served to provide a real economic incentive for pipeline systems to seek out vigorously nonowner shippers in order to realize the economies of scale made possible by these "big inch" lines. Thus the forces of economics joined with the requirements of the ICA to assure that pipelines operated, both in law and in fact, as common carriers on a nondiscriminatory basis.

By 1957, there was an indication that approximately 40 percent of all crude movement ultimately reaching refineries was for consignees *other* than affiliates of the parents of the pipeline carriers.[1932] More data became available with respect to movements during 1968 to 1972 and they showed substantial increases in Badger, Kansas, Texaco-Cities Service and Texas-New Mexico.[1933] At Senator Stevenson's request, the AOPL conducted a survey covering the then latest year available, 1973, which showed that 89 percent of the total shippers on 51 products lines were nonowners, and of those shippers 71 percent were "independents," and a similar analysis for *all* pipelines for the same year was presented to the ICC, which showed that 88 percent were nonowner shippers, and 65 percent were independents.[1934] The data on the four individual pipelines were updated for the 1973-1976 period and Amoco, West Shore and Olympic statistics were added. The four lines all had increases in nonowner shipments.[1935] Marathon submitted data for the year 1976 to the Justice Department, which evinced 53 percent nonowner shipments over its gathering lines; 34 percent over its products lines; and 31 percent over its crude trunk lines.[1936]

Erickson, Linder & Peters, in a paper presented at the American Enterprise Institute's Conference on Oil Pipelines and Public Policy, held in Washington, D.C., on March 1-2, 1979, furnished data on Colonial from 1963 through 1977 (a growth of non-affiliated shippers from 6 to 25), Plantation from 1960 through 1977 (from 9 to 21) and Explorer from 1972 through 1977 (increase from 1 to 7 outside shippers).[1937] A survey by AOPL of 74 crude oil and products pipelines for the year 1976 disclosed that 89.24 percent of their shippers were nonowners and 64 percent of all such shippers were "independents."[1938] These statistics, if anything, tended to *understate* the accessibility of pipelines to "independent"

[1932] See text at note 461, *supra*.
[1933] See text at notes 462-474, *supra*.
[1934] See text at notes 475-477, *supra*.
[1935] See text at notes 478-494, *supra*.
[1936] MARATHON, *supra* note 202, at Exhibit G.
[1937] AEI, *Oil Pipelines and Public Policy, supra* note 684, at 19-22.
[1938] See text at notes 495 and 496, *supra*.

nonowner-shippers because 84 refineries representing 16.8 percent of capacity are located in areas where they are not in a position to use any existing products line; another 39 refineries, comprising 18.0 percent of the non-"major" refining capacity, are situated where the only available products line is owned by a firm that is neither a refiner nor a refined products marketer; and a third group of 28 refineries accounting for 42.3 percent of such capacity were owned by firms which were the sole owners or part owners of products lines serving the refinery.[1939] A second fact which tends to cause the statistics to understate access is that savings in logistical "place differentials" are frequently achieved by the use of exchanges, which saves the refiners, and hence the American consumer, "a lot of money because of the saving in transportation [cost]."[1940]

In addition to the statistical data, there is testimony by three different ICC Commissioners (one of them on two occasions) attesting to the absence of complaints from industry members concerning access to pipelines[1941] and evidence from independent producers' associations on four occasions to the effect that independent producers were *not* encountering any difficulty securing shipment of their oil or discriminations by pipeline companies and were taking a position in opposition to divorcement proposals.[1942] The government has attempted to characterize this testimony as being indicative of a reluctance by potential complainants to come forward, or perhaps fear of reprisal, but this appears a bit lame in the face of several complaints that rates, the most important tariff item, have been too *low;*[1943] the surge of complaints when recent increases by Williams Pipe Line were initiated in 1971; the protest on Williams' valuation; and the initial tariff rate protest in the *TAPS* proceeding.[1944] The industry views this evidence as a strong indication of the availability of its lines to bona-fide shippers, owners and nonowners alike. Charles E. Spahr, Chairman of the Board and Chief Executive Officer of Sohio, made a common sense appraisal in his testimony before the Rodino Committee in 1975: "The record in terms of complaints to the I.C.C. from shippers who believe that they were denied proper access to common carrier lines would indicate that such problems generally do not exist. Very few such claims have been registered. A more accurate comment on the lack of complaints is that there have been few problems

---

[1939] LIVINGSTON, *supra* note 228, at 70-71; *cf. S.Res.5 Hearings, supra* note 296, pt. 3 at 899 (Statement of I.C.C. Chairman George M. Stafford).

[1940] *S.1167 Hearings, supra* note 432, pt. 8 at 5931 (Statement of Robert E. Yancey, President of Ashland Oil, Inc.).

[1941] See text at notes 1502-1505, *supra*.

[1942] See text at notes 1506-1509, *supra*.

[1943] See text at note 1516, *supra*.

[1944] See text at notes 1564-1602, *supra*.

of this nature which could not be mutually worked out between the parties before reaching the formal appeal stage. Difficulties arise because of timing, quantities and other operational factors, just as they do in other aspects of any business, and they are settled to the general satisfaction of all parties. Most business problems are settled long before they reach the stage of asking for official help and pipeline problems are no different."[1945]

There are two items which bear directly on the question of access to Colonial, seemingly the "lightning rod" for all pipeline "lack of access" claims. The first item is the Report of the Lundberg Survey on the gasoline market shares in fifteen states along Colonial's route. Briefly, the Lundberg finding was that the share of the gasoline market held by non-integrated marketers in the survey area almost doubled (going from 14 percent to 27 percent) during the years 1963-1974, while that of the integrated marketers (including Colonial's owners) steadily declined during that period.[1946] The second item is the finding by the Justice Department, after 13 years of investigating Colonial, "that there were presently no meaningful denials of access."[1947]

Perhaps Deputy Assistant Attorney General Bruce B. Wilson said it all when he testified at the *Smith-Conte Hearings:* "[The oil companies] assert that access to the pipeline is open and available to all without regard to ownership interest. This is a contention difficult to refute when we have few, if any, who will testify that they have been excluded."[1948]

(2) Lines Operating Under Capacity

Pipelines require nearly full loads to achieve maximum operating efficiency. This provides a real economic incentive for pipelines to encourage all shippers to use the line; it is a simple matter of good business to encourage nonowners to avail themselves of the service of these lines in order to achieve the greatest operating efficiency, and the lowest average cost for the owner-shippers. Ideally, if a pipeline could predict traffic with pinpoint accuracy, there would not be any undercapacity or excessive overcapacity. The plain fact of the matter is that even with the most careful forecasting, conditions beyond the pipelines' control subsequently arise which render previously valid forecasts no longer reliable and this

---

[1945] Spahr, *Rodino Testimony, supra* note 255, at 19.
[1946] *Petroleum Industry Hearings, supra* note 192, at 779.
[1947] KENNEDY STAFF REPORT, *supra* note 184, at 128.
[1948] *H.Res.5 Subcomm. Hearings, supra* note 234, at 206 (Statement of Bruce B. Wilson).

may result in either excess or insufficient capacity.[1949] Marathon provided a list of crude,[1950] and products lines[1951] which compared estimated line capacities with estimated throughputs which shows that there was substantial *underutilization*. A more comprehensive picture of pipelines operating under capacity was provided by a recent survey by AOPL of 46 integrated pipeline companies, which accounted for approximately 85 percent of total barrel-miles shipped on trunk lines in 1977. Twenty-nine of these lines continuously operated under capacity; 15 more reached capacity for limited periods or only in a relatively small portion of their systems and only two continuously ran full.[1952] The degree of underutilization of capacity may even be understated somewhat because several of the older lines, when faced with continuous, substantial underusage, first took stations off the line to save power costs and later removed them permanently where there was use for the equipment elsewhere. As discussant G. S. Wolbert remarked during the recent AEI Conference on Oil Pipelines and Public Policy, "Virtually all the lines in the United States are well under capacity today, let alone prorationed. And virtually all have substantial percentages of non-owner shipments, including a substantial percentage of 'independents,' a very amorphous term."[1953]

(3) Lines at Capacity — Proration and Expansion

Before delving into the "nitty-gritty" of proration and expansion, a general comment on the subject appears to be in order. This is the economic fact, which critics seem determined stubbornly to ignore, that the economic effect of prorationing falls more heavily on the shipper-owners than upon nonshipper-owners because the former traditionally are the larger shippers. This operates either way—if, indeed, prorationing causes the displaced shippers to utilize more expensive modes of transportation, then the larger shippers (the shipper-owners) will have more to lose by being forced to use such facilities. If, as is usually the case, alternative transportation at almost comparable cost is available, the nonowner-shipper will lose little but the shipper-owner will have lost the revenue which it could have earned by adding capacity at incremental

---

[1949] See text at notes 720-1037, *supra*.
[1950] See Appendix O.
[1951] See Appendix P.
[1952] See text at the last paragraph of Section III A 3 C, "Scramble for Traffic."
[1953] AEI, *Oil Pipelines and Public Policy, supra* note 684, at 72.

cost. Finally, if the diversion of traffic causes other modes of transportation to become less costly than the prorated line, the shipper-owner stands to lose not only by the reduction in market price which the diversion over the new more favorable mode of transportation has enabled the nonowner-shippers to make, but the effect will be cumulative since the nonowner-shippers will leave in droves, thus increasing the average cost to the shipper-owners as long as they remain with the line. This was exactly what happened to Plantation's owner-shippers when Colonial was built, until Plantation expanded and modernized its line so as to regain its competitive position.[1954] A second proposition, which at least the Antitrust Division has recognized, is that the mere existence of proration on a line does not, by itself, indicate a denial of access or undersizing; a further examination into the circumstances is required.[1955] For example, if after a reasonable period of time has elapsed while a line is under proration, so that the pipeline's owners can be reasonably sure that the demand is not an anomaly but appears to be a relatively long-term proposition, and they act promptly to increase the capacity, the most likely inference to be drawn is that demand has overtaken the original forecast. Capline has expanded several times under these circumstances, one instance being in 1977.[1956] Likewise, if, during the observational period, the demand falls off and the line no longer is prorationed, it would seem probable that the forecasts were correct and some unusual circumstance, which may never recur, caused the prorationing. For the owners to have rushed into an expansion under such circumstances would have constituted economic waste.

In Section II A 3 c, "Scramble for Traffic," an examination was made of four pipelines named either by the Kennedy Staff Report, Assistant Attorney General Shenefield's Proposed Statement or Senator Kennedy's Petition filed with the FTC for a Rulemaking Proceeding. An index of 1.00 was used to indicate a ratio when throughput equaled rated capacity under normal conditions (*i.e.*, excluding such events as a shutdown due to an accident). Table I clearly shows that all of the "suspect" lines, *i.e.*, Explorer, Platte, Wolverine and Yellowstone, were well under capacity. Another named line, Plantation, had minimal prorationing in 1950, which was lifted in 1951, and thereafter had no extended prorationing which would qualify under the ground rules set

---

[1954] *Cf.* LIVINGSTON, *supra* note 228, at 54; HOWREY & SIMON, *supra* note 3, at 78-79.

[1955] AEI, *Oil Pipelines and Public Policy, supra* note 684, at 72 (LeGrange-Hay interchange), *Id.* at 126 (Professor Edward J. Mitchell). Also, the Colonial investigation of 1963-1976 evidences this point rather forcibly.

[1956] *Cf.* HOWREY & SIMON, *supra* note 3, at 79.

forth in the first paragraph of this subsection, as can be seen by glancing at Table II. Among the systems named, Colonial was the only pipeline which had sustained prorationing (See Table III). Construction of the Colonial line commenced in 1963 and was completed in early 1965. The initial capacity was 720,000 barrels per day and initial daily shipments were approximately 600,000 barrels.

The first expansion of capacity was announced in February 1966 and completed in that year, consisting of additional pumping stations, which had been anticipated but originally thought not to be needed until the mid-seventies. The second expansion was announced in November, 1966. Two more expansions, also involving additional pumping stations, were made swiftly, bringing capacity up to 1,152,000 barrels per day.[1957] In 1971, the fifth expansion, consisting of 461 miles of 36-inch line "loops" and the addition of 128 miles of spur lines, increased the capacity to 1,464,000 barrels per day. A sixth expansion, in 1976, looped 183 miles with 40-inch pipe, raising capacity to 1,620,000 barrels per day. A seventh expansion, to be completed this year, will add more 40-inch pipe to bring capacity up to 2,100,000 barrels per day.[1958] The growth of nonowner-shippers on Colonial's line from 1965, the year the construction of the original system was fully completed (0.15 percent) to 1978 (36.26 percent) was detailed in Section II A 3 c, "Scramble for Traffic,"[1959] and the number of nonowner-shippers rose from 0 in 1965 to 32 in 1978.[1960]

In an effort to provide a more comprehensive data base, AOPL recently sent a questionnaire to its members requesting information and data on prorationing during the period 1976-1978. The responses were coded so as to preclude identification of individual companies and then were turned over to the American Petroleum Institute (API) for analysis.[1961] Altogether 53 companies responded. Of these, 19 were classified as systems wholly-owned and operated by "major" oil companies (including 14 of the 16 largest producers of crude oil and natural gas liquids in 1977); 27 were jointly owned or operated pipelines which typically included at least one "major" oil company as an owner; and seven were "independent," largely non-integrated systems.[1962]

With respect to industry coverage, the 53 firms represent 45 percent

---

[1957] *H.Res.5 Subcomm. Hearings, supra* note 234, at A71 (Letter dated Jan. 27, 1972, from Jack Vickrey to Chairman William Proxmire).
[1958] LIVINGSTON, *supra* note 228, at 54.
[1959] See text at notes 416-421, *supra*.
[1960] LIVINGSTON, *supra* note 228, at 55.
[1961] The author acknowledges his debt to Donald A. Norman, Senior Economist, API, for the analysis contained in the remainder of the discussion in this subsection [VI C 3 c (3)].
[1962] See Tables IV, V, VI.

of the approximately 119 interstate pipeline companies. However, the 53 firms accounted for approximately 88 percent of total barrel-miles shipped in 1977 and the 46 integrated pipelines were responsible for about 85 percent of total barrel-miles shipped. Hence, the sample covers the bulk of the U.S. pipeline system.

Each company was asked to report on the following:

1. The extent and duration of prorationing from 1976 through 1978.

2. The percent of prorationing in the specific segment(s) affected.

For example, if each shipper had to reduce his tender by 10 percent, then the percent of proration would be 10 percent.

3. The percent of proration in the entire system.

Pipeline systems typically are divided into segments and seldom is an entire system subject to prorationing. Thus, this information shows the extent to which a company's system was prorationed.

4. The response to prorationing.

The Kennedy Staff Report asserted that integrated pipeline firms would be slow in response to capacity shortages in order to maintain a restriction on shipments. Therefore, those companies which had to proration capacity were asked how they responded, in fact, to the situation.

5. Average daily unused capacity in the entire system.

If integrated pipelines are undersized, then one would expect not only systematic prorationing but little excess capacity. The measure of unused capacity provides another indication as to the extent to which capacity is taxed to meet shipping demands. As a rule-of-thumb, optimal throughput occurs when a pipeline is operated at 95 percent of rated capacity, *i.e.,* when unused capacity is five percent.[1963] At higher rates, regular maintenance is more difficult. And at lower rates, excess capacity exists.

The results appear in Tables IV-VI. Table IV lists the responses of the wholly-owned pipeline systems of the "major" oil companies. Table V lists the responses of jointly owned or operated systems, Table VI presents the responses of the "independent," largely non-integrated pipeline systems.

[1963] The 95 percent figure is a rough rule-of-thumb. The actual optimal throughput-to-capacity ratio varies among pipelines as it is dependent on factors such as product mix or the gravity of crude oil shipped.

(A) Extent and Duration of Prorationing, 1976-1978

Of the 46 integrated pipeline systems responding to the questionnaire, 29 had no prorationing whatsoever during the period 1976-1978. Of the 17 which did have some prorationing, only two had sustained prorationing. Prorationing on the other 15 systems was temporary and/or limited to a few specific segments of the system. The evidence on prorationing in these tables is inconsistent with the hypothesis that systematic prorationing is a general characteristic of integrated pipeline systems. Out of seven "independents," two have had to proration shipments. Thus, even the "independent" pipeline systems have had to proration shippers, though here it appears that lack of terminal capacity rather than pipeline capacity was the cause of prorationing.

The data on the percent of proration gives some perspective on the relative importance of prorationing on a company's system. For example, on system 4M, two segments have been subject to significant proration at times during the 1976-1978 period. Yet, in terms of total system shipments, prorationing has been a minor factor.

(B) Responses to Prorationing

The evidence on the responses to prorationing reveals that companies typically expand capacity, by looping and/or adding pumping capacity, in order to alleviate the pressure on capacity when such expansion is warranted. In several instances, there were no expansions because demand to ship had decreased or because such expansion was not economical.

(C) Unused Capacity

The data on unused capacity show that most systems have significant capacity. Even those systems which have had to proration capacity on certain segments generally have excess capacity as a whole. The extent of unused capacity clearly raises the question as to how these systems could reasonably be construed as undersized or discriminatively restrictive of access.

(D) Conclusions

Identifying a couple of specific pipeline systems as having been under sustained proration is insufficient empirical support for the proposition that integrated pipelines systems are restricting access or are likely to be undersized. By this method, one could just as easily identify a couple of systems which have not been under continual proration to refute that contention. Instead, one must focus on the general tendencies. The data presented in Tables IV and V are difficult to reconcile either with the claim that integrated pipeline systems restrict access or that they are likely to be undersized. While some prorationing exists, it is limited generally in terms of its extent and duration. Earlier in this subsection, it was shown that the very existence of prorationing is not, in and of itself, evidence of denial of access or of deliberate undersizing.[1964] Even in a world of perfect certainty regarding the future, some pipelines can be expected to proration capacity during their existence as fields or markets they serve change. The evidence presented here indicates that integrated pipelines do not restrict access and are not prone to systematic undersizing and hence belies the notion that divestiture is a required remedy.

(4) Decisions Not to Participate in Ownership of Pipelines

Despite critics' suspicions that "majors" simultaneously deny "independents" an opportunity to participate in the ownership of the lines and require "independents" to "buy into" jointly owned or operated lines in order to secure access,[1965] there are a number of valid reasons, entirely unrelated to such suspicions, which provide satisfactory answers to why companies in the industry, "independents" and "majors" alike, do not always seek ownership in pipelines. For convenience of analysis, these reasons have been grouped into broad classifications. First, there is a group of "geographically advantaged" refiners which enjoy an

---

[1964] See text at note 1955, *supra*.

[1965] For example, see the interchange between Senator Stevenson of Illinois and John E. Green, President of Shell Pipe Line Corporation in the *Consumer Energy Act Hearings, supra* note 154, at 604-605: Senator Stevenson. "I think it might be productive to get into the terms and conditions under which shippers, refiners and marketers acquire access to the pipelines. It has been said that Clark is heavily dependent on overseas crude through the CAP line [sic throughout quotation] and that in order to obtain this crude, it was required to buy into the line. Is this true?" Mr. Green. "Senator, I think the question hinges on one word completely, the verb 'required.' Clark could have been served by CAP line without having any ownership in CAP line, but at the time the CAP line, which is a joint venture line, was conceived and organized, Clark expressed an interest to come in and own an interest in that pipeline, and they were welcomed into the pipeline. I deal with Clark daily, at least weekly, with respect to the CAP line. I knew they are a very well respected member of the company, of the organization. It was their choice as to whether they wanted in or whether they would like to have only common carrier service." Senator Stevenson. "It is your opinion, then, that the expression of interest by Clark was entirely voluntary." Mr. Green. "Entirely voluntary." Tending to confirm Mr. Green's statement is the fact that Clark, on the "downstream" side, also bought into the Wolverine line — an 11 percent interest. See Jones & Gardner Memo., *supra* note 427, at Part B, p. 7.

unassailable position in their local market, which renders unnecessary integration into pipelines.[1966] Second, the cost of acquiring or constructing pipeline assets causes it to be unattractive to many independent producers and marketers who already have all they can handle financially to maintain their position in their primary phase of operation.[1967] Third, contractual arrangements with other forms of transportation are sufficiently satisfactory that certain "independent" refiners find it unnecessary to seek pipeline ownership.[1968] Fourth, access to a common carrier pipeline, either "major" or "independent" owned, on a tariff basis, has been so convenient and satisfactory, that both "independents" and "majors" alike deem their much needed funds, which can obtain a higher rate of return elsewhere, would not be properly invested in pipeline facilities to serve certain of their facilities.[1969] Fifth, certain refiners — "independents," self-professed "independents," [*e.g.,* Ashland] and "majors" — are able to reduce the cost of any form of transportation by means of exchange agreements, which permit them to access markets which they could not otherwise have reached on an economic basis.[1970] Sixth, there are some refiners whose locations are such that they could not use any existing pipeline and their operations are too small to justify building a line themselves.[1971]

(5) Statements of Non-Pipeline Owners Concerning Access

(a) By "Major" Oil Companies

Charles E. Spahr, Chairman of Sohio, in his testimony before the Rodino Committee stated: "In the past few years, Sohio has shipped, as a non-owner, through many lines, including Buckeye, Cherokee, Interprovincial, Lakehead, Marathon, Tecumseh, Texas and Texas-New Mexico. Our records do not indicate that we have ever been denied access to a common carrier pipeline. Those of us in Sohio who have been associated with this activity for many years have no knowledge of any such refusal."[1972]

[1966] See text at notes 1233-1235, *supra.*
[1967] See text at notes 1229-1232, *supra.*
[1968] See text at notes 1236-1238, *supra.*
[1969] See text at notes 1239-1249, *supra.*
[1970] See text at notes 1250-1255, *supra.*
[1971] LIVINGSTON, *supra* note 228, at 38 and 70-71.
[1972] Spahr, *Rodino Testimony, supra* note 255, at 18-19.

(b) By "Independents"

There is substantial evidence, from independent oil producers, both individually and through their prominent trade associations, attesting to the absence of discriminatory treatment, or denial of access, and opposing proposals to divorce pipelines from major oil companies. Mr. A. D. Barnett, an independent oil producer from Wichita, Kansas stated that he had been an independent producer in that area for 15 years, selling to both "majors" and "independents" in equal numbers, and during that time, pipeline companies had never discriminated against him, nor did he know of any such discrimination. Barnett was of the opinion that any change in the present crude oil system would only tend to increase transportation cost to the producer and hinder and delay the exploration and drilling for new reserves.[1973] Ray M. Johnson, speaking for the Oklahoma Stripper Well Association, likewise testified in *favor* of current pipeline practices and stated that pipeline divorcement proposals had "no substantial support" among the "thousands of small producers" in his Association.[1974] IPAA, either through its President or its General Counsel, on four occasions spoke favorably about pipeline practices and testified that it was not aware of any producer having difficulty in selling or moving his crude oil nor did the IPAA believe any existed. Jack M. Allen, the most recent President who testified, stated that independents had been well served by the existing pipeline industry.[1975]

The situation has been substantially similar with "independent" refiners. As far back as the *TNEC Hearings,* Louis J. Walsh, Vice President of Eastern States Petroleum Company, testified concerning the position of his company's refinery in Houston. Although Eastern States had no production or pipeline facilities, it was able, due to connections with three "major"-owned pipelines, freely to purchase oil "from practically all the fields in Texas, New Mexico, and some fields in Louisiana." Although Mr. Walsh "groused" a bit about paying the pipeline tariffs which he estimated to be about twice the cost of *operating* the lines [no mention was made of amortization of the lines' capital investment], he responded forthrightly to Congressman Henderson's question, "Do you have any trouble with these three major companies with their pipelines," by replying, "None whatever."[1976] In his testimony of August 6, 1974, Robert E. Yancey, President of Ashland Oil, Inc., a self-professed "independent," while suggesting that he might have had

---

[1973] *Consumer Energy Act Hearings, supra* note 154, at 656.
[1974] See text at note 1506, *supra*.
[1975] See text at notes 1506-1509, *supra*.
[1976] *TNEC Hearings, supra* note 6, at 7337 (Statement of Louis J. Walsh).

some difficulty in the past, related his then current experience with a major oil gathering system, which he summarized: "As of now, [we] have had no problem with it,"[1977] and he expressed the view that problems of discrimination in rates, the long- and short-haul clause or freedom of access could best be dealt with by a greater concentration by the ICC on pipeline matters.[1978] Mr. Yancey's most recent statement on the subject, and his company's official statement of position, issued May 1, 1979, is that the company is "satisfied that pipelines do not deny access to any shipper, owner or non-owner, either by behavior or discretionary structure."[1979] Thomas W. diZerega, President of Apco Oil Corporation, testified on October 29, 1975, before the Senate Antitrust and Monopoly Subcommittee that he believed all the pipeline companies realized what their legal obligations were and if an independent asserted his rights, the pipeline company would comply with the ICC's rules and regulations.[1980]

David Bacigolupo, Vice President, Beacon Oil Company, a small land-locked independent refiner-marketer in Hanford, California, located in the heart of the San Joaquin Valley, testified on November 12, 1975, before the Senate Antitrust and Monopoly Subcommittee that his company "had no difficulty filling [its] crude requirements."[1981] Richard J. Boushka, President of Vickers Energy Corp. of Wichita, Kansas, which has scattered crude oil and gas production, a 60,000 barrel per day refinery in Ardmore, Oklahoma, and markets its products through 950 service stations, also testified before the same Subcommittee, on January 28, 1976. Mr. Boushka had this to say: "We feel that the pipeline subject has been overplayed and a simple business made mysterious. Alternatives are too easy to come by for anyone to be injured over a long period of time. . . . We feel the general pipeline situation has provided the logistics for effective competition, or companies such as ours would not have been able to grow as we have over the past 5 to 10 years. Much like the service station area, any disruption could have the opposite effect than is intended by divestiture. There may be those in the oil industry who would disagree with my testimony. I ask them to examine their corporate consciences and determine whether or not they might be guilty of trying to justify past poor management decisions by blaming the majors."[1982]

[1977] *S.1167 Hearings, supra* note 432, pt. 8 at 5932 (Statement of Robert E. Yancey).
[1978] *Id.* at 5916.
[1979] See text immediately following note 1819, *supra*.
[1980] *Petroleum Industry Hearings, supra* note 192, at 252 (Statement of Thomas W. diZerega).
[1981] *Id.* at 397 (Statement of David Bacigalupo).
[1982] *Id.* at 2018 (Statement of Richard J. Boushka).

(c) By Enforcement Agencies

Not only have ICC Commissioners on four separate occasions stated that there have been no evidence or complaints against pipeline denial of access or discrimination against independent shippers,[1983] but the oft-quoted testimony of Bruce B. Wilson, Deputy Assistant Attorney General, before the Smith-Conte House Small Business Subcommittee on June 15, 1972, that the oil companies' assertions that invitations to participate in pipeline ownership and that access to pipelines was open and available to all without regard to ownership, was "difficult to refute when we have few, if any, who will testify that they have been excluded,"[1984] and the closing by the Antitrust Division of its 13-year investigation of Colonial on the basis of its finding that "there were presently no meaningful denials of access"[1985] make it extremely difficult to reach any conclusion other than the reason that no substantial evidence of denial of access has surfaced is simply because there hasn't been any.

(d) By Outside Parties

The authors of the Congressional Research Service Report on National Energy Transportation observed: "Basically, the anticompetitive nature of the shipper-owner relationship is almost being exclusively alleged by persons outside the oil industry, rather than by supposedly disadvantaged 'independent' oil companies."[1986] This was demonstrated in Section VI A 2, "Sources of Complaints."[1987] As the reader will recall, there was testimony from union officials who represented union members working the coast-wise tankers who priced themselves out of the market while Colonial's rates were going down; a member of the Ralph Nader conglomerate; an embittered ex-Guy Gillette aide and ex-Antitrust Division attorney; and the Executive Director of the National Congress of Petroleum Retailers (NCPR) who did not address the pipeline issue. Only two members of the industry testified; one of whom may have felt that pipeline rates were a bit high compared to his estimate of their cost but who stated with respect to difficulty of access to pipelines that he had "none whatever," and the other adduced equivocal testimony concerning the tender of oil which his company had obtained by outbidding an affiliate of the pipeline to which the oil was tendered,

---

[1983] See text at notes 1502-1505, *supra*.

[1984] *H.Res.5 Subcomm. hearings, supra* note 234, at 206 (Statement of Bruce B. Wilson).

[1985] KENNEDY STAFF REPORT, *supra* note 184, at 128.

[1986] NET-III, *supra* note 199, at 157.

[1987] See text at notes 1470-1510, *supra*.

with the outcome that, after a brief hassle, the pipeline ran his oil as tendered.

In addition to the ICC Commissioners' testimony mentioned in the preceding subsection, there was testimony by other non-industry members. Professor Edward J. Mitchell testified on August 6, 1974, before the Senate Antitrust and Monopoly Subcommittee concerning the results of his preliminary study of the pattern of gasoline prices in the markets served by the Colonial pipeline, which study led him to conclude that the pattern of gasoline prices and gasoline market shared was consistent with the working of the normal process of competition.[1988] Mitchell's preliminary study subsequently was confirmed and quantified by the Lundberg Survey, Inc., Report, based on an in-depth examination of gasoline market shares in 15 states along the route of the Colonial Pipeline.[1989] On January 27, 1976, Professor Richard B. Mancke testified before the same Subcommittee, summing up: "To conclude, there is no evidence that the large vertically integrated oil companies are presently exercising monopoly power in any of the four major stages of the oil business: The production of crude oil, *the transportation of crude oil and refined products,* the refining of crude oil, and the marketing of refined oil products. Indeed, all available empirical evidence supports the opposite inference" (emphasis added).[1990]

Finally, Paul J. Tierney, President of the Transportation Association of America (TAA), and former ICC Chairman, testified on October 30, 1978, in the *TAPS* proceeding that "Even though TAA has several hundred industrial and carrier members that are very heavy users of petroleum products, none to date has expressed to us its oppostition to the level of oil pipeline rates *or to any practices of these carriers"*[1991] (emphasis added).

4. The Department of Justice's Circumvention of Rate Regulation or "Undersizing" Theory

The current assault on pipeline ownership by integrated oil companies is based virtually exclusively on the so-called "undersizing" theory developed by the Department of Justice. Because of the narrow base of these attacks, a detailed examination of the development of the theory

---

[1988] Mitchell, *1974 Senate Testimony, supra* note 1056, at 40; *see also,* STEINGRABER, *supra* note 193, at 135-136; Ritchie, *Petroleum Dismemberment,* 29 Vand. L. Rev. 1131, 1144 (1976).

[1989] *Petroleum Industry Hearings, supra* note 192, at 779-785 (Digest).

[1990] *Id.* at 1894 (Statement of Richard B. Mancke); *see also,* HOWREY & SIMON, *supra* note 3, at 27; text at note 1517, *supra.*

[1991] Tierney, *TAPS Rebuttal Testimony, supra* note 175, at 8.

and its premises appears to be warranted. The seminal expression of the theory appeared in the Attorney General's Deepwater Port Report of November 5, 1976, to Secretary of Transportation William T. Coleman, pursuant to Section 7 of the Deepwater Port Act of 1974. At pages 3-5 of the Attorney General's Report, the following exposition appears:

B. Economic Principles Underlying Antitrust Review

"The analysis in this report rests on several economic principles. Accordingly, before developing factual and analytical sections, it will be useful to introduce here briefly the economic principles upon which the subsequent discussion is premised.

"Deepwater ports are natural monopolies in their respective markets. In economic terms this means that they have increasing returns to scale; as the unit gets larger, there are increasing cost savings associated with it. A 2,000,000 barrel per day port is much more efficient and achieves greater cost savings than a 1,000,000 barrel per day port. With the introduction of deepwater ports, other modes of transportation in most circumstances will become higher cost alternatives. To the extent that the deepwater port is available for users, it likely will be preferred to transshipment, lightering, or any other transportation alternative, all of which will be used to an increasingly lesser degree.

"As with all monopolies, certain economic principles apply. The monopolist has the incentive and ability to raise the price, which will result in an output below the competitive level. In a competitive situation, on the other hand, the producer has little or no control over price, which is determined by the competitive interaction of all buyers and sellers. In the competitive situation, the output will be the maximum that can be produced at that price. The producer will earn some profits, but they will not be excessive. In the monopoly situation, the producer has control over how much he can charge. He therefore possesses both the means (control over price) and the incentive (excess profits) to restrict output below the competitive level. With reduced output, the nature of demand ensures that he can raise prices and earn profits above what he would obtain at the competitive level. Thus, the monopolist can and normally will earn excess profits.

"In a pipeline or deepwater port setting (pipelines have the same natural monopoly characteristics as deepwater ports) an unregulated pipeline owner (no rate regulation) can maximize profits by raising the price of pipeline transportation above the competitive level, *i.e.,* above cost plus a normal, nonexcessive return to investment. His ability to accomplish this stems from economies of scale in pipelines — making competition from other pipelines unlikely — and from the relative efficiency of pipelines as an overland transport mode — conferring absolute cost advantages on the pipeline and making competition from other modes unlikely. The same is true of deepwater ports; their efficiency makes competition from other ports unlikely and confers absolute cost advantages on the port over alternatives. These cost advantages will not be passed on to consumers in the form of lower prices; rather, they will be pocketed by the monopolist as an excessive return to investment. The resulting high price reduces the demand for pipeline transportation and this artificial restriction of pipeline throughput creates a misallocation of resources — the provision of both pipeline transportation services and delivered crude oil fall short of socially desirable levels. The resulting excess profits will not be eroded by entry into the market until demand grows sufficiently large to accommodate an at least equally efficient competing pipeline.

"Normally, regulation is the method used to pass on natural monopoly cost savings to consumers. Therefore, a pipeline or a deepwater port could be prevented from fully exploiting the potential excess profits in transportation by regulation by the Interstate Commerce Commission (ICC) and an antitrust consent decree which places a ceiling on pipeline rates of return. But when the owners of pipelines and deepwater ports are integrated petroleum companies, they can partially circumvent this regulatory effect on their profits by utilizing their transportation cost advantage to earn monopoly profits in the market for delivered crude oil or even further downstream in the market for delivered petroleum products. The source of the transportation cost advantage is, of course, the same as the source of excess profits from an unregulated pipeline — economies of scale and cost advantages relative to alternative modes of transportation. Artificial quantity restrictions and resources misallocation remain substantially uncorrected.

"This exchange of excess transportation profits for excess downstream profits by the pipeline or port owners can be seen by comparing two apparently different situations: (a) the pipeline or port is unregulated, the owners set a monopolistic transportation tariff, and the flow of product through the pipe is reduced below a competitive level, versus (b) the pipeline's profits are restricted by regulation and the owners limit throughput by denying access to nonowners in various ways. Our analysis has led to the conclusion that these two cases tend to result in the same (maximum profit) product throughput and, therefore, tend to yield identical prices and quantities of product in the downstream markets and identical excess profits for the port owners. Because of this, our review has focused on the problem of access to deepwater ports by nonowners and expansion of deepwater ports by owners and nonowners."

Although the Federal Trade Commission in its report to Secretary Coleman pointed out that the mere fact that the deepwater port facilities included pipelines *did not convert the ports into pipelines* and that the enactment of the Deepwater Port Act itself reflected Congress' recognition of the difference,[1992] somehow the distinction became obfuscated and the germination of the "undersizing" theory, as fully applicable to pipelines *per se,* commenced *sub silentio.* The FTC Report made another significant observation, namely that "[a] system of prorationing that operated perfectly might not only eliminate the problem of discrimination but also provide incentives for increases in capacity. A shareholder forced to reduce its individual throughput because of prorationing might determine that it is necessary to maintain the pre-prorationing flow of crude oil to its refinery or petro-chemical facility, in which case it might be expected to advocate port expansion to accommodate the desired volume."[1993] However, it weakened this observation by saying that it could not be *assumed* that prorationing would work perfectly or even that a perfect prorationing system would result in adequate capacity. Hence, the point was not pursued actively by the FTC and it was rejected brusquely by the Justice Department in its Supplemental Deepwater Port Memorandum: "It is absolutely necessary to put to rest the shibboleth that prorationing is the solution to expansion. Prorationing should be viewed as no more than a very short term solution until additional capacity can be added."[1994]

The "undersizing" theory, as related to pipelines, pushed its head above ground on May 6, 1977, when Assistant Attorney General Donald I.

[1992] Report of the Federal Trade Commission to the Honorable William T. Coleman, Jr., Secretary of Transportation, on the Applications of LOOP, Inc. and SEADOCK, Inc. to Own, Construct, and Operate Deepwater Ports Under the Deepwater Port Act of 1974, dated November 5, 1976, p. 37 (hereinafter cited as FTC's Deepwater Port Report).

[1993] *Id.* at 34; essence repeated, *Id.* at 39.

[1994] Supplemental Memorandum of the *Department of Justice* to the Secretary of Transportation on the Deepwater Port License Applications of LOOP, Inc. and SEADOCK, Inc., dated November 19, 1976, pp. 12-13 [hereinafter cited as DJ Supplemental Deepwater Port Report].

Baker told a Senate Judiciary Antitrust Subcommittee on Oversight of Antitrust Enforcement that the Justice Department was "convinced" that joint ownership and operation of oil pipelines by integrated oil companies permitted them to evade rate regulation by the ICC and to "capture monopoly profits downstream." He added that the Antitrust Division was "carefully evaluating whether a litigation solution [probably pipeline divestiture] is available to remedy this problem."[1995] One of the buds opened up a bit in a terse footnote contained in the Justice Department's Statement of Views and Argument in the Valuation of Common Carriers Pipeline proceeding, then ICC *Ex Parte No. 308*, wherein it was stated: "A more serious impediment to effective rate regulation, however, is the possibility that excess earnings will simply be transferred to another level of a vertically integrated firm's operations. By restricting the throughput or capacity of a pipeline, its owners can force some oil to flow to the final market by more costly, less efficient means of transportation. As the price of crude or product downstream is generally set by the cost of the incremental barrel, throughput restriction has the effect of raising the downstream profits. Vertically integrated firms can thus capture at other unregulated stages some of the potential monopoly profit on pipeline transportation denied by regulation."[1996]

The "undersizing" theory began to blossom in the Attorney General's Alaska Natural Gas Report, where the observation was made, with respect to Alaskan gas production, that "even well-designed tariff regulation by the Federal Power Commission is doomed to failure" because with relaxed well-head price regulation "producer-owners of a pipeline can circumvent tariff regulation by shifting some or all of the potential profits stemming from the transportation natural monopoly to upstream (*i.e.*, production) operations." The report continued "...this problem — the failure of tariff regulation to eliminate monopoly profits — is due entirely to vertical integration, *i.e.*, the ownership by producers of the most efficient means of transportation."[1997]

Three additional items in the Report merit attention. First, for the purpose of developing the theory, the authors *assumed ideal regulation,* that is, its tariff was designed to permit the pipeline to earn no more than a fair and reasonable return to its investors and the delivered price had to be equal to or less than that delivered to the destination market by alternative sources; and that a non-affiliated transporter, because he can maximize his profits only by maximizing his throughput at the FPC "squeezed down" tariff, would react by aggressively soliciting additional

---

[1995] 813 ANTITRUST & TRADE REG. REP. (BNA) E-11 (May 12, 1977).
[1996] D. J. Views and Arguments, *supra* note 656, at 22 n.44.
[1997] A. G.'s Alaska Natural Gas Report, *supra* note 1431, at 32.

supplies, whereas producer-owners would shift some or all of the potential profits stemming from the transportation natural monopoly to upstream (production) profits, *i.e.,* they would "undersize."[1998]

Second, the Report's authors also attempted to "defuse" the potential of prorationing as a preventive to "undersizing": "If prorationing is then fair and equitable, has the upstream profit-vertical integration problem been solved? The answer is emphatically in the negative. Ironically, attempting to resolve the capacity problem by prorationing tends to mask and thereby exacerbate it. Although shippers are treated equally, *the decision of owners to restrict pipeline capacity is the ultimate denial of access.* Prorationing is a symptom of excess demand for pipeline transportation motivated by the still extant upstream profits. These profits cannot be bid away by newly entering producers until pipeline capacity is expanded."[1999]

The third item mentioned in the Report is most intriguing. The Report states: "Let us now drop our original assumption that the Alaskan field is open to new entry and is potentially workably competitive. If present Alaskan producers have market power in that field not derived from the pipeline (*e.g.* if they already control all the high-quality, low-cost reserves to be found in the area), they can merely restrict gas supplies at the production stage rather than indirectly via restraints on transportation. There is no need in this case for them to tinker with the pipeline stage in order to exploit their market power (except, ironically perhaps, to encourage the lowest-cost line available to insure [sic] the delivered cost advantage margin of their gas). Integration by producers into the pipeline stage under these circumstances would produce no worse performance than would already result."[2000] This is sound economic theory, but as will be demonstrated shortly in the subsection on "Fatal Flaws in the Economic Theory Involved in the Undersizing Argument,"[2001] it serves to disclose the circularity in reasoning implicit in the "undersizing" theory.

On October 13, 1977, the "undersizing" tree was still just a sapling, as revealed by Assistant Attorney General John Shenefield's interview with the Bureau of National Affairs. During the course of that interview, Mr. Shenefield declined to play the role of McDuff's midwife — he was *then* unwilling to rip pipelines untimely from the womb of the integrated oil companies. He discussed the development of the "undersizing" theory as it applied to isolated facilities such as Alaska or deepwater ports and explained that the Antitrust Division was studying the theory to see how

---

[1998] *Id.* at 32-36.
[1999] *Id.* at 40.
[2000] *Id.* at 41-42.
[2001] See text at notes 2023-2041, especially note 2039, *infra.*

far it would go. Shenefield admitted, with admirable candor, that: "There is some considerable debate as to whether our theory is applicable in a situation where there are alternate routes or where the pipelines are not being used to capacity. Whether it makes sense, as we would contend, to undersize a pipe where there's an alternate route that is no more expensive, is not wholly clear. In fact, you might conclude that the answer would be it doesn't make any sense. So, as a general proposition, we're not at the point, I'm not at the point, of concluding that all pipelines ought to be divested from all integrated oil companies everywhere, regardless of where they exist."[2002]

In the relatively brief period between the October 1977 BNA interview and his June 28, 1978, proposed statement before the scheduled, but not held, Kennedy Subcommittee Hearings, Mr. Shenefield apparently decided to see how far his theory would go. In the proposed statement, he clearly advocated pipeline divestiture, prospective for all, and retrospective for some—an unnamed class "likely to represent significant concentrations of market power."[2003] He also spelled out the "undersizing" theory.

Mr. Shenefield developed his theme as follows. First, he listed three questions that he said would lead to an understanding of how and when rate regulation may be circumvented: (1) When does a pipeline have market power? (2) Is rate regulation at all effective in curbing that power? and (3) When does vertical integration enable a pipeline company to circumvent the constraints of rate regulation?[2004]

In addressing the first question, Mr. Shenefield noted preliminarily that "to some degree," all pipelines have incentive to circumvent existing regulatory constraints,[2005] but only those with "market power" could succeed. Market power is more than mere economies of scale or inherent cost advantages, declared Mr. Shenefield, it is the amount of a pipeline's [natural] monopoly advantage over its competitors, measured either at the upstream or the downstream end of the pipeline, or both. He proposed a three-fold test for downstream market power: when (1) the pipeline's throughput comprises a significant share of the downstream market; (2) it can supply the downstream market with less expensive petroleum than can

---

[2002] 834 ANTITRUST & TRADE REG. REP. (BNA) A-7 (October 13, 1977).

[2003] SHENEFIELD PROPOSED STATEMENT, *supra* note 179, at 42. Mr. Shenefield's statement subsequently has been reprinted in full in AEI, *Oil Pipelines and Public Policy*, *supra* note 684, at 191-215. However, citations in this book will continue to relate to the mimeographed version.

[2004] *Id.* at 16.

[2005] Interestingly, DEChazeau & Kahn pointed out that "independent" (*i.e.* non-affiliated with oil interests) pipelines would not have been as competitive as integrated pipelines and they would have even more of a tendency to "undersize," or at least have less temptation to overexpand pipeline investment. DE CHAZEAU & KAHN, *supra* note 30, at 337.

other suppliers, either because of lower transportation costs than alternative modes of transportation (including other pipelines) or because local producers (or refiners) cannot expand output without increasing costs; and (3) consumers in the downstream market are willing to pay a higher price to obtain petroleum if the supply is reduced. The corresponding "upstream" tests were: (1) the pipeline's throughput comprises a significant share of the upstream market; (2) increased supplies of petroleum can be absorbed upstream only at a lower price; and (3) suppliers upstream will be willing to accept lower prices to sell their crude or products, either because of a lack of transportation alternatives at costs as low as the pipeline's, or because contracting output would be difficult or unprofitable. Mr. Shenefield termed this type of market power "monopsony" power.[2006]

Mr. Shenefield's "proof" that many pipelines have substantial market power —which rate regulation has failed to curb — was that the *Division* has what *it* considered was "strong empirical evidence" that they had such power, *i.e.,* an evaluation by the Division's Economic Policy Office, embodied in the George Hay statement in *Ex Parte No. 308,* that "the rates of return for many oil pipelines [were] significantly above what [was] encountered in the *most profitable* electric utility and natural gas pipeline companies."[2007] Rates of return were examined in the earlier subsection on "So-called 'Excessive Rates' of Return."[2008] and in the AEI conference described therein. A consensus was reached at the conference that the proper base upon which rate of return should be measured was total capitalization and that an appropriate "risk premium" had to be derived before any meaningful comparison across industries could be made. One suspects that Mr. Shenefield would be happy to agree to the more objective type of enquiry discussed in the AEI Conference, cited in footnote 2008, if he would take the time to read Dr. Hay's statement upon which he had earlier built his entire case, because Dr. Hay's statement would "prove," to use Mr. Shenefield's word, that Santa Fe, Crown-Rancho, Kiantone, MAPCO, Paloma, America Petrofina of Texas, Okie, Cheyenne, Minnesota, Buckeye, Allegheny, Air Force, Acorn, Calnev, Kaneb and Southern Pacific have the requisite "market power," and Exxon, Sun Oil, Cherokee, Arapahoe, Texaco-Cities Service, Kaw, Tecumseh, Pure, Gulf, Four Corners, Phillips Pipe Line, ARCO, Sohio, Hess and Getty, all of whom return less to capital than does the *lowest* electric utility shown, would lack such power. If the *least* profitable natural gas pipeline shown, Panhandle Eastern, were used as a yardstick,

---

[2006] SHENEFIELD PROPOSED STATEMENT, *supra* note 179, at 16-19.
[2007] *Id.* at 19-20, footnoted to Hay, *FERC Direct Statement, supra* note 1689, at 6-7.
[2008] See text at notes 1675-1710, especially 1695-1700, *supra*.

Plantation, Laurel, Mid-Valley, Chase, Cook Inlet, and Sun Pipe Line also would lack the requisite "market power."[2009]

Mr. Shenefield's answer to the effectiveness of rate regulation in curbing this market power that he had conjured up was equally self-serving and lacking in substance. While not stating explicitly what was the basis for his conclusion that the 8 percent and 10 percent ICC limitations and the 7 percent Consent Decree limitations were *partially* effective, it obviously is implicit that his frame of reference against which to judge "partial" versus "full" regulatory effectiveness was the Hay testimony. Analyzing his statement, it appears evident that Mr. Shenefield wished to avoid the logic trap that if he "found" the current rate regulation to be *completely ineffective*, economic theory would force him to admit that pipelines, assuming *arguendo* that they were "natural monopolies" without effective competition, would simply take their "monopoly profits" the easiest way, *i.e.,* with excessive tariffs.

Thus, in order to "set up" his "circumvention of regulation" or "undersizing" argument, he had to create some incentive to shift profits upstream or downstream. He did this by his "finding" of "some practical bounds" set by regulation, thereby leaving himself a differential of "excessive profits" between what the rates would have been if the lines were regulated on an average cost basis and that which the "partially effective" regulation permitted (according to Dr. Hay). This maneuver set the stage for Mr. Shenefield's answer to his third question: When does vertical integration enable a pipeline company to circumvent the constraints of rate regulation? He answered that one: "As we see it, the ability to avoid regulation is rooted in the ability of the shipper-owner to limit pipeline throughput below the level an independent would size or operate the line. If the vertically integrated pipeline owner is a significant seller in the downstream market and the pipeline has market power downstream, the pipeline's throughput may be such as to ensure that at least some not insubstantial amount of supply arrives by the more expensive transport modes. Where this occurs, the downstream market price would likely reflect the cost of delivery by the next least expensive transport mode, allowing the pipeline's shippers to pocket the difference in transportation costs."[2010]

Thus, the "undersizing" tree was in full bloom. There was a pruning of one of the weakest limbs in the so-called "Sohio" Report where the theory was held to be inapplicable to Sohio's proposed Long Beach to Midland, Texas, crude line which was intended to transship Alaskan crude brought to Long Beach by tanker to Midland where connecting pipelines

---

[2009] Hay, *FERC Direct Statement, supra* note 1689, at Table I, Table II and Table III.
[2010] SHENEFIELD PROPOSED STATEMENT, *supra* note 179, at 22.

could move the crude to PADs II (Midwest) and III (Southwest). The percentage of this area's crude supply that would have gone through Sohio's line during Phase I was 5 percent and only 10 percent if Phase II expansion took place. Approximately 36 percent of the crude supply to the destination area at the time of the Sohio Report came from imports. "Thus, the incremental supply of petroleum to the midcontinent is forthcoming at the world price of petroleum," and that cost "determines the market price of oil in PADs II and III." Sohio could not, by restricting throughput, increase the price of the delivered oil above the world price and to pursue a policy of throughput restriction would merely cut Sohio's throat in PADs II and III and benefit the foreign producer.[2011]

Arizona markets were served productwise by the Southern Pacific pipeline and crudewise by ARCO's reversed Four Corners line, so no market power was present. New Mexico's indigenous crude production exceeded local refinery capacity and the excess entered the pipeline network near Midland, so Sohio could not raise the price of New Mexican crude by undersizing.[2012] Nor could Sohio discourage bidding for leases or developing new crude reserves in PAD V (West Coast). If it produced a surplus on the West Coast, prices and profits for all sellers including Sohio in PAD V would be forced downward. Then Sohio would have to expand the line or lose profits. Moreover, prorationing of the line would "back out" more Sohio crude which, instead of being sold in PADs II and III, would have to be "dumped" in PAD IV.[2013] Based on this analysis, the SOHIO Report concluded that since Sohio had no incentive, in fact an economic disincentive, to restrict the line, the "undersizing" theory was not applicable to the proposed Long Beach-Midland line.[2014]

The Division's new theory rapidly became the "in" theme throughout the Washington circuit. The Kennedy Staff Report,[2015] the Bureau of Competition of the FTC,[2016] the Policy and Evaluation Section of the DOE,[2017] and, of course, Senator Kennedy's Petition to the FTC for a Rulemaking Proceeding to divest pipelines of certain classes,[2018] all adopted the "undersizing" argument, in most cases as their principal argument.

The Department perhaps began to get a little uneasy about its theory

---

[2011] D. J. SOHIO REPORT, *supra* note 633, at 40-41.
[2012] *Id.* at 44-45.
[2013] *Id.* at 48.
[2014] *Id.* at 49.
[2015] KENNEDY STAFF REPORT, *supra* note 184, at 15, 63-65.
[2016] DOUGHERTY, REVISED PROPOSED STATEMENT, *supra* note 656, at 3-8.
[2017] ALM PROPOSED STATEMENT, *supra* note 646, at 1-2.
[2018] Kennedy FTC Rulemaking Petition, *supra* note 1641, at 7-10.

due to some of the criticism. When Donald L. Flexner, Deputy Assistant Attorney General, delivered his paper "Oil Pipelines: The Case for Divestiture" before the AEI Conference on Oil Pipelines and Public Policy on March 1, 1979, he first "tidied up" the third market power prerequisite to: "alternative supply is less than perfectly elastic, that is, any increase in the supply required from alternative sources can be attained only at a higher price."[2019] He next attempted to inflate the downstream market shares of the owners by reasoning that while: "In many areas of the country, where workable competition may exist within other sectors of the petroleum industry, the downstream market shares of individual companies are often not large. However, the combined downstream market shares of the joint owners of a pipeline can present a very different picture.[2020] Decision making by a joint venture, especially where unanimous or near unanimous consent of the members is required, is necessarily a collective process. That collectivity, in turn, reflects the fact that the individual interests of participants in the joint venture are maximized in accordance with the cumulative downstream market share of the members. In short, the joint venture will operate as though the sum of the downstream market shares of all the individual firms is affected by each collective decision. Accordingly, the market shares and resulting market effects are much more significant than the corresponding effects of a decision made by one firm with a relatively small market share."[2021] Mr. Flexner's footnote discloses that his purpose in attempting to engraft a "group monopoly" concept onto the "undersizing" theory was to try to mitigate the devastating effect of the Mitchell-Lundberg evidence of absence of anticompetitive effect which should have attended the construction of Colonial's line had the "undersizing" theory been valid.

Thus, at least, the "undersizing" theory has proven to have a high degree of elasticity. Each time an objection or impediment to its application has been voiced, or the Antitrust Division itself has detected an internal inconsistency, the theory has been stretched, amended, or

---

[2019] D. FLEXNER, OIL PIPELINES: THE CASE FOR DIVESTITURE in AEI, OIL PIPELINES AND PUBLIC POLICY 5 (E. Mitchell ed. 1979) [hereinafter cited as FLEXNER, *The Case for Divestiture*].

[2020] *Id.* at 9. Flexner's footnote reads, "For example, in 1973, in any one of the states of Virginia, Maryland, Delaware, New Jersey, and in the District of Columbia, no single company among the group of integrated oil companies which jointly own Colonial Pipeline Company appeared to have a retail market share greater than 19 percent (Lundberg Survey, published by *National Petroleum News*). The sum of the retail market shares of all ten owners is much greater. In addition, in 1976 the owners of Colonial, taken collectively, owned about 49 percent of the refining capacity in those same states (*Oil & Gas Journal,* 1976). In those states, then, it would be appropriate to take into account their collective share of the supply, including the impact of their ownership of refineries, in determining the downstream effects of the joint ownership of Colonial." *Id.* at 9 n.11.

[2021] *Id.* at 9.

reformed in shape so as to retain the appearance of viability. The closest anyone has come to establishing a finite limitation to its extent was in the open discussion at the AEI Conference in Oil Pipelines and Public Policy. The dialogue ran about like this. Dr. George Hay, in responding to Ulyesse LeGrange's comment that operators who were attempting to abide by the law had to have some ground rules in advance so that they could make a business judgment on risk, replied that the Department recognized that many things determine how large a pipeline should be including whether the demand is going to be there and whether there might be a new source of supply. But, he added, "there may be an additional consideration: an incentive to make the pipeline a little smaller out of fear of 'spoiling' the market." If that fundamental incentive existed, then he felt it was useless to worry about details of expansion rules and ownership rules; a structural solution [*i.e.*, divestiture] may be required. Whereupon, G. S. Wolbert commented that it seemed to him that if there were an incentive to undersize, it would have to show up in a pattern of constant prorationing of a sizable number of lines, which just wasn't the fact. Dr. Hay's answer to that was that the nonowner-shipper's fair share under the prorationing rules might be so small, or they might be so uneasy about whether they would get access to the pipeline and unwilling to commit themselves to the supply which would make the pipeline a good idea in the first place, that they wouldn't make the effort to get in.

Professor Edward Mitchell couldn't see that argument at all. He believed that if the theory were sound, prorationing would be operating continuously, all over the place. He believed that if someone were really setting to undersize, he would do so sufficiently to prevent the pipeline from completely serving the market. Therefore, if for substantial periods of time pipelines were not full and were not prorating, this fact would seem to him to undercut seriously the undersizing theory. It would have to be set aside with respect to a line that was rarely prorated.

At this point, Wolbert remarked that virtually *all* the lines in the United States were operating well under capacity today, let alone prorated. Moreover, all had substantial percentages of nonowner shipments, including a substantial percentage of "independents." He then addressed the following question to Dr. Hay. "If we can assume for the sake of argument that this is so — investigation could verify it — then can these lines be undersized?" To which question, Dr. Hay replied: "If there are many nonowner-shippers, and if they have access to more cheap oil than they can bring in, and there is empty space on the pipeline which they can get access to, then, I would agree that the pipeline could not be undersized with respect to current demand facing it."[2022]

[2022] AEI, *Oil Pipelines and Public Policy, supra* note 684, at 70-72.

a. Fatal Flaws in the Economic Theory Involved in the Undersizing Argument

In order to facilitate the discussion of the economic flaws in the "undersizing" theory, a recapitulation of the theory will save the reader from having to reread the previous subsection. *First:* pipelines are "natural monopolies" because of their cost efficiency and economies of scale.[2023] *Second:* By restricting the throughput or capacity of a pipeline, its owners can force some oil to flow to the final market by more costly, less efficient, means of transportation. As the price of oil is generally set by the cost of the incremental barrel, the output restriction has the effect of raising the downstream profits. Vertically integrated firms can thus capture at other unregulated stages some of the potential monopoly profit on pipeline transportation denied by regulation.[2024] *Third:* Other stages of the industry are assumed to be workably competitive.[2025] *Fourth:* Pipeline tariff policy is assumed to be ideal, *i.e.,* the tariff is designed to permit the pipeline to earn no more than a fair and reasonable return to its investors.[2026] *Fifth:* Fair and equitable prorationing will not solve the problem.[2027]

The Department of Justice has not presented any data relating to its undersizing theory, nor has it exposed any formal economic model which would support its undersizing allegation.[2028]

On the other hand, the American Petroleum Institute has introduced into the public domain a rigorous mathematic model which demonstrated, even taking the assumptions used by the Justice Department as factual (several of which will be demonstrated in a subsequent subsection to be incompatible with the real world facts), that the Department's undersizing theory is a myth and that the Department's conclusions regarding pipeline

---

[2023] FLEXNER, *The Case for Divestiture, supra* note 2019, at 4.

[2024] D. J. Views and Arguments, *supra* note 646, at 22 n.44.

[2025] A. G.'s Alaska Natural Gas Report, *supra* note 1431, at 30-31. To assume otherwise would remove the incentive for integrated firms to undersize pipelines. *Id.* at 41-42: "If present Alaskan producers have market power in that field not derived from the pipeline. . . . [t]here is no need. . . for them to tinker with the pipeline stage in order to exploit their market power (except, ironically perhaps, to encourage the lowest-cost line available to insure the delivered cost advantage margin of their gas)."

[2026] *Id.* at 32-33. This assumption becomes necessary ". . . because with completely ineffectual rate regulation there would be no need to use vertical integration to reap the pipeline's monopoly profits. Such profits could be obtained directly with excessive tariffs." SHENEFIELD, PROPOSED STATEMENT, *supra* note 179, at 20.

[2027] A. G.'s Alaska Natural Gas Report, *supra* note 1431, at 40; D. J. Supplemental Deepwater Port Report, *supra* note 1994, at 12-13; *but cf.* FTC's Deepwater Port Report, *supra* note 1992, at 34, 39.

[2028] PIETTE, *U.S. Oil Lines — Public Policy, supra* note 1492, at 13.

integration do not follow from their premises.[2029]

In an earlier section of this book, III D, "Similiarities and Dissimilarities to a Natural Monopoly,"[2030] the application of the economists' natural monopoly appellation to pipelines was questioned severely. Mr. Shenefield's expression in his BNA interview of his doubt whether "undersizing" was applicable where there were alternate routes and/or lines not used to capacity appear to reflect some of the same points made in Section III D.

Another economic fallacy is contained in the Justice Department's second point, *i.e.*, that by restricting the capacity of the pipeline, its owners will capture the difference between the downstream price set by "the incremental barrel" reaching the market by "more costly, less efficient means of transportation" and the owners' cost of shipping through the (undersized) line. While the "higher cost" of the alternate mode of transportation will be demonstrated in the following subsection to be an incorrect assumption in fact, for the purpose of showing the economic error in the argument, let us accept, *arguendo,* the fact asserted. The error in the economic theory is the predicate that the higher cost unit required to balance supply and demand in the destination market will come *solely* from the cost of shipping to that point by the assumed higher cost mode of transportation. This is a misuse of the concept of marginal cost. Remember that Shenefield postulated perfect prorationing and no lack of access. Under these conditions, the marginal cost is *not* the cost of the assumed higher-cost mode of transportation. The marginal cost is a weighted average of the pipeline tariff and the assumed higher cost of the alternative mode of shipment. Each shipper, owner and nonowner alike, has the same average cost and there is no "monopoly profit."[2031]

This can be illustrated by a simple example. Assume that a company builds a pipeline designed to serve 90 percent of an area's needs and that the remaining 10 percent must employ a more costly mode of transportation. What will the shipper do who is using the more costly mode of transportation? Clearly, he will tender his material to the pipeline for shipment. If the common carrier statutes are working correctly, it will not be long before every shipper is moving 90 percent of his material by pipeline and 10 percent by the higher cost mode. Shenefield sees this equilibrium but says the owner-shipper is simply sharing his monopoly profits with non-owner shippers, so all are content.[2032]

[2029] P. KOBRIN, A FORMAL CRITIQUE OF THE JUSTICE DEPARTMENT'S PIPELINE UNDERSIZING ASSERTION (A.P.I. Critique #004, August 21, 1978).

[2030] See text at notes 657-701, especially 672-701, *supra*.

[2031] LIVINGSTON, *supra* note 228, at 43; AEI, *Oil Pipelines and Public Policy, supra* note 684, at 67 (Comment by Dr. Morris Livingston).

[2032] SHENEFIELD PROPOSED STATEMENT, *supra* note 179, at 27.

There are two things wrong with the conclusion. First, and less significant, it is not likely that investors would continually put up the huge sums of money involved and let others be "free-riders." If there were real monopoly profits, all would want to be "free-riders" and no one would want to be an investor.

Of greater importance, however, is the fact that Mr. Shenefield has analyzed the problem incorrectly. Consider the situation if demand rises by one barrel. Everyone who perceives that increase will tender one more barrel to the pipeline and he will move 0.9 barrel by pipeline and 0.1 barrel by the higher cost mode. In this case the 0.1 barrel moving by the high cost mode does not set the price. The incremental barrel in this case is a weighted average of the two costs.

The least expensive way to move a barrel of oil into the market is to move 0.9 barrel by pipeline and 0.1 by the high cost mode. Note that the 0.9 low cost barrel automatically drags along 0.1 high cost barrel; one cannot move the 0.9 without moving the 0.1. Given these fixed relations, the minimum incremental cost is the weighted average of the two modes. It is this weighted average incremental cost which will set the market price. And since everyone has the same costs — the very same costs which set the market price—there is no excessive return for anyone.

Consider the possibility that one ingenious shipper decides to supply his incremental barrel solely by the high cost mode, thereby setting a higher market price and creating excessive profits for all. Every other shipper then has the incentive to tender a barrel to the pipeline, ship part by each mode to obtain a lower delivered cost than that of the high cost shipper, shade the new price somewhat, take this incremental sale away from the original shipper and thereby increase his profits still further. Indeed, no price above that which is based on weighted average cost can be an equilibrium price unless there is some form of collusion or lack of competition in the final market. But if there is collusion or lack of competition, the problem has nothing to do with the size of the pipeline and there are ample remedies for collusion already on the books.

In short, if pipelines are in fact common carriers and if they are, in fact, undersized, the final market price is set by the weighted average cost of the two modes — the actual costs incurred by all shippers — and there are no downstream monopoly profits. The reader will recall that the FTC recognized this in its report to the Secretary of Transportation on LOOP and SEADOCK but it assumed, without any supporting evidence, that oil pipelines were not common carriers in fact and that downstream monopoly profits therefore could occur. In order to provide an intellectually honest argument that there is an incentive to undersize pipelines, it must first be demonstrated that oil pipelines are not common carriers in fact. No such documented argument to this effect has been advanced.

Canes & Norman also have spotted a weakness in the Justice Department's economic theory underlying the undersizing argument.

As they phrase it: "That something is amiss in the Justice analysis is easily seen by recalling that the Department assumes workable competition in non-pipeline industry stages *and* asserts that monopoly profits are earned there. The simultaneous existence of competition and monopoly profits violates an elementary principle of economic analysis .... Suppose for the sake of argument that a group of integrated firms indeed sought to restrict pipeline throughput so as to earn monopoly profits at an adjacent industry stage. By assumption, the adjacent stage is competitive and there is common carrier, non-discriminatory access to the pipeline. Because the monopoly profits are gained via access to the restricted-throughput pipeline, firms will be motivated to compete for this access. Some set of rules will govern the obtaining of access, but regardless of the form of these rules, existing firms and entrants will devote resources to obtaining more pipeline access so long as extra-normal profits can be earned thereby. But the competitive devoting of resources to acquiring access will affect both costs and returns from such access, and this competitive process will result in an equilibrium in which no extra-normal profits will be earned. Hence, contrary to the Department's assertion, its own assumptions concerning competition at the adjacent stage will ensure that no monopoly profits are earned."[2033]

There was an additional facet of the "circumventing regulation" argument advanced by Justice Department representatives at the AEI Conference. In Thomas Spavins' paper, "The Regulation of Oil Pipelines," he wrote: "In the case of petroleum pipelines, the economic benefit earned by a shipper-owner from the movement of petroleum through a pipeline is not measured by the tariff, but by the difference in the value of the petroleum between origin and destination. This difference can be estimated by the market prices of the petroleum at those points, if that information is available. If the owner can influence the throughput of the pipeline, and thereby the price differential between origin and destination points, then the possibility exists for the subversion of any pipeline regulatory scheme that is concerned solely with pipeline tariffs."[2034] Parenthetically, it might be noted that the nonowner likewise receives the economic benefit of the difference in value between origin and destination, less the tariff rate. Hence the only *additional* benefit to the

---

[2033] M. CANES & D. NORMAN, PIPELINES AND ANTITRUST 14-15 (API Critique #005, November, 1978) (An enlarged and updated version of this paper appears *sub. nom. Pipelines and Public Policy* in AEI, *Oil Pipelines and Public Policy, supra* note 684, at 141-163.) [hereinafter cited as CANES & NORMAN, *Pipelines & Antitrust*].

[2034] SPAVINS, PIPELINE REGULATION, *supra* note 676, at 9-10. Spavins' paper is reprinted in its entirety in AEI, *Oil Pipelines and Public Policy, supra* note 684, at 77-105.

shipper-owner would be the difference between the cost and the tariff, which, under the undersizing premises restated at the beginning of this subsection, was deemed to be no more than a fair and reasonable return to the investor, and which was demonstrated to be so in fact in the discussion in Section VI C 1, above.

Spavins raised the point as a problem of regulatory boundaries — the "tar-baby" syndrome discussed in an earlier section,[2035] but Dr. George Hay brought it squarely into the "undersizing" context when he replied to Dr. Michael Canes' question concerning a proper empirical test of the undersizing hypothesis, that "[t]he ideal empirical test would be a comparison of the value of the oil at one end of the pipeline—less the market value of the oil at the other end of the pipeline compared with the actual economic cost of transportation. Where that exceeds the actual cost of transportation, then one can infer that the size of the pipeline—." Dr. Hay never got a chance to finish his sentence before he was interrupted by Professors Edmund Kitch, remarking that such a difference would exist in *every* case where some of the oil is moving to the destination market by an inferior technology, and Professor Teece, saying "It is a quasi-rent." Professor Kitch bored in a little harder: "That doesn't show the pipeline is undersized since the sizing decision was made when it was built. It may be simply proof of an error." Dr. Hay: "Okay, I accept that." Professor Kitch: "And if you found a dispersion in a set of markets, where some pipelines seem to be oversized and some seem under, you would just be showing dispersion of errors around some mean." Dr. Hay retreated: "Suppose I refine my testimony to say, at the time pipeline is built, we observe a difference between the market value of the oil at either end compared with the cost of building the pipeline?"

This "fall back" position didn't satisfy Professor Teece, who asked where would lie the incentive for anyone ever to build a pipeline if there was not such a discrepancy, *i.e.*, if the value of the oil that came out of the other end was exactly the same as the value of what went in, plus transportation costs. Dr. Hay then qualified his statement by saying, "transportation costs" as he was using it, included a competitive rate of return. But Professor Teece felt that still did not provide an incentive; there had to be something more — "epsilon" greater than just a competitive rate of return. Dr. Hay was willing to allow "epsilon," but he stuck to his guns and said, "Then the question is how big to build it." Professor Teece agreed. At that point, Dr. Canes, who had initiated the line of questioning, attempted to get a definitive answer by framing, with an amendment by Professor Edward Mitchell, a test that before there could be any question of undersizing at least there must be systematic

---

[2035] See text at notes 1005 and 1006, *supra*.

prorationing. Mr. Spavins, limiting his remark to an attempt to answer Dr. Canes' question of what was a set of necessary and/or sufficient conditions to indicate pipeline "undersizing," stated: "Assume there is a pipeline that has unused capacity in the sense that it is continuously willing to offer transportation, and can pump more petroleum through the line. This fact is a valid test of throughput restrictions and vertical restrictions and is only valid if access is equal in the sense that there are no differentials in the access cost of getting into the pipeline."[2036]

Finally, a problem in logic arises when undersizing drifts completely into a vertical integration argument, namely the transfer of assumed "monopoly power" from the transportation stage to another level. There is a respectable body of economic thought which discounts the "spill-over" theory. In their view, vertical integration does not impair competition nor can it enhance monopoly power. In fact, they argue that where a vertically integrated firm includes one phase with greater monopoly power than the others, vertical integration may serve to increase output, and where the final market is also monopolistic, to reduce market price.[2037]

This school of thought builds upon the generally accepted economic theory that a monopolist can extract only so much profit from its position, which it takes at the one level in which it has a monopoly; it cannot increase its monopoly returns by obtaining another monopoly in any other level of the industry. It thus perceives the monopoly problem as a *horizontal* issue, rather than a *vertical* one. The Justice Department's theory attempts to "wire around" this problem by saying, even though the ICC rate regulation has been ineffective it has prevented "natural monopoly" pipelines from taking all the monopoly profits they could have taken in that phase, hence the "undersizing" theory, which presumably relates only to the differential amount of "monopoly profit" which remains untaken after the ICC/FERC rate of return limit is reached, does not run afoul of the "take-it-all-in-one-stage" theory.

It is rather hard to tell whether Justice's argument is ingenious or disingenuous. Even if one were willing to grant, purely for the sake of argument, that the Department had threaded its way successfully around the problem of taking all the profit which arises from an alleged "natural monopoly" power in the pipeline phase, there remain two other obstacles.

[2036] AEI, *Oil Pipelines and Public Policy, supra* note 684, at 123-126 (General Discussion).

[2037] SPENGLER, *Vertical Integration and Antitrust Policy*, 58 J. Pol. Econ. 347 (1950); BORK, *Vertical Integration-Economic Misconception, supra* note 106, at 196-197; *cf.* LIEBELER, *Integration and Competition, supra* note 976, at 23; DE CHAZEAU & KAHN, *supra* note 30, at 45, 553-554; ROSTOW & SACHS, *Entry into the Oil Refining Business: Vertical Integration Re-Examined,* 61 Yale L.J. 856, 876 (1952). This is not to imply that there are not economists who disagree with this thesis, but the Justice Department apparently agrees with it. See notes 2025 and 2026, *supra.*

First, there is the fact that the assumption that pipelines are "natural monopolies" is extremely tenuous.[2038] Second, again assuming, *arguendo*, that pipelines *do* have such power, there is the admitted prerequisite for the undersizing theory to apply that not only must the pipeline's throughput comprise a significant share (according to the SOHIO Report more than 10 percent) of the downstream market's supply, and be the lowest cost method of supply, but the pipeline's owner must be a significant seller in the downstream market (the upstream market, for all practical purposes in today's shortage of supply situation, is merely a symbol of theoretical symmetry). If the pipeline's shipper-owner already has market power downstream, one questions whether the same economic theory doesn't hold true, *i.e.,* why should it fiddle around with pipeline undersizing when it can simply take its profit in the marketing stage based on its power in that market.[2039]

The undersizing theory bears a strong resemblance to the now thoroughly discredited charge of transfer of monopoly profits backward to the production phase so as to reap a "tax-advantage" profit,[2040] and to a lesser degree, the "subsidization" of shipper-owners by nonowner shippers which, although popular politically, received a very frigid reception from academicians, even those traditionally critical of the industry.[2041]

b. Legal Imperfections in the Undersizing Theory

The Justice Department initially couched its undersizing argument in terms of a single firm pipeline owner. It used (or as contended in this book, misused) the term "natural monopoly" as a springboard from which to develop the concept of transferral of "monopoly power" profits from pipelines to adjacent phases of the industry. Assuming, for the sake of this discussion, that pipelines' economies of scale *did* provide a monopoly power, the very use of the adjective "natural" indicates that such a monopoly would be one that was "thrust upon" its holder by superior efficiency. In the famous *Alcoa* decision, Judge Learned Hand, in a much quoted statement, said that certainly thirty-three percent of the relevant market was not enough to constitute an illegal monopoly; that it was doubtful whether sixty or sixty-four percent would be enough;[2042] but

---

[2038] See text at notes 672-701 and 2030, *supra*.
[2039] PIETTE, *U.S. Oil Lines - Public Policy, supra* note 1492, at 13. This takes us back to the Alaska Natural Gas Report's third item mentioned in text at note 2001, *supra*.
[2040] See text at notes 1167-1175, *supra*.
[2041] See text at notes 1293-1296, *supra*.
[2042] The Supreme Court held in United States v. International Harvester, 274 U.S. 693 (1927) that 64.1 percent was not enough.

ninety percent was enough.[2043]

The penalization of a "natural monopoly" would pervert antitrust law and policy which *seeks* competition based on efficiency.[2044] The *Brown Shoe*[2045] market foreclosure doctrine, upon which the Department may be relying implicitly, has two strands, one of which is purely protection from competition,[2046] which has no place in any public policy designed to increase the well being of consumers,[2047] and the other is an "entry barrier" concept which is factually inapposite both to the refinery market and to the petroleum wholesale and retail market, as will be demonstrated in the next subsection. The only other plausible legal theory upon which to build an antitrust complaint about a single owner pipeline would have to be denial of access. After having read through Subsection 3 of this Section, "Asserted Denial of Access to Pipelines,"[2048] the reader will appreciate how unpromising would be a case on that ground. Moreover, despite the disparagement of ICC/FERC regulation by the Justice Department, if conditions should change in the future there are adequate means already in place to remedy any such future denial,[2049] and the Draconian measure of divestiture of an industry in order to correct one, or even a few instances, appears, as ICC Commission Charles Webb once remarked, comparable to the use of a sledge hammer to drive a tack.[2050]

Because most of the large-diameter, big-volume lines that have been constructed over the past three decades have been jointly owned and operated lines (and around 40 percent of oil pipelines reporting to the Commission are jointly owned), any deliberate undersizing of the line at the time of construction would be fully actionable under Section One of the Sherman Act and the Department of Justice could proceed directly against the owners without going through the convolutions of the "circumvention of regulation" theory. One has to suspect that the reason no

---

[2043] United States v. Aluminum Company of America, 148 F.2d 416, 424 (2d Cir. 1945).

[2044] Connell Construction Co. Inc. v. Plumbers & Steamfitters Local Union No. 100, 421 U.S. 616 (1975); United States v. Aluminum Company of America, 148 F.2d 416, 430 (2d Cir. 1945) (". . . the [Sherman] Act does not mean to condemn the resultant of those very forces which it is its prime object to foster: finis opus coronat. The successful competitor, having been urged to compete, must not be turned upon when he wins."); Adelman, *Effective Competition and the Antitrust Laws*, 61 Harv. L. Rev. 1289, 1310 (1948); WOLBERT, *supra* note 1, at 78-79.

[2045] Brown Shoe Co. v. United States, 370 U.S. 294, 328 (1962).

[2046] See text at notes 1514-1519, *supra*.

[2047] LIEBELER, *Integration and Competition*, *supra* note 976, at 25.

[2048] See text at notes 1807-1991, *supra*.

[2049] See text at notes 1520-1540, *supra*.

[2050] C. WEBB, I.C.C. REGULATION OF THE OIL PIPELINE INDUSTRY 7 (Address Before the National Petroleum Association, September 17, 1959).

such action has been brought simply is because there have been no such happenings.

There are many self-interest reasons, in addition to the undesirability of breaking the law, why this would be so. First, the "undersizers" would be locking themselves for 30 to 40 years into a self-imposed inefficient facility which would result in higher transportation costs. Second, if the owners of such a line undersized it even just a little too much, they would be inviting a new, more efficient form of transportation to enter the field and "eat their lunch" in the market place. Third, as was explained in detail in Section III F, "Risks Involved in Pipeline Ventures,"[2051] forecasting the accurate volumes for the optimal size line — which would be a necessary starting point from which to undersize — is just not precise enough an art to commence such a plan. In addition, there has to be an economic cost in determining just how much the undersizing should be and the more companies that were involved in the assumed collective non-competitive behavior, the greater the cost would be. This would suggest minimizing the owner group. But that is inconsistent with the well-known practice of giving wide publicity to the line and multiple solicitations of prospective owners to join.[2052] Finally, there is a practical difficulty in agreeing on a common course of action that increases sharply with the number of parties seeking to follow such a course of conduct.[2053] Considering the size and complexity of the industry, ease of entry, the divergence of interests, both upstream and downstream, the fact that, unlike what one would expect of colluders, the rates of return of the joint owners are widely divergent and the overall return on the line is below that which could be attained from alternate investment opportunities, the chance of collusion among the owners is remote.[2054]

[2051] See text at notes 720-917, especially 727-734, *supra*.
[2052] CANES & NORMAN, *Pipelines & Antitrust*, supra note 2033, at 10-11.
[2053] E. ERICKSON, G. LINDER & W. PETERS, THE PIPELINE UNDERSIZING ARGUMENT AND THE RECORD OF ACCESS AND EXPANSION IN THE OIL PIPELINE INDUSTRY in AEI, *Oil Pipelines and Public Policy*, supra note 684, at 20 [hereinafter cited as Erickson et al., *The Pipeline Undersizing Argument*].
[2054] *Cf.* Ikard, *Petroleum Industry Competition*, supra note 1055, at 952; TEECE, supra note 329, at 101-102; MANCKE, *Oil Industry Competition*, supra note 674, at 42 (nearly impossible); CANES & NORMAN, *Pipelines and Antitrust*, supra note 2033, at 10-11; DE CHAZEAU & KAHN, supra note 30, at 28.

c. Undersizing Theory is Unsupported by Evidence and is Contrary to the Real World of Pipelines

(1) No Credible Evidence has been Adduced by Proponents of "Undersizing"

One thing that sticks out like a wart on a spinster's nose is the utter lack of *any* credible evidence on the part of the Justice Department in support of its undersizing theory. The sole item advanced by Assistant Attorney General Shenefield was his self-serving statement that the *Department* was convinced that it had what "we consider to be strong empirical evidence that many pipelines have substantial market power," citing the Hay statement in *Ex Parte No. 308.*

The vacuous quality of that "proof" was demonstrated earlier in this chapter.[2055] Its contrast with the standard contained in Deputy Assistant Attorney General Bruce Wilson's testimony before the Smith-Conte House Small Business Subcommittee[2056] is striking, which provides a clear explanation why Mr. Shenefield, whose business is to prosecute cases, had attempted to dump the problem into the willing arms of a friendly Congressional Subcommittee rather than bring his complaint before a Federal judge who would require proof of the allegation. Professor Edward Erickson has analogized Mr. Shenefield's theory to the economics of the carnival midway in which the vertical stages of the petroleum industry cause "monopoly profits" earned through the exercise of market power at one stage to be transferred to, and presumably hidden in, other stages of the industry, in a manner similar to the shell game where the pea (of monopoly profits) is passed and hidden among the walnuts (of the vertical stages), regardless of the appearance of effective competition or effective regulation at any or all stages."[2057] As Professor Erickson pointed out, in addition to the various problems of economic logic, sooner or later, this argument must bear the burden of showing where the monopoly profits are hidden.[2058]

The evidence on overall industry profits certainly does not reveal any excessive profits; the overall average return on invested capital over the 10 year period 1964-1973 for the petroleum industry was only 10.3 percent as opposed to all manufacturing industries of 9.8 percent, and over the 16 year period 1958-1973 was only 10.0 percent compared to the average of

---

[2055] See text at notes 2007-2009, *supra.*

[2056] *H.Res.5 Subcomm. Hearings, supra* note 234, at 206-207 (Statement of Bruce B. Wilson).

[2057] Erickson et al., *The Pipeline Undersizing Argument, supra* note 2053, at 16-17.

[2058] *Id.* at 17 n.5.

all manufacturing industries of 9.3 percent.[2059] Thomas Duchesneau, in a research project for the Ford Foundation, found that for the period 1967-1973, in spite of the 1973 results, oil company reported profitability had not been generally above the average profitability of all manufacturers, *i.e.* 10.8 percent for each.[2060] The "big eight" (*FTC v. Exxon* defendants), influenced by foreign operation, had a net after tax income over equity ratio of 12.0 percent as opposed to an average all manufacturing ratio of 10.9 percent for the period 1951-1971.[2061] If the period is extended to 1951-1973, only in 1951 and 1973 (both unusually good years for the petroleum industry), did the most profitable of the eight oil companies equal or exceed the five year(1969-1973) average ratio of return earned by a group of non-petroleum firms thought to represent oligopolistic industries.[2062] Individual companies within that group did not fare as well. Shell's return on total capital from 1961 through 1973 was less than that of all manufacturing corporations for all years except 1963 and 1967 and less than that of utility companies from 1969 to 1973. Even its return on shareholders' investment was exceeded by all manufacturing companies

---

[2059] Ritchie, *Petroleum Dismemberment,* 29 Vand. L. Rev. 1131, 1148 (Table 6, tabulation from data contained in the testimony of Secretary of the Treasury William Simon in the *Hearings on Oil Profits and Their Effect on Small Business and Capital Investment Needs of the Energy Industries Before the Subcomm. on Government Regulation of the Senate Select Comm. on Small Business,* 93d Cong., 2d Sess. 145 (1974). See also First National City Bank, Monthly Letter April each year, 1965-1974: 13.4 percent average for petroleum industry, slightly above all manufacturing (13 percent), but below average for mining companies (14.7 percent); 1964-1973 and 1963-1972: lower for petroleum industry than for manufacturing generally and about the same for mining. *Petroleum Industry Hearings, supra* note 192, at 338 (Statement of W. T. Slick, Jr.) ("Historically, profits in the petroleum industry have been about equal to the average for all of U.S. manufacturing"); MANCKE, *Oil Industry Competition, supra* note 674, at 61 (American petroleum companies were significantly less profitable than the S&P 500 over the 1953 to 1972 period. Not one equalled it.) Johnson et al., *Oil Industry Competition, supra* note 1055, at 2451.

[2060] T. DUCHESNEAU, COMPETITION IN THE U.S. ENERGY INDUSTRY 155, 157 (1975).

[2061] Ikard, *Petroleum Industry Competition, supra* note 1055, at 588 n.24, citing the FEDERAL TRADE COMMISSION, PRELIMINARY FEDERAL TRADE COMMISSION STAFF REPORT ON ITS INVESTIGATION OF THE PETROLEUM INDUSTRY 280 (July 2, 1973) [hereinafter cited as FTC Preliminary Staff Report]. *See also Petroleum Industry Hearings, supra* note 192, at 489 (Statement of Joseph Tovey, Vice President of Faulkner, Dawkins & Sullivan, Inc., Stockbrokers and Investment Bankers) (Rate of return for the five big U.S. based internationals is less than for a comparable group of well managed billion dollar U.S. firms in other industries during 1967-1974); Mitchell, *1974 Senate Testimony, supra* note 1056, at 9 ("The FTC Eight — the five internationals plus Atlantic Richfield, Shell and Standard of Indiana — were less profitable than S&P [Stock Composite] from 1953 to 1972 and from 1953 to 1974.").

[2062] Ikard, *Petroleum Industry Competition, supra* note 1055, at 589, citing E. ERICKSON and R. SPANN, An Analysis of the Competitive Implications of the Profitability of the Petroleum Industry with Special Reference to the Comparison Between the Cost of Capital and the Rate of Return Earned on Stockholders' Equity 7 (February 1974) (unpublished paper on file with the Oregon Law Review).

from 1969 through 1973 and by utility companies from 1970 to 1973.[2063]

Professor Erickson summed this up by saying "There are two problems with such an accommodation [*i.e.,* hiding the profits], to the shell game theory of monopoly profits as applied to pipeline undersizing. First, the DOJ posits effective regulation of the pipeline sector so that there is no monopoly pea to be shunted from this stage to other stages. Second, according to the DOJ hypotheses, the exercise of market power through undersizing of pipelines should result in a recorded profit enhancement downstream in marketing. To the extent that profit data by stage of the industry are available and reliable, the marketing activities of the industry are, as a class, the least profitable component among a set of activities whose overall returns are only normal. Thus, by the DOJ assumption, there is no pea, and were there a pea, the shell under which it is alleged to be hidden is, upon inspection, empty."[2064]

(2) Assumptions Underlying Theory are Mistaken

One of the key assumptions of the undersizing theory is that downstream prices are set by alternate, *more expensive,* modes of transportation. The theory postulates that the owners of a pipeline, by undersizing it, can force enough crude or products to reach the destination market by more costly means of transportation so as to be the "marginal cost" which sets the downstream price and thereby the integrated pipeline owners can capture the difference in transportation costs.[2065]

Strangely, both Mr. Shenefield and Mr. Flexner, who espoused this theory, made reference to the Antitrust Division's SOHIO Report, which appears to refute completely this assumption. In Shenefield's proposed statement, less than three pages prior to his articulation of the "Capture of Profit" thesis, Mr. Shenefield stated: "...the proposed Sohio Pipeline...would not have market power in the Gulf Coast/Midwest crude market, because its input of crude into that market would not be substantial and the delivered cost via that pipeline of oil in that market is comparable to that for foreign sources of supply."[2066] This author has

---

[2063] *Hearings on Profitability of Domestic Energy Company Operations Before the Senate Committee on Finance,* 93d Cong., 2d Sess. 41 (1974) (Statement of G. S. Wolbert, Jr. formerly Vice President - Finance, Shell Oil Co.); Mobil lost money on its domestic refining, marketing and transportation assets in 1974 and during the first half of 1975. *Petroleum Industry Hearings, supra* note 192, at 1823 (Statement of William P. Tavoulareas, President, Mobil Oil Corp.).

[2064] Erickson et al., *The Pipeline Undersizing Argument, supra* note 2053, at 17 n.5.

[2065] SHENEFIELD PROPOSED STATEMENT, *supra* note 179, at 22; FLEXNER, *The Case for Divestiture, supra* note 2019, at 7.

[2066] SHENEFIELD PROPOSED STATEMENT, *supra* note 179, at 18-19.

difficulty in reading that to mean anything other than an acknowledgement, contrary to the assumption that alternate modes of transportation of crude from the same source would determine the downstream price, but that alternate sources of crude, mostly foreign, would determine the downstream price. It likewise would follow that movements from different producing areas into the questioned line's destination market, via pipeline, barge or any other means which would result in a *lower* laid-down price would determine the market price level.[2067]

Pursuing the import point, the SOHIO reasoning would seem to eliminate most of the nation's crude lines because about 90 percent of domestic refining capacity is located on a coast, and supplied not only by domestic pipelines but also, increasingly, by imported crude oil brought in by tanker. Since an extensive worldwide crude market exists in which U.S. firms purchase over six million barrels per day, it seems highly unlikely that a single crude line supplying a coastal refinery center would have the price effect that the undersizing theory requires.[2068] As the SOHIO Report put it: "These imports will still be the incremental supply to the mid-continent throughout the range of throughput possible for this pipeline. SOHIO could not, by restricting throughput, increase the price of the delivered oil above the world price. To pursue a strategy of throughput restriction would be merely to invite greater imports of foreign crude at the world price, a result which could not benefit SOHIO in PADs II and III."[2069]

Most inland refineries have multiple sources of crude.[2070] The Chicago area, which has about 50 percent of the inland refining capacity, is supplied from Canadian fields by Interprovincial-Lakehead; from Mid-Continent/Gulf coast areas by ARCO, Amoco and Texaco-Cities Service pipelines; and imported crude by the Capline-Chicap systems.[2071] The St. Louis/Wood River refining area receives Montana, Wyoming and Rocky Mountain crude via Platte and Amoco; and West Texas/Mid-continent crude through Basin-Ozark, Shell and Mobil; and imported crude by Capline.[2072]

Similarly, few if any, products pipeline would meet Shenefield's

---

[2067] MARATHON, *supra* note 202, at 65.

[2068] CANES & NORMAN, *Pipelines & Antitrust, supra* note 2033, at 11; API's Comments on Kennedy FTC Petition, *supra* note 1879, at 26-27.

[2069] D. J. SOHIO REPORT, *supra* note 633, at 41.

[2070] See text at notes 214-224, and 402-406, *supra*. See also Appendices G & J. For a detailed analysis of crude oil modes of transport see LIVINGSTON,, *supra* note 228, at 44-49.

[2071] CANES & NORMAN, *Pipelines and Antitrust, supra* note 2033, at 12 n.1; API's Comments on Kennedy FTC Peitition, *supra* note 1641, at 27.

[2072] CANES & NORMAN, *Pipelines and Antitrust, supra* note 2033, at 12 n.1.

market power requisite for undersizing. Most large inland markets are served by more than one product line and these lines must compete with product outturn by refineries served by crude lines originating from the same or different sources. For example, Salt Lake City is supplied product by Pioneer and local refineries to which crude is supplied by the Amoco and Chevron lines. Gulf Coast refineries may ship products to the Chicago area via Explorer, Williams Pipe Line or Texas Eastern. When the product arrives there it must face competition from Kansas and St. Louis refiners and local refineries such as Amoco's Whiting refinery.[2073]

Readers will remember Explorer's misfortunes with product transportation competition from barges and Texas Eastern's line and the two new crude lines, Texoma and Seaway, which brought crude to Cushing where local refiners' products not only competed with Explorer's Tulsa marketers' shipments but also in Chicago which these same refiners reached via Williams Pipe Line. Small wonder that Assistant Attorney General Shenefield in his philosophically relaxed BNA interview soliliquized on the soundness of his "undersizing" theory. Perhaps he had in mind the foregoing multiple competitive line situation when he remarked, "In fact, you might conclude that the answer would be it doesn't make any sense."[2074]

Even the most criticized line, Colonial, not only competes with Plantation a good part of the way, but products going through its lines have to meet competition at destination markets on the East Coast, which is provided by refineries running imported crude. As Professor Erickson commented at the AEI Conference, "I am not sure that I would identify the incremental source of supply for refined products to the East Coast as shipments over Colonial or Plantation." It seemed to him that the incremental source was imported crude that came in and was refined for example, at [Bayway], New Jersey.[2075] The reader will recall Amoco's construction of a refinery at Yorktown, Virginia, and Gulf and ARCO have huge refineries on the Delaware River, at Philadelphia. Nearby are Mobil, Getty and Sun, all of which are geared to run on imported crude.

A second key assumption is that prorationing will not be effective. According to Mr. Shenefield, "...effective prorationing would not alter basic throughput restriction incentives. Prorationing would merely require the monopolist to dilute its monopoly return by sharing it with others."[2076] Under the Interstate Commerce Act, a pipeline is obliged to

---

[2073] See Appendices H & K.
[2074] 834 ANTITRUST & TRADE REG. REP. (BNA) A-7 (October 13, 1977).
[2075] AEI, *Oil Pipeline and Public Policy, supra* note 684, at 66 (Comment by Professor E. Erickson during general discussion).
[2076] SHENEFIELD PROPOSED STATEMENT, *supra* note 179, at 27.

serve all shippers (and prospective shippers) on equal and nondiscriminatory terms at reasonable rates. If excess demand exists, everyone's shipments—owners' and nonowners' alike—must suffer proportionately from the higher cost of using alternative modes. As has been demonstrated, this would harm the owner-shippers the most because, contrary to the Department's implication that it could force "others" to use higher cost alternatives, owner-shippers would simply have their average cost raised along with the prorated nonowner-shippers, gaining nothing, while losing both the advantage they would have received from the lower incremental cost of a properly sized line for their own shipments plus the profit, meager though it is, that increased shipments by nonowners would return to them via the allowed tariff rate. In addition, the owners would be risking civil and criminal penalties under both the Interstate Commerce Act and the antitrust laws by any such denial of proportional access to competitors, and the chance that they would end up with a non-competitive line with a 30-40 year debt obligation.

Given the Department of Justice's assumption that pipeline "tariff policy is ideal, *i.e.*, the tariff is designed to permit the pipeline to earn no more than a fair and reasonable return to its owners'...investment,"[2077] the theory can be depicted by a simple graph (Figure 1). In the absence of any rate regulation, the monopoly rate would be charged. In Figure 1 this would be Pm, and the monopoly profit would be represented by the area PmABC (the difference in price being Pm-C, the price levels at the respective intersections of the quantity qm with the demand curve D [at A] and with that which would pertain under effective rate regulation, the long run average cost curve LRAC [at B]). The shortrun average cost curve of the "undersized" line that would achieve this maximization of monopoly profits under the Department's theory is the curve AC. The price-demand equilibrium intersection under the "regulatory optimized" curve LRAC (we will disregard the theoretical "consumer optimized" long run marginal cost curve [LRMC] because it would be impossible to achieve from a regulatory standpoint and would require government subsidization) would produce a throughput of qr (at a tariff rate of Pr) and short-run average cost curve would be ACr. Under the Department's theory, because of its postulated effective rate regulation, the "price" (pipeline tariff) of the assumed undersized line (held back to qm) would be C, below that necessary to clear the market. The effect of this is that there is a permanent long-run shortage of capacity.[2078]

It would seem to follow that for the undersizing theory to be valid,

---

[2077] A. G.'s Alaska Natural Gas Report, *supra* note 1431, at 32-33.
[2078] CANES & NORMAN, *Pipelines and Antitrust*, *supra* note 2033, at 7.

there would have to be a pattern of constant prorationing of a number of lines.[2079] However, on the other hand, the mere fact that a line did have constant, or frequent prorationing would not, in and of itself, prove undersizing. It would simply call for investigation, because there are many reasons unrelated to anticompetitive behavior, why a line, properly sized, could be forced into prorationing,[2080] such as miscalculation of the market or subsequent events unforseen, and some unforseeable, at the time the line was sized.[2081]

(3) Empirical Evidence Refutes or Makes Inapplicable to Pipelines the Undersizing Theory

(A) Comparison of Values at Ends of Pipeline

Readers will recall that Dr. George Hay, Director of Economics, Economic Policy Office at the Antitrust Division, stated at the AEI Conference that "the ideal empirical test [of undersizing] would be a comparison of the value of the oil at one end of the pipeline—less the market value of the oil at the other end of the pipeline compared with the actual economic cost of transportation."[2082] This, of course, assumes instantaneous adjustment of prices and capital flows, which do not occur in the real world. Although he qualified the statement to allow for an incentive, *i.e.,* a competitive rate of return plus an "epsilon" risk element[2083] and Scott Harvey, an economist with the Bureau of Economics, Federal Trade Commission, cautioned about the distortional effects of DOE regulations,[2084] it would appear that the test has some limited relevance, if nothing more than an exclusionary test, *i.e.,* if the value figures do not show a difference, the validity of the theory would seem to be weakened.

Mr. Shenefield's proposed statement used Hay's approach to "prove" that product pipeline capacity from the Gulf Coast to the East

---

[2079] *Id.* at 21; AEI *Oil Pipelines and Public Policy, supra* note 684, *Id.* at 71 (G. S. Wolbert), *Id.* at 72 (Professor Edward Mitchell).

[2080] CANES & NORMAN, *Pipelines and Antitrust, supra* note 2033, at 21-22; AEI *Oil Pipelines and Public Policy, supra* note 684, at 124 (Professor Edmund Kitch; Dr. George Hay, Director of Economics, Economic Policy Office, Antitrust Division).

[2081] See text at notes 720-1037, *supra*; API's Comments on Kennedy FTC Petition, *supra* note 1879, at 25-26.

[2082] AEI, *Oil Pipelines and Public Policy, supra* note 684, at 123.

[2083] *Id.* at 124.

[2084] *Id.* at 172.

Coast (*i.e.*, Colonial) was restricted so that movements would have to be made by higher cost coastwise tanker transportation. He even went so far as to estimate the cost to the consumer of this alleged restriction by comparing the pipeline rates with the tanker rates appearing in the May 25, 1978, issue of *Platt's Oilgram Price Service*. He multiplied the "about 37 cents per barrel" differential by the 1976 tanker volumes reported by the Bureau of Mines, and by the pipeline volumes for 1973, estimated from National Energy Transportation System Map No. 12, "Total Petroleum Products Movement," accompanying the Senate Report entitled "National Energy Transportation," to get $44.6 million of "dead weight loss" (unnecessary additional transportation costs) and $93 million "increased profits downstream," respectively. The sum of these two items ($137.6 million) was discounted by a "Kentucky windage" factor of 2/3 for "errors that might be contained in such rough estimates" and he came up with an estimated consumer cost of $45 million annually.[2085] Testing this approach by using the same method and the same source date, *i.e.*, the May 25, 1978, issue of *Platt's Oilgram Price Service* for four key products: premium gasoline, regular gasoline, unleaded gasoline and home heating oil (No. 2 distillate), the high and low New York harbor spot cargo prices were determined for these products (and for the foreign equivalent price for regular gasoline). The appropriate transportation costs were added to the U.S. Gulf spot cargo prices for waterborne and pipeline movements to obtain the laid-in spot costs of products in New York. Tanker costs are based on the clean products rates from the U.S. Gulf to New York. All grades of gasoline are costed at the regular gasoline rate and No. 2 heating oil at the gas oil rate. The pipeline rates are for Colonial to Newark, New Jersey, and are volume weighted 75 percent from Pasadena and 25 percent from Lake Charles to reflect different origin points. The laid-in spot prices from the U.S. Gulf were then compared with New York spot prices using the domestic low as a base. The results are shown on Table VII. As can be noted, the New York spot cargo prices are generally lower than the laid-in costs from the U.S. Gulf except for pipeline movements of No. 2 heating oil and they more closely approximate the prices of products out of the U.S. Gulf by pipeline. As can also be noted, foreign supplies may have influenced the price of regular gasoline. If the Justice Department's assumptions were correct, New York prices should have been the same as the spot laid-in price of products from the U.S. Gulf by tanker, *which they are not.*

Certainly, a single data point is not conclusive on the question. Therefore, a similar analysis was made for the year 1978. The basic data

---

[2085] SHENEFIELD PROPOSED STATEMENT, *supra* note 179, at 28-29.

source was again *Platt's Oilgram Price Service Report.* Market price data were developed in the same way as for Table VII using the low spot prices. Daily plots for each of the products shown in Table VII were developed and are attached as Figures 2 through 5. These data are plotted by the issue date of *Platt's* and show the domestic New York spot cargo price (also foreign equivalent when available) on the dashed (- - -) line and the laid-in U.S. Gulf spot cargo prices by tanker on the dotted (. . .) line and by pipeline on the solid (—) line.

The plots show the absolute spot cargo prices of products from each source. As can be noted, all prices follow the same general trend reflecting changes in product prices at the source which in turn primarily reflect changes in raw material (crude oil) costs. Transportation costs included in the laid-in spot cargo prices from the U.S. Gulf by tanker or pipeline are generally 10 percent or less of the total shown. The prime purpose of the plots is to show the differential between the New York spot cargo prices and the laid-in spot cargo prices from the U.S. Gulf by pipeline and tanker. Under DOJ's assumptions, there should be no differential between the New York spot cargo prices shown on the dashed line and the laid-in spot cargo prices from the U.S. Gulf by tanker shown on the dotted line. The dashed and dotted plots of prices should coincide or track each other.

Figures 2-5 also show, as did the single point data of Table VII, that New York spot cargo prices are not the same as laid-in spot cargo prices for products moved in from the U.S. Gulf by tanker as believed by the DOJ. As can be noted, the dotted plot of laid-in spot cargo prices for products moved from the U.S. Gulf by tanker are generally higher, indicating spot cargos are available at lower prices in New York or from the U.S. Gulf delivered by pipeline. During some brief period, New York spot cargo prices approached either the laid-in prices of products from the U.S. Gulf by tanker or pipeline and at other times were either above or below delivered Gulf prices. One of the major influences indicated is the availability of lower priced foreign products. This is notable only for regular gasoline because this is the sole product for which prices were quoted. Even then, foreign prices were not quoted for the entire year. During the May through September period, New York spot cargo prices were generally lower than the laid-in price of products from the U.S. Gulf, indicating foreign products may be influencing the New York spot cargo prices for regular gasoline. Also, substantial volumes of foreign crude are refined on the East Coast, and products from the U.S. Gulf transported either by tanker or pipeline must compete with products from these sources. These plots indicate that New York spot cargo prices are not the same as laid-in spot cargo prices for products moved from the U.S. Gulf by tanker and that there is no discernable correlation between these two prices.

Although secondary in importance to the pricing assumption, it appears that the 37 cents/barrel differential quoted by the Justice Department is high in that it does not take into account several factors.

One factor the Justice Department neglected to take into account is the pipeline costs required to move from a refinery to the entry point on Colonial. As can be noted in the May 25, 1978, issue of *Platt's*, U.S. Gulf Coast spot cargo prices for pipeline shipments range from 0-0.5 cents / gallon (0-21.00 cents/barrel) higher than for waterborne shipments to reflect the cost of this move. During some periods of 1978, apparently depending on the source of the products, this differential is 0.75-1.00 cents/gallon (31.5-42.0 cents/barrel). The range of these costs are such that their inclusion could reduce significantly the differential between tanker and pipeline costs calculated by the Justice Department.

Using the same data source for pipeline movements from the U.S. Gulf (National Energy Transportation System Map No. 12), it is indicated that 75-80 percent of product movements originate in the Houston/Pasadena area instead of coming entirely from the Lake Charles area. The May 25 *Platt's* rate for Colonial from Lake Charles was reported as 1.26 cents/gallon (52.92 cents/barrel) while the rate from Pasadena was reported to be 1.35 cents/gallon (56.70 cents/barrel) which, if volume weighted 75 percent Pasadena/25 percent Lake Charles, would be 1.33 cents/gallon (55.86 cents/barrel) thus reducing the differential used by about 3 cents/barrel.

The clean tanker rates from the U.S. Gulf to New York in the May 25, 1978, issue of *Platt's* are quoted as 2.17 cents/gallon (91.14 cents/barrel) gas-oil and 1.90 cents/gallon (79.80 cents/barrel) for motor gasoline. Using the Justice Department's differential of 37 cents/barrel between the pipeline rate from Lake Charles (52.92 cents/barrel), the indicated average tanker rate would be 89.92 cents/barrel. The Bureau of Mines reports of tanker movements for 1976 (120,656,000 barrels), referred to earlier, indicate a product distribution of 39 percent motor gasoline, 48 percent distillate and 13 percent kerosene and jet fuel. Assuming all products except motor gasoline move at the gas-oil rate, the average product weighted tanker rate would be 86.72 cents/barrel, some 3 cents/barrel less than used by the Justice Department, further reducing the differential.

Thus, it appears that the Justice Department simply overstated the differential used in its calculations and inclusion of all of the above factors could significantly reduce this differential between tanker and pipeline costs.

While Colonial shippers undoubtedly see a benefit over tanker movements, it should not be assumed that tanker movements are made only because capacity is not available in Colonial. The tanker rates quoted in *Platt's* may not reflect the tanker costs of all operators. There is a wide range in the size of tankers used to make these product movements, differences in operating efficiencies and other factors which determine the actual average costs of tanker movements from the U.S. Gulf.

The statement concerning unnecessary tanker transportation costs implies that a pipeline should be built or expanded in size to cover all future requirements. However, it is extremely difficult to forecast requirements for pipelines because they usually have a life of about 30 years. As has been shown, pipelines are highly capital intensive requiring major expenditures early in the life of the project. Overinvestment could result in higher rates and corresponding higher costs of products to the consumer. A decision to expand an existing pipeline also is difficult. Perhaps one of the biggest problems facing a pipeline company is the difficulty of determining the probable actions government will take. Changes in the pipeline rate system, which has been established for about 30 years, now are being considered. This will influence decisions to build or expand. Pipeline projects are being delayed and may, in some cases, be cancelled because of ever-increasing difficulties in obtaining permits. Effects of government controls on crude and product pricing are unclear. Also, the actions of foreign governments could influence future supply sources and, for example, could result in greater imports of foreign products on the East Coast, reducing the need for products from the U.S. Gulf. The location of new domestic refinery capacity on the East Coast would have the same effect as foreign product imports, and the two effects would be additive. These are just a few of the problems facing pipeline companies considering new construction or major expansions.

In summary, it appears that the Justice Department's theory concerning restriction of pipeline capacity to the East Coast cannot be supported by the data. New York spot cargo prices are not the same as laid-in product prices from the U.S. Gulf by tanker and there is no discernable correlation. Although secondary, it appears the differential between tanker and pipeline transportation rates from the U.S. Gulf to the East Coast simply is overstated.

(B) Pipelines Mostly Are Running Under Capacity

In Section II A 3 C, "Scramble for Traffic," in the last paragraph; in Tables I and II; and in this chapter in subsection C 3 c (2), "Lines Operating Under Capacity,"[2086] it was shown that Explorer, Platte,

---

[2086] See text at notes 1949-1953, *supra; see also* Erickson et al., *The Pipeline Undersizing Argument, supra* note 2053, at 19-22.

Wolverine and Yellowstone ran well under capacity for the years in which data were available. There was also reported an AOPL survey which showed that out of a total of 46 integrated pipeline companies, accounting for approximately 85 percent of the total barrel-miles shipped in trunk lines in 1977, 29 had no prorationing whatsoever, and, on 15 of the remaining 17, prorationing was limited in duration and/or confined to a relatively small part of the system (see Tables IV-VI). The only lines under fairly constant proration, Colonial, or more than incidental proration, Plantation, will be discussed in the following subsection. The degree to which this undercapacity exists is demonstrated by the DOJ's own statement in *Ex Parte No. 308* wherein it was shown that, during 1974-1975, 9 pipelines earned 5 percent or less on capital invested, and an additional 7 pipelines actually operated at a loss.[2087] This condition would seem to negate an intent to undersize, or at the very least, to make it inapplicable to all but two pipelines.[2088]

(C) Prorationing is Limited in Extent and Administered Equitably

As has been noted previously, it would appear that for the undersizing theory to have any validity, there would have to be a consistent pattern of prorationing;[2089] moreover, the mere presence of a continuing prorationing situation does not, of itself, prove undersizing because there are many reasons for the existence of such a situation which are not anticompetitive in nature.[2090] In the preceding subsection, it was observed that incidents of prorationing were sparse with the exception of two lines, Colonial and Plantation. In Subsection VI C 3 b (8), "Alleged Discriminatory Prorationing,"[2091] and Subsection VI C 3 c (3), "Lines at Capacity—Proration and Expansion,"[2092] of this chapter, the empirical evidence was examined and note was taken of the legal requirements for equitable prorationing and the presence of prorationing principles in tariffs filed with the Commission. A legitimate complaint might be lodged that not all carriers spell out their prorationing policies as

---

[2087] Hay, *FERC Direct Statement, supra* note 1689, at Table I.
[2088] See text at note 2036, *supra*.
[2089] See text at note 2022, *supra*.
[2090] AEI, *Oil Pipelines and Public Policy, supra* note 684 at 126 (Professor Edward Mitchell); *Id.* at 57-58 (G. S. Wolbert: unforeseeable growth in demand); *Id.* at 72 (U. LeGrange: came about over time), *Id.* at 127 (Professor Edmund Kitch, Dr. George Hay: error in estimation); see text at notes 742-746 (sizing dilemma), 753-755 (difficulties in estimating field reserves), 780-785 (market miscalculations), 786-799 (fluctuation in throughput), *supra*.
[2091] See text at notes 1916-1926, *supra*.
[2092] See text at notes 1954-1964, *supra*.

fully as, say, Colonial. This can be remedied easily by the Commission or by the carriers on their own volition and avoid thereby the type of innuendos which have been raised in Congressional hearings. These subsections reported the absence of formal complaints, with the exception of the rather unusual *Powder River Pipeline* case, which appears to be a "frontier" situation involving the unique question whether a carrier is responsible for the actions of one of its shippers. The operation of the two principal methods of proration was examined; the owner-shipper's economic disincentive to permit its line to get into, or remain in, a prorationed operation mode was explored; and the corrective expansion actions taken by Capline and Colonial were detailed (Plantation's expansion moves were examined in Section II A 3 C, "Scramble for Traffic").[2093] Colonial's prorationing treatment and expansion actions were given special attention; first, because it has been the line with the greatest continuity of prorationing and second, because it has somehow become the prime target of the industry's critics. It was recorded that Colonial started out on a risky basis, but because increasing union labor scales initially diverted tanker business to Colonial and the Consent Decree forced Colonial to reduce its tariffs while tanker rates continued to increase, thereby widening further the transportation differential, Colonial began to get an increasingly larger share of an unexpectedly rapidly growing market and filled its line to capacity. Section II A 3 C, "Scramble for Traffic,"[2094] documented the treatment of nonowner shippers over Colonial's line from 1967, when the line first was prorationed, until 1978, during which time nonowner shipments rose from 0.37 percent to 36.26 percent, not in spite of, but *because* of, Colonial's equitable administration of its prorationing obligations. It was established that from 1969 to 1978, owners' shipments *decreased* from 1.1 million barrels per day to 1.05 million barrels per day, while nonowners' shipments during the same period *increased* from 37,000 barrels per day to 595,000 barrels per day.

The other two subsections chronicled Colonial's efforts to get off prorationing, expanding seven times and tripling its initial capacity. These same subsections observed that Colonial's practices were investigated thoroughly by the Justice Department over a period of time from 1963 to 1976 (with the investigation "oversighted" by virtually every session of Congress and by multiple committees in several of them). The investigation was closed in 1976 on the finding by the Department that there

---

[2093] See text at notes 424-426, *supra*.

[2094] See text at notes 416-423, *supra*.

was no meaningful denial of access to nonowner shippers. The conclusion reached after this examination was that the empirical data were difficult to reconcile with the claim that integrated pipelines restricted nonowner access or that they were likely to be undersized.

(D) Access to Pipelines by Nonowners

In the outline of the evolution of the DOJ's undersizing theory, Dr. George Hay's statement on access was set forth: "If there are many nonowner-shippers, and if they have access to more cheap oil than they can bring in, and there is empty space on the pipeline which they can get access to, then, I agree that the pipeline could not be undersized with respect to current demand facing it."[2095] In Sections II A 3 C, "Scramble for Traffic,"[2096] and VI C 3 c (1): "Nonowner Usage,";[2097] a detailed examination of the empirical evidence was made, starting with specific lines identified by critics as being examples of pipelines which presented problems of access, *viz.*, Colonial, Plantation, Wolverine, Yellowstone, Explorer, Platte and Texoma. Colonial's nonowner usage was shown to have increased from 0.62 percent in 1965, its first full year of operation, to 36.26 percent in 1978. Plantation's nonowner shipments rose from 9.0 percent in 1946 to 67 percent in 1964, when it lost a number of them to Colonial which had started operations. Although the wholesale flight of nonowner-shippers to Colonial pulled Plantation down to 35 percent in 1965, and further erosion took place until Plantation undertook a vigorous expansion and modernization program, it rebuilt its nonowner shipment percentage back up from its low of 30 percent in 1967 to 36 percent in 1974, the last year data were available. Explorer, through its extreme efforts to build traffic, including adjusting its minimum tender requirements to meet small shippers' needs and the unprecedented step of "batching" crude in its refined products system, built its nonowner traffic from 0.9 percent in the start-up year of 1972 to 28.9 percent in 1977; 26 of the 29 "outside" shippers were "independents."

A detailed examination was made of the Meyer Kopolow claim that Explorer discriminated against his terminal company in requiring a throughput guarantee to construct a spur line underneath the Mississippi River and up through a heavily congested area of St. Louis, and the invalidity of this claim was established by the testimony of Kopolow's own nominated referral witnesses. The Kennedy Staff's Report of the inability of an unnamed marketer in the Dallas area to obtain a terminalling connection was

---

[2095] AEI, *Oil Pipelines and Public Policy*, supra note 684, at 72.
[2096] See text at notes 412-496, *supra*.
[2097] See text at notes 1929-1948, *supra*.

discussed to the extent possible considering the vagueness of the allegation and the concomitant circumstances at the time — *i.e.*, Explorer was losing $46 million and making cash calls on its owners to the tune of $42 million, making it unlikely that it would willingly rebuff *any* customers and Vernon T. Jones, President of Explorer at the time of his testimony before the Senate Antitrust and Monopoly Subcommittee, stated unequivocally that Explorer had never been approached by any independent and asked to provide terminal facilities.[2098] The conclusion was reached that if, in fact, there had been a denial of terminal space, it was due to a unilateral refusal by a competitor to lease space to the unknown complainant and no action or inaction by Explorer was involved.

The growing trend of nonowner shippers was traced from the 1930's, when over 60 percent of the large integrated companies carried no oil for nonowners in their gathering lines, and about one-half of them did not transport nonowners' oil in their trunklines, up to the point where a survey for the year 1976 of 74 crude oil and products pipelines owned or controlled by "majors" reporting to the ICC disclosed that 89.24 percent of their shippers were nonowners, and 64 percent were "independents." It was noted that, if anything, these statistics *understated* the case for access because of the locational impossibility of independent refiners representing 16.8 percent of independent refining capacity to make use of the lines, 39 independent refineries, comprising 18.0 percent of such capacity had access to non-oil-interest-related pipelines ("independent" pipelines as defined by Mr. Shenefield); and 28 independent refineries, with 42.3 percent of such capacity, were owned by firms which were the sole, or part, owners of products lines serving their refineries. A second "understating" factor was the extensive use of exchanges, which saved the refiners concerned, and hence the consumer, "a lot of money" because of the saving in transportation cost.

There was presented in these sections testimony by ICC Commissioners attesting to the absence of industry members' complaints, and evidence from independent producers' associations that independent producers had no difficulty in securing shipment of their crude nor had they been subjected to discrimination by pipeline companies. There was testimony by Professor Edward Mitchell that the pattern of gasoline prices and gasoline market shares along Colonial's line was consistent with the normal processes of competition,[2099] Mitchell's preliminary study was reinforced by a subsequent Lundberg Survey which showed that the share of the gasoline market in the 15-state area involved held by nonintegrated marketers almost doubled (from 14 percent to 27 percent)

[2098] See text at notes 1872-1973, *supra*.
[2099] Mitchell, *1974 Senate Testimony, supra* note 1056, at 40.

during the years 1963-1974, while that of the integrated marketers, including Colonial's owners, steadily declined during that period. Again, the finding of "no meaningful denials of access" by DOJ after 13 years of investigation is relevant to this question. The second cited section closed with the famous [Deputy Assistant Attorney General] Bruce Wilson's statement that it was difficult to refute the oil companies' assertion that access to pipelines was open and available to all without regard to ownership interest, "when we have few, if any, who will testify that they have been excluded."

(E) Competition at the Transportation Stage

In Section III F 8, "Competition,"[2100] a rigorous examination was made of the competition faced by oil pipelines. There were multiple instances of point-to-point competition; many examples of area competition; instances of competition from lines passing through the destination of existing lines; and intermarket competition, *i.e.*, traffic over products lines going to market areas faced competition from crude lines going to refineries located either at such market areas or in such manner as to reach them by products lines. Instances were given where competition was created by the reconfiguration of existing facilities, usually crude lines converted to products service, in some cases with reversed flow. There was a portion of this subsection devoted to competition from tankers, barges, trucks and railroads.

The wealth of this material makes it difficult, if not impossible, to say that, except in very isolated situations, which are not competitively significant, the requisite "market power" of any single line to "prop up" market prices exists so as to make "undersizing" a rational strategy. Moreover, there is freedom of access to pipeline ownership, not only because of Congress' repeated actions in opting not to impose a requirement for certificates of convenience and necessity, but also because of the growing number of large diameter lines, which tend to bring in smaller shippers who would not otherwise have the financial ability or traffic to justify a line on their own. Between 1951 and 1972, the number of firms owning interstate pipelines increased from 65 to about 90. Nor is the ownership static; five of the top 20 firms in 1972 were not even among the top 20 firms in 1951, and yet they have since captured at least 1 percent of the market.[2101] The trend has continued since then.

Professor Edward Erickson and his colleagues showed that new

[2100] See text at notes 814-887, *supra*.
[2101] CANES, *An Economist's Analysis, supra* note 1021, at 48.

products pipeline entities entering the market during the period 1947-1976 represented 48.7 percent of the value of carrier property in 1976 and 79.2 percent of the total barrel miles in that year;[2102] the corresponding figures for crude oil pipelines were 28.9 percent and 50.6 percent.[2103] The authors noted that of the 21 new entity jointly owned products lines, 16 had ownership participation by firms other than the 18 largest oil companies, and of the 42 new entity sole ownership product pipelines, 38 were built by firms other than the 18 largest oil companies.[2104] The corresponding figures for new entity crude oil pipelines were: 21 out of the 26 new jointly owned lines had ownership participation by firms other than the 18 largest oil companies, and 21 out of the 29 new solely owned crude lines were built by firms other than the largest 18 oil companies.[2105] The authors commented: "As with the data on access to pipelines, this pattern of diverse, nonexclusionary ownership is inconsistent with the behavioral implications of the exploitive pipeline undersizing argument."[2106]

(F) The Existence of Competitive Markets Both Upstream and Downstream

Although the Justice Department in developing its undersizing theory has assumed the other stages to be workably competitive,[2107] it has shown a willingness to change assumptions abruptly when an inconsistency shows up. Hence, before the Department attempts to change its position on the competitive nature of adjacent markets, it appears sound to nail down this point: there definitely is an absence of concentration in the other stages of the oil business. In the crude oil and natural gas production end the figures on market concentration for the largest firm in the 1970's vary from 8 to 10.7 percent; for the top 4 firms, from 23.9 to 31 percent; and top 8 firms, from 42.2 to 49 percent. Refining concentration figures are: top firm, from 8.3 to 8.4 percent; top 4 firms, from 27.2 to 33 percent; and top 8 firms, from 51.1 to 57 percent. Gasoline marketing shares are: top firm from 8.1 to 8.2 percent; top 4 firms from 29 to 31

[2102] Erickson, et al., *The Pipeline Undersizing Argument, supra* note 2053, at 31, Table 9.
[2103] *Id.* at 34, Table 12.
[2104] *Id.* at 34. *See also* his Tables 13 and 14.
[2105] *Id.* at 34. *See also* his Tables 15 and 16.
[2106] *Id.* at 36.
[2107] A. G.'s Alaska Natural Gas Report, *supra* note 1431. Indeed, if they were not, there would be no need for the integrated firms "to tinker with the pipeline stage in order to exploit their monopoly power" *Id.* at 41-42.

percent; and top 8 firms from 51.8 to 53 percent.[2108]

Professor David Teece, who has been a close observer of the industry, summed up the matter: "The data on concentration and changes in concentration in all stages of the industry in no way make an *a priori* case for monopoly power by any one of the [FTC v. Exxon] respondents. No matter how the relevant market is defined, market share statistics suggest that the industry is competitive."[2109]

(G) Facts of Life in Adjacent Stages

(i) The Domestic Crude Oil Market [2110]

The misconception that there is no effective market for crude oil is one that is widely held in some circles. Senator Birch Bayh, for example, recently said that: "From the time the drill bit enters the ground to develop a new well, until the oil from that well is pumped into the consumer's gasoline tank, it is controlled by a single company."[2111]

If this quotation represented an isolated view, it would not be cause for concern — even considering its highly visible source. Unfortunately, this view is reflective of a substantial school of thought that has a significant impact on the nation's emerging energy policies. Since the nation's future well-being requires intelligent energy policies it is necessary to dispel this misconception.

Before quantifying the dimensions of the domestic crude oil market it may be useful to describe this complex market in general terms.[2112]

There are several reasons why producers, refiners and others come

---

[2108] *Petroleum Industry Hearings, supra* note 192, at 43 (Statement of Senator Roman L. Hruska of Nebraska); *Id.* at 345 (Exhibit 2) (Statement of W. T. Slick, Jr.); *Id.* at 444-445 (Statement of L. C. Soileau III); *Id.* at 1821 (Statement of William P. Tavoulareas); Ikard, *Petroleum Industry Competition, supra* note 1055, at 598; CANES, *An Economist's Analysis, supra* note 1021, at 61.

[2109] TEECE, *supra* note 329, at 100.

[2110] The author acknowledges a heavy debt to Dr. John Ryan for the contents of this subsection which came from a paper, entitled "Domestic Crude Market," presented before the Southwest Economics Association at Fort Worth, Texas on March 30, 1979.

[2111] *Petroleum Industry Hearings, supra* note 192, at 3 (Opening Statement of Senator Birch Bayh).

[2112] For present purposes, the discussion will be limited to the market for crude oil produced in and largely consumed in PAD Districts I-IV, *i.e.*, the area East of the Rockies. This area accounted for 82.7% of the nation's indigenous crude oil and condensate production in 1977 and 84.1% of the nation's crude oil input to refineries (Bureau of Mines, "Petroleum Statement," Annual, 1977). More recently, significant volumes of Alaskan North Slope crude oil have begun to move into the Gulf and East Coast areas. This movement could alter significantly the balances described below.

together to buy and sell crude oil. In the first place, most producers do not own refinery capacity; they are looking for an outlet for their crude oil production. Secondly, nearly all refiners are crude deficient and are looking for raw material for their facilities. Indeed, among the top twenty refiners in the United States, only Getty/Skelly was self-sufficient in 1973 while the other 19 had self-sufficiency ratios ranging from 6.4 percent to 67.8 percent.[2113] On balance, the smaller refiners are probably even less self-sufficient that the larger firms. There is, therefore, a major imbalance between the producing sector and the refining sector and substantial volumes of crude oil must flow from one to the other.

A related need for a crude oil market arises from the uncertain timing of crude oil receipts or requirements; for example, the planned balance of a refinery's supply may be interrupted by a delay in the arrival of a vessel. Conversely, an unanticipated event, such as a refinery shutdown, may reduce sharply the need for crude oil and pose a containment problem. In these continually reoccurring events, the refiner turns to the crude market to get back into balance.

Another major reason for crude oil trading by refiner/producers is to upgrade crude oil streams. The value of a particular grade of crude oil to a refiner will depend primarily on the chemical characteristics of the crude oil, the configuration of the refiner's plant, the location of the crude oil relative to that plant, and the prices at which he can sell the final refined products. Since there are significant differences among refiners as to plant configurations, plant locations, and potential markets, different refiners will generally impute different values to the same crude oil streams. A refiner with facilities for converting the heavier fractions of the barrel into gasoline will generally impute a higher value to a barrel of heavy crude oil than will a refiner without such conversion equipment. Similarly, a refiner who does not have the equipment to handle high sulfur crude oil will generally impute a lower value to such crude oil than will a refiner who is equipped to handle it.

The crude oil produced by a given refiner seldom will be optimal for that refiner. Prior to successful exploration, he cannot know the quality of any crude oil he may discover and he must search for oil where it is thought to be, not where he would wish it to be. Furthermore, he is likely to have access to a far greater number of types of crude oil than he can handle conveniently. As a result, he will seek to upgrade his crude oil stream by selling off those grades worth the least to him and replacing them with those worth more to him. Since different refiners place dif-

---

[2113] *Petroleum Industry Hearings, supra* note 192, at 345 (Statement of W. T. Slick, Jr.) With declining domestic crude oil production and increased refinery runs, domestic refiners must be even less self-sufficient today than they were in 1973.

ferent values on the same grades of crude oil, such an upgrading should be possible. (This difference in perceived values for a given grade of crude is also a fundamental driving force behind crude oil exchanges, which are explained later.)

Generally, a refiner will designate a crude oil which may be bought or sold freely in large volumes as a "reference" crude. Today, such a crude oil might be Arabian Light. Then the other crude oil streams tributary to his plants will be evaluated in terms of differences in value, or "deltas," from his reference crude; *e.g.,* crude oil A might be worth 50 cents a barrel more to him than the reference crude (in terms of the value of the products it will yield and the costs to him of processing it) whereas crude oil B might be worth 30 cents less. The company's crude oil traders then are instructed to dispose of crude oil with negative deltas and to replace it with reference crude or better. As a result of this process, each of the crude oils being retained in his final crude oil "slate" should be worth more to the refiner than its delivered cost at his refinery gate.[2114]

The traders carry out their function largely by telephone, balancing the refinery's crude oil supply and demand, while upgrading the crude stream. The telephone verbal agreements generally are followed by written agreements, but in many cases the crude oil flows before the paper does.

There are two distinct kinds of transactions in crude oil, depending on whether the purchase/sale is at the "wellhead" (actually the lease tank) or downstreamn from the "wellhead." The so-called "lease transaction" is the first purchase or sale after the crude oil leaves the well. The contract to purchase lease production is generally referred to as a "division order" because all of the interest owners in the lease —including royalty owners — must execute the contracts which also provide for the appropriate division of the payments among the interest owners. Such a purchase contract normally allows for termination after notice of 24 to 48 hours.

On the other hand, a "contract purchase" provides for the more generalized transfer of title for any quantity of crude oil by any means at any point specified after the initial well head acquisition. Such contracts today are generally "evergreen," (automatically renewed) with a 30-day cancellation provision.

[2114] The market price of any commodity is established by the consumer who wants it least, who is indifferent to whether he purchases that unit or not. All other consumers would have been willing to pay more than the market determined price if they had been forced to do so and, as a result, they command a consumer surplus. As a corollary, it follows that crude oil posted prices are determined by the refining values of the marginal refinery using that crude oil; the other users obtain a consumer surplus. The spread between the value of a crude oil to a refiner and its delivered cost to him must cover his out-of-pocket refining cost and, in the long run, his fixed costs including return on investment.

The Texas Railroad Commission requires the operator of every lease in Texas to file a monthly report for that lease showing oil and gas production, along with the identities of the operator and the gatherer, as well as certain other information. These voluminous data are stored on magnetic tape by the Commission and the tapes are made available to anyone who desires access to them. A number of commercial organizations, such as Petroleum Information Corporation, obtain these tapes and summarize them in a variety of ways for their individual clients. Table VIII, which shows the identity of the 20 largest crude oil "producers" (lease operators) in Texas in 1977, is abstracted from a tabulation of the 200 largest producers prepared by one such organization.

It is clear from Table VIII that the number of companies producing crude oil in Texas is quite large. The top 8 firms account for only 54.43 percent of the state's production. This figure is somewhat larger than the more common nationwide 8-firm oil industry concentration ratio, but is still small in relation to other industries. The 20-firm concentration ratio is 78.06 percent.[2115] Finally, below the top 20 firms, there is an extremely large number of companies producing significant volumes of crude oil. The structure of the crude oil producing sector in Texas comes about as close to meeting the criteria for a competitive market on the supply side as one is likely to encounter in an industrial field.

Reference to Bureau of Mines data on refinery ownership shows that most of the smaller producing companies do not have any refining and that some of the larger firms — such as General Crude and Texas Pacific — also have no refining interests. Quintana is a recent new entrant into refining.

The data in Table VIII are on a "custody" basis rather than on the more common "working interest" basis. This means that the 60 million barrels reported by Chevron, for example, was the volume that it produced as an operator in Texas. This sum includes volumes owned by others, including royalty interests, which were produced by and disposed of by Chevron as an operator. Production owned by Chevron where others were operators is excluded from this figure.

Now consider the demand side of the crude oil market: Who are the purchasers of crude oil? As noted above, there are two types of purchases, those at the lease and those downstream from the lease. Lease purchases will be considered first, using Texas as an example.

[2115] The reader is reminded of the limited value of concentration ratios in appraising market power. See text at notes 1055-1057, *supra*. The use of data from the State of Texas is purely one of convenience due to their availability and in no way implies that Texas, or any other state, is a "relevant market" for competitive purposes.

Traditionally in Texas (and in other so-called "market demand" states), anyone wishing to purchase crude oil at the wellhead will "nominate" for the level of purchases he desires. The regulatory agency, in this case the Texas Railroad Commission, then sets an allowable level of production for the ensuing period such that there will be sufficient production to meet the demand, *i.e.,* total nominations. Actual purchases also are reported to assure that individuals do not consistently thwart the goals of the system by nominating for significantly more or less than they purchase subsequently.[2116]

As spare producing capacity in the United States dried up in 1972, the regulatory agencies no longer had the ability to set an output equal to the demand for domestic crude oil; since that time, all domestic fields (with the exception of a very few fields where technical problems exist) have been producing at capacity and nominations have exceeded lease purchases.

Each month, the Texas Railroad Commission issues a "Recapitulation" statement, which shows nominations by companies for coming months and actual purchases by those companies in the latest past month. Figure 6, entitled "Crude Oil Nominations and Purchases," is the recapitulation sheet for May, 1976, a randomly selected month.

Several conclusions are apparent from Figure 6. First, the purchasing companies nominated for 3,233,703 B/D in May and would have nominated for another 658,075 B/D if they had thought it would have been available. In March 1976, the companies made nominations for 3,260,588 B/D but purchased only 3,185,042 B/D, for an underpurchase of 75,546 B/D. The preponderance of companies were unable to purchase as much crude oil in March as they had nominated even though production was virtually at capacity.

In all, 75 entities nominated for lease purchases in Texas in May of 1976, but elimination of affiliated organizations pares the list to about 70. The number of purchasers at the lease, though fairly large, is still far smaller than the number of companies selling at the lease. Lease purchasers include large and small refiners, pipelines, producers, traders and others. In terms of size, nominations that month ran from a high for Exxon of 532,000 B/D to a low of 2 B/D for Stannar Oil Corp. It is clear from Figure 6 that the opportunity to purchase crude oil at the lease in Texas is not limited to refiners or to giants in any category.

The top 4 producers in Table VIII purchased 39.3 percent of the Texas crude oil purchased in the representative month, slightly above their

---

[2116] Historically, the function of the system was to ensure that non-refiner (generally small) producers had an outlet for their crude and that crude-short refiners had access to supply. The system forced a kind of *de facto* vertical integration on a largely unintegrated industry.

36.3 percent share of 1977 Texas production. The top 8 purchased 54.6 percent of all Texas purchases, almost exactly the same as their share of production (54.4 percent). Marathon, which is one of the largest producers in the state, has a medium-sized refinery in Texas but still sold essentially all of its Texas crude oil production to others.

It will be recalled that sales of crude oil also take place downstream from the lease, after the crude oil has entered a company's supply system. Unfortunately, there are no public data on such transactions, which are generally arm's length deals between unrelated companies. In order to fill this gap, a study of Exxon's total crude oil sales and other deliveries was conducted for the years 1970 and 1977. The year 1970 was chosen as representative of the period prior to today's pervasive governmental regulations on petroleum transactions. The year 1977 was the last full year for which data were available at the time of the study.

The first conclusion of the Exxon study was that about twice as much crude oil was sold by Exxon under contract, downstream from the lease, as Exxon sold at the lease. There is reason to believe that the other large companies also sell significantly more crude oil beyond the lease than they do at the lease, but this presumption cannot be documented.

In all, Exxon was able to identify 181 companies (after elimination of duplications) to which it sold crude oil in 1970 and 213 in 1977. A summary of the results of this study is as follows:[2117]

	Share of Exxon USA's Total Crude Oil Sales		*Share of Nation's Refining Capacity*	
	1970	**1977**	**1/1/70**	**1/1/77**
Seven Other Largest	34.3%	26.6%	49.3%	43.8%
Thums	1.9	----	0	0
Second Tier	25.0	22.8	21.3	20.2
Others	38.8	50.6	20.4	27.8

Memo:
Exxon USA Net Crude Oil Production (MB/D)			761.5	631.0
Exxon USA Crude Oil Sales and Deliveries (MB/D)			957.1	855.5

[2117] Exxon USA Study. Refining capacity, from U.S. Bureau of Mines, "Petroleum Refineries in the United States and Puerto Rico," *Petroleum Refineries,* Annual, January 1, 1970, 1977. (Exxon's share is excluded from this table which, therefore, adds to less than 100.0%.)

In analyzing the data of the above Table, it is reasonable to presume that if a company has x percent of the nation's refining capacity and if it buys more than x percent of the crude oil that Exxon sells, then it is accounting for a disproportionately large share of Exxon's sales. Using this presumption, it follows that the seven largest firms other than Exxon actually bought a disproportionately *small* share of Exxon's crude oil while the smaller companies bought a disproportionately *large* share of Exxon's sales. Again, it cannot be proved, but there is no reason to believe that Exxon is atypical in this matter. The above Table also notes that Exxon's domestic crude oil and sales and deliveries were substantially in excess of its domestic production in both years. This, too, probably is a fairly common situation among the larger firms.

By 1977, the share of Exxon's crude oil going to its largest seven competitors had declined appreciably while the share going to the smaller companies had risen. The significance of this shift is offset somewhat by the fact that the share of refining capacity of the top seven companies had declined by 1977 while that of the smaller companies had risen. Furthermore, some of the change undoubtedly was caused by the Federal Government's Mandatory Crude Oil Allocation Program and subsidies for small refiners. Nevertheless, the company was still selling a disproportionately large share of its crude oil to smaller companies in 1977.

It is alleged frequently that numbers of buyers and sellers, concentration ratios, data on sales and other quantitative measures are meaningless in the oil industry because the major oil companies control crude oil gathering and thus excercise effective control over the industry. If, indeed, the large companies had erected road blocks between independent refiners on the one hand and producers on the other, and if independent refiners can only obtain crude oil at the sufferance of the majors then there could be some substance to the allegation. Hence the ownership and control of the crude oil gathering and trunk lines merits some study in any analysis of the crude oil market. For this purpose it is fortunate that, in addition to reports by operators, the Texas Railroad Commission requires gatherers and transporters to file monthly data on receipts and deliveries and also requires refiners to file data on their crude oil receipts. These data are available to the public, and organizations, such as Petroleum Information Corporation, collect these data and summarize them for their clients.

Before turning to the Texas Railroad Commission data, it is useful to note a major source of error which permeates most available analyses of oil pipelines. The most convenient data on oil pipelines and that most commonly employed have been published by the Interstate Commerce Commission. Unfortunately, the ICC data cover only those interstate pipelines reporting to the ICC; intrastate common carriers and private pipelines are excluded from the ICC reports. This fact turns out to be quite significant in that the larger companies are more likely to own ICC pipelines than the smaller companies, which are more likely to own non-ICC lines. The situation is particularly troublesome in the case of gathering lines, where the non-ICC lines are probably about as significant as those reporting to the ICC.

Although there are no reliable data on the extent of non-ICC gatherers, it is illustrative of their importance that the two largest gatherers in the big East Texas field are Scurlock and Matador, the latter owned by Koch Industries. Neither of these pipelines reports to the ICC; each is a Texas, intrastate, common carrier. It is apparent, therefore, that when concentration ratios of crude oil gathering line ownership is calculated on the basis of ICC data, serious errors arise. These errors can be avoided, in Texas at least, by using Railroad Commission data.[2118]

Data on crude oil gathering receipts in Texas for the first six months of 1978 were calculated from Petroleum Information Corporation summaries of Railroad Commission data. These data are shown in Appendix D.[2119]

The largest single gatherer, Exxon, accounted for over 17 percent of all the crude oil gathered in Texas during the period. Concentration in gathering — while still low relative to other industies — was somewhat higher than concentration in production, but the participants were different. Whereas Texaco and Socal ranked quite high in production, they were rather far down the list of gatherers. Furthermore, some new faces show up among the larger gatherers, namely Permian, Scurlock and Koch. In fact, Permian is the fifth largest gatherer in Texas; larger than ARCO, Gulf, Chevron, or Texaco; and was about twice as large as Marathon.

During the period studied, it was possible to identify 125 different crude oil gatherers (after eliminating affiliated companies) reporting to the Railroad Commission. Those firms below the top 20 accounted for 6.4 percent of all crude oil gathered in Texas. Most of these smaller gatherers were not engaged in any refining activities.

The eight largest refiners on a national basis (*i.e.*, the so-called "integrated majors") accounted for 57 percent of the gathering in Texas, a figure very near their share of production as seen in Table VIII.

Finally, Texas Railroad Commission data on refinery receipts as compiled by Petroleum Information Corporation cast further light on the

---

[2118] See text following note 371, *supra*.

[2119] The peculiar time period was necessary because of a change in definitions. Prior to 1978 all of the production of some of the larger producing units was attributed to one gatherer when there were, in fact, several. This inconsistency was corrected starting with 1978 data.

structure of the Texas crude oil market. Appendix E shows the source of crude oil for every refinery in Texas during the first six months of 1978 grouped by Railroad Commission Districts.[2120] Several interesting factors emerge from study of this Appendix.

The larger refiners, such as Exxon, tend to rely primarily on their own pipelines to supply their own refiners but they also supply those of others as well. Some of the smaller refiners rely almost exclusively on the pipelines of the major companies (Quintana-Howell Joint Venture, for example). Others rely heavily on the independent transporters (*e.g.*, Adobe and Flint Ink). Still other smaller refiners have developed their own transportation systems in large measure (Dorchester Refining and Diamond-Shamrock).[2121] Some refiners, such as Champlin and Texas City, use a combination of suppliers while others, such as Winston Refining, rely only on one supplier pipeline. It would appear that there is no uniformity, and that each of the refiners in Texas has adopted the mode of transportation that best suits its requirements.

Finally, one other fact is clear from Appendix E: It is necessary to import crude oil into the United States to balance supply and demand. From Appendix E, it is apparent that the smaller companies have differentially *favorable* access to domestic crude but, beyond this, that the "majors" are not reluctant to supply them with foreign source oil. See, for example, Southwestern Refining which receives virtually its entire supply of crude oil through major company pipelines or in the form of imports supplied by majors.[2122]

While the subject of exchanges was discussed generally in Section IV B 4 [2123] they recently have assumed greater importance in the crude oil market because the effects of the changeover from domestic crude surplus to deficiency and government regulation favoring smaller refiners have been to dislocate logistical patterns of supply. Exchanges have grown in number as a means of accommodating to these circumstances. As noted previously, the typical refiner is continually engaged in upgrading his crude oil stream; he is constantly selling off those grades least attractive to him and replacing them with those of greater value. But, so long as he is

---

[2120] These data represent trunk line deliveries to refineries plus deliveries from barges and ships across the docks of refiners. There also may be some truck or rail deliveries included as well. Obvious misspellings or multiple listings for individual companies have not been changed.

[2121] Dorchester, interestingly, was the 21st largest gatherer and just barely missed in Appendix D.

[2122] This dependence on major companies has not inhibited Southwestern's growth. Between 1970 and 1977, Southwestern's crude oil distilltion capacity grew from 43,500 B/D to 120,000 B/D or by 176%. During the same period total refining capacity in Texas grew only by 31%. (U.S. Bureau of Mines, "Petroleum Refineries in the United States and Puerto Rico," *Petroleum Refineries,* Annual, January 1, 1970 and January 1, 1977.).

[2123] See text at notes 1250-1255, *supra*.

crude oil deficient, as has increasingly been the case, he will have to replace each barrel he sells. In arranging for the replacement barrel, it sometimes happens that an exchange is more attractive than an outright purchase.

One reason that an exchange may be attractive is that it is not as vulnerable to cancellation as is a purchase. The typical purchase contract usually allows cancellation on fairly short notice; a rational seller would not normally write a long term crude oil sales contract without some kind of reopener. Thus, replacement barrels obtained under pure purchase contracts are subject to potential loss.

Replacement barrels obtained on exchange (or back-to-back purchases/sales, which accomplish the same result) may offer greater security than those obtained by outright purchase. There are two basic reasons for this added security. First, a general increase in crude oil prices could lead to a cancellation of a purchase contract and possible loss of the replacement material. In the case of the exchange, on the other hand, the relative values of the two crude oils to each refiner could remain essentially unchanged with the result that cancellation would not be triggered.

The second reason has to do with the fact developed earlier that different refiners impute different values to the same crude stream. When refiner A imputes a higher value to a crude in refiner B's possession than B does and refiner B imputes a higher value to a stream in A's possession than A does, there is a clear basis for a mutually beneficial trade. Under an exchange agreement, A would be less likely to withdraw the crude oil he was supplying to B than he would if he were making an outright sale. The reason, of course, is that B could retaliate at some cost to A. Thus a refiner is likely to view crude oil obtained on exchange as relatively more secure than that obtained by purchase on the open market. And, since market disruptions have not been uncommon in the crude oil market (the OPEC embargo of 1973-74 and the 1979 Iranian shutdown are just the two most recent occurrences), more security is preferable to less. Thus, exchanges have tended to come about as refiners have attempted to establish some kind of security in an increasingly uncertain market.

An entirely different kind of exchange is found in the so-called "time trade." Here a refiner may find that he is temporarily short of raw material due to some event such as tanker delay resulting from bad weather. If he can locate a supplier who is temporarily in long supply, he will try to negotiate a swap. The advantage of the swap transaction is that it is beneficial to both parties; it solves the supply problem of one while reducing the storage costs of the other. Furthermore, in contrast to a sale and a subsequent repurchase, the transaction is perfectly hedged; the party supplying the crude oil knows that he will receive back the equivalent of the material he has delivered, no matter what has happened to prices in the interim.

Finally, in connection with exchanges, it might be useful to lay to rest another common myth. It is apparently believed on a fairly wide scale that crude oil exchanges are used in some way by larger refiners to deny small refiners access to crude oil.[2124]

The only publicly available data on exchanges were provided by Exxon in its testimony before the Antitrust and Monopoly Subcommittee of the Senate Judiciary Committee in 1975. In that testimony the following data on Exxon USA's 1970 crude oil exchanges were submitted:[2125]

Exxon USA Crude Oil Exchanges, 1970

	With Others in Top 8	With 9th Through 16th Largest	With Non-Majors
Percent by Volume	35.5	26.8	37.3
Percent by Number	23.1	17.1	59.8

Using the same criterion that was developed earlier for analyzing crude oil sales, one would have to conclude that Exxon USA had a disproportionately *large* number and volume of crude oil exchanges with the independent sector. Although the volume of crude oil moving on exchange has increased recently, there is no evidence that there has been a substantial change in these patterns.

In any event, it would be improper to dwell at more length on crude oil exchanges. They have not been a predominant form of doing business in the past; they are just one of several types of crude oil transactions that may occur; they are the preferred form of transaction only when there is some economic advantage in the transaction itself. In contrast to critics' charges, there is no evidence that exchanges have ever been used in an anti-competitive or exclusionary manner.

[2124] *See e.g.*, FTC Preliminary Staff Report, *supra* note 2061, at 30.
[2125] *Petroleum Industry Hearings, supra* note 192, at 346 (Statement of W. T. Slick, Jr.).

(ii) The Refining Sector

Much of the criticism of the industry has involved the refining stage of the business. The S.2387, or PICA, Report cited the standard allegation of too high a concentration.[2126] (James T. Halverson, of the Federal Trade Commission told a Senate Committee about that time that "The industry is highly concentrated at the refinery level."[2127])

The Report also alleged that independent refiners were inhibited by control of crude by "majors" (based on independent refiner lack of crude self-sufficiency and a "dependence" on "majors" arising from an average purchase of about 50 percent from "majors" and an additional 12 percent on exchange from the eight largest "majors.")[2128] Edwin J. Dryer, General Counsel and Executive Secretary of the Independent Refiners Association of America (IRAA), built on that theme and charged that the integrated "majors" as "owners of most domestic crude oil and as purchasers of much of the balance, have set crude oil prices which are artificially high in relation to the prices set by them for refined petroleum products."[2129] The FTC Staff in its Preliminary Report (later incorporated into the *FTC vs. Exxon, et al.* proceeding) claimed that barriers to entry by "independents" had been erected and maintained in order to preserve their profits.[2130] Mr. Halverson, in his statement mentioned above, characterized such barriers as "overwhelming." In addition to control of the crude oil market, among the devices alleged to build these entry barriers were control of products pipelines and exchanges.[2131] Finally, a statement by Professors Fred C. Allvine and James

---

[2126] S.2387 REPORT, *supra* note 49, at 49. The Report presented a table, giving the Bureau of Census as the source, showing the following:

	Percent of value of shipments		
	Top 4	Top 8	Top 20 Firms
1972	31	56	84
1967	33	57	84
1963	34	56	82
1958	32	55	82
1954	33	56	84
1947	37	59	83
1935	38	58	NA

[2127] Halverson's Statement is cited in *Consumer Energy Act Hearings, supra* note 154, at 1332.
[2128] S.2387 REPORT, *supra* note 49, at 52.
[2129] *Petroleum Industry Hearings, supra* note 192, at 240 (Statement of E. J. Dryer).
[2130] FTC Preliminary Staff Report, *supra* note 2061, at 25.
[2131] S.2387 REPORT, *supra* note 49, at 54-56.

M. Patterson, in their book *Highway Robbery*, that "[s]ince 1950 the integrated oil companies have taken over several of the important independent refineries and there have been built no new independent refineries with over 50,000 barrel per day capacity,"[2132] was picked up by Senator Philip Hart and inserted in the *Congressional Record*,[2133] and echoed by Mr. Halverson, who went even further and said there had been virtually no new entry into the industry since 1950.

As the late Al Smith used to say, "Let's look at the record." First, as to concentration, there is a plethora of evidence on the subject, which runs contradictory to the critics' assertions. There were two compilations of identical concentration figures, one entered as an exhibit to the testimony of Jack Vickrey in the *Consumer Energy Act Hearings*[2134] and the other in Thomas Duchesneau's *Competition in the U.S. Energy Industry*[2135] which were as follows:

Percentage of Total Certified Refinery Inputs of Total U.S. Refineries
[PAD] Districts I-IV

Class of Refiners	1960	1966	1970
Independents	16.2	16.5	17.1
Majors (including Ashland)	83.8	83.5	82.9

District V

Class of Refiners	1964	1966	1968	1970
Independents	11.1	11.5	11.6	13.6
Majors	88.9	88.5	88.4	86.4

It can be seen that the independents had commenced to *increase*, rather than decrease their share of the refining business, albeit, slightly during this time.[2136] Professor Richard Mancke who has been a close

---

[2132] F. ALLVINE & J. PATTERSON, HIGHWAY ROBBERY 216 (1974).

[2133] 121 Cong. Rec. S.17690 (daily ed. Oct. 7, 1975).

[2134] *Consumer Energy Act Hearings, supra* note 154, at 652 (Statement of Jack Vickrey, Exhibit I). Source was Office of Oil & Gas, Department of the Interior.

[2135] T. DUCHESNEAU, COMPETITION IN THE U.S. ENERGY INDUSTRY 300 (Table F-2) (1975).

[2136] *S.1167 Hearings, supra* note 432, pt. 9 at 594 (Statement of Fred F. Steingraber); STEINGRABER, *supra* note 193, at 134 (Working from figures of Staff Analysis, Office of the Energy Advisor, Department of the Treasury, dated August 27, 1973).

observer of the industry, provided a list of the top 20 refiners in 1970, and one notes the names of Ashland (above Continental, Marathon, and Getty), Tenneco, Clark and American Petrofina, which have not been on any Congressional "hit" parades. The Energy Resources Council in its *Analysis of Vertical Divestiture* of May 1976, in its Executive Summary, page iv, concluded that "[c]oncentration levels for petroleum refining have been less than the average for all U.S. manufacturing." Professor Richard Mancke observed: "In comparison with other major American heavy industries, oil refining is not highly concentrated."[2137] W. T Slick, Jr., testifying before the Senate Antitrust and Monopoly Subcommittee on November 12, 1975, observed that not only had the concentration ratios of "major" refiners been declining since 1950, but he pointed out that the trend had accelerated since 1970.[2138] By 1974 the top refiner had only 8.4 percent, the top 4 had 31 percent and the top 8 had 55 percent.[2139] By 1977, although the top firm had a fraction of a percent more, the top 4 were exactly the same and the top 8 had declined to 52.7 percent.[2140]

Also, of significance was the fact, noted earlier, that the membership of the group and their relative positions changed during this period. From January 1, 1951, to January 1, 1975, Mobil slipped from 2nd to 7th, Texaco from 3rd to 4th, Gulf from 5th to 6th, Cities Service from 9th to 17th and Texas City from 20th to 31st, while Shell moved up from 7th to 2nd, Standard (Indiana) from 4th to 3rd, Ashland from 18th to 15th, and Amerada-Hess, which was not even in the top 20 in 1951, came on with a rush to grab the 8th spot; and new names like Coastal States and Petrofina moved into the largest 20 refiners.[2141] Donald C. O'Hara refined this point in his testimony of November 12, 1975, before the Senate Antitrust and Monopoly Subcommittee when he displayed the top 20 refiners of January 1, 1951, which had 80.68 percent of U.S. refining capacity, and carried the same refiners over to January 1, 1975 when they had only 71.47 percent of capacity.[2142] It is only by "filling in" with Amerada-Hess, Marathon, Coastal States and American Petrofina that the "top 20" get built back up to 88 percent. The same type of reshuffling took place between 1975 and 1977, with firms such as Champlin and

---

[2137] MANCKE, *Oil Industry Competition,* supra note 674, at 47.

[2138] *Petroleum Industry Hearings, supra* note 192, at 335 (Statement of W. T. Slick, Jr.).

[2139] *Id.* at 345.

[2140] API, Market Shares and Individual Company Data for U.S. Energy Markets: 1950-1977 (Discussion paper #014, November 1978) p. 93. Source: NPRA "U.S. Refining Capacity," July 1978; Company Annual Reports and 10-K's.

[2141] *Petroleum Industry Hearings, supra* note 192, at 392 (Statement of Donald C. O'Hara, President of National Petroleum Refiners Association, "NPRA").

[2142] *Petroleum Industry Hearings, supra* note 192, at 392 (Exhibit A) (Statement of Donald C. O'Hara).

Tosco Corp. breaking into the select group. These compilations show two things, both antithetical to the critics' assertions, *i.e.*, concentration ratios in the refinery phase are "moderately low" using the 8 firm test, and "low," using the standard 4 firm test normally used to measure concentration in American manufacturing industries;[2143] and the "barriers to access" didn't keep firms like Ashland, Coastal States, Champlin, Tosco, Kerr-McGee and American Petrofina, among others, from working their way up to the "big boys." As Donald C. O'Hara, President of the National Petroleum Refiners Association (NPRA), stated in his famous "Let's Look at the Record" address before the Symposium on the Energy Crisis sponsored by the Oil, Chemical, and Atomic Workers Union (OCAW), in Washington, D.C., on March 14, 1974, "The fact is that no other basic industry even comes close to having the [large] number of strong, independent competitors that we have in the oil refining business."[2144]

Since the discussion already has entered into the topic of barriers of entry, let's look again at the record. The allegation of control of crude oil has already been discussed rather thoroughly in the immediately preceding subsection (see also the discussion following footnote 371 and Appendix E). Exchanges have been examined in previous sections and their procompetitive and consumer-benefiting nature has been established; all that needs to be added at this point is that "majors" conduct approximately 37 percent of their total exchanges with independent refiners, which is more than three times the percentage differential between their respective refining capacities.[2145] The allegation that independents are dependent on majors for about 50 percent of their crude from "majors" appears to cut even more strongly the other way, as pointed out in the preceding subsection,[2146] *i.e.*, not only do independents have more favorable access to domestic crude because of governmental programs biased in their favor,[2147] but the "majors" are not reluctant to supply them with domestic crude and foreign crude as well.[2148] The contention of the FTC staff that "majors" have kept independents out of refining by controlling gasoline marketing through the device of segmenting the market for retail gasoline into two markets, and then restricting the

---

[2143] See text at notes 1060-1062, *supra*.

[2144] *Consumer Energy Act Hearings*, *supra* note 154, at 1332 (Reprint of Mr. O'Hara's address).

[2145] See text at notes 1250-1255 and 2123-2125, *supra*, for earlier discussion. The 37 percent exchange with independents comes from ERC REPORT, *supra* note 184, at 15 (Table 13); the capacities are taken from API, Market Shares and Individual Company Data for U.S. Energy Markets: 1950-1977 (Discussion Paper #014, December 1978) p. 93.

[2146] See text at note 2122, *supra* and Appendix E.

[2147] See text at notes 1022-1027, *supra*. MANCKE, *Oil Industry Competition*, *supra* note 674, at 64 (MOIP).

opportunity to supply branded outlets to the "majors' " refineries and "cooperating refineries," borders on being a ludicrous statement. In the next subsection, the obvious fact — that "majors" hardly control gasoline marketing — will be documented. The division into branded and non-branded markets is a result primarily of two factors: chronic (before recent OPEC-induced shortages) excess refinery capacity; and the FTC itself, which always has covertly, and since its 1967 policy statement has openly proceeded against any action by a branded marketer to price its gasoline competitively with unbranded gasoline.[2149] Alleged denial of access by independents to pipelines, both products and crude, has been laid to rest in Section VI C 3, "Asserted Denial of Access to Pipelines."[2150] There are, indeed, barriers to entry in refining: capital requirements of over $2,500 per barrel per day of capacity,[2151] and the current uncertainty in obtaining a secured supply of crude.[2152] The FTC implies that the high capital requirements constitute a "significant" barrier to new entry[2153] but, as will be shown in the following treatment of new entrants into the market, this implication appears to be dispelled by the empirical evidence.

There have been some significant deterrents to new entry, but they have come from exogeneous sources, operating against "majors" and independents alike. These are the efforts of local and environmental groups to block the construction of refineries.[2154] Donald O'Hara named some of these in his speech: Metropolitan Petroleum Co., was prevented from building a 250,000 B/D refinery at Eastport, Maine, by the Maine Environmental Commission; Maine Clean Fuels was turned down in an attempt to construct a 200,000 B/D refinery at Searsport, Maine;

[2148] FTC Preliminary Staff Report, *supra* note 2061, at 8 (eight largest "majors"). For an example of the sales and purchases by the largest company in the industry, Exxon, to and from independents in years 1972-1973, by quarter, see Appendix Q. Another interesting item is the fact that when Hudson Oil Company, one of the most aggressive "independent" refiner-marketers in the business, was looking for a refinery site in the Houston area, it was able to purchase a site for a 200,000 B/D refinery in the Bayport Industrial Development Area in Harris County, Texas, in February, 1975 from the Friendswood Development Company, which is 100 percent owned by Exxon Company, U.S.A., Houston Post, Feb. 22, 1975, p. 6-D.

[2149] JOHNSON et al., *Oil Industry Competition*, *supra* note 1055, at 2448.

[2150] See text at notes 1807-1991, *supra*.

[2151] J. HASS, E. MITCHELL & B. STONE, FINANCING THE ENERGY INDUSTRY 32 (1974). *See also* text at note 1180, *supra*.

[2152] *Cf.* note 1183, *supra*; TEECE, *supra* note 329, at 99.

[2153] MANCKE, *Oil Industry Competition*, *supra* note 674, at 47. But Professor Teece does not believe that adverse capital costs and scale economies are barriers to entry. TEECE, *supra* note 329, at 99.

[2154] JOHNSON, et al., *Oil Industry Competition*, *supra* note 1055, at 2443 give the number of new refineries blocked by such groups as 16. *See also* Ritchie, *Petroleum Dismemberment*, 29 Vand. L. Rev. 1131, 1152 (1976); TEECE, *supra* note 329, at 99.

Aristotle Onassis (Olympic Refining) was refused permission by local authorities to build a 400,000 B/D refinery at Durham, New Hampshire; Northeast Petroleum was rejected by the City Council of Tiverton, Rhode Island, in its attempt to build a 65,000 B/D refinery; Commerce Oil, whose President sunk most of his own money in fighting unsuccessful court battles to build a 45,000 B/D refinery on Jamestown Island, Rhode Island; Shell Oil, which fought local environmentalists for years in order to build on Blackbird Hundred, a mosquito sanctuary in Delaware, and won its battle with them only to fall prey to the Delaware legislature which enacted a 10-mile strip restriction which brought down its plans to build a 150,000 B/D refinery. (As an item of wry amusement, during the gasoline shortage of 1973, the Governor of Delaware contemplated limiting the sale of gasoline to cars with Delaware license plates.) The reader who wishes to pursue the rest of this sorry story will find it chronicled in O'Hara's speech, reprinted in the *Consumer Energy Act Hearings* at pages 1334-1344.

Before moving on, one ghost should be laid to rest, *i.e.*, that the majors prevented Occidental from building a 300,000 B/D refinery at Machiasport, Maine. The fact is that what the majors (and independents alike) objected to was not the construction of the refinery but Armand Hammer's attempt to envelope his refinery within a one refinery "foreign trade zone" and to move products from this refinery, which would run on foreign crude, into the United States under an exemption from the oil import regulations. It is noteworthy that 62 "small business" independent refiners also objected to this "loaded dice" deal. When the Hawaiian Independent Refinery applied for a permit to build a refinery in a foreign trade zone without these exemptions, no one in the industry filed any objections. Apparently, Machiasport citizens changed their minds when Atlantic Richfield attepmted to locate a 200,000 B/D refinery there because it was turned back by the local authorities. Messrs. Johnson, Messick, Van-Vactor and Wyant, who were authors of the Report, "Competition in the Oil Industry," which grew out of the George Washington Uniersity's Energy Research Project, concluded:

> "In sum, there have been and continue to be barriers to entry in refining. However, these barriers do not appear to have been insurmountable. Nor have they been the result of the sinister workings of the major oil companies. Instead, they are the result of technological and economic factors inherent in the refining business. Or, they are a product of often misguided or uninformed government regulation. None of these barriers will be lowered by breaking up or otherwise prosecuting the major oil companies."[2155]

Turning to the famous Allvine and Patterson statement, echoed by James T. Halverson of the FTC, and included in the *FTC vs. Exxon, et al* proceding, that since 1950 no new independent refineries with over 50,000 B/D capacity have been built, one finds it to be contradicted squarely by the facts. Professor Mancke found that there were 47 new entrants into U.S. refining between 1950 and 1972 and 13 new entrants between 1972 and 1975. Of the 23 largest new entrants between 1950 and 1972, 8 were crude producers, 6 had marketed refined products, 8 were of unknown antecedent, and in one case, the entry represented acquisition of an integrated operation.[2156] Donald O'Hara, in his testimony in the *Consumer Energy Act Hearings*, listed the independent refineries as of January 1, 1973, who were not in the refining business in 1951 (*) or who built 50,000 B/D or more capacity since 1951(#). Where both items were true, the company is marked by both symbols. Amerada-Hess (*,#): 510,000 B/D; Union Pacific (*,#): 140,000 B/D; Commonwealth (*,#): 140,000 B/D; Coastal States (*,#): 135,000 B/D; Petrofina (*,#): 108,500 B/D; Murphy (*,#): 105,000 B/D; Clark (#): 104,500 B/D; Koch (*,#): 97,900 B/D; Crown Central (#): 96,000 B/D; Tenneco (*,#): 87,000 B/D; Charter (*): 82,900 B/D; Tosco Petro (*): 62,500 B/D; and while not independent, Marathon (#): 223,000 B/D.[2157]

William P. Tavoulareas, President of Mobil Oil Corporation, testified on January 21, 1976, before the Senate Antitrust and Monopoly Subcommittee that 10 companies that were not even in the refining business at all in 1950, had refineries with capacities of over 50,000 B/D by 1973.[2158]

In the same Hearings, on November 12, 1975, Mr. O'Hara quoted Professors Allvine and Patterson and then contrasted their statement with the facts, *i.e.*, that the following new refineries with a capacity of 50,000 B/D or more capacity had been built by independents since 1950: Amerada-Hess, St. Croix, Virgin Islands — 700,000 B/D; Commonwealth, Puerto Rico —161,000 B/D; Koch Industries, St. Paul, Minnesota — 109,800 B/D; Amerada-Hess, Perth Amboy, New Jersey — 67,900 B/D; Hawaiian Independent, Honolulu, Hawaii — 60,000 B/D; United Refining, Warren, Pennsylvania — 58,000 B/D; and Ecol, New

---

[2155] JOHNSON, et al., *Oil Industry Competition, supra* note 1055, at 2448; *see also* CANES, *An Economist's Analysis, supra* note 1021, at 48; *Petroleum Industry Hearings, supra* note 192, at 336 (Statement of W. T. Slick, Jr.).

[2156] MANCKE, *Oil Industry Competition, supra* note 674, at 47-48, citing original source material collected by The Chase Manhattan Bank, reprinted in BNA, *Energy User's Report*, November 13, 1975, p. A-20.

[2157] *Consumer Energy Act Hearings, supra* note 154, at 1333 (Exhibit A) (Statement of Donald C. O'Hara).

[2158] *Petroleum Industry Hearings, supra* note 192, at 1821 (Statement of William P. Tavoulareas).

Orleans, Louisiana —200,000 B/D (under construction) (Marathon has subsequently purchased this refinery). In addition, 15 other independent companies each had built 50,000 B/D or more of new capacity since 1950, either by adding to existing refineries or by building on a site occupied previously by an older refinery, which meant that the equivalent of 22 additional refineries of 50,000 B/D or more had been built by independent companies since 1950.[2159]

Frank N. Ikard, writing in the Oregon Law Review, updated some of this information by showing new U.S. refineries built and planned since January 1, 1965 with 739,000 B/D aggregate new capacity; substantial (45,000 B/D+) refinery expansions since January 1, 1965 in the aggregate of 1.3 million B/D and a whole host of announced expansions and new constructions as of August 9, 1973.[2160] Professor Mancke listed 16 U.S. refinery expansions set for 1975-1977 aggregating 1.35 million barrels per day.[2161] Thomas Duchesneau noted that from the beginning of 1966 to the end of 1972, independents built a total of 18 new refineries with a combined capacity of 146,000 B/D and expanded the capacity of existing refineries by 260,000 B/D.[2162] The Energy Resources Council commented in 1976 that: "There has been significant entry or expansion by independent refiners in the past 15 years. Table 17 shows that 22 refiners have grown to more than 50,000 B/D of capacity during the period of 1951 to 1975. These companies have built or acquired 2,950,000 B/D of refining capacity during this period. As of January 1, 1975, these companies accounted for 20 percent of domestic refining capacity."[2163]

At the request of the Senate Antitrust and Monopoly Subcommittee, the API prepared and forwarded to the Subcommittee a list of new refineries built between 1950 and 1970 in the United States, and what their subsequent history has been. The list identified new entrants ("N"), defined as firms which had not previously owned domestic refinery capacity and "refineries" used the Bureau of Mines definition. It is attached herein as Appendix R. According to these data, 129 new refineries were built in the United States between 1950 and 1976. These new refineries represented 35 percent of the refineries in the United States and accounted for about 20 percent of domestic refining capacity in 1976. New entrants accounted for over two-thirds of the new refineries, but only

---

[2159] *Id.* at 386 (Statement of Donald C. O'Hara); *see also* Ritchie, *Petroleum Dismemberment,* 29 Vand. L. Rev. 1131, 1151 (1976).

[2160] Ikard, *Petroleum Industry Competition, supra* note 1055, at 593-596.

[2161] MANCKE, *Oil Industry Competition, supra* note 674, at 49 (Table 6).

[2162] T. DUCHESNEAU, COMPETITION IN THE U.S. ENERGY INDUSTRY 299 (1975).

[2163] ERC REPORT, *supra* note 184, at 35; *see also Petroleum Industry Hearings, supra* note 192, at 373 (Exhibit D) (Statement of Walter R. Peirson); *cf. Id.* at 335 (Statement of W. T. Slick, Jr.).

24 percent of the initial capacity of the new refineries. The rate of expansion by the new entrants has been slightly greater than that of the other new refineries built between 1950 and 1977, and these new entrants now hold about 28 percent of the capacity attributed to new refineries and about 5 percent of total domestic refining capacity. These data on new entrants understate the degree of entry in the 1950-1976 period for two reasons. First, the data exclude non-refiners which bought domestic capacity and subsequently expanded it significantly (as in the case of Coastal States). Similarly, firms which were relatively new entrants prior to 1950 have been excluded.

To sum up, the empirical evidence seems convincing that entry by "independents" has not been barred by "majors" to preserve monopoly profits,[2164] or to put it affirmatively, the diversity of the firms entering the refinery phase of the business is such that large firms lack the protection of high entry barriers that sometimes has been alleged.[2165] For those readers who might have been misled in the past by versions of the Allvine-Patterson allegations, one would hope that their reaction would be similar to that of Professor Leonard Weiss in his deposition as an expert witness for the Antitrust Division of the Department of Justice in the IBM case, when he stated: "I mentioned...the number of firms, including some independents I have never heard of, who set out to build refineries between May and July of 1973, and it is just astounding — and these were one hundred million dollar investments — that shook my belief in capital requirements as a high barrier to entry quite a bit."[2166]

(iii) Gasoline Marketing

As the 1973 Preliminary Staff Report of the Federal Trade Commission acknowledged: "...gasoline marketing is the most competitive area of the petroleum industry and has the largest number of independent companies."[2167] There are a number of submarkets, some of which are on a "bid" basis, such as government business, large corporate accounts and railroads. There is a cargo lot market, which exists in the gasoline market but is more prevalent in the residual fuel business; for example, sales to utilities, etc. The gasoline market is itself divided into the direct consumer

---

[2164] Ikard, *Petroleum Industry Competition, supra* note 1055, at 597.

[2165] MANCKE, *Oil Industry Competition, supra* note 674, at 48; *cf. Petroleum Industry Hearings, supra* note 192, at 1063 (Statement of Professor E. Erickson).

[2166] United States v. I.B.M., S.D.N.Y., Deposition of Leonard Weiss 354-355, June 11, 1974.

[2167] FTC Preliminary Staff Report, *supra* note 2061.

market, the jobber market, the service station market, and the direct service station sales to automotive customers. According to the S.2387, or PICA, Report in 1975, about two-thirds of gasoline sold was delivered at "dealer tank wagon" prices through refinery-owned terminals and bulk plants. The balance moves through jobbers, who purchase at a discount (the jobber "margin"), transport it to bulk storage facilities, and complete the distribution process either through their owned stations or through dealers who own or lease their stations. Jobbers are either branded jobbers, who sell their product under their refiner-supplier's brand, or unbranded jobbers, who purchase products on specification and sell under their own brand or completely unbranded.[2168] The decision by a refiner as to whether he will utilize the jobber channel is a matter of economics, in line with the discussion in Section IV B "Vertical Integration."[2169] It is the usual choice to distribute through local jobbers where the station density is slight and average throughput volumes are low. In some instances, it is uneconomic for the refiner to do business there and the jobber can do a better job of serving the customers.[2170]

Concentration ratios for the motor gasoline sales market for 1977 were 29.1 percent for the top 4, 49.2 percent for the top 8, and 72.4 percent for the top 20, as compared with 1954 figures of 31.2, 54.0 and 80.4 percent, respectively.[2171] Once again, one notes that under the 4-firm test, this concentration would be rated "low," and under the 8-firm test, it is "moderately low." There is also the change in the "cast of characters" among the top twenty, with Getty, Clark, Tosco and Kerr-McGee, all names absent from the 1954 lineup. Incidentally, although Kerr-McGee edged out American Petrofina and Tenneco for the 20th spot, it was by the narrowest of margins — their percentage of the market was exactly the same when carried out to one significant digit in the tenths of one percent, and Amerada-Hess and Diamond Shamrock were extremely close to Tenneco.

There are some 15,000 wholesale dealers and 300,000 retailers in the marketplace,[2172] most of which are independent businessmen[2173] who are

---

[2168] S.2387 REPORT, *supra* note 49, at 15.

[2169] See text at notes 1256-1264, *supra*.

[2170] *Petroleum Industry Hearings, supra* note 192, at 2083 (Statement of Charles R. Jackson, a branded jobber in Chesterfield County, South Carolina.).

[2171] API, Market Shares and Individual Company Data for U.S. Energy Markets: 1950-1977 (Discussion Paper #014, November 178) pp. 151, 160. The intervening years appear on pages 152 through 159. *See also* ERC REPORT, *supra* note 184, at 17 (1972-1975 ratios for Middle Distillates, Residual Fuels and MoGas).

[2172] CARD, *Industrial Organization, supra* note 1152, at 40.

[2173] *Petroleum Industry Hearings, supra* note 192, at 336 (Statement of W. T. Slick, Jr.).

vigorously competitive.[2174] As William P. Tavoulareas testified before the Senate Antitrust and Monopoly Committee on January 21, 1976: "Indeed at the retail level, most of the comments I hear criticize the large number of service stations available to the public. I must say I don't hear communities complaining because there is insufficient oil company competition for the customer's business."[2175]

The most significant item besides the relatively large share of the market held by the independent marketer is the fact that it has been growing steadily since 1954, and, except for a brief dip during the 1973-1974 crude oil and product shortage,[2176] has grown by leaps and bounds in recent years.[2177] The Johnson study compared the top 5 firm and 8 firm shares in 1954 to 1972, reflecting declines by both.[2178] Jack Vickrey, testifying before the Senate Commerce Committee in the *Consumer Energy Act Hearings*,[2179] and Fred Steingraber, appearing before the Senate Antitrust and Monopoly Subcommittee in the *S.1167*, or *Energy Industry Hearings*,[2180] both showed increases in independents' share of the market from 1968 to 1972. William P. Tavoulareas, also appearing before the Senate Antitrust and Monopoly Subcommittee on January 21, 1976, testified concerning the "significant achievement" by independent marketers in increasing their market share (from 25 percent to 33 percent) between 1968 and 1975.[2181] Professor Richard B. Mancke, presented a Table showing market shares of the top 4, top 8, and independent firms for the years 1964, 1970 and 1974 which showed a steady decline by the top 4 and top 8 during this period and a substantial growth by independents from 1970 to 1974.[2182] William T. Slick, Jr., appearing at the same hearings as Tavoulareas, noted the almost

---

[2174] *Id.; Id.* at 2088 (Statement of John R. Johnson, a Shell jobber from Morristown, Tennessee).

[2175] *Id.* at 1821 (Statement of William P. Tavoulareas).

[2176] JOHNSON et al., *Oil Industry Competition, supra* note 1055, at 2364 ("In 1973 and early 1974 the market shares of the sample firms [the narrow FEA definition of independents — Independent Gasoline Marketers' Council (IGMC) members] fell from about 5.2 percent to about 4.5 percent of total gasoline sales. This was during the period when crude oil and product shortages first occurred. As one might expect, these shortages hit hardest the nonbranded segment of the industry. (Some nonbranded distributors have bought heavily on the spot market. By definition, branded dealers have not.) However, by the fourth quarter of 1974 the nonbranded independent marketers had more than restored their 1972 market shares. By then, they accounted for 5.5 percent of total gasoline sales." *See also Petroleum Industry Hearings, supra* note 192, at 202 (reprint of excerpt from Testimony of John Hill, Deputy Administrator, FEA before Senate Interior Committee on Sept. 4, 1975).

[2177] See note 2171, *supra*.

[2178] JOHNSON et al., *Oil Industry Competition, supra* note 1055, at 2343 (Table 1).

[2179] *Consumer Energy Act Hearings, supra* note 154, at 615, and Exhibit J, p. 652. (Statement of Jack Vickrey.)

[2180] *S.1167 Hearings, supra* note 432, pt. 9 at 594 (Statement of Fred F. Steingraber).

[2181] *Petroleum Industry Hearings, supra* note 192, at 1821 (Statement of William P. Tavoulareas).

[2182] MANCKE, *Oil Industry Competition, supra* note 674, at 51 (Table 8).

50 percent growth in independents' share of the market between 1968 and 1975.[2183] In addition, there was some localized information adduced. The Johnson paper listed the top 15 increases in market shares at the state level from 1970-1973. The highest and 3rd, 4th, 5th, 9th, and 12th greatest increases at the state level were by independents.[2184] Professor Edward J. Mitchell testified on August 6, 1974 before the Senate Antitrust and Monopoly Subcommittee concerning the results of his preliminary study of the effect on gasoline prices and market shares of independents in the markets along the route of the Colonial Pipeline. His graphs show marked declines in prices after shipments commenced through the line and declines in market shares of Colonial's owners and other major oil companies during that period of time, leading him to conclude on the basis of such studies that the pattern of market shares and gasoline prices were consistent with the workings of competition.[2185] Professor Mitchell's findings were confirmed by a subsequent more detailed report by Lundberg survey, Inc., introduced in *The Petroleum Industry Hearings.*[2186]

The increase in market share of the "independents" obviously has been at the expense of the "major" oil companies.[2187] One of the principal reasons for this phenomenon has been the long term trend of the public to shift from branded dealers to cut-rate, self-service outlets operated by both independents and major oil companies,[2188] which mode of operation, as Professors Allvine and Patterson pointed out, was initiated by "independents" at a time when the "majors" were attempting to operate their marketing properties at a profit and because of these large investments in the traditional friendly full-service stations were caught "running counter to the trend."[2189] Many of the more progressive majors swung over to the self-serve style of marketing but the impetus remained with the independents. However, as an independent jobber from Collins, Mississippi, remarked: "If the larger, integrated companies had the control in this industry that is purported, self service would not exist, for it is mainly the large oil company that has an investment in properties costing $100,000 or more. These high-cost stations are the ones suffering large gallonage losses to independent wholesalers and retailers, branded

---

[2183] *Petroleum Industry Hearings, supra* note 192, at 336 (Statement of W. T. Slick, Jr.).

[2184] JOHNSON et al., *Oil Industry Competition, supra* note 1055, at 2373 (Table 10).

[2185] Mitchell, *1974 Senate Testimony, supra* note 1056, at 40. See Mitchell's Exhibits 5-8.

[2186] *Petroleum Industry Hearings, supra* note 192, at 779.

[2187] CARD, *Industrial Organization, supra* note 1152, at 40, citing the August 27, 1973 Staff Analysis of the Office of the Energy Administrator, Department of the Treasury.

[2188] JOHNSON et al., *Oil Industry Competition, supra* note 1055, at 2366.

[2189] F. ALLVINE & J. PATTERSON, HIGHWAY ROBBERY 77-79 (1974).

and unbranded."[2190]

Out of this examination of the marketing segment of the business and the earlier section on pipeline access seems to flow the following conclusions: there is no evidence of restrictions on entry in the marketing sector of the industry,[2191] and the charge that "major" companies dominate or have increased their domination in particular regions is not supported by the facts.[2192] Once again, as with the other segments of the industry, the large number of companies, the low levels of concentration and the continual growth of "independents" belie a market structure dominated by a few.[2193] The second conclusion interacts with the first, namely, that refiners and marketers of petroleum products, as well as consumers, have benefited materially from the network of oil pipelines. More sellers of more products are able to reach more different markets more efficiently at a lesser cost and with greater quantities of petroleum than would otherwise be possible.[2194]

[2190] *Petroleum Industry Hearings, supra* note 192, at 2086 (Statement of Pat Green, Jr., Independent Jobber).

[2191] JOHNSON et al., *Oil Industry Competition, supra* note 1055, at 2443.

[2192] *Id.* at 2374.

[2193] *Petroleum Industry Hearings, supra* note 192, at 336 (Statement of W. T. Slick, Jr.).

[2194] *Consumer Energy Act Hearings, supra* note 154, at 596 (Statement of Jack Vickrey).

TABLE VIII

TOP CRUDE OIL PRODUCERS IN TEXAS, 1977

Rank	Company	Million Barrels	Cumulative Share of Total
1	Exxon	137.4	12.48
2	Amoco	130.5	24.33
3	Shell	71.9	30.86
4	Chevron	60.2	36.32
5	Texaco	58.8	41.66
6	Gulf	56.5	46.79
7	Arco	43.3	50.73
8	Marathon	40.8	54.43
9	Mobil	39.6	58.03
10	Sun	34.8	61.19
11	Amerad Hess	33.8	64.26
12	Union	23.8	66.42
13	Quintana	23.1	68.52
14	Getty	19.5	70.29
15	Continental	18.0	71.92
16	Cities Service	17.9	73.55
17	General Crude	16.6	75.05
18	Phillips	12.9	76.23
19	Texas Pacific	12.7	77.38
20	Hunt Oil	7.5	78.06
21-200	Next 180	149.1	91.60
200+	All Others	92.5	100.00
	Total	1101.2	

Source: Petroleum Information Corporation, *Company Production, Crude and Casinghead Gas by Company, State of Texas,* 1977.

Chapter VII

Proposed Solutions to an Assumed Problem

A. Divestiture of Oil Pipelines from Oil Company Ownership

The political view, expressed by the staff of the Senate Antitrust and Monopoly Subcommittee, which has invested innumerable hours in hearings and investigations of most of the recent bills on the subject, probably epitomized the view of the industry's Congressional critics and, for that reason, the following lengthy quotation is taken from the Kennedy Staff Report's Conclusions:

"Petroleum pipelines are natural monopolies. As natural monopolies they have been treated in the law as regulated common carriers. But unlike other common carriers they have escaped the traditional prohibition on shipper ownership. The exemption of oil pipelines from this prohibition is without justification. In pipeline transport, as elsewhere, there exists a basic conflict between the interest of a firm as owner of a common carrier facility and the interest of a firm as shipper. The common carrier obligation to treat all shippers equally is simply inconsistent with a firm's practical motivation to advance its own commercial interests. It is unreasonable to expect a firm to resist the myriad of incentives and opportunities to provide itself with advantages over its competitors who also must use its facilities.

"One of the most important disadvantages the integrated pipeline owner can inflict on its rivals is the requirement that they pay excessive tariffs. The integrated owners must, of course, also pay the high tariffs. But the high tariffs are only transfer payments for the integrated owners who, in an economic sense, still benefit from the low transportation costs. For the nonowner shipper, however, the high tariffs represent a real cost disadvantage. In effect, the nonowner must subsidize the transportation of its competitors.

"The integrated owners of pipelines have in fact exploited their advantages. Integrated company ownership of petroleum pipelines has had a substantial impact on the ability of nonintegrated refiners and marketers to compete with the pipeline owners in the marketplace for petroleum products. Control of crude oil pipelines has enabled these vertically integrated oil companies to gain control of crude oil production greatly exceeding their own refinery needs and has worked to prevent the formation of a domestic crude oil market. Their operation of petroleum pipelines allows them to control the distribution and flow — and consequently influence the price — of refined petroleum products. The lower real costs of pipeline transportation have not been translated into lower consumer prices, but merely into higher oil company profits. Oil company ownership of petroleum pipelines has demonstrably failed to make petroleum pipelines a practical transportation alternative for many smaller refiners and marketers trying to compete with the pipeline owners.

"The current system of common carrier regulation has totally failed to deal with these problems. Moreover the regulatory agencies lack the legal authority to prevent integrated oil companies from restricting pipeline throughput capacity in order to maximize downstream profits. It is, perhaps, conceivable that a regulatory scheme could be devised which in theory could control the behavior of shipper owners. But it would necessarily involve massive regulatory intervention in every aspect of pipeline operations. It is improbable that such a scheme could be effective in practice. This is particularly true with regard to the construction and expansion of throughput capacity. Neither the history of pipeline regulation nor the history of regulation in other industries suggests that such a massive and complex system could successfully serve the public interest.

"Subjecting petroleum pipelines to the commodities clause of the Interstate Commerce Act cannot effectively resolve the problems created by oil company ownership of pipelines. Such an approach would be hard to police, given the prevalence of exchange within the industry, and in fact, would encourage reciprocal dealing among pipeline owners. More importantly, a commodities clause would still leave many refiners and marketers dependent upon competitors for a service essential to their ability to compete.

"There thus remains only one practical and effective solution to the grave competitive problems inherent in oil company ownership of petroleum pipelines: to prohibit oil companies from owning petroleum pipelines and to require divorcement of existing pipelines from oil company ownership."[2195]

Assistant Attorney John Shenefield, in his Proposed Statement which was to be submitted to the Kennedy Subcommittee on June 28, 1978, which hearings in fact were not held, was much more cautious. For one thing, he drew a sharp distinction between retrospective divestiture and a prospective ban against the ownership of oil pipelines to be constructed in the future. In his words: "Because of its much clearer potential

---

[2195] KENNEDY STAFF REPORT, *supra* note 184, at 151-152.

for disruptive effects, we have greater concern about across-the-board restructuring of the entire existing interstate pipeline industry by retrospective divestiture. Retrospective divestiture, therefore, requires careful analysis of its potential impact on the industry as it exists today."[2196] Without any discussion of the constitutionality of legislation discriminating between classes of pipelines, he declared that a class of pipelines "likely to represent significant concentrations of market power may properly serve as the exclusive target of such structural relief," thereby minimizing industry disruption, "although there would be significant legislative and administrative costs with respect to both identifying the class of pipelines to be divested and supervising the divestiture." Mr. Shenefield summed up his remarks on the subject by saying if such a selective approach were followed, he believed that a strong case for divestiture could be made, adding, however, "In reaching this conclusion, I certainly agree with divestiture *opponents* that, before such restructuring is imposed, it is up to divestiture *proponents* to make a convincing case that there is a problem and that there is no effective, less drastic solution."[2197] Since the time of Mr. Shenefield's proposed statement, the Antitrust Division appears, to this author at least, to have concentrated its efforts in searching for alternative, less drastic solutions, and in making a more careful examination of some of the premises that it apparently previously had assumed were true without much depth of examination. Examples are: Thomas Spavin's work on rates of return; Dr. Reynold's tentative approach to a "joint venture rules" adaptation of the Deepwater Port Report's "Competitive Rules"; active participation in the FERC *Valuation of Common Carrier Pipelines* proceedings and in the FTC *Rulemaking* proceeding being held as a result of Senator Kennedy's Petition; and its attempts to grapple with the "free rider" and "risk premium" problems arising from the open ownership and expansion aspects of the "Competitive Rules," as evidenced by its submission in the FERC proceeding on the Texas Deepwater Port Authority's petition for a "two tier" tariff. The Division's participation in the American Enterprise Institute's Conference on Oil Pipelines and Public Policy held in Washington, D.C., on March 1st and 2nd, 1979, evidenced a sincere desire to communicate with the industry and to discover wherein its theories raised difficulties which it had no way of visualizing in the absence of such conversations.

[2196] SHENEFIELD PROPOSED STATEMENT, *supra* note 179, at 42.
[2197] *Id.* at 42-43.

1. Propositions Common to Both Prospective and Retrospective Divestiture

a. Heavy Burden of Proof on Divestiture Proponents

There seems to be clear agreement between the critics,[2198] defenders[2199] and observers[2200] of the industry that the proponents of pipeline divestiture must bear a heavy burden of persuasion. This burden can *only* be discharged by a clear and convincing showing[2201] that (1) a serious problem exists; (2) divestiture will solve the problem; (3) that the anticipated benefits will outweigh expected costs; and (4) there is no effective, less drastic solution.[2202]

(1) To Show that a Serious Problem Exists

The reader will recall Assistant Attorney Shenefield's words in his proposed statement that it was up to divestiture *proponents* to make a convincing case that there is a problem.[2203] Senator Clifford P. Hansen elaborated somewhat on this theme when he stated in *The Petroleum Industry Hearings* on September 23, 1975, that he "wanted to hear the evidence;" he didn't want "vague allegations" or a "long list of complaints from those who are really suing for less competition under the guise of promoting competition." He wanted to have witnesses explain *how* the public would be better served and he wanted to make sure that witnesses in support of the legislation [S.2387; S.739; S.745; S.756; S.1137 and S.1138] didn't have the wrong idea of what kind of an oil industry would be forthcoming under the bills. He felt that it was the duty of the [Senate Antitrust and

---

[2198] *Petroleum Industry Hearings, supra* note 192, at 165 (Senator Birch Bayh, introducer of S.2387); SHENEFIELD PROPOSED STATEMENT, *supra* 179, at 43.

[2199] *Petroleum Industry Hearings, supra* note 192, at 41 (Opening Statement of Senator Roman Hruska) ("the heavy burden, a very responsible burden"); *Id.* at 72 (Statement of Senator Clifford P. Hansen ("a great burden"); *Id.* at 2322 (Letter to Senators Eastland and Hart from Charles E. Spahr enclosing a copy of a guest lecture delivered by Spahr to Mount Union College students.).

[2200] *Id.* at 1957-1958 (Reply by Peter A. Bator, Davis, Polk & Wardwell, to question by Senator Hruska) ("a very heavy burden of proof"); *Id.* at 2040 (statement by Otis H. Ellis, Private Citizen, Consumer and Farmer) ("beyond a reasonable doubt").

[2201] SHENEFIELD PROPOSED STATEMENT, *supra* note 179, at 43 ("convincing case") *Petroleum Industry Hearings, supra* note 192, at 71 (Statement of Senator Clifford P. Hansen) ("I want to hear the evidence. I don't want vague allegations."); *Id.* at 199 (Statement of Senator Dewey F. Bartlett ("Obligation to learn the answers"); *Id.* at 1109 (Statement of Charles E. Spahr) ("substantive evidence", not "vague, unsupported assertions"); ERC REPORT, *supra* note 184, at v ("confirmation of any such abuse [prior to change]"); see note 1266, *supra*.

[2202] See text at notes 1264-1268, *supra*; NET-III, *supra* note 199, at 156; API's Comments on Kennedy FTC Petition, *supra* note 1879 at 17.

[2203] See text at note 2198. *supra*.

Monopoly] Subcommittee to probe the proponents of such legislation to make sure that they understood that the changes the bills would bring would indeed remedy their complaints and not, in fact, have quite different effects.[2204] At the time of the precursor *Consumer Energy Act Hearings* in 1973-1974 before the Senate Commerce Committee, there was a flat statement by an industry witness that "The broad claims of discriminatory treatment of nonowner shippers made by some pipeline critics has never been substantiated by reliable and factual evidence." This witness continued, "The actual facts, as contrasted with conjecture or 'mere suspicion' show that the non-owner shippers are not denied access to petroleum pipelines but, on the contrary, account for a substantial portion of the shipments over many of these lines."[2205] In support of his statement, this witness submitted tables showing, by years 1968-1972, nonowner shipments over several lines, both crude and products; data on Explorer, which was then getting underway; a letter from ICC Chairman Stafford to Congressman Joe Evins, a letter from ICC Chairman Stafford to Congressman Joe Evins, Chairman of the House Small Business Committee, attesting to the lack of complaints about discriminatory rates and practices of pipelines; price trends of oil pipelines compared with rail, truck and air transportation, showing a markedly better performance by oil pipelines in lowering revenues per ton-mile than the other carriers during the period 1947 to 1971; market share data for refiners between 1960 and 1970, showing steady increases in the shares held by independents, and a 1968-1972 yearly time series of percentage shares of the gasoline market, also showing a substantial and continuous increase by the independents' share of that market.[2206] At the same hearings, Keith Clearwaters, Deputy Assistant Attorney General, appeared and, in a very equivocal statement, said that "We believe that there *may be* sound reasons for enacting legislation which would require that oil pipelines be independently owned, free from control by persons engaged in any other phase of the petroleum business...." (emphasis added). His stated reasons were that the shipper-owners of a jointly owned pipeline could determine the route, size of the line and location of input and delivery points; that such lines "almost invariably" did not operate input or output terminals but simply provided connections to shippers' terminals. One page later in his statement, he said "Since we are not justified in instituting antitrust cases upon the mere suspicion of possible anticompetitive purposes or effects, we have attempted to examine a number of joint-venture pipelines for specific evidence that the

---

[2204] *Petroleum Industry Hearings, supra* note 192, at 71 (Statement of Senator Clifford P. Hansen).

[2205] *Consumer Energy Act Hearings, supra* note 154, at 617 (Statement of Jack Vickrey).

[2206] *Id.* at 634-635 (nonowner shipments); 641-643 (Stafford letter); *Id.* at 643 (revenue levels of rail, truck, air, and oil pipelines; *Id.* at 652 (market share data).

general concerns I enumerated are in fact applicable in the particular case." [Note, at this point, the Antitrust Division had been investigating Colonial for ten years, Olympic for eight, Explorer for six, and TAPS for four years, without finding anything that would justify a complaint, civil or criminal]. In response to Senator Stevenson's question whether he was supporting pipeline divestiture legislation, Clearwaters replied: "We certainly lean toward that way, Mr. Chairman. However, I think that at this point at least I am not in a position, based upon what we know about the efficiencies and the economy of scale of these pipelines, to categorically state that oil companies should be denied entry in all cases from the business of pipeline transportation."[2207] ICC Chairman George M. Stafford also appeared at the same hearings and after reviewing the pros and cons of divestiture stated that "...it would appear that such far-reaching changes [divestiture] are premature in light of the limited facts known to date."[2208]

In introducing S.2387, Senator Birch Bayh stated the theory which had inspired his proposal, *i.e.,* "The lack of competition in the oil industry is the result of the unique convergence of two factors: intense concentration and vertical integration." Senator Phillip Hart, a cosponsor of the bill, offered a somewhat different version. His reasons were the degree of cooperation between the participants in joint arrangements, including jointly owned pipelines; and that ownership and control of gathering lines, products lines, and crude trunk lines by "majors" effectively foreclosed transportation to new entrants and made "smaller companies stay in line."[2209]

A broad-scale reaction to the underlying theories expounded by Senators Bayh and Hart was expressed by witness Otis Ellis, who styled himself as a "Private Citizen, Consumer and Farmer," but who had served as Staff Director of the House Small Business Committee in 1948 and 1949, and in private law practice had represented the National Oil Jobbers Council, a federation of independent jobbers, subsequently represented hundreds of independent jobbers, service station dealers and held interests in 13 oil jobberships, and prior to retirement, had served as a consultant to the Ministry of Mines and Hydrocarbons of Venezuela. Mr. Ellis testified before the Senate Antitrust and Monopoly Subcommittee on January 28, 1976, as a concerned citizen. Because of his

---

[2207] *Id.* at 1023-1024 (believes sound reason for legislation); *Id.* at 1026 (routing and terminals); *Id.* at 1027 (can't act upon mere suspicion); *Id.* at 1029 (leaning toward, but not stating).

[2208] *Id.* at 1268 (Statement of ICC Chairman George M. Stafford).

[2209] *Petroleum Industry Hearings, supra* note 192, at 2217 (Statement of Professor Neil H. Jacoby).

legal training, he had examined S.2387's "Findings" which he characterized as follows: "The language under findings ranges all the way from that which might be contained in a mother's prayer to that which might be used in describing an economist's nightmare."[2210] Mr. Ellis' comments on the merits were that, on the basis of his 25 years experience with the petroleum industry, he could not agree that the "findings" were correct. In fact, his experience had been exactly the opposite.[2211]

Examining more analytically Senator Bayh's basic premises, his characterization of "intense concentration" was shown in Section IV A 1, "Presumed Concentration in Hands of 'Major' Oil Companies,"[2212] to be completely mistaken;[2213] likewise, vertical integration in the oil industry was examined in depth in Section IV B. "Vertical Integration,"[2214] wherein it was found that (1) the U.S. petroleum industry had relatively little integration compared with U.S. manufacturing in general;[2215] (2) the advantages and disadvantages of vertical integration in the oil industry on balance favored integration as being a more efficient mode of effecting transactional economies than was the open marketplace in most instances;[2216] and (3) that in those instances where it was not as efficient, oil companies tended volunatrily to refrain from integrating, or if integrated, to disintegrate on their own initiative, basing their decisions on sound economic principles rather than acting pursuant to antitrust motivations.[2217] Senator Hart's premises likewise were demonstrated to be groundless. First, "joint ventures," as he called them, were scrutinized minutely in Section IV C, "Jointly Owned and Jointly Operated Pipelines,"[2218] wherein such lines were found to meet fully Dr. Walter Mead's four-fold test for legality, *i.e.*, these "industry lines" were brought about by the capital investment required to realize the economies of scale; they spread risks that were too great for any single, and very few small combinations of existing firms, to be able practically, or prudently willing, to assume the risk; separate operations by competing lines would be economically wasteful and economies were produced by them which tended to be *external* to the owner-shippers and to accrue to *all* firms in the relevant area. Nor

---

[2210] *Id.* at 2031 (Statement of Otis H. Ellis, Esq., Private Citizen, Consumer and Farmer).
[2211] *Id.*
[2212] See text at notes 1051-1082, *supra*.
[2213] *See* Bock, *The Lizzie Borden Solution,* XIV CONFERENCE BOARD MAGAZINE 50 (No. 3, March 1977) (Citing testimony of Thomas E. Kauper, Assistant Attorney General, Antitrust Division, before the Senate Judiciary Committee in June 1976) [hereinafter cited as Bock, *Lizzie Borden*].
[2214] See text at notes 1125-1128, *supra*.
[2215] See text at note 1155, *supra*.
[2216] See text at notes 1176-1228, *supra*.
[2217] See text at notes 1229-1360, *supra*.
[2218] See text at 1269-1333, *supra*.

was there any empirical evidence supporting the existence of any of Dr. Mead's three possible anti-competitive effects. The evidence disclosed no slackening of competition among the owners, nor any refraining from competing in the same market as the joint enterprise. Rather than "market foreclosure" by the large lines, the evidence supported the opposite result, *i.e.*, (a) independent marketers along the Colonial line grew and prospered at the expense of Colonial's owners,[2219] and (b) the economic incentives behind the larger industry lines enabled smaller companies to participate in operations from which they would have been excluded had it been necessary for them to accumulate the resources needed to finance the entire investment required.[2220] Senator Hart's second premise fared no better than his first when exposed to real world facts. Section II A 3 c, "Scramble for Traffic,"[2221] documented thoroughly the rise in nonowner shippers over oil pipelines culminating in a 1976 survey which showed that in such year, 84.24 percent of the shippers were nonowners and 64 percent of the total shippers were "independents." Moreover, Section VI C 3, "Asserted Denial of Acess to Pipelines,"[2222] reviewed painstakingly the absence of complaints by industry members; the empirical evidence on each of several alleged "indirect" methods of denial; the use of the lines by nonowners; lines operating under capacity; prorationing and expansion practices of the two lines which had more than sporadic prorationing; the reasons why certain parties declined to participate in ownership of the lines; and statements by persons who were not owners concerning access to pipelines. The sweep of this evidence overwhelmingly was contradictory to Senator Hart's premises. Three items of evidence by the Department of Justice confirm this finding: Deputy Assistant Attorney General Bruce B. Wilson's testimony before the House Small Business Committee on June 15, 1972, that the Department could not refute the industry's testimony that access to pipelines was open and available to all without regard to ownership interest;[2223] Assistant Attorney General Thomas B. Kauper's testimony of June 1976 before the Senate Judiciary Committee that the industry was neither monopolistic nor highly oligopolistic and that its conduct, including cooperative pipeline ownership, was not anticompetitive,[2224] and the Department's conclusion that the "lightning-rod" pipeline, Colonial, which has been the critic's archetype of all that possibly could be evil in a

---

[2219] See text at note 1327, *supra*.

[2220] Bock, *Lizzie Borden, supra* note 2213, at 50, citing Assistant Attorney General Thomas E. Kauper's testimony of June, 1976 before the Senate Judiciary Committee.

[2221] See text at notes 407-496, *supra*.

[2222] See text at notes 1807-1991, *supra*.

[2223] See text at note 1808, *supra*.

[2224] Bock, *Lizzie Borden, supra* note 2213, at 50.

Proposed Solutions to an Assumed Problem 447

pipeline, was, after a 13-year investigation, found not to have committed any meaningful denial of access.[2225] In short, the finding of Section VI was that "major" oil company-owned pipelines are readily accessible to nonowners including other "majors," "independent" crude gatherers, "independent" producers, "independent" refiners and "independent" marketers alike.[2226]

Despite the comments of writers,[2227] government officials,[2228] and witnesses in the *S.1167 Hearings*[2229] and in *The Petroleum Industry Hearings,*[2230] to the effect that no hard data or facts had been adduced to support a proposal for pipeline divestiture legislation, the staff of the Senate Antitrust and Monopoly Subcommittee came out with a Staff Report in June 1978 which advocated such legislation. In order to save the reader the trouble of turning back to the beginning of this section, the Staff Report's points will be discussed seriatim. It starts out with the sweeping statement that pipeline are "natural monopolies." In Section III D, "Similarities and Dissimilarities [of pipelines] to a Natural Monopoly,"[2231] this assertion was examined and its evolution was traced from the Department of Justice's Deepwater Port Report with its *assumed* equation of pipelines to Deepwater Ports, the stretching of the classical economic definition of a "natural monopoly" in an attempt to enshroud pipelines therein; the failure to recognize the effect of rapidly growing markets; and, most importantly, the failure to mention the existence of

---

[2225] See text at note 1813, *supra*.

[2226] See text at notes 1824-1828, *supra*.

[2227] CRUDE OIL PIPELINES, *supra* note 158, at 125 (Leslie Cookenboo: "This demonstration [sufficient to warrant pipeline divorcement] has not yet adequately been made by the proponents of divorcement"); *cf.* SPECTRE, *supra* note 104, at 110 (G.S. Wolbert: "In the absence of *proof* of these assumed facts, the basic argument comes down to the contention that separation of pipelines from major oil companies would [proponents of divorcement say] promote competition, thus there should be a crime *implied* in order to fit or justify the punishment").

[2228] *Cf.* Frank Zarb, Administrator of the Federal Energy Administration (FEA): "Those who want divestiture just haven't generated enough hard data to tell people whether they are better off with that kind of legislation. We need to get a full understanding of the issue first." The Oil Daily, November 11, 1975, p. 2.

[2229] *S.1167 Hearings, supra* note 432, pt. 9 at 610 (Statement of Vernon T. Jones) ("Absolutely no hard facts have been made available to support a change from the present and historic regulation and structure of the oil pipeline industry.").

[2230] *Petroleum Industry Hearings, supra* note 192, at 319 (Statement of J. Donald Durand) ("We have never seen any hard evidence that pipelines are not operating as common carriers, as they are required to under the law."); *Id.* at 2046 (Statement of Otis H. Ellis) ("We have had an FTC that, for most of the years I have known them, have had a gleam in their eye to tear up anything big. They have not torn [the oil industry] up. Was it because the laws were inadequate? Not in my judgment. It was because there were inadequate facts to support prosecution of these laws effectively against this industry. It is inadequacy of existing facts warranting a breakup of the so-called oil monopoly, not inadequate laws").

[2231] See text at notes 656-701, *supra*.

intra- and inter-modal transportation and multiple source competition. These factors combined to lead to a conclusion that the "natural monopoly" doctrine probably was inapplicable entirely to oil pipelines and absolutely limited to at most one or two specific instances.

Point two — there is a basic conflict between the interest of a firm as owner of a pipeline and a common carrier furnishing services to a competitor. One has to agree that there is a degree of tension in these two roles; which is why industry members concede that human nature being what it is, some form of regulation is necessary to monitor those who cannot successfully "wear the two hats." However, the obvious, and well documented, advantages of shipper-ownership well outweigh the *potential* for wrongdoing, as the record of the past three decades has demonstrated. The conflict makes a point for regulation, but not for divestiture. This point will be addressed in more depth in the cost/benefit subsection below.

The third point was two-fold: excessive rates, and subsidization by the non-owner of its shipper-owner competitor. The question of excessive rates was scrutinized minutely in Section VI C 1, "So Called 'Excessive Rates' of Return,"[2232] in which the strawman of "Return on Equity" was shown to be what it really is — an item of political campaign rhetoric;[2233] the constructive work of the Justice Department and industry, in dialogue, hammering out an acceptable method was discussed, and a conclusion reached that additional research was required to determine an appropriate "risk premium" that would permit meaningful comparisons to be made across industry lines as a forerunner to deciding the appropriate level of returns to be allowed by regulatory agencies to pipelines. At the very least, the conclusion has to be that the jury must remain out on this question, and the existence of the outstanding question of fact precludes the Kennedy Staff's "summary judgement" of excessive rates. Moreover, the discussion in Section V A 1, "The Required Rate of Return: The Modern Day Cost of Capital,"[2234] sheds additional light on the subject of the level of rates that the knowledgeable investor expects in order to invest in pipelines, which is a key item in *Hope Natural Gas'* requirement that the allowed rate must be such as to assure confidence in the financial integrity of the pipeline venture, maintain its credit standing and enable it to attract new capital. The second part of the staff's third conclusion, the exhumed charge of a shipper-owner being subsidized by the tariffs paid by a nonowner shipper, was examined in Section IV C,

---

[2232] See text at notes 1675-1720, *supra*.

[2233] See text at notes 1768-1776 and 1392-1394, *supra*.

[2234] See text at notes 1346-1394, *supra*.

"Jointly Owned and Jointly Operated Pipelines;"[2235] and found to be without merit.

The fourth point was an accusation that integrated company ownership of petroleum pipelines has had a substantial impact on the ability of non-integrated refiners and marketers to compete with the pipeline owners in the market place for petroleum products. Chapter VI, subsections C 3 C (3), (g), (ii) and (iii), "The Refining Sector"[2236] and "Gasoline Marketing"[2237] respectively, documented the entry into, and steady growth of refining capacity of independent refiners from 1955 to 1977, during which time not only had the share of the top 4 firms declined from 33.1 percent to 31.0 percent, the top 8 from 57.7 percent to 52.7 percent and the top 20 from 84.8 percent to 78.0 percent; but many firms which were "independents," or not even refiners in 1955 were in the top 20 by 1977, *i.e.,:* Coastal States, Champlin, Getty and Tosco. Donald O'Hara, President of the NPRA, made the point dramatically by taking the top 20 refiners as of January 1, 1951, which had 80.68 percent of "U.S. refining capacity" [the foregoing term included both U.S. and Puerto Rico], and looking at the same list of refiners as of January 1, 1975, he found their shares to have declined to 71.47 percent of U.S. capacity. Professor Richard Mancke found 47 new entrants into U.S. refining between 1950 and 1972 and 13 new entrants between 1972 and 1975. A quick look at Appendix R will give the reader an idea of how lacking in factual validity is the Kennedy Staff's contention with respect to independent refiners. In the marketing sector, the cited subsection shows that concentration ratios for motor gasoline sales dropped from 1954 figures of Top 4 — 31.2 percent, Top 8 — 54.0 percent, Top 20 — 80.4 percent to 1977 figures of Top 4 — 29.1 percent, Top 8 — 49.2 percent and Top 20 — 72.4 percent, accompanied by the same phenomenon of independents and non-existing firms in 1954 being found in the top 20 in 1977, *i.e.,* Getty, Clark, Tosco, Kerr-McGee (with American Petrofina and Tenneco so close to Kerr-McGee that they had the same rounded off percentage share). There are some 15,000 wholesale dealers and 300,000 retailers in the market place, most of which are independent businessmen. In the light of these documented statistics, the staff's allegation of pipeline exploitation of independent refiners and marketers appears to be clearly contrary to available facts.

The Staff Report's fifth "conclusion" was that the major oil control of crude oil pipelines has enabled them to gain control of crude oil production and to prevent the formation of a free domestic crude oil

---

[2235] See text at notes 1293-1296, *supra.*

[2236] See text at notes 2126-2144, *supra,* API Market Shares and Individual Company Data for U.S. Energy Markets: 1950-1977 (Discussion Paper #014, November 1978) pp. 93 (1977) and 124 (1955).

[2237] See text at notes 2167-2194, *surpa.*

market. Section I B 6 "[Pipeline System] Flexibility and Alternatives to Shippers,"[2238] describes the various alternative pipeline transportation routes available to crude oil shippers, noting that, with very minor exceptions due to locational peculiarities, most regions are served by one or more pipelines and that pipelines extend from every producing region to a number of refining areas and conversely, every refinery area is connected to a number of producing regions. Section VI C 4 c (3) (G) (i), "The Domestic Crude Oil Market,"[2239] contains a detailed description of the domestic crude oil market and how it works. A look was taken at the State of Texas [which has detailed reports available on magnetic tape and hence readily available] and Table VIII shows the Top Twenty crude producers to include Quintana, Getty, General Crude, Texas Pacific and Hunt Oil. There is an extremely large number of companies producing significant volumes of crude oil. The structure of the crude oil producing sector in Texas comes about as close to meeting the criteria for a competitive market on the supply side as one is likely to encounter in an industrial field. On the purchasing side, Figure 6, which is a "Recapitulation Statement" issued by the Texas Railroad Commission each month [Figure 6 was a randomly selected month May, 1976], shows that the opportunity to purchase crude oil at the lease in Texas is not limited to refiners or to large companies in any category. Mention was made of an Exxon study [which is in the public domain] of the "aftermarket," *i.e.*, sales downstream from the lease, which shows that the seven largest firms other than Exxon actually bought a disproportionately *smaller* [than their refinery capacity] share of Exxon's crude oil while the smaller companies bought a disproportionately *larger* share of Exxon's sales. The question raised by the Staff Report encompassed also the issue whether "major" oil company control of crude oil and gathering trunk lines was used to erect "road blocks" between independent refiners and producers. Data on crude oil gathering receipts in Texas [again selected because of the data availability] for the first 6 months in 1978 are shown in Appendix D. Once again, some new faces showed up in the list of larger gatherers: Permian, Scurlock and Koch. Texas Railroad Commission data were also used to compile data on refinery receipts. See Appendix E which shows the source of crude for every refinery in Texas during the first 6 months of 1978. This Appendix makes it apparent that smaller refiners may have differentially *favorable* access to domestic crude and also that the "majors" are not reluctant to supply them with foreign oil. The Kennedy Staff Report apparently did not pay much heed to the FTC's Bureau of Competition and Economics September, 1975, Report on the Effects of Decontrol on

---

[2238] See text at notes 212-224, *supra*.
[2239] See text at notes 2110-2125, *supra*.

Competition in the Petroleum Industry wherein it was concluded that: (1) the supply and price of imported oil was in the hands of OPEC rather than the domestic oil companies; (2) the pricing of such crude by OPEC would be a major if not dominant factor affecting the domestic oil industry; (3) if prices were decontrolled, there would not be a "squeeze" by integrated oil companies upon refining profit margins in favor of higher crude prices; and (4) independent refiners would have no difficulty in satisfying all their requirements from a combination of domestic companies and foreign sources.[2240] Once again, the preconceived theory does not stand up well when confronted with the facts.

Conclusion Number 6 was that the lower real costs of pipeline transportation have not been translated into lower consumer prices, but merely into higher profits. Although the Report cited Professor Arthur Johnson's book *Petroleum Pipelines and Public Policy: 1906-1959* several times, the staff must have overlooked his statement made after reviewing the growth of pipelines and the accelerating changes that took place between 1931-1941: "The savings that accrued to oil companies by virtue of owning gasoline lines appear to have been passed on to consumers of the product."[2241] In the old Temporary National Economic Committee (TNEC) days of 1939-1940, one of the industry's most severe critics, John E. Shatford, who had been the sole witness in the early *Reduced Pipeline Rates and Gathering Charges* proceeding, testified that it was "indisputable" that both crude oil and gasoline pipelines had resulted in lower costs to consumers.[2242] In Section I B 2, "Anti-Inflationary Rate Trends,"[2243] data from 1945 to 1964 showed that the average rate for all pipelines remained virtually constant and in 1971 it was actually 5 percent *lower* than in 1947. Paul J. Tierney, President of the Transportation Association of America, submitted data for the *TAPS* record showing that oil pipeline rates, as reflected by the generally accepted Average Revenue Per Ton Mile yardstick used in the transportation industry, had remained well below general price levels throughout the 1947-1976 period. During this period, consumer prices rose 155 percent, producers' (wholesale) prices rose 138 percent while oil pipeline levels increased only 40 percent — and all of that increase occurred from 1973 onward.[2244] In Section VI C 2 c, "Where Did the 'Monopoly Profits' Go — Why Isn't

---

[2240] *Petroleum Industry Hearings, supra* note 192, at 1032 (Letter to Chairman Philip Hart from Stark Ritchie, attaching a copy of the DECONTROL REPORT).

[2241] PETROLEUM PIPELINES, *supra* note 45, at 258, 259 (Chart I); see text at note 146, *supra*.

[2242] *TNEC Hearings, supra* note 6, vol. 6 at 429. For a modern comment to the same effect, see *Petroleum Industry Hearings supra* note 192, at 2224 (Statement of Professor Neil H. Jacoby).

[2243] See text at notes 188-197, *supra*.

[2244] See table in note 194, *supra*.

Everyone Clamoring to Get In?,"[2245] there was an examination of this point. Due note was taken of Dr. George Hay's statement in the *Valuation of Common Carrier Pipelines* proceedings, then ICC *Ex Parte No. 308,* which showed on its face that only 26 of the oil pipelines listed exceeded the rate of return to capital earned by the highest natural gas line listed, and 9 of the 26 (including the 1st, 2nd, 4th and 5th highest) were "independents." It also showed that 7 of the lines had *negative* rates of return and 15 made less than 5%, well below the 12.1 percent earned by the *lowest* of the five natural gas lines shown. There was no attempt to make a "risk premium" adjustment, although the dispersion of the natural gas line rates was only 1.6 percent, whereas the range of oil pipeline rates was from a positive 305.6 percent (by Santa Fe, an "independent") to a negative 10.2 percent (by Getty, a "major") which certainly would appear to indicate a difference in risk. Professor Stewart Myers, who testified before the Senate Interior Committee Special Subcommittee on Integrated Oil Operations, derived a weighted average rate of return using 1971 ICC "Form P" reports, adjusted for "normalized" accounting, of 7.7 percent, which he compared with AT&T's 7.8 percent. LeGrange and Spavins at the AEI Conference, using the same method except that Spavins did not take into account the tax effect on the interest which he added back to after-tax net income, calculated 7.6 percent and 13.0 percent, respectively, mean average weighted returns. As witness E. P. Hardin testified in the *Petroleum Industry Hearings,* a 7 percent return on valuation (even after deducting interest) "is not a red-hot business deal." Despite Mr. Shenefield's imaginative attempt to explain why there wasn't a stampede of investors into this "fantastically lucrative" business by saying that potential shippers would not want to pay the excessive tariffs that an independent would charge, the basic fact is just as Hardin put it: no big deal. This was corroborated by Robert Yancey's (President of Ashland) testimony that he would rather ship over Colonial's line and put his company's money where it would earn a good return; by the "opting out" of ownership in lines by so many substantial entities, such as Standard of California and the State of Alaska in the TAPS deal (and Mobil's cut-back to a 5% interest at its first opportunity) and by all the "drop-outs" shown in connection with the discussion on jointly owned and jointly operated lines.[2246]

Count Seven is the "undersizing" argument borrowed from the Justice Department, Having just devoted about 25 percent of the preceding Chapter and over 200 footnotes to this subject under the

---

[2245] See text following note 1799 to note 1806, *supra.*
[2246] See text at notes 1331-1332: Yellowstone (3 dropouts out of 7); Capline (15 dropouts out of 23); Eugene Island (49 dropouts out of 56).

heading Section VI C 4, "The Department of Justice's Circumvention of Rate Regulation or 'Undersizing' Theory,"[2247] it is hard to say anything more than (1) there are serious internal inconsistencies and flaws in the economic theory; (2) its legal premises are subject to questions; (3) no credible evidence has been produced in support of the theory by its proponents; (4) the key assumptions underlying the theory are mistaken; and (5) there is substantial empirical evidence of industry conditions which make the theory probably invalid and certainly inapplicable to virtually all of the industry's pipelines. One knowledgeable commentator has analogized it to the economics of the midway — a shell game theory of monopoly,[2248] and its principal proponent publicly prophesied that where there are alternate routes that are no more expensive than the subject pipeline, "...you might conclude that [the undersizing theory] doesn't make any sense."[2249] There can be absolutely no question but that clear and convincing evidence is lacking on this count.

Senator Kennedy's petition to the FTC for a rulemaking proceeding has as its "factual basis" the Kennedy Staff Report,[2250] and its pivotal rationale is the Department of Justice's Undersizing Theory.[2251] With both feet thus planted firmly in mid-air, there is no purpose to be served by belaboring the issue of proof any longer.

To sum up, every allegation in the documents which have urged, or been used to urge, pipeline divestiture not only have failed to come even close to meeting the consensus test of "clear and convincing" proof, but appear to lack even respectable evidence.

(2) To Show That Divestiture Will Solve the Problem

In the light of the findings that no showing has been made that a problem in fact does exist, it is somewhat difficult to treat this second prerequisite. Moreover, the task is rendered even more complex by the fact that the proponents were exceedingly vague about *what* pipeline

---

[2247] See text at notes 1992-2194, *supra*.

[2248] ERICKSON et al., *The Pipeline Undersizing Argument, supra* note 2053, at 17. As Professor Erickson has noted: "In addition to various problems of economic logic, this conception must ultimately bear the burden of identifying where the monopoly profits are hidden and specifying their effect upon overall petroleum company rates of return. Petroleum company rates of return are, however, generally regarded to be within the range consistent with effective competition." *Id.* at 17 n.5.

[2249] 834 ANTITRUST & TRADE REG. REP. (BNA) A-7 (October 13, 1977).

[2250] Kennedy FTC Rulemaking Petition, *supra* note 1641, at 3 (Harm to the Public Interest), 4-7 (Factual Background — specifically referenced to the Kennedy Staff Report).

[2251] *Id.* at 7-10 (The Case for Prohibiting Shipper Ownership).

divestiture was going to accomplish. The Kennedy Staff Report stated: "It is equally clear that corrective measures are required if non-owners are to take advantage of the sizeable economies provided by pipeline transportation and if these economies are to be passed along to the consumer in the form of lower prices. Regulation has failed. Antitrust efforts have failed. Congressional oversight has failed. It is essential that a solution be found that will solve this longstanding problem. Some form of divestiture of pipelines from the remaining segments of the industry is necessary. Anything short of this solution will be doomed to failure, since the integrated oil companies will adapt it to their goals and use it as a shield to ward off more drastic attempts to make their pipelines truly available to all potential users."[2252] No mention is made as to *how* divestiture will solve "the problem" (which itself really is not clearly spelled out).

Assistant Attorney General John Shenfield was equally vague. In fact, he devoted most of his discussion to an attempt to "defuse" the social costs of divestiture and a discussion of alternative solutions. He also used the technique that other solutions, such as the "Competitive Rules," merely "treated the symptoms" whereas divestiture ". . . attacks the problem directly because the problem is clearly rooted in the structure. What might appear to be a drastic approach, therefore, is in reality a clean and decisive break with the burdensome requirement of continuous, pervasive regulatory scrutiny."[2253] Again, a nebulous reference was made to "the problem" and there was no discussion of *how* divestiture would solve such assumed problem. This is a combination of *Maxwell Smart's* old "throw your hands up in the air" trick and the familiar Army adage of "Do something even if it's wrong". The closest Mr. Shenefield came to spelling out a theory of accomplishment was his assertion that divestiture would transform present pipeline owners into ratepayers and thus create an adversary relationship between carrier and ratepayer, which would aid the regulatory process.[2254] Senator Kennedy's Petition for Rulemaking by the FTC simply echoed Mr. Shenefield's statement. In fact, he gave as his rationale for divestiture the very language quoted above at note 2253.[2255] Straining to do the proponents' job for them, the most that this writer can develop — within the framework of proponents' documents and prior Congressional Committee Reports — is a claim that there will be (1) an increase in "competitiveness" in the pipeline industry; (2) a pressure for rate reductions arising from a "new" adversary stance between shippers and owners; (3) "independent" pipelines provide more open access to

---

[2252] KENNEDY STAFF REPORT, *supra* 184, at 150.
[2253] SHENEFIELD PROPOSED STATEMENT, *supra* note 179, at 37.
[2254] *Id.* at 37-38.
[2255] Kennedy Rulemaking Petition, *supra* note 1641, at 10.

shippers; and (4) that pipelines will not "undersize" their lines because, given effective rate regulation, the only way for an independent operator to make money is to expand his throughput.

Passing for the moment the key point that no *"how* this is going to come about" evidence has been forthcoming from proponents, let's look at the hypotheses. The first one, the textbook paradigm dealing with increased competitiveness, has a certain surface appeal — who can be opposed to increased competition? However, when exposed to analysis, it does not weather well. Economists who are knowledgable in the field have eschewed it. Edward Erickson, Professor of Economics and Business at North Carolina State University, filed a statement with the Senate Antitrust and Monopoly Subcommittee which pointed out that the petroleum industry already was effectively competitive. Entry and exit are every day phenomena. Resources are mobile. Long-run profitability indicated competitive performance. The rate of technical advance was impressive. He continued: "If the major benefit of divestiture is presumed to flow from the breakup of an alleged monopoly, that benefit is illusory. It is not too strong a figure of speech to say that the contribution of divestiture to the effectiveness of competition in the U.S. petroleum industry would be equivalent to the contribution of another ton of salt to the salinity of the oceans."[2256] Dr. Leslie Cookenboo, Jr., who, at the time he wrote his book on Crude Oil Pipelines and Competition in the Oil Industry, was an Assistant Professor of Economics at Rice, stated: "In sum, then, the economic case for pipe-line divorcement, insofar as refineries are concerned, may be said to stand on the (unknown) degree of freedom of entry into refining which would be created, and the (unknown) extent of effective new competition which would be occasioned by such entry. In this writer's view it is only by demonstrating that divorcement would appreciably assist those independents large enough to promote more vigorous price competition in the refined [products] market that a program of dis-integration can be justified — especially when such a program carries with it the danger of higher transportation costs. This demonstration has not yet adequately been made by the proponents of divorcement."[2257] Lawyers have expressed severe questions that it would accomplish what the proponents of divestiture think it will do.[2258] The

---

[2256] *Petroleum Industry Hearings, supra* note 192, at 1053 (Letter from Professor Edward Erickson to Senator Birch Bayh, Senate Judiciary Committee, transmitting two chapters of his book on the Energy Question).

[2257] CRUDE OIL PIPELINES, *supra* note 158, at 125.

[2258] *Petroleum Industry Hearings, supra* note 192, at 1957 (Answer of Witness Peter A. Bator, Davis, Polk & Wardwell, to question of Staff Counsel Henry Banta); Bork, *Vertical Integration-Economic Misconception, supra* note 106, at 197 (". . . dissolving the vertical integration accomplishes precisely nothing"); *cf.* Ritchie, *Petroleum Dismemberment,* 29 Vand. L. Rev. 1131, 1132 (1976) (". . . the grounds used publicly to justify divestiture efforts appear to be specious").

leading business historian, whose two-volume work is the bible of pipeline history, Professor Arthur M. Johnson, has pointed out that the arguments for the benefits of divorcement rest on analogies, hypotheses and predictions.[2259] Writers have commented that divestiture would *not* make the industry more competitive and would raise prices.[2260] They noted also that anyone who thinks that the preservation of small companies automatically aids the competitive atmosphere is confusing preservation of competitors with preservation of competition,[2261] and protectionism has no place in any public policy that is designed to increase the well-being of consumers.[2262] Even if the theory were to survive this wave of broad-scale criticism, one may ask how it would work out in the "oil patch." Several observers have commented and/or testified to the fact that divestiture of pipelines will not change the physical location of the present lines,[2263] connections to the lines,[2264] or the number of terminals.[2265] Service to shippers will not improve since currently the most efficient form of transportation known for the movement of petroleum liquids is being provided on a non-discriminatory basis.[2266] The purchaser of a line running from a producing field to a refinery could not buy that line or operate it without a long-term contract from the refiner, and he would be just as dependent upon the refinery as if the pipeline were owned by it.[2267]

The second hypothesis, concerning rate reductions, likewise crumbles like a cookie in the hands of a two-year old. Its first weakness is the implied assumption that the present shipper-pipeline relationship has been the cause for absence of complaint. The willingness of shippers to take rate cases to the ICC has been demonstrated amply in recent years. The fact that all protests have been resolved at staff levels for several decades, except for one against Williams Pipe Line Company, shows that the system works. The case against Williams arose not, as alleged, because Williams Pipe Line was an "independent" but because technological

---

[2259] JOHNSON, Lessons, *supra* note 108, at 191.
[2260] E. MITCHELL, INTRODUCTION, VERTICAL INTEGRATION IN THE OIL INDUSTRY 3 (E. Mitchell ed. 1976) (refering to a consensus of the five authors in the book).
[2261] Comment, *Making the Oil Industry Competitive — Problems and Solutions,* 5 U. Hous. L. Rev. 315, 319 (1967).
[2262] LIEBELER, *Integration and Competition, supra* note 976, at 24-25.
[2263] *E.g.,* WOLBERT, *supra* note 1, at 100; *S.1167 Hearings, supra* note 432, pt. 9 at 638 (Statement of Fred F. Steingraber); HOWREY & SIMON, *supra* note 3, at 73; TEECE, *supra* note 329, at 120.
[2264] *S.1167 Hearings, supra* note 432, pt. 9 at 638 (Statement of Fred F. Steingraber); *Consumer Energy Act Hearings, supra* note 154, at 632 (Statement of Jack Vickrey).
[2265] TEECE, *supra* note 329, at 120.
[2266] STEINGRABER, *supra* note 193, at 153.
[2267] *Petroleum Industry Hearings, supra* note 192, at 383 (Statement of Donald C. O'Hara, President of NPRA).

improvements, competition and the Consent Decree had kept pipeline rates steadily reducing while all other costs were going up until the last surge of inflation and the probable reaching of an end point of cost savings; at which point Williams was forced to raise its rates to cover its added costs, and shippers descended on Williams like the Pittsburgh Steelers in the luggage commercial. If the cause were really that Williams was an "independent," a projection of the number of cases against "independent" lines from January 1, 1969 to February 1975 into the future would lead to an estimated additional cost of regulating divested interstate oil lines in excess of $49 million annually.[2268] In addition, "independent" lines perhaps are more apt to succumb to the Averch-Johnson "gold-plating" syndrome because, unlike present shipper-owners, they would stand to benefit rather than suffer from the added costs of the unnecessary capital additions.[2269] There is no indication that new (independent) owners will be any less eager to make money than would an integrated owner.[2270] Moreover, the 7 percent consent decree limitation would not be applicable to such lines,[2271] and their financing costs would be higher, thus the impetus would be upwards, not downwards, as hoped by divestiture proponents.

The third hypothesis is that pipelines would provide more open access to nonowners. This is difficult to visualize as the examination of access discussed above demonstrated current free access.[2272] What would change? The lines would be located in the same location as present lines and the criteria for routing new lines will be the same no matter who owns the line: economically, it must follow the traffic, modified by exogenous forces such as zoning, government permits, right-of-way availability and environmental considerations.[2273] Minimum tenders,[2274] "batch" positioning and cycle scheduling are based on technological considerations which obey the laws of physics and not of Congress. Extensions will still be subject to payout considerations and, if anything, an independent pipeline company might be somewhat more cautious about

---

[2268] E. THOMPSON, R. ROONEY & R. FAITH, COMPETITION AND VERTICAL INTEGRATION IN THE OIL INDUSTRY (Unpublished study February, 1976) [hereinafter cited as THOMPSON et al., *Competition*].

[2269] *Id.* The authors estimate social costs of this to be several billions of dollars per year if retrospective divestiture of all lines were to take place; lesser amounts if the divestiture were only prospective or the retroactive divestiture were only partial.

[2270] WOLBERT, *supra* note 1, at 100; HOWREY & SIMON, *supra* note 3, at 86.

[2271] Waidelich, Rodino Testimony, supra note 197, at 8; STEINGRABER *supra* note 193, at 153; HOWREY & SIMON, *supra* note 3, at 86.

[2272] See text at notes 1807-1991, especially 1824-1828, *supra*.

[2273] See text at notes 1831-1834, *supra*.

[2274] TEECE, *supra* note 329, at 120.

extension until it had seen the traffic materialize,[2275] if for no other reason than it would lack the "observational economies" that the present shipper-owner enjoys and would suffer from the "information impactedness" that the transaction analysis school of economists talks about.[2276] Operating solely for the transportation profit, its management would be obliged to cut off marginal production in a situation where the integrated owner might give more weight to the need for the crude in its business.[2277]

The fourth hypothesis rests entirely on the validity of the undersizing theory, which was demonstrated to be imaginary in Section VI C 4, "The Department of Justice's Circumvention of Rate Regulation or 'Undersizing' Theory."[2278] Hypothesis four thus is left in the posture of a solution with the Sisyphean task of looking for a problem. Ironically, even if the "undersizing" theory were more than an economist's fantasy, there is a basic fallacy in Mr. Shenefield's "leaping-before-looking" assumption that an independent would not be susceptible to "undersizing" temptations. While it could indeed only *make* an incremental *profit* by expanding throughput, the business and financial risks of oversizing would weigh more heavily on it than would the loss of the possible additional revenue because its *only* risk in undersizing would be the unlikely possibility that another line would be built. Thus, it would be more apt to be conservative in order to guarantee full utilization of the pipeline because it would suffer no detriment from prorationing, as do the present shipper-owners. This possibility is increased by the "independent's" difficulties in obtaining financing and the higher cost of such financing to it than to the present lines, whose credit is backed by shipper-owner's throughput deficiency agreements or direct guarantees.[2279]

Summarizing this subsection, there appears to be a complete failure by proponents to make the requisite showing that divestiture will solve the "problem" — in fact, one is left in doubt as to what *is* the problem.

[2275] HOWREY & SIMON, *supra* note 3, at 89.
[2276] TEECE, *supra* note 329, at 123-128 for a Glossary explaining this wonderful world of new terms used by transaction analysts.
[2277] Consumer Energy Act Hearings, *supra* note 154, at 632 (Statement of Jack Vickrey); *S.1167 Hearings, supra* note 432, pt. 9 at 638 (Statement of Fred F. Steingraber).
[2278] See text at notes 1992-2194, *supra*.
[2279] MARATHON, *supra* note 202, at 94-95.

(3) To Show that the Anticipated Benefits Will Outweigh the Expected Costs

An indication of the Department's increased awareness, since the issuance of Mr. Shenefield's Proposed Statement to the Kennedy Subcommittee, of the complexity of the problems that divestiture would cause was displayed in Deputy Assistant Attorney General Donald Flexner's answer to a question put to him by this author at the AEI Conference on Oil Pipelines and Public Policy. In his words: "Before we support a proposal for a flat across-the-board divestiture of existing pipelines we will carefully weigh the benefits against the potential costs. We are sensitive to the potential costs that might attend if we make a mistake. For example, after a close look at Sohio [the proposed Long Beach, California, to Midland, Texas, line] we concluded that there would not be a competitive problem, and there may be other cases like that. We want to approach that part of the issue rather carefully."[2280] Unfortunately, this careful, exploratory *modus operandi* was not present at the earlier hearings, reports or statements, so instead of a detailed list of concrete benefits that proponents anticipate will be achieved by pipeline divestiture, the closest one can come are vague allusions to increased competitiveness, greater access to lines by smaller shippers and exactitude in optimal sizing of lines. Presumably, the hope is that a "ripple," or second-level, effect will be lower costs to consumers. These are all wonderful aspirations and devoutly to be desired, but the gap between the desire and the deed makes the Grand Canyon look like a pot hole in a city street. Unfortunately for the country, there does not appear to be much support for the chance that the hope will become a reality. In fact, the commentators on the subject assign scant odds to its achievement. This author, writing in *Oil's First Century,* a seminar sponsored by the Harvard Graduate School of Business Administration, bringing students of the oil industry, oilmen and government representatives together for a mutual exchange of views on significant aspects of the industry's development on the occasion of the industry's centennial, expressed the view that: "those who have speculated on how [the problems faced by the pipeline industry] would be handled by nonintegrated pipelines appear unanimous that the results reasonably to be anticipated are well short of the actual performance record established by integrated pipelines."[2281] Dr. Michael Canes in his paper, *"An Economist's Analysis,"* after reviewing the arguments of economists who supported integration and those who argued that it was a strategy to discourage entry or exit from

---

[2280] AEI, *Oil Pipelines and Public Policy, supra* note 684, at 69 (General Discussion).
[2281] SPECTRE, *supra* note 104, at 118.

the industry, as well as the empirical evidence that he developed from his own research, summed up as follows: "The major finding of the chapter is that there are no discernible consumer benefits from oil industry divestiture whereas costs are certain to be large. Since a relevant criterion for social policy is whether consumers are made better off, the major implication is that no case can be made for the vertical divestiture of U.S. oil firms."[2282] The Energy Resources Council (with participation by the FEA, Treasury Department, Commerce Department, and the State Department) in its report of May 1976, entitled "Analysis of Vertical Divestiture," after reviewing the benefits that would stem from the alleged increased competition that might be fostered by divestiture, had this to say: "The realization of any of these potential benefits depends upon the ability of divestiture to explicitly effect requisite changes. If anticompetitive behavior is partly responsible for the status quo situation, then it would be incumbent on divestiture to rectify this behavior enroute to bringing the desired benefits. Whether divestiture could accomplish either the intermediate (reducing anticompetitiveness) or the ultimate goal is open to substantial question."[2283] Mr. Charles P. Siess, Jr., President of Apco Oil Corp., in his 1974 *S.1167* testimony, expressed some support for pipeline divestiture, but that same testimony showed that, after outbidding a major company for some crude oil, he was able to get the major's pipeline affiliate to transport it.[2284] No other evidence of lack of access surfaced and the overwhelming weight of evidence attesting to the current freedom of access which was marshalled in Section VI C 3, "Asserted Denial of Access to Pipelines," virtually compelled the conclusion that there was, in fact, no such benefit to be derived.

This leaves only the so-called conflict of interest between the role of shipper and owner. The only evidence found on that point was the judgement of business historian Arthur Johnson, who, after examining in minute detail the evolution of the pipeline industry from 1862 to 1959 in a two-volume work which is the classic in the field, reached a conclusion favorable to pipeline management as responsible for conducting a soundly managed common carrier operation and, although undoubtedly spurred by governmental intervention, emerged with a solid record of accomplishment and a high value on pipelines' public image which approximated the goals sought by the early critics of the industry.[2285] In the final analysis, the answer to the conflict of interest question has to come from the same type of balancing that this chapter established as the

---

[2282] CANES, *An Economist's Analysis, supra* note 1021, at 46-47.
[2283] ERC REPORT, *supra* note 184, at 43-44.
[2284] *S.1167 Hearings, supra* note 432, pt. 8 at 6235, 6240-6242 (Statement of Charles P. Siess, Jr.).
[2285] PETROLEUM PIPELINES, *supra* note 45, at 477-478.

overall test, *i.e.,* is the public better served by an efficient, well financed, knowledgeable pipeline company which is affiliated with other oil interests or by someone who has no relationship with another phase of the business but whose performance may be nowhere near as good as the integrated company? As Donald Baker testified before the Senate Judiciary Committee when he was in charge of the Antitrust Division. "...I'd like to see more effort to clearly understand the problem and less rhetoric about simple solutions."[2286]

As might be expected, there is a wealth of evidence on expected costs, so much so that it required organizing into subheadings.

(a) Interim Chaos and Uncertainty

From a practical standpoint, there are only two methods to accomplish pipeline divestiture: sale of assets or a "spin-off." The sale of assets, if it followed the pattern envisioned by *S.2387,* would involve at least 84 firms: 18 "major" oil companies, 45 other firms having a partial ownership in interstate pipelines, and 21 additional firms if applied to intrastate pipelines, altogether worth over $14 billion in 1977-1978,[2287] today probably closer to $16 billion. There is a substantial question as to where investors could be found who would buy these lines.[2288] The size of these assets caused one witness, who represents daily the investment banking industry, to conclude that it was "extraordinarily unlikely" that the new capital could be found to finance their purchase.[2289] Another witness, not similarly conditioned by constant dealing with the cautious wording of registration statements filed with the SEC, simply said it was practically impossible.[2290] Thus the choice narrowed down to a "spin-off," *i.e.,* a pro-rata distribution of the shares of a present pipeline company which owned the assets, or of a company formed to receive them in a tax-free reorganization under Section 351 of the Internal Revenue Code, to the shareholders of the parent oil company as a tax-free distribution (dividend) under Section 355 of the Internal Revenue Code. Immediately, there would be a problem with the covenants contained in many, if not most, existing financing documents forbidding the disposition of assets of the borrower, or restricting the amount of

---

[2286] 813 ANTITRUST & TRADE REG. REP. (BNA) E-9 (May 12, 1977).

[2287] NET-III, *supra* note 199, at 152-153.

[2288] *Petroleum Industry Hearings, supra* note 192, at 166 (Peter N. Chumbris, Minority Chief Counsel reading Senator Hruska's questions); *cf. Id.* at 2016 (Statement of Richard J. Boushka, President of Vickers Energy Corp.).

[2289] *Id.* at 1929 (Statement of Peter A. Bator, of Davis, Polk & Wardwell).

[2290] *S.1167 Hearings, supra* note 432, pt. 9 at 602-603 (Statement of Fred F. Steingraber).

dividends that a borrowing company could pay, or requiring the maintenance of a certain net worth, or prohibiting the disposal of the stock of certain subsidiaries of the borrowing company. The problem would spread even wider because many lending instruments contain "cross-default" clauses whereby a technical default in one loan will accelerate the unpaid indebtedness in *all* outstanding loans.[2291] Litigation by lenders in order to protect themselves almost certainly would ensue,[2292] and there would be an upset of the money markets for years.[2293]

Moreover, contracts with third parties which prohibited the assignment or transfer of rights without the approval of such parties would be breached and more litigation would result.[2294] There would be a swarm of cases testing the bounds of the "commercial frustration" provisions of the Uniform Commercial Code. If companies attempted to forestall "events of default" by renegotiating their agreements, lenders and other parties could pressure them into curtailing capital investment programs so as to provide cash to repay the loans over the shortest term possible or make more costly arrangements to see that their committments were met.

Some federal agency would have to oversee and administer the divestiture to ensure that the provisions of the law were executed in accordance with the intent of the divestiture law. For example, under *S.2387*, the FTC was assigned that task.[2295] There are bound to be questions of interpretation, and consultations, rulings and litigation reasonably can be anticipated.

This leads again to more social costs: *e.g.*, the maintenance of a large organization requiring countless man-hours of effort by both the administering agency and by the subject companies. As one jobber put it "...what we need in our industry is less regulation, and not more... We have dealt with FEA now for awhile and I will tell you for sure, I will be happier with my future in the hands of a major oil supplier than I am with someone up here [Washington], very frankly. We have a fellow in Mississippi that they say can mess up a one-car funeral. I think that we do not have a monopoly on these types of individuals."[2296] A leading in-

---

[2291] ERC REPORT, *supra* note 184, at 53.
[2292] DEPARTMENT OF THE TREASURY, IMPLICATIONS OF DIVESTITURE 201 (1976).
[2293] *Cf. Petroleum Industry Hearings, supra* note 192, at 204. (Note by Richard C. Sparling, Energy Economist, Chase Manhattan Bank).
[2294] *Id.* at 1932 (Statement of Peter A. Bator, of the law firm of Davis, Polk & Wardwell).
[2295] ERC REPORT, *supra* note 184, at 48-49. For a list of some of the activities S.2387 would require of the FTC, see *Petroleum Industry Hearings, supra* note 192, at 1933-1934, (Statement of Peter A. Bator).
[2296] *Petroleum Industry Hearings, supra* note 192, at 2093 (Statement of Pat Green, Jr., a jobber from Collins, Miss.).

vestment banker expressed the concern that every party who had an interest would want to be represented before such an administrative agency and the courts in order to protect those interests. This would mean a multitude of litigants and years of uncertainty and confusion as plans were made, revised, opposed, and changed again. While all this was going on, no investor was going to be very anxious to put up any capital as investors have no liking for uncertainty.[2297]

A reasonable estimate of the period of chaos[2298] and litigation is from 10 to 20 years,[2299] or perhaps even a generation.[2300] The El Paso divestiture of Northwestern took 17 years.[2301] The Public Utility Company Holding Company Act was passed in 1935; it was 1946 before the constitutionality of the Act was determined, and it was in the 1950's before most of it was concluded and even then there were some remnants yet unfinished.[2302] Proponents of divestiture have attempted to minimize the dislocations of divestiture by analogizing the situation to the Pennzoil spin-off of United Gas Pipeline;[2303] to the court-ordered divestiture of Peabody Coal Company by Kennecott Copper and the spin-off of Louisiana-Pacific by Georgia-Pacific;[2304] to the Standard Oil divestiture of 1911; to the Public Utility Holding Company Act of 1935;[2305] the Bank Holding Company Act of 1956;[2306] or the voluntary "divestitures" made by major oil companies as they adapted to changing circumstances.[2307]

Each of these suggested analogies was methodically shot down by the more reasoned replies of divestiture opponents or committee members or their counsel. Pennzoil was a voluntary, shareholder-voted spin-off of a single entity with no impairment of obligations.[2308] The Peabody Coal divestiture involved a single company which was not a functional part of Kennecott and its financial and operational aspects were easily ascertainable.[2309] The Standard Oil Trust divestiture took place in an industry

---

[2297] *Id.* at 1964 (Statement of Raymond B. Gary, Managing Director, Morgan Stanley & Co., Inc.).

[2298] *Id.* at 1839 (Statement of William P. Tavoulareas, President, Mobil Oil Corp.).

[2299] *Id.* at 1937 (Statement of Peter A. Bator).

[2300] *S.1167 Hearings, supra* note 432, pt. 9 at 639 (Statement of Fred F. Steingraber).

[2301] *Id.,* pt. 8 at 6233 (Statement of Charles P. Siess Jr., President Apco Oil Corp.).

[2302] *Petroleum Industry Hearings, supra* note 192, at 2039 (Senator Roman Hruska).

[2303] *Id.* at 190 (Answer by Ronald J. Peterson, President, Independent Terminal Operators' Association to question by Peter N. Chumbris, Minority Chief Counsel).

[2304] S.2387 REPORT, *supra* note 49, at 133-135.

[2305] *H.Res.5 Subcomm. Hearings, supra* note 234, at A66 (Reprint of Beverly Moore's *Proxmire Hearings* Testimony).

[2306] *Petroleum Industry Hearings, supra* note 192, at 49-50 (Senator Robert Packwood).

[2307] *Id.* at 493 (Senator John Tunney).

[2308] *Id.* at 190 (Peter N. Chumbris, Minority Chief Counsel); *cf. Id.* at 197 (Statement of Senator Dewey Bartlett); *Id.* at 367-368 (Statement of Walter R. Peirson, President, Amoco Oil Co.).

[2309] *Id.* at 197 (Statement of Senator Dewey Bartlett).

setting that bears little or no resemblance to today's economy or to the industry structure today and it was a divestiture ordered by a court after a complete examination of both the facts and the remedy with full opportunity for opponents to present their case, and the division was along corporate, not functional lines.[2310] Witness Joseph Tovey, Vice President of Faulkner, Dawkins & Sullivan, Inc., whose speciality was the petroleum business, explained to Senator Tunney that there was a substantial difference to the consumer between the "pruning" by a major oil company of an uneconomic portion of its business, which was done in a carefully planned manner, with contracts being honored, people being retrained suitably and notice given to interested parties and a "wham, bam, thank you ma'm," precipitous rout of the entire industry.[2311] With respect to Senator Tunney's Public Utility Holding Company Act analogy, Senator Roman Hruska pointed out that it had been a relatively small operation, easy to deal with as there were geographical units, separately incorporated, each of which had its own profit and loss statement, its assets and liabilities and its functioning organization. Simple as it was compared to the present proposal, it still took 20 years or more. Witness Ellis felt that the relative simplicity of the Public Utility Holding Company Act of 1935 to the complexity of the current proposal was like comparing a molecule to a dinosaur.[2312] Moreover, both it and the Bank Holding Company Act were directed solely to holding companies and not to operating entities that were the product of normal business growth.

Raymond Gary, Managing Director of Morgan Stanley & Co., expressed concern about the reluctance of investors during the period of turmoil. Mr. Gary's worry was echoed by the Energy Resources Council ("ERC") with respect to the affected companies' incentives to make investments during this time. As the ERC Report noted, new investments in all but the most profitable areas would tend to be curtailed because of the financing difficulties and this would be true especially of investments in large projects.[2313] Some examples of specific projects come quickly to mind. TAPS would not only need to establish new owners and operators but had divestiture come before the project was underway, it would have

---

[2310] *Id.* at 324 (Statement of W. T. Slick, Jr.); JOHNSON, *Lessons, supra* note 108, at 191-198; H. WYMAN, *The Standard Oil Breakup of 1911 and Its Relevance Today* in WITNESSES FOR OIL 63-74 (1976).

[2311] *Petroleum Industry Hearings, supra* note 192, at 493 (Answer by Joseph Tovey to question by Senator Tunney).

[2312] *Id.* at 2039 (Comment by Senator Hruska; comment by Otis H. Ellis, Private Citizen, Consumer and Farmer).

[2313] ERC REPORT, *supra* note 184, at 54; NET-III, *supra* note 199, at 155; *Petroleum Industry Hearings, supra* note 192, at 493 (Statement of Joseph Tovey, Vice President, Faulkner, Dawkins & Sullivan, Inc.).

required new financial backers as well. The pipeline and the development of the Prudhoe Bay reserves could have been shut down indefinitely until new arrangements could have been made, if they even could have been at all.[2314]

Even though the project is now in place, many of the same questions, such as finding new owners and managers for the pipeline and the key question of refinancing over 8 billion dollars remain. Indeed, the mere threat of divestiture discourages oil companies from making investments of the magnitude necessary if the United States is going to reduce its dependence on oil imports.[2315] Moreover, costs would be raised by the turmoil and added uncertainty.[2316] In addition, there would also be costs which would be incurred by the companies in designing and effectuating plans, such as categorizing and separating assets, arranging for, and establishing of, new managements, and the expenses of litigation.[2317] In this environment of confusion and uncertainty, with its concomitant impairment of the affected companies to finance, it is obvious that operating and planning decisions of the industry will be affected substantially,[2318] or, as one witness put it, "we are in sort of a holding pattern" until things get sorted out.[2319]

(b) Decrease in Efficiency of Capital Formation — Increased Cost of Financing

Just as in the case of companies postponing making investments in the face of an impending threat of divestiture, the mere prospect of pipeline divestiture pursuant to generally applicable formulae would have a profoundly depressing effect on capital markets for pipeline obligations.[2320] An example of this was the experience of Sohio, which was coming to market with a $1.75 billion private placement to finance its

---

[2314] *Petroleum Industry Hearings, supra* note 192, at 326-327 (Statement of W. T. Slick, Jr.).

[2315] MANCKE, *Oil Industry Competition, supra* note 674, at 71.

[2316] MITCHELL, *Capital Cost Savings, supra* note 1088, at 101.

[2317] Ritchie, *Petroleum Dismemberment,* 29 Vand. L. Rev. 1131, 1156-1157 (1976); *Petroleum Industry Hearings, supra* note 192, at 339 (Statement of W. T. Slick, Jr.) (the immediate impact would be to suspend capital spending programs and focus top management attention on developing its dismemberment plan and on reevaluating and restructuring short and long-range investment plans.).

[2318] *Petroleum Industry Hearings, supra* note 192, at 1972 (Statement of Raymond B. Gary, Managing Director of Morgan Stanley & Co.).

[2319] *Id.* at 405-406 (Statement of Robert J. Welsh, Jr., President & Owner of Welsh Oil Inc., at independent Jobber in Gary, Ind.).

[2320] *S.1167 Hearings, supra* note 432, pt. 8 at 5916 (Statement of Robert E. Yancey, President, Ashland Oil, Inc.).

portion of TAPS at a time when the Senate was considering *S.2387* and related divorcement bills.[2321]

If a pipeline divestiture bill were passed and signed into law, during the period of chaos and uncertainty described above, investors would not be eager to put up any capital. Raymond B. Gary, Managing Director of Morgan Stanley & Co., who has placed as much or more pipeline debt as any investment banker, testified as to his belief that in the case of complete divorcement of all phases of the industry, investors, large and small, would "dump" their oil securities — almost certainly at a substantial loss — and billions of dollars of security values would have been wiped out. While the magnitude of this flight of capital could be expected to be less in the case where only pipelines were divested, investors clearly dislike uncertainty, and what could be today's pipeline divestiture could be tomorrow's total dismemberment of the oil industry. Hence, much of Gary's testimony is applicable. He speculated about the difficulties of a company trying to sell debt after the required splitup, but before the manner of effecting it and the legal questions had been resolved. As he noted, investors would lack their traditional yardsticks to measure assets back-up and debt equity ratios; earnings predictions would be impracticable; the after-market could not be gauged and, moreover, the investors would be so busy litigating that they wouldn't have time for investing. The market for the company's equity securities would be one of depressed disarray. In short, he concluded that a major portion of the industry might be effectively foreclosed from capital markets for an indefinite period.[2322] There is some possibility that this disenchantment on the part of the investor could spread to other parts of the market, which is noted for its psychological instincts.[2323]

What will happen after the dust has settled down? The very factors discussed in Chapter V, "Financing,"[2324] come into play. The cost of financing pipelines will rise sharply for the following reasons: (1) the reduction in size and operational stability will give investors pause concerning the security of repayment;[2325] (2) loss of diversification, placing all the investor's eggs in one basket;[2326] (3) the need to establish a "track record" of profitable operations *before* investors will consider it as an

---

[2321] *Petroleum Industry Hearings, supra* note 192, at 198 (Statement of Senator Dewey F. Bartlett).

[2322] *Id.* at 1964, 1975-1976 (Statement of Raymond B. Gary).

[2323] *Cf. Id.* at 341-342 (Statement of W. T. Slick, Jr.).

[2324] See text at notes 1346-1453, *supra*.

[2325] ERC REPORT, *supra* note 184, at 45, 71; *cf. Petroleum Industry Hearings, supra* note 192, at 205 (Letter of Gerald L. Parsky, Department of the Treasury).

[2326] ERC REPORT, *supra* note 184, at 70-71.

Proposed Solutions to an Assumed Problem 467

investment possibility;[2327] (4) investors' perception of loss of operational efficiencies;[2328] (5) the new company will have to finance on a "stand-alone" basis, without the traditional throughput and deficiency agreements or guarantees;[2329] and (6) it must compete for funds in the speculative money market.[2330] Assistant Attorney General Shenefield, in his Proposed Statement to the Kennedy Subcommittee, stated that an independent pipeline company would have no intrinsic difficulty in obtaining financing if the oil industry were willing to make throughput commitments to it on the same basis that they are now willing to do for their own pipeline subsidiaries and that he had "...every reason to believe that once divestiture forecloses equity participation by shippers, their throughput commitments to independent carriers would be much more readily forthcoming than they are now."[2331] This bland assertion was shown in Section V B, "How Pipelines are Financed,"[2332] to be awry. Apparently the Department has come to realize this, judging by its "two-tier" pricing suggestion in the *Texas Deepwater Port Authority* proceedings before FERC. There has also been some speculation that existing throughput and deficiency agreements or guarantees might follow the assets of a divested pipeline, but there are substantial constitutional questions raised by such suggestion, and it would also run afoul of existing definitions of "control" forbidden by divestiture statutes.

The greater cost of financing and the resulting higher tariffs which would result were first suggested *a priori* by Professor Steward Myers in his 1974 testimony before the Senate Interior Committee.[2333] His testimony was reinforced by Raymond Gary's testimony in the *Valuation of Common Carrier Pipelines* and *TAPS* proceedings.[2334] Fred Steingraber, then Vice Chairman of Colonial, testified before the Senate Antitrust and Monopoly Subcommittee in January 1975, that if Colonial, *which had a very good track record of earnings,* had been a "stand-alone," it would have required a 38 percent increase in its tariff rate to

---

[2327] *Id.* at 55, 70.
[2328] *Petroleum Industry Hearings, supra* note 192, at 205 (Letter of Gerald L. Parsky, Department of the Treasury); *cf.* Ritchie, *Petroleum Dismemberment,* 29 Vand. L. Rev. 1131, 1158 (1976).
[2329] *Consumer Energy Act Hearings, supra* note 154, at 597, 616, 632, (Statement of Jack Vickrey); STEINGRABER, *supra* note 193, at 151; *cf.* Myers, *supra* note 198, at 22.
[2330] *Id; see also Petroleum Industry Hearings, supra* note 192, at 302 (Statement of E. P. Hardin.).
[2331] SHENEFIELD PROPOSED STATEMENT, *supra* note 179, at 40-41.
[2332] See text at notes 1428-1442, *supra.*
[2333] Myers, *supra* note 198, at 22.
[2334] Gary, *FERC Direct Testimony, supra* note 613, at 32-33; Gary, *Direct TAPS Testimony, supra* note 634, at 51-52.

achieve a comparable return on total investment.[2335] Professor Edward Mitchell approached the subject from the cost-of-capital vantage point, examining investor expectations in the cases of "stand-alones" such as Williams Pipe Line during the period 1964-1974. He concluded that the cost of capital to nonintegrated pipelines was substantial.[2336]

(c) Chilling Effect on Construction of New Lines, Expansion and Continuation of Service by Present Lines

It is, of course, impossible to adduce positive proof that divestiture would discourage the construction of new pipelines, especially the most economic large diameter pipelines with economies of scale, or would have a deleterious effect on existing lines. It should be sufficient to show that there is a real and substantial risk of that result.[2337] It appears clear beyond a peradventure of doubt that, at the very least, expansion of existing lines and construction of new lines would be slowed down substantially because of funding problems and the difficulty of private interests in obtaining the necessary support to justify the huge investments involved.[2338] Stronger expressions of doubt have been voiced by witnesses in Congressional hearings and commentators to the effect that it is very doubtful that any new pipelines would be built by private industry and the continued operation of many existing lines would be imperiled, particularly the marginal lines serving depleted oil fields, stripper wells or refineries suffering reduced throughput.[2339]

When the scope of the statements is narrowed, the degree of probability increases. Thus, it has been said that, in the event of pipeline divestiture, it is unlikely that economic and efficient large diameter lines would ever again be built by private enterprise, either within or without the oil industry.[2340] Another expression was that, in all probability, in-

---

[2335] *S.1167 Hearings, supra* note 432, pt. 9 at 600 (Statement of Fred F. Steingraber) (For the year 1973 this would have increased the cost to its shippers, and on to the consuming public, close to $52 million).

[2336] Mitchell, *Capital Cost Savings, supra* note 1088, at 98.

[2337] MARATHON, *supra* note 202, at 97.

[2338] *S.1167 Hearings, supra* note 432, pt. 9 at 659 (Answer by Fred F. Steingraber to Senator Strom Thurmond's question); Ritchie, *Petroleum Dismemberment,* 29 Vand. L. Rev. 1131, 1158 (1976); *cf. Petroleum Industry Hearings, supra* note 192, at 208 (Reprint of *S.1167 Testimony* of Gary L. Swenson, Vice President, The First Boston Corp.).

[2339] *Consumer Energy Act Hearings, supra* note 154, at 579, 616 (Statement of Jack Vickrey); *S.1167 Hearings, supra* note 432, pt. 9 at 638 (Statement of Fred G. Steingraber); HOWREY & SIMON, *supra* note 3, at 88-89.

[2340] HOWREY & SIMON, *supra* note 3, at 91; CRUDE OIL PIPELINES, *supra* note 158, at 127.

dependently owned capacity would be built only in the face of sustained, severe shortages of pipeline capacity and that new projects would be postponed or abandoned and expansion of existing lines would be in limbo.[2341]

There are some empirical data in support of these judgements. In Sections IV A 2, "Paucity of Pipelines Operated by Non-Oil-Related Owners"[2342] and V B 3, " 'Stand-Alone' Pipeline Ventures,"[2343] an examination was made of the fortunes (and misfortunes) of pipelines promoted and operated by owners who had no other affiliation with the oil business. Even operating in a period of growth, relatively smaller investments and a stable long-term debt market, the record was quite unimpressive. When the "price of poker" went up, as in the case of TAPS, it was found that the line could not be built on a "stand-alone" basis or even with the credit of the State of Alaska behind it.[2344]

Nor is it without significance that in the Deepwater Port situation, despite considerable political pressure to exclude oil companies from deepwater port ownership, the concern that no entity other than oil companies would be willing, or able, to provide the necessary financial support for the ports caused the Congress to decide not to exclude them.[2345] When most of SEADOCK's "major" oil company owners, already faced with a marginal proposition, were unwilling to go forward under the added weight of overregulation, and withdrew from the project,[2346] SEADOCK lost its viability. The State of Texas Deepwater Port Authority is attempting to perform a salvage job (with the benefit of all the work done by the companies) but the outcome is, to say the least, problematical.

(d) Loss of Integration Efficiencies — Resultant Higher Costs to Shippers and Higher Fuel Prices to Consumers

Perhaps the most concise statement on this topic is found in the Congressional Research Service Report to Senators Jackson and Cannon on Issues and Problems in National Energy Transportation:[2347] "Since it

---

[2341] *Petroleum Industry Hearings, supra* note 192, at 302 (Statement of E.P. Hardin); *cf.* HOWREY & SIMON, *supra* note 3, at 64-65.

[2342] See text at notes 1083-1124, *supra*.

[2343] See text at notes 1449-1453, *supra*.

[2344] See text at notes 1450-1451, *supra*.

[2345] WOFFORD PROPOSED STATEMENT, *supra* note 656, at 2-3.

[2346] AEI, *Oil Pipelines and Public Policy, supra* note 684, at 183 (Comment by John Shamas, Director of Transportation, Getty Oil Co. in general discussion).

[2347] NET-III, *supra* note 199, at 154.

is generally recognized that vertical integration confers substantial cost savings, dis-integration could therefore reimpose those costs without necessarily providing any offsetting cost savings to the firm. The precise magnitude of these costs is difficult to quantify, but, whatever their size they would either have to be passed on to the consumers in the form of higher prices and reduced supplies, or have to be absorbed by the firms and their stockholders, or both."

The reader will recall the extended discussion in Section IV B 3, "Reasons for Vertical Integration"[2348] which catalogued, among others, the following advantages/savings: security of supply; avoidance of the inadequacies and higher costs of contractual arrangements; coordination of planning of huge, interdependent capital investments and execution of smooth, continuous operation; lower inventories and reduced working capital; reduction of risk and the cost of raising capital; the reduction of cost achieved by common use of information, training, experience and managerial guidance across the various stages of the industry.

The witnesses and commentators are virtually unanimously of the view that, if adopted, divestiture is likely to result in higher fuel prices because the elimination of some or all of these real integration economies will raise costs.[2349]

(4) To Show That There Is No Effective, Less Dramatic Solution

In his "series of gates" testimony before the Antitrust and Monopoly Subcommittee, Assistant Attorney General Donald I. Baker stated:"It is a series of gates. If you decide the case for breaking up isn't strong, you forget about it. If you can't devise a statute that will do the job in a workable way, forget about it. If the cost is excessive, you forget about it."[2350] Even John Shenefield, in his Proposed Statement to the Senate Antitrust and Monopoly Subcommittee, unbent enough to concede that ". . . it is up to divestiture *proponents* to make a convincing case . . . that

---

[2348] See text at notes 1156-1228, especially 1176-1221, *supra*. See also *Petroleum Industry Hearings, supra* note 192, at 203 (Statement of Richard C. Sparling, Energy Economist, Chase Manhattan Bank).

[2349] MANCKE, *Oil Industry Competition, supra* note 674, at 71; Mitchell, *Capital Cost Savings, supra* note 1088, at 101; *Petroleum Industry Hearings, supra* note 192, at 131 (Statement of Frank N. Ikard, President, API); *Id.* at 327 (Statement of W. T. Slick, Jr., Senior Vice President, Exxon, USA); *Id.* at 466 (Statement of John Winger, Vice President, Chase Manhattan Bank); *Id.* at 2224 (Statement of Professor Neil H. Jacoby, Graduate School of Management, UCLA); *S.1167 Hearings, supra* note 432, pt. 8 at 5916, 5932 (Statement of Robert E. Yancey, President, Ashland Oil, Inc.); CANES, *An Economist's Analysis, supra* note 1021, at 51; Ritchie, *Petroleum Dismemberment,* 29 Vand. L. Rev. 1131, 1154 (1976); *cf.* TEECE, *supra* note 329, at 105.

[2350] See note 1267, *supra*.

there is no effective, less drastic solution."[2351] Courts, in nonmerger cases, have eschewed "surgical ruthlessness" and treated divestiture as a remedy of last resort to be used only where a clear record of abuse exists and no lesser remedy will suffice.[2352] They deem themselves to be charged with "...a heavy responsibility to tailor the remedy to the particular facts of each case."[2353]

Betty Bock, a former FTC economist, now with The Conference Board, wrote a delightful article entitled, "The Lizzie Borden Solution,"[2354] dealing with *S.2378* and related petroleum industry dismemberment bills.[2355] In her article, Ms. Bock pointed out that "breakup" was not simply a concept or word; it was a drastic, highly specialized, unpredictable, and irreversible remedy being proposed for the energy problems of the last quarter of the 20th Century. She noted also that "...despite the widespread use of the 'monopoly' label for the oil industry, neither Congress nor any serious research body has developed sufficient facts to show how the industry operates or what would happen if the major companies were broken up."[2356] The article closed with a searching question — "Do we understand that dismemberment is irreversible; that mistakes cannot be rectified; and that if we weaken our industrial and defense base, we can become a second United Kingdom — with a strong bureaucracy and an economy that no longer can meet the needs of its people?"[2357] Although Ms. Bock wrote about complete dismemberment, her words are applicable equally to pipeline divestiture. Courts,[2358] commentators[2359] and industry critics of integration[2360] all agree that one is talking about a "drastic," "radical," "ruthless surgical" action when contemplating divestiture; something not to be taken lightly

---

[2351] SHENEFIELD PROPOSED STATEMENT, *supra* note 179, at 43.

[2352] United States v. United Shoe Machinery Corp., 110 F.Supp. 295, 348 (D.Mass 1953), *aff'd*, 347 U.S. 521 (1954).

[2353] Gilbertville Trucking Co. v. United States, 371 U.S. 115, 130 (1962).

[2354] BOCK, *Lizzie Borden*, *supra* note 2213.

[2355] Hence, the name of her article as she was reminded of the old (ca. 1893) rhyme: "Lizzie Borden took an ax — And gave her mother forty whacks; When she saw what she had done — She gave her father forty-one!".

[2356] *Id.* at 52.

[2357] *Id.* at 53.

[2358] United States v. United Shoe Machinery Corp., 110 F.Supp 295, 348 (D.Mass. 1953), *aff'd*, 347 U.S. 521 (1954); Gilbertville Trucking Co. v. United States, 371 U.S. 115, 130 (1962).

[2359] *Petroleum Industry Hearings*, *supra* note 192, at 2217 (Statement of Professor Neil H. Jacoby); Harmon, *Effective Public Policy to Deal with Oil Pipelines*, 4 Am. Bus. L. J. 113, 121 (1966).

[2360] *Petroleum Industry Hearings*, *supra* note 192, at 182 (Statement of Ronald J. Peterson, President of Independent Terminal Operators' Association); *Id.* at 240 (Statement of Edwin J. Dryer, General Counsel and Executive Secretary of the Independent Refiners Association of America).

nor to be used indiscriminately.

Because of the Draconian nature of the remedy, the factual basis for the need especially is important. As has been demonstrated in the foregoing subsections, proponents have come nowhere near discharging their duty of proving there was in fact a problem; or that, if had there been a problem, divestiture would have solved it; and, moreover, the benefit/cost equation was so heavily loaded on the cost side and so scant of benefit that each of the "gates" would have precluded even considering the question of remedies. As Charles Spahr put it: "To attempt a reorganization of an industry that has performed as well as this one and to do so with a complete lack of substantive evidence of any wrongdoing is simply not warranted."[2361] Even industry witnesses who were critical of alleged "major" oil company practices, were disinclined to support divestiture proposals. Edwin J. Dryer, General Counsel and Executive Secretary of the Independent Refiners Association of America, while complaining about the alleged absence of a free and open crude market, opposed divestiture as a "drastic surgery" and suggested in lieu thereof amendments to the antitrust laws to require each phase to be operated as a profit center and a "federally chartered crude oil market" (perhaps this was the genesis of the "Open Market Credit" proposal).[2362] Likewise, David Bacigalupo, Vice President of Beacon Oil Company, a small, landlocked "independent" refiner-marketer in the San Joaquin Valley, California, upon "serious reflection," concluded that divestiture would "[create] more problems than presently exist." He suggested that legislators should "...explore very carefully all of the available options open to insure [sic] the existence of a petroleum industry that operates in the public interest because hasty action could be disastrous."[2363]

2. Additional Reasons Why Retrospective Divestiture Is Inappropriate

There are serious constitutional questions which would be raised by any retrospective divestiture. Although the Contract Clause (Article 1, Section 10) is not applicable to Federal legislation, impairment of contract concepts have crept into the Due Process Clause, perhaps because the *Ex Post Facto Clause* was held inapplicable to civil legislation in the *Calder* case.[2364] At any rate, where business investments representing billions of

---

[2361] *Id.* at 1109 (Statement of Charles E. Spahr).

[2362] *Id.* at 240-241 (Statement of E. J. Dryer).

[2363] *Id.* at 395-396 (Statement of David Bacigalupo).

[2364] Calder v. Bull, 3 U.S. (Dall.) 386 (1798).

dollars have been made in good faith pursuant to a normal business growth in order efficiently to conduct a legitimate business in full compliance with legal requirements in effect at the time they were made, retrospective divestiture, especially in the light of the paucity of evidence of any wrongdoing, need, benefit, or inappropriateness of alternate remedies, would appear to be "harsh and oppressive," and hence invalid under the Due Process Clause.[2365] There would also seem to be a serious question under the "Taking Clause" of the Fifth Amendment,[2366] especially if the divestiture removed the right of ownership from the oil company owners of integrated pipelines without providing adequate relief from its existing burdens under throughput and deficiency agreements and guarantees.[2367] There is a separate Due Process argument, based on an overinclusive statute if it purports to be across-the-board, and an unlawful discrimination violation unless substantial evidence is produced concerning the appropriateness of application to particular pipelines.[2368] With a petition already drawn, it would be marginally worthwhile to see if, because of the unsubstantiated anticompetitive behavior base for legislatively singling out certain oil companies and their affiliated pipelines, the "sleeping giant" of Article I, Section 9, Clause 3's prohibition against "Bills of Attainder" might not be revitalized.[2369]

In bringing the discussion of divestiture to a close, it appears crystal clear that pipeline divestiture is not warranted. It has been demonstrated that there has been a complete failure to produce the requisite proof that a serious problem exists; that divestiture would have solved the problem if it had existed; the benefit/burden balance was demonstrated to be lopsidedly against divestiture proponents, and there has been no really serious effort, until after proponents recently realized glaring omissions in their case, to explore less drastic remedies which are capable of solving any difficulties that might exist. In a nutshell, pipeline divestiture is not a responsible proposal at this junction.

[2365] United States Trust Co. v. New Jersey, 431 U.S. 1, 17 (1977); Hochman, *The Supreme Court and the Constitutionality of Retroactive Legislation,* 73 Harv. L. Rev. **692 (1960)**.

[2366] *Petroleum Industry Hearings, supra* note 192, at 303 (Statement of E. P. Hardin); *S.1167 Hearings, supra* note 432, pt. 9 at 638 (Statement of Fred F. Steingraber); *Consumer Energy Act Hearings, supra* note 154, at 632 (Statement of Jack Vickrey).

[2367] HOWREY & SIMON, *supra* note 3, at 95.

[2368] ALM PROPOSED STATEMENT, *supra* note 646, at 3; *see* SHENEFIELD PROPOSED STATEMENT, *supra* note 179, at 37. Even Dean Rostow, who was no admirer of the industry, opted for antitrust proceedings rather than statutory relief because of his concern about the difficulty in achieving a supportable discrimination under a statute. E. ROSTOW, A NATIONAL POLICY FOR THE OIL INDUSTRY 123 (1948).

[2369] *Cf.* JOHNSON et al., *Oil Industry Competition, supra* note 1055, at 2403.

B. Alternative Solutions which Are Being Discussed

Although it is beyond the scope of this book to examine in any depth proposals for alternative relief, a brief catalogue of some of the more prominent ideas which are being discussed may be of benefit to the reader.

1. Commodities Clause

No time should be spent on this oft-raised and equally oft-rejected remedy, other than to say that, as Champlin's pipeline was found to be by the United States Supreme Court in *Champlin II*,[2370] this remedy is unused,[2371] unsought after,[2372] and unneeded by independents.[2373] The reader will remember that this subject was discussed in Section VI A 1, "Brief History of Public Policy Controversies."

2. Government Operated Oil Function

The extreme parameter on one side of the suggested alternatives is the Government Operated Oil Function (GOOF). Historian Arthur Johnson chronicles an early suggestion in the form of a bill introduced by Senator Robert Owen of Oklahoma on May 13, 1914 calling for government ownership and operation of trunk pipelines, pumping stations, and terminal facilities.[2374] There were two suggestions in the 1940's by writers in the field, one suggesting limited scale involvement, *e.g.*, one of the crude lines running from the Mid-Continent area to the Chicago area as a "yardstick" which would serve independents and keep the industry owned systems "in line" *à la* TVA operation.[2375] The other writer was more concerned with provision of a pipeline system adequate to meet the needs of the national economy and the national defenses and deemed the availability to independent shippers as a "by-product" benefit.[2376]

[2370] United States v. Champlin Refining Co., 341 U.S. 290 (1951).
[2371] United States v. Elgin, Joliet & Eastern Railroad, 298 U.S. 492 (1936); United States v. South Buffalo Ry. Co., 333 U.S. 771 (1948).
[2372] KENNEDY STAFF REPORT, *supra* note 184, at 152.
[2373] *Consumer Energy Act Hearings, supra* note 154, at 601, 616 (Statement of Jack Vickrey); WOLBERT, *supra* note 1, at 56.
[2374] PETROLEUM PIPELINES, *supra* note 45, at 107
[2375] Comment, *Public Control of Petroleum Pipe Lines,* 51 Yale L. J. 1338, 1357 (1942).
[2376] Whitesel, *Recent Federal Regulation of the Petroleum Pipe Line as a Common Carrier,* 32 Corn. L. Q. 337, 375-376 (1947).

Senator Birch Bayh, in a colloquy with Frank Ikard during *The Petroleum Industry Hearings*, suggested that the logical extension of Mr. Ikard's argument that continued integration of pipelines was necessary to preserve the efficiencies of integration for the consuming public would be for the Federal Government to take over all the lines and have one unified management.[2377] The immediate response by witnesses[2338] and commentators on the subject[2379] was that post office or railroad type operation by the government was counterproductive and greatly to be avoided. They were quick to point out that in 1952 a gallon of oil and a post card could both be sent from Texas to New York for a penny; 22 years later, a gallon of oil could, under oil company operation, still be transported between these points for a penny but the cost of delivering post cards, under government operation, had risen to 8 cents.[2380] Public reaction to the proposed Federal Oil and Gas Corporation (FOGCO) proposal appears to be strongly on the industry's side of the argument.

3. "Do Nothing" Approach

At the other extreme of the spectrum, there are certain people who, based on the complete failure of proof of any systematic wrongdoing on the part of the oil industry, suggest that the nation would be served better by leaving the industry as it is and turning attention to other matters urgently requiring solution. This would appear to be a bit cavalier, in the light of the fact that there have been some complaints. However, occasional specific complaints can and should be dealt with by the regulatory authorities within the present framework. If violations are found, appropriate remedies are available at their hands. Such complaints should not be used to justify sweeping structural changes of an industry which the evidence shows, on balance, to be highly efficient, flexible, and to have served well producers and refiners, independents as well as majors, with the result that the transportation element has remained a very small percentage of the delivered price to the consumer.

[2377] *Petroleum Industry Hearings, supra* note 192, at 174.
[2378] *Consumer Energy Act Hearings, supra* note 154, at 1307-1308 (Statement by John Winger, Vice President of Chase Manhattan Bank); *Id.* at 1040 (Senator Dewey F. Bartlett).
[2379] STEINGRABER, *supra* note 193, at 153-154; *cf.* Bock, *Lizzie Borden, supra* note 2213, at 52.
[2380] *Consumer Energy Act Hearings, supra* note 154, at 1307-1308 (Statement of John Winger); STEINGRABER, *supra* note 193, at 153.

4. "Competitive Rules" and their Progeny

In its *Deepwater Port Report*[2381] the Justice Department laid down four "ground rules" which it felt were required to avoid the potential competitive problems which it visualized could develop from the Port's construction, ownership and operation by major oil companies. Despite the objections of the parties to the onerous burden imposed by the rules which led to the "cratering" of the SEADOCK proposal; the refusal of the Secretary of Transportation to adopt them without modification,[2382] and the substantial differences between deepwater ports and pipelines, Assistant Attorney General John Shenefield, in his Proposed Statement to the Kennedy Subcommittee, suggested that the original "Competitive Rules" suggested in his Deepwater Port Report be applied without change to oil pipelines.[2383] In doing so, however, Mr. Shenefield surfaced candidly the problems of dilution in quality of issued debt security and compensating the original owners for having carried the risk up to the point of expansion or entry of new owners.

Mention has been made previously of the "Offshore Competitive Principles" written into the Outer Continental Shelf Lands Acts Amendments of 1978, which amended Section 5 of the OCS Lands Act, to provide for FERC regulation of pipelines operating under an OCS right-of-way permit to ensure open and nondiscriminatory access to such lines and mandatory expansion at the request of one or more owners or shippers upon the requisite showing of financial responsibility, technological capability and economic feasibility.[2384] The original paper by Ms. Lucinda Lewis and Dr. Robert Reynolds, presented at the AEI Conference on Oil Pipelines and Public Policy, contained an outline of another adaptation of the "Competitive Rules" called "Joint Venture Rules." It is interesting to note that the authors suggest that "Competitive Rules" are an alternative to rate regulation as well as to divestiture. The idea ought not to be disparaged. All of these proposals represent worthwhile attempts to assuage the Department's concern that pipelines operate in the public interest. This author is of the belief that the Department is on the right track in attempting to improve the current system by isolating those factors and practices it considers to be contrary

---

[2381] A.G.'s Deepwater Port Report, *supra* note 446, at 106-108.

[2382] See HOWREY & SIMON, *supra* note 3, at 102-117 for a critique of these rules, the reasons for the collapse of SEADOCK and the modifications made by the DOT in its Deepwater Port Licenses. See also AEI, *Oil Pipelines and Public Policy, supra* note 684, at 175-187 for a lively discussion of the Rules.

[2383] SHENEFIELD PROPOSED STATEMENT, *supra* note 179, at 33.

[2384] See text at notes 1012-1013, *supra*.

to the public interest and to propound their suggested remedies both in dialogue with the industry and by participating in the administrative process before FERC. Judging from discussions at the AEI Conference, Justice has raised basically three kinds of proposals: One, conditions or rules which the industry, if it will review and evaluate in good faith, will find understandable and liveable, and which could be adopted without undue harm to the industry although their advantages might be debatable. Two, proposals which are clear enough but which are counterproductive or unworkable, to which the industry should respond by giving detailed explanations why they are unacceptable. The third, and last category, are proposals which are unclear or not sufficiently well thought out so that industry is unable to reach an intelligent decision whether it can or cannot live with them. These should be the subject of further mutual discussion so that, ultimately, they can be placed in the first or second categories. By doing this, the public interest will be served. This approach would seem to be precisely in accord with Donald Kaplan's sentiments expressed during the general discussion at the AEI Conference.[2385] It likewise would go far toward solving the practical problem, voiced by Ulyesse LeGrange, at the same conference, that a businessman needs to know in advance how much these additional safeguards are going to cost and how they will affect his operation so that he may make an intelligent "going-in" decision whether the additional risks posed by such conditions make his projected investment uneconomic.[2386] Much work needs yet to be done to resolve the questions and concerns raised by the Justice Department's apparent insistence on some form of "competitive rules" as a condition for permitting new facilities to be placed into operation.

[2385] AEI, *Oil Pipelines and Public Policy, supra* note 684, at 129 (Comment by Donald Kaplan, Chief, Energy Section, Antitrust Division, U.S. Department of Justice).

[2386] *Id.* at 70, 130, 131, 180-181 (Comments by Ulyesse J. LeGrange, Controller, Exxon Corporation); *see also Id.* at 187. (Summation by Donald I. Baker, former Assistant Attorney General, Antitrust Division, now Partner in law firm of Jones, Day, Reavis, and Pogue).

Appendixes, Figures and Tables

Appendix A

RANGE OF PETROLEUM TRANSPORTATION COSTS / CENTS PER 100 BARREL MILES

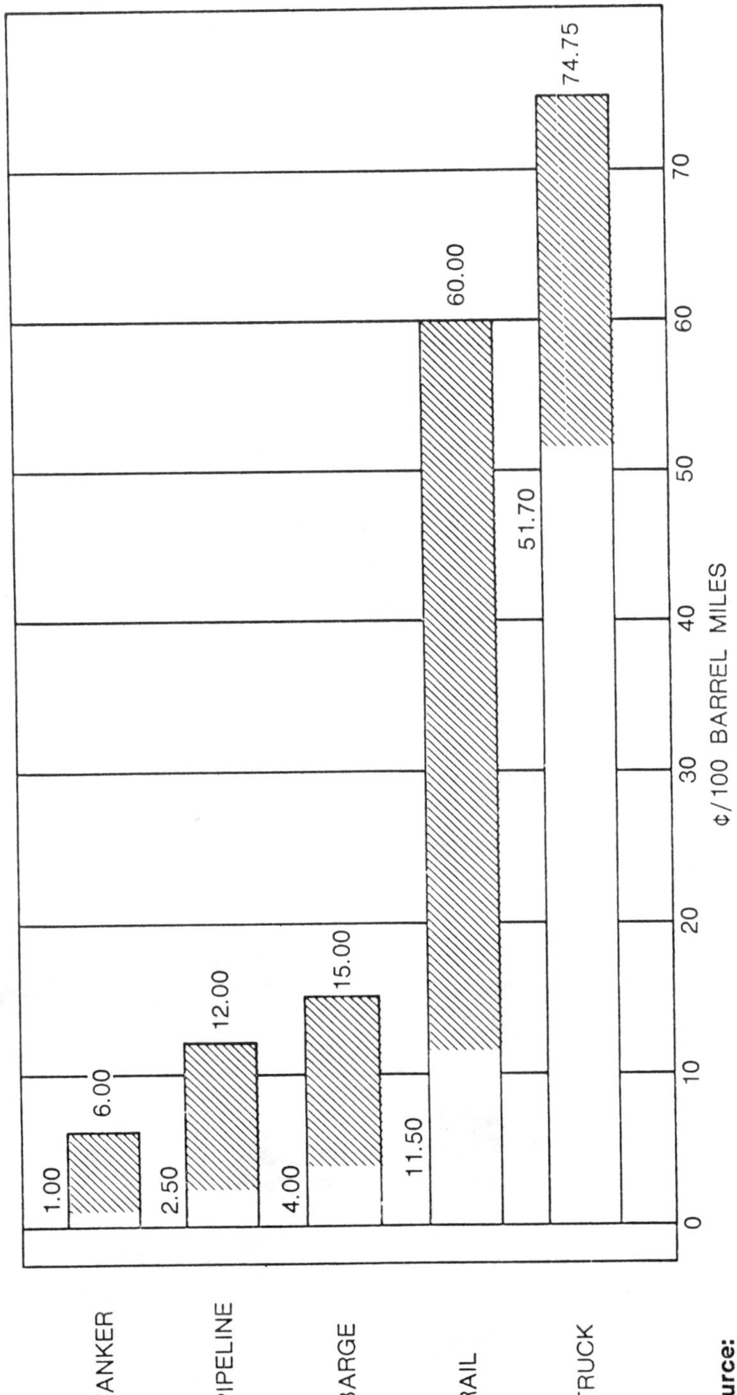

Source:
Exhibit 29, p. 13
Statement of Ulyesse J. LeGrange

Appendix B

SHELL PIPE LINE CORPORATION
TEXAS LOCAL GATHERING TARIFF No. 728
Cancels Tariff No. 598

The rates named in this tariff are for gathering of crude petroleum and petroleum products within the areas of Shell Pipe Line Corporation's gathering systems in the State of Texas named below. The rates are expressed in cents per barrel of 42 U. S. gallons each, subject to regulations of the Railroad Commission of Texas and to the other rules herein contained, and to change without notice, except as otherwise provided by law.

GATHERING SYSTEM	Gathering Charge
Cleveland - North Cleveland ....Liberty County	15
Includes the following fields as designated by Railroad Commission of Texas: Cleveland; Cleveland Cockfield A, Segment E; Cleveland, North; Cleveland Wilcox 9000 (Condensate); Cleveland Yegua 5800 (Condensate)	
Kilgore .... Gregg, Rusk, Smith and Upshur Counties	7
Includes the following field as designated by Railroad Commission of Texas: East Texas Field	
Livingston ....Polk County	10
Includes the following fields as designated by Railroad Commission of Texas: Livingston; Livingston Cockfield; Livingston McElroy; Livingston Sparta; Livingston Wilcox; Livingston Wilcox 1, FB-A; Livingston 4050 Yegua; Livingston 4100 Yegua; Livingston 4950 Yegua (Condensate); Livingston 5000 Yegua	
Mercy .... San Jacinto County	25
Includes the following fields as designated by Railroad Commission of Texas: Mercy; Mercy Mt. Selman (Condensate)	
Cass .... Cass County	10
Includes the following fields as designated by Railroad Commission of Texas: Carbondale North Smackover (Condensate); Carbondale 9660 Smackover; Frost Smackover	

NOTE

Pending reissuance of this tariff, any newly designated field that may be connected to an above-named gathering system will be subject to the applicable gathering charge named for that gathering system.

CANCELLATION NOTICE

Effective with June 10, 1978, McCoy Gathering System, Liberty County, Texas, is eliminated from this tariff. Thereafter, no rate for gathering within this system will be in effect.

ISSUED JUNE 9, 1978   EFFECTIVE JUNE 10, 1978

J. R. HURLEY, President
SHELL PIPE LINE CORPORATION
P. O. Box 2648
Houston, Texas 77001

RULES AND REGULATIONS OF RAILROAD COMMISSION OF TEXAS

1. **All Marketable Oil To Be Received for Transportation.** — By the term "marketable oil" is meant any crude petroleum adapted for refining or fuel purposes, properly settled and containing not more than two per cent (2%) of basic sediment, water, or other impurities above a point six (6) inches below the pipe line connection with the tank. Pipe lines shall receive for transportation all such "marketable oil" tendered; but no pipe line shall be required to receive for shipment from any one person an amount exceeding three thousand (3,000) barrels of petroleum in any one (1) day; and, if the oil tendered for transportation differs materially in character from that usually produced in the field and being transported therefrom by the pipe line, then it shall be transported under such terms as the shipper and the owner of the pipe line may agree or the Commission may require.

2. **Basic Sediment, How Determined—Temperature.** — In determining the amount of sediment, water or other impurities, a pipe line is authorized to make a test of the oil offered for transportation from an average sample from each such tank, by the use of centrifugal machine, or by the use of any other appliance agreed upon by the pipe line and the shipper. The same method of ascertaining the amount of the sediment, water or other impurities shall be used in the delivery as in the receipt of oil. A pipe line shall not be required to receive for transportation, nor shall consignee be required to accept as a delivery, any oil of a higher temperature than ninety degrees Fahrenheit (90° F) except that during the summer oil shall be received at any atmospheric temperature, and may be delivered at like temperature. Consignee shall have the same right to test the oil upon delivery at destination that the pipe line has to test before receiving from the shipper.

3. **"Barrel" Defined.** — For the purpose of these rules, a "barrel" of crude petroleum is declared to be forty-two (42) gallons of 231 cubic inches per gallon at sixty degrees Fahrenheit (60° F).

4. **Oil Involved in Litigation, Etc.—Indemnity Against Loss.** — When any oil offered for transportation is involved in litigation, or the ownership is in dispute, or when the oil appears to be encumbered by lien or charge of any kind, the pipeline may require of shippers an indemnity bond to protect it against all loss.

5. **Storage.** — Each pipe line shall provide, without additional charge, sufficient storage, such as is incident and necessary to the transportation of oil, including storage at destination or so near thereto as to be available for prompt delivery to destination point, for five (5) days from the date of order of delivery at destination.

6. **Identity of Oil, Maintenance of.** — A pipe line may deliver to consignee, either the identical oil received for transportation, subject to such consequences of mixing with other oil as are incident to the usual pipe line transportation, or it may make delivery from its common stock at destination; provided, if this last be done, the delivery shall be of substantially like kind and market value.

7. **Minimum Quantity To Be Received.** — A pipe line shall not be required to receive less than one (1) tank carload of oil when oil is offered for loading into tank cars at destination of the pipe line. When oil is offered for transportation for other than tank car delivery, a pipe line shall not be required to receive less than five hundred (500) barrels.

8. **Gathering Charges.** — Tariffs to be filed by a pipe line shall specify separately the charges for gathering of the oil, for transportation, and for delivery.

9. **Gauging, Testing and Deductions.** — All crude oil tendered to a pipe line for transportation shall be gauged and tested by a representative of the pipe line prior to its receipt by the pipe line. The shipper may be present or represented at the gauging and testing. Quantities shall be computed from correctly compiled tank tables showing one hundred percent (100%) of the full capacity of the tanks, and adjustments shall be made for temperature from the nearest whole number degree to the basis of sixty degrees Fahrenheit (60° F.) and to the nearest whole API degree in accordance with the volume correction Table No. 6 contained in the American Society for Testing Materials and the Institute of Petroleum's Petroleum Oil Measurements Tables, ASTM Designation ASTM: D-1250-IP:200, issued January, 1952. A pipe line may deduct the basic sediment, water, and other impurities as shown by the centrifugal or other test agreed upon. The net balance shall be the quantity deliverable by the pipe line.* In allowing the deductions, it is not the intention of the Commission to affect any tax or royalty obligations imposed by laws of Texas on any producer or shipper of crude oil.

The gauging and testing of oil by the pipe line representative is directed toward and intended to require tank gauge measurement of produced crude prior to the transfer of custody to the initial transporter from a producing property. A transfer of custody of crude between transporters is subject to measurement as agreed upon by the transporters.

---

* This deviates from Rule 71(A)(9) of the General Conservation Rules, in that no deduction will be made for evaporation and loss during transportation.

10. **Delivery and Demurrage.** — Each pipe line shall transport oil with reasonable diligence, considering the quality of the oil, the distance of transportation, and other material elements, but at any time after receipt of a consignment of oil, upon twenty-four (24) hours' notice to the consignee, may offer oil for delivery from its common stock at the point of destination, conformable to Section 6 of this rule, at a rate not exceeding ten thousand (10,000) barrels per day of twenty-four (24) hours. Computation of time of storage (as provided for in Section 5 of this rule) shall begin at the expiration of such notice. At the expiration of the time allowed in Section 5 of this rule for storage at destination, a pipe line may assess a demurrage charge on oil offered for delivery and remaining undelivered, at a rate for the first ten (10) days of one-tenth of one cent per barrel; for the next ten (10) days at a rate of two-tenths of one cent per barrel; and thereafter at a rate of three-tenths of one cent per barrel, for each day of twenty-four (24) hours or fractional part thereof.

11. **Unpaid Charges, Lien For and Sale to Cover.** — A pipe line shall have a lien on all oil to cover charges for transportation, including demurrage, and it may withhold delivery of oil until the charges are paid. If the charges shall remain unpaid for more than five (5) days after notice of readiness to deliver, the pipe line may sell the oil at public auction at the general office of the pipe line on any day not a legal holiday. The date for the sale shall not be less than forty-eight (48) hours after publication of notice in a daily newspaper of general circulation published in the city where the general office of the pipe line is located. The notice shall give the time and place of the sale, and the quantity of the oil to be sold. From the proceeds of the sale, the pipe line may deduct all charges lawfully accruing, including demurrage, and all expenses of the sale. The net balance shall be made to the person lawfully entitled thereto.

12. **Notice of Claims.** — Notice of claims for loss, damage or delay in connection with the shipment of oil must be made in writing to the pipe line within ninety-one (91) days after the damage, loss, or delay occurred. If the claim is for failure to make delivery, the claim must be made within ninety-one (91) days after a reasonable time for delivery has elapsed.

13. **Telephone-Telegraph Line—Shipper to Use.** — If a pipe line maintains a private telegraph or telephone line, a shipper may use it without extra charge, for messages incident to shipments. However, a pipe line shall not be held liable for failure to deliver any messages away from its office or for delay in transmission or for delay in transmission or for interruption of service.

14. **Contracts of Transportation.** — When a consignment of oil is accepted, the pipe line shall give the shipper a run ticket, and shall give the shipper a statement that shows the amount of oil received for transportation, the points of origin and destination, corrections made for temperature, deductions made for impurities, and the rate for such transportation.

15. **Shipper's Tank, etc.—Inspection.** — When a shipment of oil has been offered for transportation, the pipe line shall have the right to go upon the premises where the oil is produced or stored, and have access to any and all tanks or storage receptacles for the purpose of making any examination, inspection, or test authorized by this rule.

16. **Offers in Excess of Facilities.** — If oil is offered to any pipe line for transportation in excess of the amount that can be immediately transported, the transporation furnished by the pipe line shall be apportioned among all shippers in proportion to the amounts offered by each; but no offer for transportation shall be considered beyond the amount which the person requesting the shipment then has ready for shipment by the pipe line. The pipe line shall be considered as a shipper of oil produced or purchased by itself and held for shipment through its line, and its oil shall be entitled to participate in such apportionment.

17. **Interchange of Tonnage.** — Pipe lines shall provide the necessary connections and facilities for the exchange of tonnage at every locality reached by two or more pipe lines, when the Commission finds that a necessity exists for connection, and under such regulations as said Commission may determine in each case.

18. **Receipt and Delivery—Necessary Facilities For.** — Each pipe line shall install and maintain facilities for the receipt and delivery of marketable crude petroleum of shippers at any point on its line if the Commission finds that a necessity exists therefor, and under regulations by the Commission.

19. **Fires, Lightning and Leakage, Reports of Loss From.**

(a) Each pipe line shall immediately notify the Commission, by telegraph, telephone, or letter, of each fire that occurs at any oil tank owned or controlled by the pipe line, or of any tank struck by lightning. Each pipe line shall in like manner report each break or leak in any of its tanks or pipe lines from which more than five (5) barrels escapes. Each pipe line shall report in writing to the Commission, by the fifteenth (15th) day of each calendar month, the estimated amount loss of oil by fire or leakage from its tanks and pipe lines for the preceding month; but not including leakage or evaporation ordinarily incident to transportation.

(b) No risk of fire, storm, flood or act of God, and no risk resulting from riots, insurrection, rebellion, war, or act of the public enemy, or from quarantine or authority of law or any order, requisition or necessity of the government of the United States in time of war, shall be borne by a pipe line, nor shall any liability accrue to it from any damage thereby occasioned. If loss of any crude oil from any such causes occurs after the oil has been received for transportation, and before it has been delivered to the consignee, the shipper shall bear a loss in such proportion as the amount of his shipment is to all of the oil held in transportation by the pipe line at the time of such loss, and the shipper shall be entitled to have delivered only such portion of his shipment as may remain after a deduction of his due proportion of such loss, but in such event the shipper shall be required to pay charges only on the quantity of oil delivered. This rule shall not apply if the loss occurs because of negligence of the pipe line.

20. **Printing and Posting.** — Each pipe line shall have Sections 1 through 19 of this rule printed on its tariff sheets, and shall post the printed sections in a prominent place in its various offices for the inspection of the shipping public. Each pipe line shall post and publish only such rules and regulations as may be adopted by the Commission as general rules or such special rules as may be adopted for any particular field.

SPECIAL RULES COVERING PETROLEUM PRODUCTS

This tariff applies on shipments of: (1) Any petroleum product, or mixture of petroleum products, which by A. S. T. M. distillation methods substantially distills below seven hundred degrees (700°) and above four hundred degrees (400°) Fahrenheit, and which has a Reid vapor pressure not exceeding twenty (20) pounds at one hundred degrees (100°) Fahrenheit; or (2) Any petroleum product, or mixture of petroleum products, which by A. S. T. M. distillation methods substantially distills below four hundred degrees (400°) Fahrenheit, if the same shipper also offers for shipment to the same destination a sufficient quantity of crude petroleum to produce a mixture which has a Reid vapor pressure not exceeding that permitted by carrier's facilities and operating conditions.

Appendix C

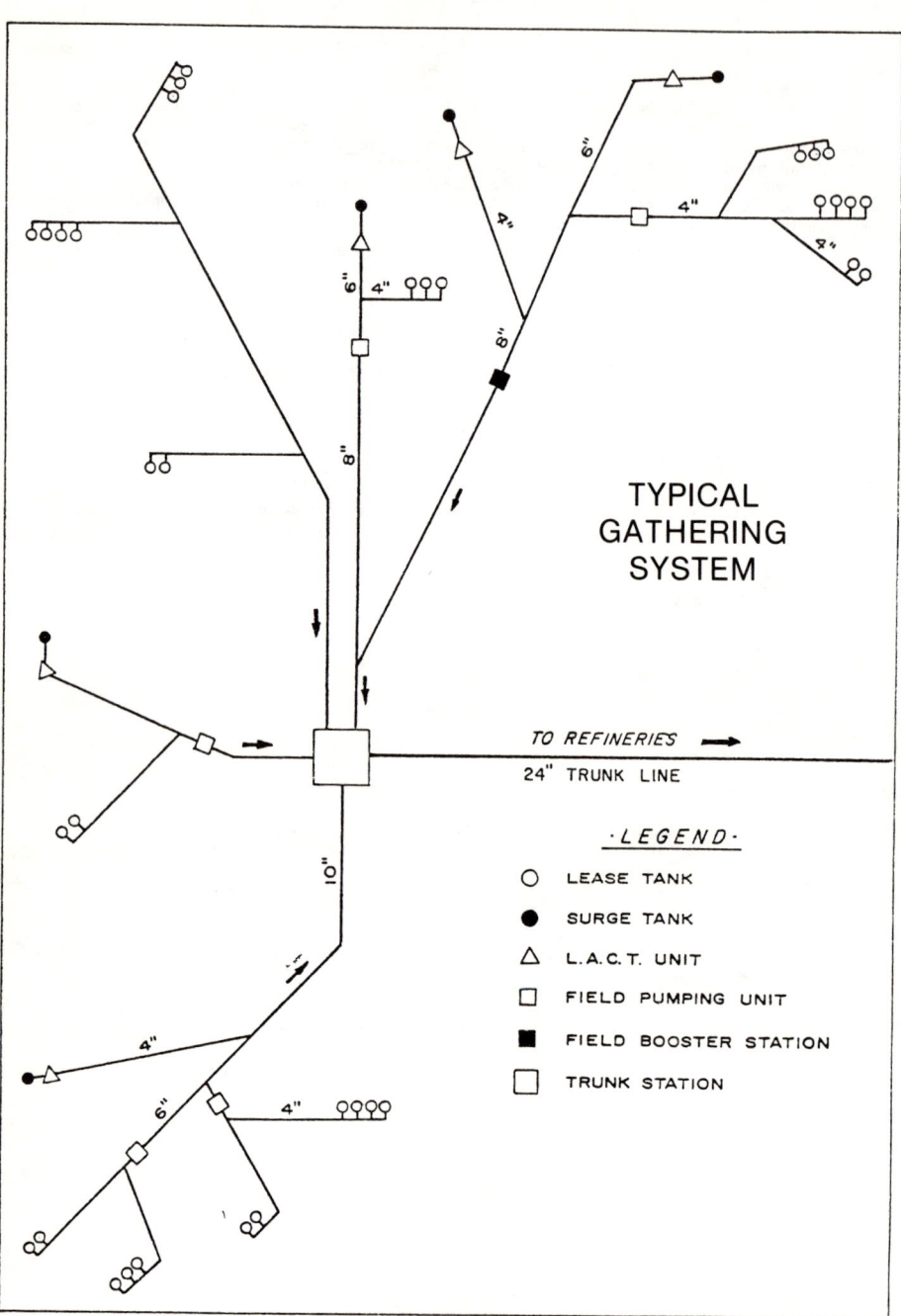

CRUDE OIL GATHERING IN TEXAS
December 1978

Rank	Company	B/D Gathered	Cumulative Percent
1	Amoco	586,152	21.8
2	Exxon	405,660	36.9
3	Permian (Inc. Western Oil & Trns)	179,699	43.6
4	Mobil	168,708	49.9
5	Arco	157,359	55.7
6	Marathon	125,940	60.4
7	Gulf	119,023	69.1
8	Shell	113,888	69.1
9	Sun	108,427	73.1
10	Texas-New Mexico Pipeline*	97,557	76.8
11	Koch (Inc. Matador)	86,253	80.0
12	Scurlock	84,236	83.1
13	Phillips	71,582	85.8
14	Conroe Corp.	58,962	88.0
15	Cities Service	54,087	89.0
16	American Petrofina (Inc. Cosden & Amdel)	45,466	91.7
17	Texaco	34,023	92.9
18	Union of Cal.	31,946	94.1
19	Tesoro (Inc. Hondo)	28,330	95.2
20	Sohio	17,945	95.8
21-145		111,485	100.0
	Total	2,686,728	

*Undivided joint interest pipeline: Texaco (45.3%), Shell (31.3%), Cities Service (8.9%), and Arco (14.5%).

Source: Petroleum Information Corporation, *Company Production, Crude and Casinghead Gas,* State of Texas, December 1978.

REFINERY RECEIPTS OF CRUDE OIL AND CONDENSATE TEXAS, FIRST HALF 1978

REFINER Crude Source	Thousand Barrels	REFINER Crude Source	Thousand Barrels
ADOBE REF. CO		**COASTAL STATES**	
Crystal Oil	320	**PETROCHEMICAL**	
Permian Corp.	519	Coastal Sts. Mktg. (I)	22,160
Koch Oil Co.	30	Coastal Sts. Crude Gath	2,252
		Texaco Inc. (I)	504
CHAMPLIN PETROLEUM		Wholesale & Retail Outlets	609
Agip (I)	1,222		
Amoco (I)	2,221	**FLINT INK CORP.**	
Arco	329	Permian Corp.	87
Champlin PL	1,453	Scurlock Oil Co.	32
Ashland Oil Co. (I)	1,325		
Chevron Oil Co. (I)	381	**HOWELL HYDROCARBONS**	
Coastal States Gath Co.	482	Ensearch Exploration	1
Coastal Sts. Crude (I)	575	Howell Hydrocarbons	435
Coastal Sts. PL	578	Koch Oil Co.	25
Conoco (I)	414	Permian Corp.	114
Crown Central (I)	358	Tesoro Petroleum	18
Coral Petroleum Co. (I)	155		
Derby & Co.	461	**PIONEER REFINING LTD.**	
Derby	341	Basin Inc.	33
Ekofisk (I)	200	Coastal Sts. Trading Inc.	5
Exxon PL Co.	2,517	Compton Corp.	310
Kerr McGee	396	P&O Falco Inc.	a
Koch Oil Co. (I)	543	Permian Corp	179
La Gloria (I)	900	Producers Crude	25
Mesa PL Co.	38	Scurlock Oil Co.	58
Mobil PL Co.	43	Summit Gas Co.	109
Permian Corp.	37	Union Oil Inc.	62
Powerine Oil Co. (I)	133		
Rittenbury Oil Co. (I)	1,131	**QUINTANA-HOWELL**	
Saber Petroleum	4	**JOINT VENTURE**	
Scurlock Oil Co.	166	Exxon PL Co.	6,166
Sonatrak (I)	2,575	Koch Oil	110
Sun Oil Co. (I)	477	SWTL	76
Summit Oil Co.	21	Unknown Receipts	33
Tesoro Crude Oil Co.	19		
Texaco Oil Co. (I)	1,649	**RAYMAL REFINING LTD.**	
Total Intl. (I)	733	Petroleum Mgt. Inc.	9
Union Oil Co. (I)	2,770	Tesoro Petroleum Co.	5
Wesco (I)	314	Union Oil Inc.	163

REFINER Crude Source	Thousand Barrels	REFINER Crude Source	Thousand Barrels
SABER REFINING INC.		**TIPPERARY REFINING CO.**	
American Petrofina	14	Compton Corp	2
Carbonit	444	Petroleum Mgm. Inc.	6
Derby & Co.	541	Sun PL Co.	655
Saber Refining	1,526	Union Oil Co.	312
Southwestern	150	Hondo PL Co.	5
Sun Oil Co.	799		
Sigmor	83	**AMERICAN PETROFINA CO. OF TX.**	
		Amdel PL Co.	12,764
SIGMOR REFINING CO.		Amoco PL Co.	1,877
Basin Inc.	9		
Energy Distributing Co.	33	**AMOCO TX. REFINING CO.**	
Falco Inc.	22	Amoco Production Co.	688
Gulf Energy	a	Amoco Production Co. (I)	36
PMI	1	Amoco PL Co.	21,519
Permian Corp.	92	Amoco Oil Co.	43
Saber Petroleum	14	Amer. Overseas Oil (I)	32,588
Scurlock	11	American Petrofina	1,137
Sigmor PL Co.	1,220	Scurlock Oil Co. (I)	94
Sigmor Refining Co.	52	Scurlock Oil Co.	55
Tesoro Petroleum	217	Shell Oil Co. (I)	542
		Sohio (I)	5,111
SENTRY REFINING INC.		Sun Oil Co.	102
Coastal Sts. Crude Gath	62	Sun PL Co.	81
Coastal Sts. Trading	54	Texas PL Co.	65
Coral Petroleum	88	Intracompany Receipts	4,847
Import frm. Venezuela (I)	625		
		ATLANTIC RICHFIELD CO.	
SOUTHWESTERN REFINING CO.		Act	3,148
Agip (I)	1,909	Amoco (I)	122
Arco (I)	766	Amoco Intl. (I)	1,409
BP Trading (I)	3,005	Arco PL Co.	32,124
Exxon PL Co.	2,422	Bonaire Petroleum (I)	1,824
Exxon PL	3,336	British Petroleum (I)	371
Mobil PL	203	BP Trading (I)	1,625
Permian Corp.	1,675	Chevron (I)	2,089
Shell Intl. (I)	3,267	Derby (I)	1,451
Shell (I)	683	ECI (I)	1,164
		Exxon PL Co.	838
SUN OIL CO. OF PA		Getty	307
American Petrofina PL	63	Gulf Trading	352
Compton Corp.	60	IPT	551
Exxon Pipeline Co.	734	Mark Rich (I)	3,420
PMI	81	Mobil	326
Sun Oil Co.	944	Petroleo Mexicanos (I)	1,825
Sun PL Co.	8,499	Phillips Petroleum (I)	253
		Rancho PL	14,314
TESORO PETROLEUM CORP.		Rittenberry (I)	979
Hondo PL	2,093	Scurlock Oil Co.	262
Tesoro Crude Oil Co.	460		

REFINER Crude Source	Thousand Barrels	REFINER Crude Source	Thousand Barrels
ATLANTIC RICHFIELD CO. (con't)		**MARATHON OIL CO.**	
Shell	137	Amerada Hess	390
Shell Oil Co. (I)	1,423	Cities Service Oil Co.	188
Sohio (I)	4,685	Exxon PL	2,029
Texaco Exporting Co. (I)	608	Exxon PL Co.	7,260
Texas PL Co.	217	Marathon Oil Co. (I)	102
Texas-NM PL Co.	403	Mobil Oil Co. (I)	489
Various Companies (I)	563	Permian Oil Co.	195
		Shell Oil Co. (I)	214
CHARTER INTL. OIL CO.		Texaco Inc. (I)	780
Isthmus Crude (I)	334	Texas PL Co.	261
Texas New Mexico PL	1,448	Texas City Refining (I)	370
		Western Crude Oil Co. (I)	206
CROWN CENTRAL PETROLEUM CORP.		To Balance Exxon PL Reports	75
Algerian National (I)	5,511	**MOBIL OIL CORP.**	
American Petrofina (I)	978	Ensearch Expl. Inc.	6,683
Amoco Oil Co. (I)	103	Mobil Oil Corp.	11,954
Amoco Production Co. (I)	418	Mobil PL Co.	27,651
Aminoil (I)	73	Wholesale/Retail	8,003
Atlantic Richfield (I)	324		
Champlin	80	**PHILLIPS PETROLEUM CO.**	
Crown Central PL	689	Amoco Production Co. (I)	276
Cities Service	382	Phillips Petro. Co (I)	11,305
Derby & Co. (I)	383	Phillips Petroleum Co.	803
Exxon PL Co.	682	Phillips PL Co.	3,955
Mark Rich (I)	138		
Nigerian National (I)	1,754	**RANCHO REFINING CO. OF TX.**	
Occidental	1,118	Coastal Sts. Gath	152
Pemex	246	Summit Gas Co.	4
Rancho PL	3,085		
Scurlock	35	**SHELL OIL CO.**	
		Barges from Buccaneer	42
EDDY REFINING CO.		Barges from Wks. Island	613
Scurlock	493	Shell Oil Co.	12,834
		Rancho PL Co.	22,407
EXXON CO. USA		Shell PL Co.	7,046
Arco PL Co. (I)	378	To Balance Report	9
Champlin Petroleum	445		
Exxon Intl. (I)	37,441	**SOUTH HAMPTON CO.**	
Exxon Co. USA (I)	644	Gulf Sts. PL Co.	1,574
Exxon PL Co.	46,128	Hill Brothers	2
Pemex (I)	2,373	Permian Corp	120
Texaco (I)	422	Petran Inc.	46
To Balance Report	2	Scurlock Oil Co.	57
Mobil Oil Co.(I)	324	South Hampton Co.	14
		Western Crude Oil Co.	1
GULF OIL CO. USA			
Gulf Refining Co. PL	37,810		
Gulf Oil Co. (I)	17,120		

REFINER Crude Source	Thousand Barrels	REFINER Crude Source	Thousand Barrels
TEXACO INC.		**J & W REFINING INC.**	
(Port Arthur)		Basin Inc.	10
East Tx. Main PL	1,055	Falco	137
Gulf PL Co.	434	Intercoastal	a
Texas PL Co.	40,527	J&W Refining	502
Texaco Inc. Refinery	20,735	Lubrizoil	a
(Port Neches)		Permian Corp.	66
Coastal States (I)	491	Petroleum Carriers	22
Texaco Export Inc.(I)	24,477	Placid/Scurlock	1
Texas Pipeline Co.	3,795	Scurlock Oil Co.	88
		Ted True Inc.	5
TEXAS CITY REFINING INC.		UPG	1
Amoco Oil Overseas (I)	548	Intracompany Receipts	6
Amoco Production Co.	110		
Cajun Marine/Shell	272	**LA GLORIA OIL & GAS CO.**	
Chevron	37	McMurrey PL Co.	4,380
Companie de Francais de Petrol (I)		Scurlock Oil Co.	691
	468		
Crown Petroleum Co.	128	**LONGVIEW REFINING CO.**	
Derby & Co.	426	Abney Oil Co.	54
Exxon PL Co.	1,200	Crystal Oil Co. (I)	43
To Balance Report	267	Permian Corp.	1,349
Koch Oil Co. (I)	282	Scurlock Oil Co.	173
Koch Oil Co.	233	Ted True Inc.	9
Marathon Oil Co. (I)	178		
Permian Corp.	39	**MID-TEX REFINERY**	
Petrobarge	39	Exxon PL	60
Rittenbury & Assoc. (I)	202	Dynamic Industries	66
Scurlock Oil Co. (I)	102	Falco Inc.	50
Scurlock Oil Co.	249		
Scurlock Corp.	29	**PETROLITE CORP BARCO DIV.**	
Shell Oil Co. (I)	2,181	Hill Bors. Transport	a
Sonatrack (I)	750	Petrolite Corp.	70
Texas PL Co.	2,578	Wholesale/Retail	5
UNION OIL CO. OF CA.		**TEXAS ASPHALT & REFINING CO.**	
Union 76 Div/Union Oil	15,694	Armada Petroleum Corp.	6
Union 76 Div/Union Oil (I)	2,707	Big-Tex Crude	2
		Compton Corp.	23
DORCHESTER REFINING CO.		Geer Tank Trucking	10
Dorchester PL Co.	3,980	Gratex	67
		Hill Brothers	1
GULF STATES OIL AND REFINING CO.		Petroleum Industries Inc.	132
Falco Inc.	a	**WINSTON REFINING CO.**	
Lajet Inc.	1	Permian Corp.	2,466
Permian Corp.	16		
Scurlock Oil Co.	408		
Western Crude Oil	228		

REFINER Crude Source	Thousand Barrels	REFINER Crude Source	Thousand Barrels
PRIDE REFINING INC.		**THRIFTWAY INC.**	
Big-Tex	19	Encorp	I
Doma Corp.	65	Falco	I
Falco	a	Geer Tank Trucks	27
Holtman Tank Trucks	8	Gough Tank Trucks	78
Kerr-McGee	22	Gratex Corp.	35
Mid-Tex Refining	2	Mobley Transport Inc.	1
Permian Corp.	1,310	Permian Corp.	25
Pride Transport Co.	7	Thriftway Inc.	19
Pride PL Co.	1,014	Teague Tank Trucks	2
Saber	39		
Scurlock Oil Co.	50	**DIAMOND SHAMROCK CORP.**	
Sigmor	2	Diamond Shamrock Corp.	10,749
Union Oil Co.	12		
West Tx. Gulf PL	186	**PHILLIPS PETROLEUM CO.**	
		Phillips Petroleum Co.	17,332
CHEVRON USA INC.			
Chevron PL Co.	13,773	**TEXACO INC.**	
		Texas PL Co.	32,591
COSDEN OIL & CHEMICAL CO.			
Cosden PL Co.	9,143		
SHELL OIL CO. Shell PL Co.	4,210		
TEXACO INC.			
Chevron PL Co.	3,136		

a = Less than 500 barrels
I = Import

Source: Petroleum Information Corporation, *Receipts, Deliveries and Stocks, Crude and Condensate, State of Texas* (Houston: January-June 1978). Based on reports to Texas Railroad Commission by gatherers, transporters, and refiners of crude oil in Texas.

Appendix F

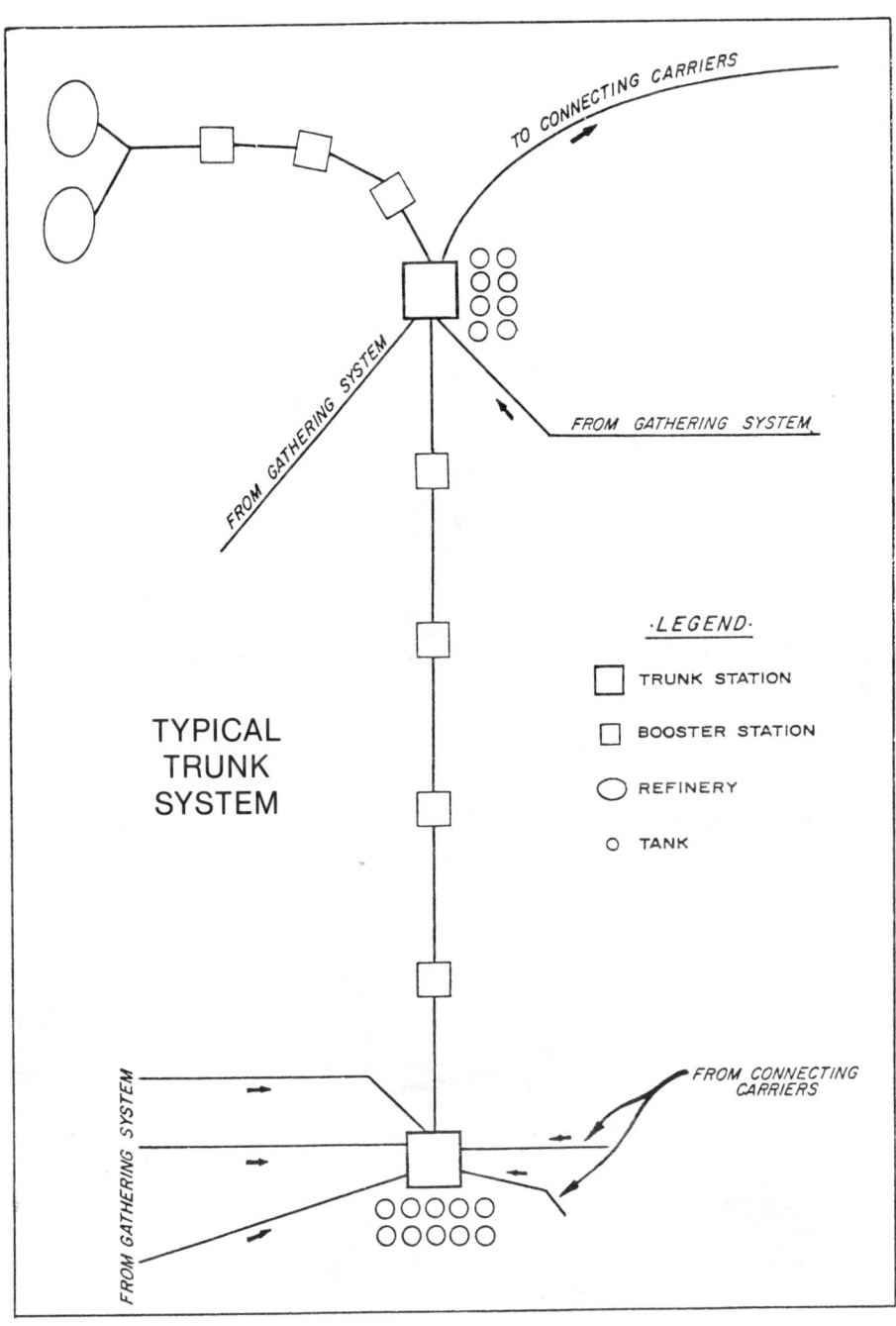

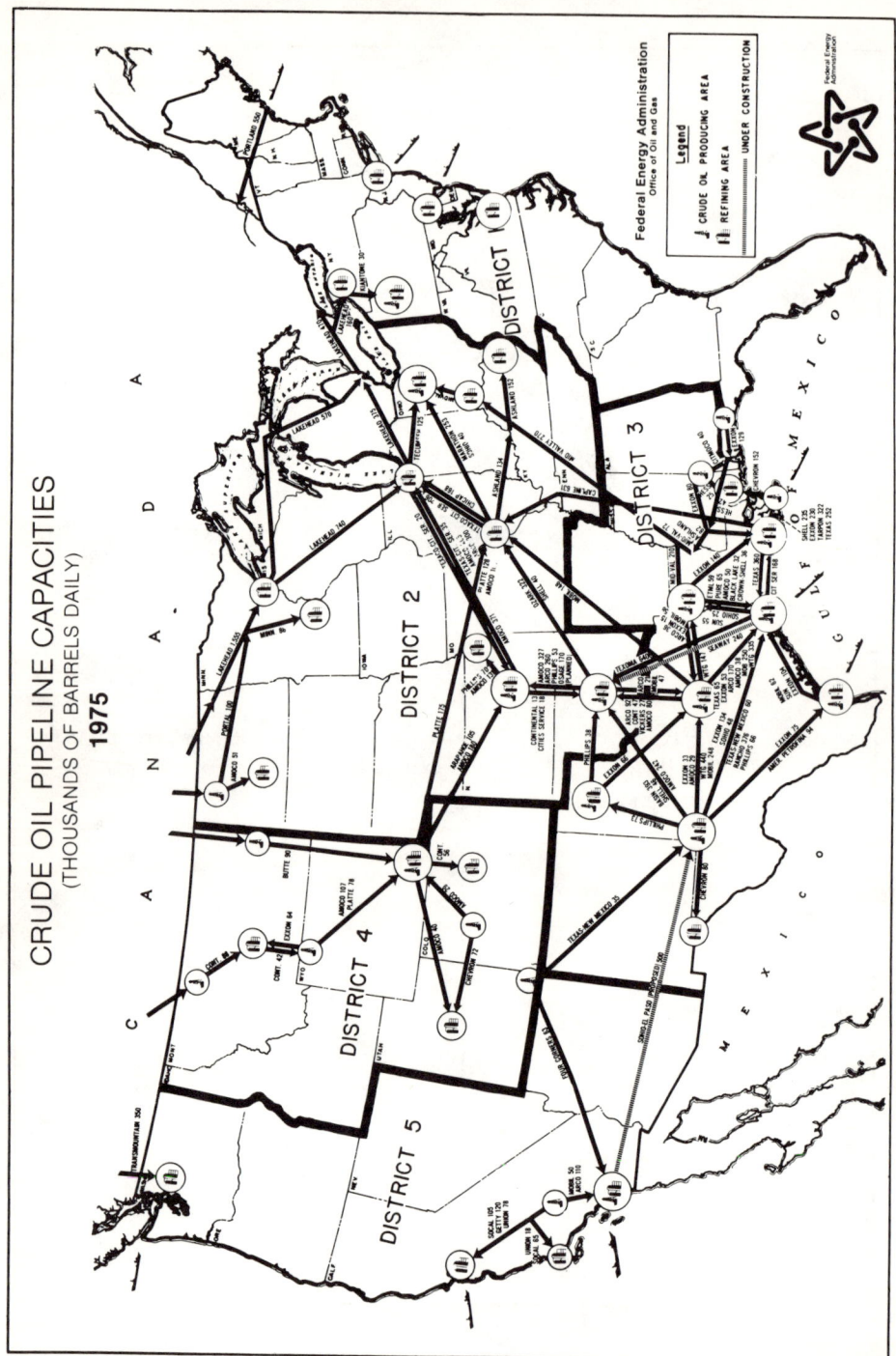

Appendix H

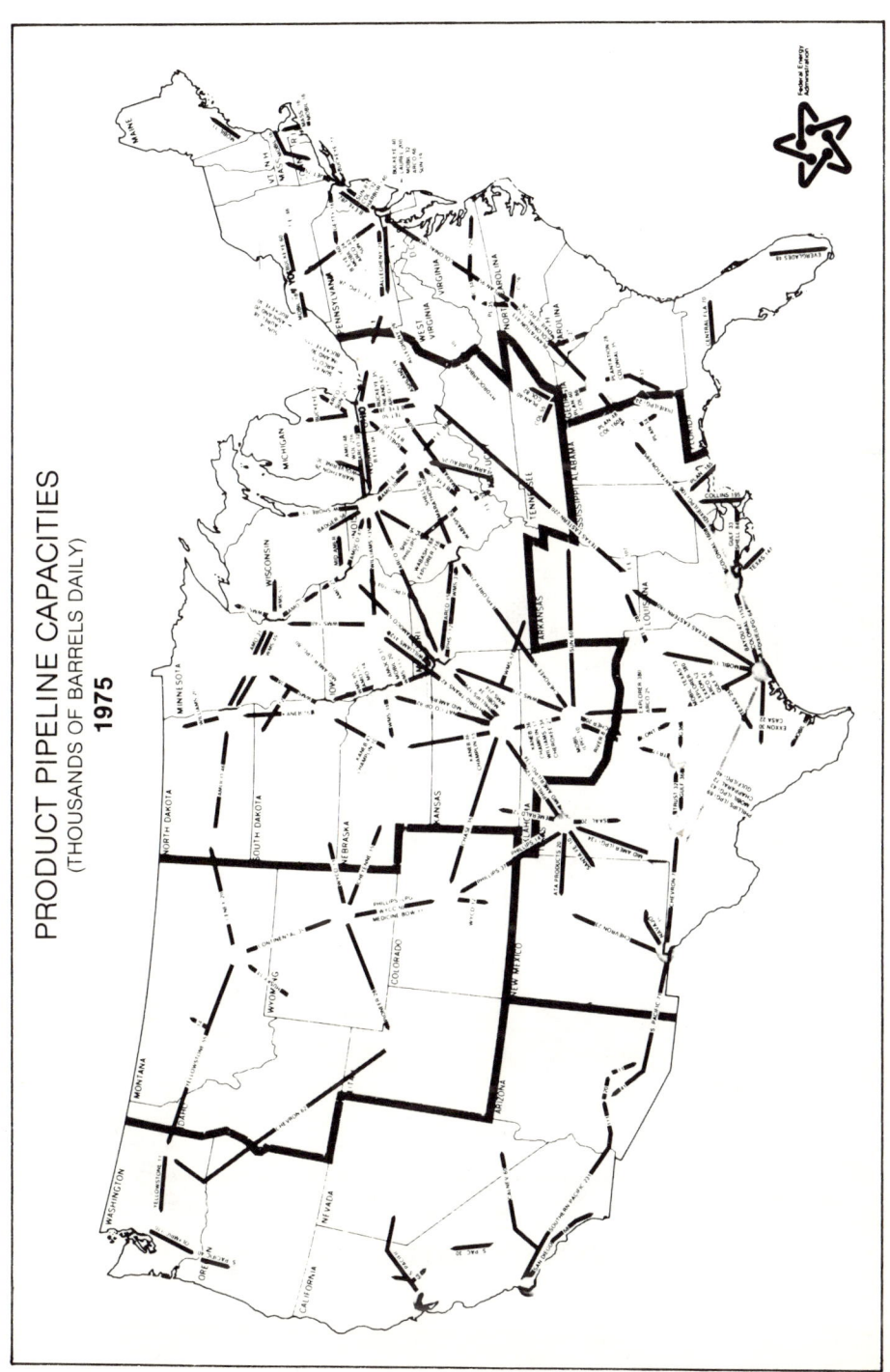

Appendix I

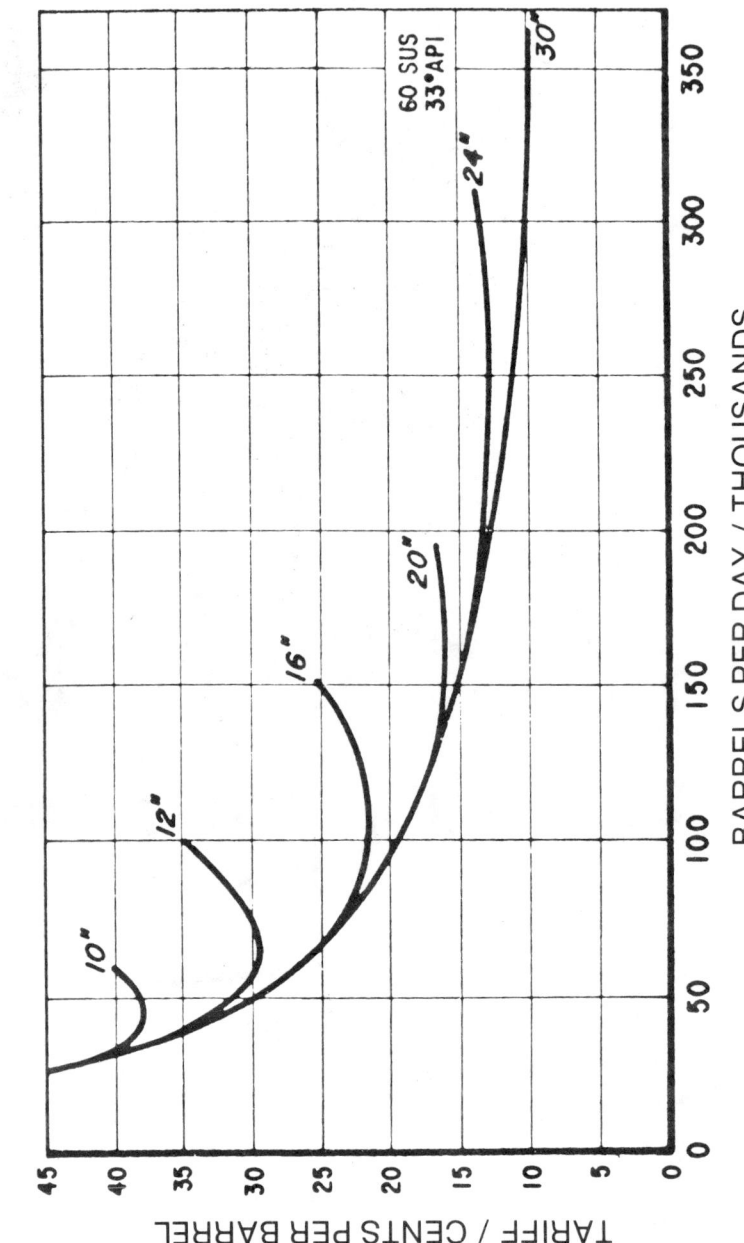

Source: Daniels, L.C. "Economics of Pipeline Industry." Paper presented to Association of Oil Pipe Lines, 1976 Educators Conference, Houston, Texas.

Appendix J

CRUDE REFINING AREAS
AND
CRUDE PIPE LINES
December 1977

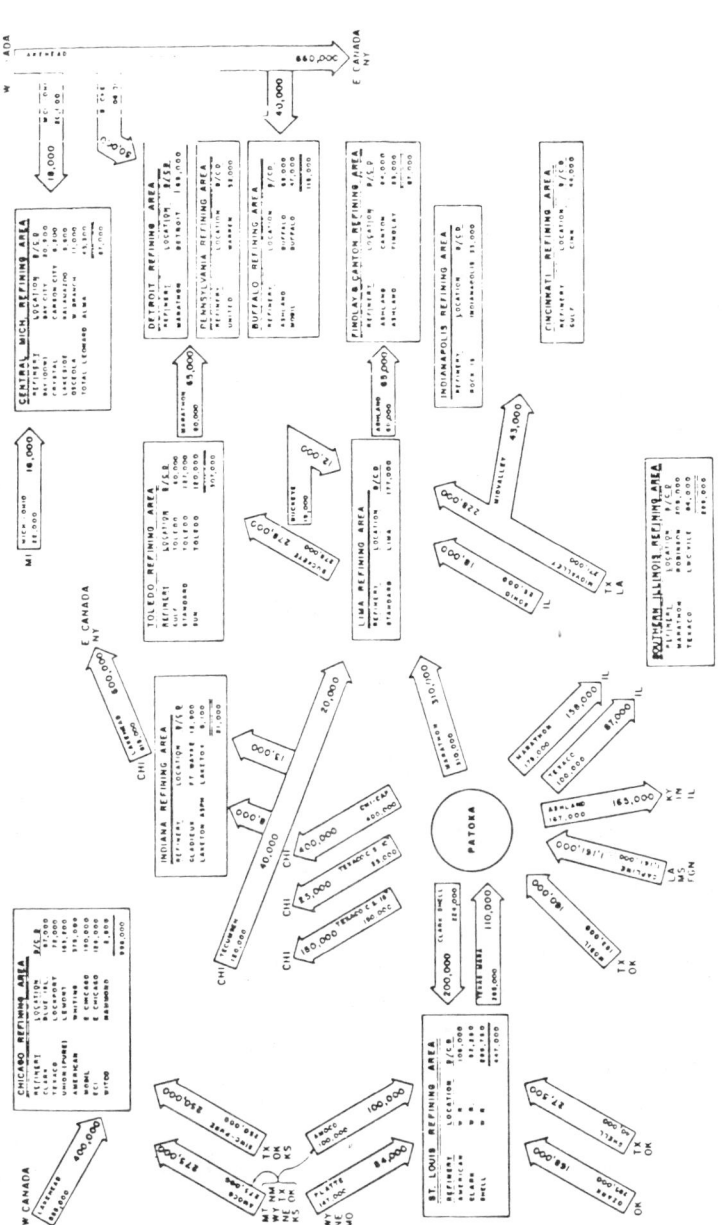

Notes:
1. This chart is not geographically accurate.
2. Flow figures are averages based on Marathon's estimates; certain lines may reach capacity on a seasonal basis.
3. The origins and destinations listed are for the pipeline systems — it should be noted that they could receive crude from many geographic areas depending upon connections. (Source: Marathon Appendix B)

Appendix K

PRODUCTS PIPE LINES
December 1977

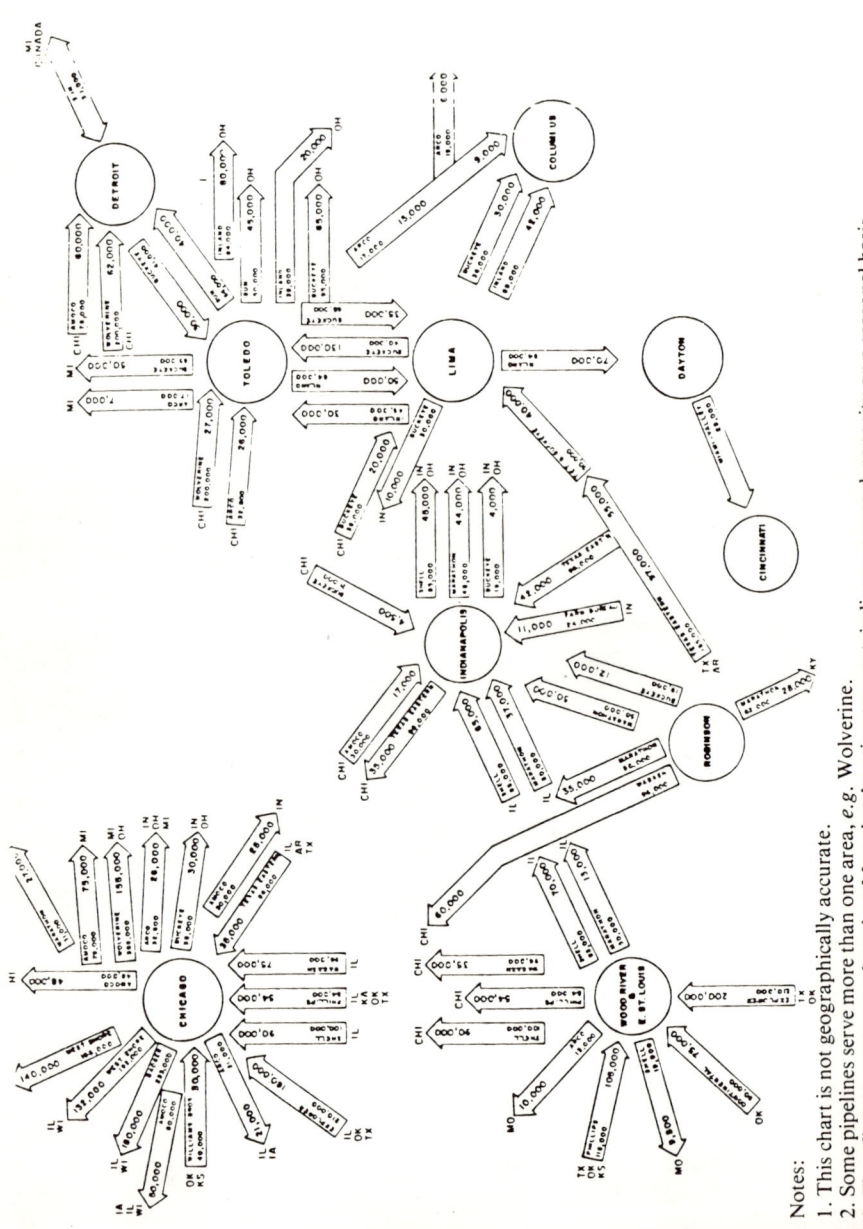

Notes:
1. This chart is not geographically accurate.
2. Some pipelines serve more than one area, *e.g.* Wolverine.
3. Flow figures are average based on Marathon's estimates: certain lines may reach capacity on a seasonal basis.
4. The origins and destinations listed are for the pipeline systems — it should be noted that they could receive product from many geographic areas depending upon connections. (Source: Marathon Appendix D)

TOTAL PETROLEUM PRODUCTS CARRIED IN DOMESTIC TRANSPORTATION AND PERCENT OF TOTAL CARRIED BY EACH MODE OF TRANSPORTATION

In Tons of 2,000 Pounds

Year	Total Petroleum Products Carried	PIPELINES		WATER CARRIERS		MOTOR CARRIERS		RAILROADS	
		Tons Carried	Percent of Total	Tons Carried	Percent of Total	Tons Carried	Percent of Total	Tons Carried	Percent of Total
1968	988,583,300	300,606,600	30.41	253,992,300	25.69	408,800,000	41.35	25,184,400	2.55
1970	1,070,468,000	333,085,000	31.12	286,367,000	26.75	425,200,000	39.72	25,816,000	2.41
1971	1,103,555,900	346,810,800	31.43	302,071,300	27.37	429,900,000	38.96	24,773,800	2.24
1972	1,199,710,500	388,641,400	32.39	322,930,400	26.92	462,500,000	38.55	25,638,700	2.14
1973	1,282,527,200	419,827,600	32.74	330,687,300	25.78	504,177,000	39.31	27,835,300	2.17
1974	1,253,462,500	420,375,600	33.54	323,868,200	25.84	481,993,000	38.45	27,225,700	2.17
1975	1,219,899,100	424,759,300	34.82	326,077,900	26.73	444,398,000	36.43	24,663,900	2.02
1976	1,336,604,000	475,600,300	35.58	349,947,400	26.18	486,615,700	36.41	24,440,600	1.83

TOTAL PETROLEUM PRODUCTS CARRIED IN DOMESTIC TRANSPORTATION AND PERCENT OF TOTAL CARRIED BY EACH MODE OF TRANSPORTATION

In Billions of Ton Miles

Year	Total Petroleum Products Ton Miles	PIPELINES		WATER CARRIERS		MOTOR CARRIERS		RAILROADS	
		Ton Miles	Percent of Total	Ton Miles	Percent of Total	Ton Miles	Percent of Total	Ton Miles	Percent of Total
1972	476.8	191.3	40.1	254	53.3	22	4.6	9.5	2.0
1973	480.4	205	42.7	238	49.5	23.7	4.9	13.7	2.9
1974	488.8	203	41.5	244	49.9	27.7	5.7	14.1	2.9
1975	515.2	219	42.5	257.4	50.0	26.2	5.1	12.6	2.4
1976	526.9	215	40.8	269.1	51.1	30.4	5.8	12.4	2.3

Source: Exhibit K of Tierney TAPS Rebuttal Testimony.

Original Source: "Shifts in Petroleum Transportation," Association of Oil Pipe Lines.

Appendix M

EUGENE ISLAND PIPELINE SYSTEM
PROJECT DEVELOPMENT CHRONOLOGY

1974

8/23/74	Representatives of three companies — Exxon Pipeline Company, Exxon Company, U.S.A., and The Texas Pipe Line Company met to discuss the need for a new pipeline from the Eugene Island and South Marsh Island areas. It was decided that a joint industry study was required to solve this transportation problem.
9/17/74	Approximately 56 companies were invited to participate in the Eugene Island study. All companies having production or production interest in the Eugene Island Block 332 area were invited to participate in the project.
9/18/74	As an alternative to a new pipeline project, the possibility of expanding the capacity of the Bonito Pipeline was examined. The conclusion was to continue to pursue a joint-interest pipeline.
10/30/74	The first steering committee meeting was scheduled for 11/14/74. A summary of the results of the inquiry of 9/17/74 indicated that 15 producers did not respond to the inquiry.
11/14/74	The first steering committee was held. Of the 56 companies initially contacted, 14 companies attended the first steering committee meeting.
12/11/74	The second steering committee was held and Exxon Pipeline Company was selected to perform the feasibility study.

1975

1/2/75	Sun Oil Company submitted a letter of intent to participate in the project. Ten companies are participating in the project.
2/11/75	C&G Producing Company was extended an invitation to participate in the project as a result of a telephone inquiry.
2/14/75	Forrest Oil Corporation was extended an invitation to participate in the project as a result of a telephone inquiry.
4/3/75	Seven companies expressed interest in continuing with the project as shipper-owners. Three of the ten companies — Cities Service Oil Company, Sun Oil Company, and Tenneco Oil Company — have withdrawn from the project as owners. Tenneco and Sun did express interest as shippers only.
4/28/75	An invitation was extended to Mesa Petroleum Company to participate in the project. Mesa joined the project on 4/29/75.

4/30/75	An invitation was extended to Chevron Oil Company to participate in the project.
5/12/75	Burmah Oil Development Inc. joined the project.
6/2/75	Amoco withdrew from project as owner but indicated that volumes may be available for thruput.
6/3/75	The feasibility study of the Eugene Island Pipeline was declared.
6/11/75	Mesa withdrew from project as owner but expressed an interest to use the line as a shipper.
6/23/75	The total volume nominations for the pipeline decreased from a peak of 185,500 B/D in April, 1975, to 174,000 B/D in May, 1975, to 136,000 B/D in June, 1975.
7/17/75	Burmah Oil Development, Inc. withdrew from project as owner but expressed interest as a shipper.
8/75	Williams Pipe Line Company and Kaneb Pipe Line Company had been contacted verbally to determine if they were interested in ownership participation in the Eugene Island pipeline. Both companies would participate if they could arrange for ship-or-pay agreements. Neither company finally participated. Contacting Williams and Kaneb was an effort to solicit additional ownership participation in the pipeline even though neither company had any production in the Eugene Island area.
8/15/75	Oxy Petroleum, Inc. expressed an interest in the EIPS but did not join the project.
8/20/75	Amerada-Hess expressed an interest in the EIPS but did not join the project.
9/8/75	As of September, 1975, only 93.5% of the ownership in the pipeline could be placed. Several of the owner companies that had initially indicated ownership interest —Amoco, Sun Pipe Line Company, Amerada-Hess, and Oxy Petroleum — were contacted again to try to fill the open ownership.
9/8/75	Sun Pipe Line Company rejoined the project as a 7% owner.
10/28/75	The delay in finalizing the ownership nominations was jeopardizing the project. Therefore, the tentative owners of the pipeline were forced to accept larger ownership percentages than each wanted to finalize the project.
5/7/76	Construction began on the 20" pipeline.
11/1/76	The Eugene Island Pipeline was officially "started up".

Source: Exxon Pipeline Company.

TEXOMA PIPELINE SYSTEM
PROJECT DEVELOPMENT CHRONOLOGY

December, 1972 - February 1973	Information concerning the Texoma project and inviting participation was sent to all refineries and pipeline companies which conceivably could have an interest in the proposed system. Of the 23 companies who received invitations, 10 companies expressed some interest.
March, 1973	Meetings with interested companies were held to determine which were willing to participate in and support a final feasibility study and preliminary engineering design.
June, 1973	Crude Oil International Company, Kerr-McGee Corporation, Lion Oil Company, Sun Pipe Line Company, and Western Crude Oil, Inc., had executed a Letter of Intent to proceed with the preliminary design of the pipeline.
September, 1973	Crude Oil International Company decided to participate with the Apco Oil Corporation, Continental Oil Company, Diamond Shamrock Oil and Gas Company, and Phillips Petroleum in a project to construct a competing pipeline from Freeport, Texas, to Cushing, Oklahoma, and therefore withdrew from the Texoma project.
Late September, 1973	Skelly Oil Company and United Refining Company expressed their desire to participate in the Texoma project.
October 18, 1973	The Shareholders' Agreement was executed by Kerr-McGee Pipeline Corp., Lion Oil Company, Skelly, Sun Oil, United Refining, and Western Crude Oil, Inc. Subsequent to the execution of the Shareholders' Agreement, additional companies requested an opportunity to participate in Texoma Pipe Line Company and were allowed to do so. Rock Island Refining Corp., Texas Eastern Transmission Corp., Mobil Oil Corp, and Vickers Petroleum Corp.; received ownership interest in the company.

Source: Texoma Pipe Line Company.

Appendix O

CURRENT CAPACITY — CRUDE OIL LINES

Circa 1977
EASTERN MIDWEST

Pipeline	Destination	Estimated Capacity B/D	Recent Estimated Flow B/D
Amoco	East Chicago	275,000	275,000
Arco	East Chicago	250,000	250,000
Ashland	Canton	65,000	65,000
Buckeye	Toledo	106,000	50,000
Buckeye	Toledo Samaria	278,000	278,000
Buckeye	Findlay	19,000	12,000
Capline	Patoka	1,161,000	1,161,000
Clark-Shell	Wood River	224,000	200,000
Lakehead	Sarnia	615,000	600,000
Lakehead	East Chicago	855,000	400,000
Marathon	Detroit	80,000	65,000
Marathon	Lima	310,000	310,000
Marathon	Robinson	175,000	158,000
Michigan-Ohio South Line	Alma	22,000	16,000
Michigan-Ohio North Line	Alma	20,000	18,000
Mid Valley	Lima	271,000	271,000
Mobil	Patoka	163,000	160,000
Ozark	Wood River	285,000	168,000
Platte	Wood River	147,000	84,000
Shell	Wood River	40,000	27,500
Sohio	Lima	25,000	18,000
Tecumseh	Cygnet	120,000	40,000
Texas	Lawrenceville	100,000	87,000
Texaco-Cities Service 18"	East Chicago	150,000	150,000
Texaco-Cities Service 10"	East Chicago	25,000	25,000
Texas-Marathon	Patoka	285,000	110,000

Source: Marathon Appendix F

Appendix P

CURRENT CAPACITY — PRODUCT LINES
Circa 1977
EASTERN MIDWEST

Pipeline	Destination	Estimated Capacity B/D	Recent Estimated Flow B/D
Chicago Area			
Amoco	Chicago-West	50,000	50,000
Amoco	Chicago-North	48,000	48,000
Amoco	Chicago-East	75,000	75,000
Amoco	Chicago-Indianapolis	30,000	25,000
Arco	Chicago-East	32,500	26,000
Arco	Chicago-West	21,000	21,000
Badger	Chicago-North	233,000	180,000
Buckeye	Chicago-East	35,000	30,000
Explorer	Chicago	210,000	160,000
Marathon	Champaign-Chicago	96,000	75,000
Marathon	Muskegon	31,000	27,000
Phillips	Chicago	54,000	54,000
Shell	Chicago	100,000	90,000
Texas Eastern	Chicago	96,000	35,000
West Shore	Chicago-North	380,000	272,000
Williams	Chicago	40,000	30,000
Wolverine	Chicago-East	365,000	155,000
Wood River Area			
Arco	Missouri Points	13,000	10,000
Continental	Wood River	90,000	75,000
Explorer	Wood River	210,000	200,000
Marathon	W. River-Champaign	96,000	35,000
Marathon	W. River-East	50,000	13,000
Marathon	Robinson-Champaign	96,000	35,000
Phillips	Chicago	54,000	54,000
Phillips	Wood River	115,000	105,000
Shell	Wood River-West	19,000	9,500
Shell	Wood River-East	85,000	70,000
Shell	Chicago	100,000	90,000
Indianapolis-Ohio Area			
Amoco	Indianapolis	30,000	17,000
Amoco	Detroit	75,000	60,000
Arco	Toledo	32,500	26,000
Arco	Toledo-North	17,000	7,000
Arco	Toledo-South	17,000	15,000
Buckeye	Lima-North	140,000	130,000
Buckeye	Robinson-East	19,000	12,000
Buckeye	Lima-West	30,000	10,000
Buckeye	Lima	35,000	20,000

Pipeline	Destination	Estimated Capacity B/D	Recent Estimated Flow B/D
Indianapolis-Ohio Area (Continued)			
Buckeye	Lima-South	36,000	30,000
Buckeye	Detroit	65,000	50,000
Buckeye	Toledo-South	160,000	100,000
Buckeye	Toledo	41,000	30,000
Farm Bureau	Indianapolis	24,000	11,000
Inland	Toledo-South	64,000	50,000
Inland	Lima-South	149,000	112,000
Inland	Lima-North	45,000	30,000
Inland	Toledo-East	119,000	100,000
Marathon	Robinson-Lima	48,000	44,000
Marathon	Indianapolis	100,000	85,000
Marathon	Louisville	62,000	28,000
Shell	Indianapolis-East	85,000	45,000
Sun	Toledo-East	50,000	45,000
Sun	Detroit	54,000	40,000
Sun	Detroit	30,000	Unknown
Texas Eastern	Indianapolis	96,000	42,000
Wolverine	Toledo	200,000	27,000
Wolverine	Detroit	200,000	62,000

Source: Marathon Appendix E.

EXXON COMPANY U.S.A.
Crude Oil Sales (and Purchases)
By Quarter, 1972, 1973 (P.A.D. I-IV Only) Listed by Buyer/(Seller): M Bbls.

Sales to:	1972				1973			
	1st Qtr	2nd Qtr	3rd Qtr	4th Qtr	1st Qtr	2nd Qtr	3rd Qtr	4th Qtr
Adobe Refining Company	-	-	-	-	-	-	595	16
Alabama Refining Company	1,593	748	585	780	926	709	1,115	4,473
Amerada Hess	1,338	612	591	929	990	1,327	1,285	1,667
American Petrofina Company of Texas	-	-	577	382	157	747	480	369
Amoco Oil Company	-	-	-	-	-	-	284	-
Amoco Production Company	6,629	7,649	7,129	6,148	5,658	5,408	5,688	5,122
Ashland Petroleum Company	635	477	564	646	627	180	201	286
Ashland Oil & Refining Company	-	-	-	55	-	-	224	319
Atlantic Richfield Company	1,359	1,963	2,322	2,391	5,903	6,100	4,757	2,053
Atlas Processing Company	554	-	134	102	-	-	-	-
Perry R. Bass	57	92	87	94	84	83	89	80
Canadian Hydrocarbons, Ltd.	-	-	-	-	-	-	1,725	56
Carson Oil Company	180	-	500	-	-	-	-	-
Champlin Petroleum Company	264	3,344	3,216	3,194	2,349	2,177	2,317	2,250
Charter International Company	230	137	287	462	190	189	232	227
The Chevron Oil Company	1,695	1,877	1,844	1,951	2,309	1,469	2,562	2,750
Cities Service Oil Company	2,313	1,912	1,199	1,222	957	-	977	-
Clark Oil & Refining Company	3,135	2,662	3,072	3,027	2,053	1,505	632	636
Coastal States Crude Gathering Company	3,247	5,628	3,516	3,019	1,896	1,265	2,144	1,013
Coastal States Gas Producing Company	306	1,021	-	1,429	-	-	-	-
Continental Oil Company	2,027	944	934	984	886	1,377	1,444	3,961
Cosden Oil & Chemical Company	368	-	-	-	-	-	-	-
Cotton Valley Solvents	67	71	68	68	56	61	51	52
CRA, Inc.	-	-	-	-	-	-	25	12
Cross Oil & Refining Company	-	-	-	-	30	-	-	223
Crown Central Petroleum Corporation	560	1,448	747	1,456	1	-	778	786
Deepwater Oil Terminals	-	-	101	-	-	-	-	-
Delta Refining Company	2,877	1,577	2,058	1,989	1,835	4,674	2,310	1,920
Diamond Shamrock Corporation	286	296	452	898	264	244	925	330
Edgington Oil Company	499	1,075	590	226	-	-	-	-

Sales to:	1972				1973			
	1st Qtr	2nd Qtr	3rd Qtr	4th Qtr	1st Qtr	2nd Qtr	3rd Qtr	4th Qtr
El Paso Natural Gas Products Company	107	-	-	30	-	-	-	-
Falco, Inc.	615	742	902	1,092	973	975	1,146	1,091
Farmers Union Central Exchange, Inc.	-	-	-	-	61	100	-	-
J. E. Fowler Petroleum Products	-	-	-	56	134	175	169	121
General Tire and Rubber Company	30	-	-	-	-	-	-	-
Getty Oil Company	556	293	571	526	738	944	748	734
Goodrich Chemical Company BF	-	-	-	7	-	-	-	-
Great Lakes Chemical Company	-	-	-	6	-	2	-	-
Gulf Oil Corporation.	3,152	2,239	2,258	2,900	2,920	3,092	3,273	2,975
Hercules, Inc.	-	-	-	86	-	-	-	-
Howell Corporation	-	-	-	-	-	218	1,505	304
J. M. Huber Corporation	-	75	-	39	-	-	-	-
Hunt Oil Company	84	137	143	126	143	158	182	206
Kewanee Oil Company	506	253	-	-	-	-	-	-
Koch Industries	-	-	-	-	536	-	-	-
La Gloria Oil & Gas	499	680	644	1,129	537	1,102	820	489
Lion Oil Company	-	-	-	131	665	620	886	551
Marathon Oil Company	4,331	4,497	4,009	4,181	3,942	4,023	3,836	3,732
Mobil Oil Corporation	1,902	2,552	-	2,969	2,375	-	-	2,471
Moncrief W. A. Jr.	-	-	210	251	100	-	-	-
Monsanto Chemical Company	1,202	994	576	1,964	71	-	-	-
Osceola Refining Company	-	-	-	-	-	-	-	-
Pennzoil United	-	-	-	-	-	-	2,257	1,003
The Permian Corporation	2,442	1,326	1,328	1,775	1,253	1,149	1,357	1,854
Petroleum Refining Company	195	103	77	144	50	78	37	1,658
Petro-Tex Chemical Corporation	71	-	-	76	-	-	-	-
Phillips Petroleum Company	3,949	4,349	5,662	5,675	2,602	2,388	1,648	2,030
Plateau, Inc.	-	-	-	-	-	-	-	765
Powerine Oil Company	-	-	719	697	-	-	-	-
Pride Refining Company	-	-	-	-	600	-	1,035	54
Quintana Petroleum Company	-	-	-	-	-	-	540	-330
Quintana Refinery Company	-	-	-	-	-	-	375	2,242
Refinery Corporation, The	699	37	-	-	-	-	-	-
Sid Richardson Carbon & Gasoline Company	174	-	-	-	-	-	-	2,210
Rock Island Refining Company	260	91	134	434	72	70	212	57

	1972				1973			
Sales to: *(Continued)*	1st Qtr	2nd Qtr	3rd Qtr	4th Qtr	1st Qtr	2nd Qtr	3rd Qtr	4th Qtr
Scurlock Oil Company	1,323	2,420	891	320	223	232	350	376
Seminole Asphalt Refining Company	-	-	-	43	-	-	-	-
Shell Oil Company	5,540	6,233	7,213	6,329	5,103	5,154	5,667	5,332
Signal Marine Company	-	-	325	285	-	-	-	-
Skelly Oil Company	86	-	-	-	-	-	-	-
Sohio Petroleum Company	4,426	4,712	3,844	4,080	1,888	2,172	1,760	883
Southland Oil Company	-	-	-	266	-	-	312	-
Southwestern Oil & Refining Company	3,074	3,333	3,703	2,743	2,899	3,490	7,501	3,731
Space Petroleum, Inc.	-	-	-	-	-	100	-	-
Standard Oil Company of California	354	340	363	352	3,909	2,493	2,713	3,185
Sun Oil Company	2,285	2,814	3,251	3,869	451	660	407	271
Tenneco Oil Company	699	405	705	1,591	-	-	-	185
Tesoro Crude Oil Company	-	-	-	-	-	-	-	-
Texaco, Inc.	3,875	4,538	5,857	4,329	3,759	3,123	4,542	4,678
Texas City Refining, Inc.	1,469	2,182	1,846	1,754	1,726	1,188	1,256	1,277
Texas Eastman Company	-	128	-	109	90	46	-	-
Toro Petroleum Company	-	182	-	-	-	-	-	-
Triangle Refineries, Inc.	-	94	60	-	-	1,167	1,708	790
Union Carbide Corporation	-	243	-	-	-	-	2,957	43
Union Oil Company of California	2,005	1,795	4,398	5,679	3,291	3,470	3,156	2,419
Union Texas Petroleum Inc.	-	-	-	104	-	-	-	188
Uniroyal Inc.	34	-	-	7	-	-	-	-
Valcan Asphalt Company	-	-	-	39	-	-	-	-
Webber Tanks, Inc.	-	-	398	-	-	-	-	-
Witco Chemical Corporation	-	-	-	258	-	-	-	-
All Others	3,589	3,182	5,438	2,878	2,855	6,160	5,084	3,658
	82,752	84,512	86,098	90,781	71,097	72,064	88,279	79,834

Purchases from:

	1st Qtr	2nd Qtr	3rd Qtr	4th Qtr	1st Qtr	2nd Qtr	3rd Qtr	4th Qtr
Adobe Refining Company	-	-	-	-	-	-	595	-
Alabama Refining Company	-	-	-	-	-	-	907	5,094
Amerada Hess	1,831	1,835	1,174	1,698	1,147	708	1,276	1,626
American Petrofina Company of Texas	1,952	871	727	493	390	550	796	383
Amoco Oil Company	-	-	-	-	-	-	508	-
Amoco Production Company	4,358	5,184	5,181	4,888	4,538	4,837	4,766	4,488

Purchases from: (Continued)	1972				1973			
	1st Qtr	2nd Qtr	3rd Qtr	4th Qtr	1st Qtr	2nd Qtr	3rd Qtr	4th Qtr
Ashland Petroleum Company	-	-	259	77	-	-	719	310
Ashland Oil & Refining Company	-	-	-	38	-	-	-	40
Atlantic Richfield Company	1,765	1,632	1,316	1,746	4,984	5,041	5,825	2,104
Atlas Processing Company	704	-	-	251	-	-	-	-
Perry R. Bass	290	299	280	278	257	264	263	201
Canadian Hydrocarbons, Ltd.	-	-	-	-	-	-	1,966	60
Champlin Petroleum Company	40	38	36	36	139	30	111	635
Charter International Oil Company	-	-	-	-	-	-	127	32
The Chevron Oil Company	2,432	2,198	2,092	2,267	1,609	1,863	2,107	2,757
Cities Service Oil Company	2,145	1,808	856	886	184	-	340	-
Clark Oil & Refining Company	2,280	1,198	1,664	924	1,005	1,626	690	755
Coastal States Crude Gathering Company	765	735	1,383	595	148	51	735	997
Coastal States Gas Producing Company	96	617	-	1,614	-	-	-	-
Continental Oil Company	1,950	1,799	1,931	2,014	1,692	1,624	2,082	5,192
Cotton Valley Solvents	150	221	188	184	182	122	105	127
Cross Oil & Refining Company	-	-	-	-	-	-	865	220
Crown Central Petroleum Corporation	9	14	12	477	12	-	-	-
Delta Refining Company	1,773	-	528	426	-	3,101	611	334
Diamond Shamrock Corporation	5	5	4	608	88	4	819	114
Dow Chemical	-	-	-	8	-	-	-	-
Falco, Inc.	687	735	900	745	678	943	1,153	1,021
Getty Oil Company	497	480	469	513	331	330	321	307
Gulf Oil Corporation	2,390	3,049	2,721	3,037	2,362	2,676	2,605	1,549
Howell Corporation	-	-	-	-	-	-	43	1,192
J. M. Huber Corporation	-	10	-	9	-	-	-	-
Hunt Oil Company	49	53	49	51	40	47	36	38
Kewanee Oil Company	117	117	-	-	-	-	-	-
Koch Industries	-	-	-	-	441	-	-	-
La Gloria Oil & Gas	-	-	-	-	-	508	383	-
Lion Oil Company	-	-	-	536	-	-	771	61
Marathon Oil Company	2,848	2,862	2,713	3,016	3,025	3,164	2,800	2,500
Mobil Oil Corporation	741	861	-	981	718	-	-	2,108
Moncrief W. A. Jr.	-	-	2	2	2	-	-	-
Monsanto Chemical Company	136	428	64	1,210	48	-	-	-
Osceola Refining Co.	-	-	-	-	-	-	-	1,083

Purchases from: (Continued)	1972				1973			
	1st Qtr	2nd Qtr	3rd Qtr	4th Qtr	1st Qtr	2nd Qtr	3rd Qtr	4th Qtr
Pennzoil United	1,481	1,002	1,001	1,154	744	852	2,840	2,502
The Permian Corporation	54	51	46	109	39	61	1,096	1,776
Petroleum Refining Company							17	
Phillips Petroleum Company	3,291	3,182	3,528	4,413	1,348	1,015	1,517	1,067
Plateau, Inc.								1,015
Powerine Oil Company			353	697				3
Pride Refining Company							1,155	5,652
Quintana Petroleum Company							5,630	2,781
Refinery Corporation, The	874	41						
Rock Island Refining Company	189	32		285			163	
Scurlock Oil Company	2,284	3,018	2,396	1,829	1,715	1,930	2,225	2,239
Shell Oil Company	1,710	2,197	2,350	2,472	2,158	2,051	2,128	1,812
Skelly Oil Company	581							
Sohio Petroleum Company	570	588	551	769	1,147	1,864	1,145	837
Southland Oil Company				169			247	
Southwestern Oil & Refining Company							3,457	700
Sun Oil Company	1,817	1,832	2,365	2,716	1,989	2,500	2,131	2,210
Tenneco Oil Company	247	249	232	652	191	108	172	518
Texaco Inc.	1,643	1,576	1,525	2,088	1,666	1,651	2,269	2,789
Texas City Refining, Inc.				114	117			137
Triangle Refineries, Inc.						909	578	338
Union Carbide Corporation							2,957	43
Union Oil Company of California	2,074	3,083	3,059	4,029	3,132	2,726	3,419	3,596
Union Texas Petroleum, Inc.				99				205
All Others	21,488	25,385	25,076	23,967	24,370	25,198	17,924	16,277
Exxon Operated	81,082	87,465	88,285	86,794	83,799	82,830	84,652	74,707
Total Purchases	149,395	156,750	155,316	161,964	146,435	151,184	170,047	157,532
Net	66,643	72,238	69,218	71,183	75,338	79,120	81,768	77,698

Source:
Exxon Company, USA submission to U.S. Senate, Committee on Government Operations, Permanent Subcommittee on Investigations, June 27, 1974.

NEW U.S. REFINERIES BUILT SINCE JANUARY 1, 1950

Year Constructed	Location	Refiners	Total for State	New Entrants Indicated by "N"	Initial Operating Capacity (000 b/d) Year[1] / Capacity		1977 Operating Capacity (000 b/d)
	Alabama						
1953	Cordova	Vulcan Asphalt Refining Co.		N	1954	1.6	5.0
1954	Holt	Warrior Asphalt Corp.		N	1954	1.0	3.0
1967	Theodore	Alabama Refining Co., now Marion Corp.		N	1968	10.0	19.2
1967	Moundville	Cracker Asphalt Co.			1968	2.0	-0-
1975	Mobile	Louisiana Land & Exploration		N	1976	30.0	38.4
			5	4		44.6	65.6
	Alaska						
1964	Kenai	Standard Oil of California			1964	20.0	22.0
1970	Kenai	Tesoro-Alaska Petroleum Corp.			1970	17.0	38.0
			2			37.0	60.0
	Arizona						
1971	Fredonia	Arizona Fuels Inc.	1	N	1972	3.0	5.0
	California						
1950	Bakersfield	Norwalk (then Bankline, Signal) now Lion, subsidiary of Tosco.			1951	5.0	40.0
1951	Oxnard	Superior Asphalt Co.		N	1952	1.0	-0-
1955	Santa Maria	Union Oil Co. of California			1956	19.0	41.0
1956	Long Beach	Manespo (later NISH-PAUL, then U. S. Oil & Refining)		N	1957	5.0	-0-
1963	Long Beach	Western Oil Reduction		N	1964	<.1	-0-
1967	Hercules	Sequoia Refining (then Gulf Oil), now Pacific Refining, subsidiary of Coastal States			1968	25.0	53.3
1967	South Gate	Luday Thagard Oil Co.			1968	3.0	3.2
1968	Sangus	Lubrication Co. of America		N	1969	.8	-0-
1970	Benicia	Exxon			1970	72.0	88.0

[1] "Year" indicates the first year in which operating capacity is shown by the Bureau of Mines (as of January 1).

Year Constructed	Location	Refiners	Total for State	New Entrants Indicated by "N"	Initial Operating Capacity (000 b/d) Year	Capacity	1977 Operating Capacity (000 b/d)
1970	Bakersfield	Sabre Oil & Refining, Inc.		N	1972	2.0	3.5
1970	Wilmington	Champlin Petroleum Co., subsidiary of Union Pacific			1971	30.0	30.7
1976	Ventura	U.S.A. Petroleum Corp.		N	1977	15.0	15.0
1976	Long Beach	Basin Petroleum		N	1977	3.0	3.0
1976	Signal Hill	ECO Petroleum		N	1977	3.0	3.0
1976	Compton	Demenno Resources		N	1977	2.5	2.5
			15	9		186.3	283.2
	Colorado						
1952	Rangely	Wesco Refining (later Unita, then LUBCO)		N	1953	.8	-0-
1964	Fruita	American Gilsonite Co., now Gary Western Co.			1965	6.0	*
1965	Grand Junction	Morrison Refining Co.		N	1966	.7	-0-
			3	N			-0-
				3		7.5	
	Delaware						
1957	Delaware City	Tidewater Oil Co., now Getty	1		1958	120.0	140.0
	Florida						
1955	St. Marks	Seminole Asphalt & Refining, Inc.	1	N	1956	2.0	6.0
	Georgia						
1956	Douglasville	Cracker Asphalt & Refining, now Young Refining Co.	1		1957	1.1	5.0
	Hawaii						
1960	Barbers Point	Standard Oil of California		N	1961	32.0	40.0
1972	Oahu, Ewa Beach	Hawaii Ind. Ref. **		N	1973	** 29.5	** 59.0
			2	2		61.5	99.0

* Not given a rated capacity as this plant does not process crude oil.

** Not included in U.S. statistics by Bureau of Mines because it is located in Foreign Trade Zone.

513

Year Constructed	Location	Refiners	Total for State	New Entrants Indicated by "N"	Initial Operating Capacity (000 b/d) Year / Capacity		1977 Operating Capacity (000 b/d)
	Illinois						
1958	Crossville	Richards, T. M. Inc.			1959	.6	.1
1972	Joliet	Mobil Oil Corp.		N	1973	160.0	180.0
1976	Pana	Bio-Petro		N	1977	1.0	1.0
			3	2		161.6	181.1
	Indiana						
1954	Gary	Berry Asphalt Co.			1955	5.0	-0-
1956	Laketon	Laketon Asphalt Refining Co.		N	1957	4.0	8.5
1956	Hammond	Allby Asphalt & Refining Co., now Witco Chemical		N	1957	7.0	-0-
1960	Ft. Wayne	Gladieux Refinery		N	1961	2.0	12.2
			4	3		18.0	20.7
	Kentucky						
1964	Betsy Lane	Kentucky Oil & Refining Co.	1	N	1965	.5	3.0
	Louisiana						
1951	Ida	Great National (later Ida Gas), now Breaux Bridge		N	1952	.5	2.0
1951	Church Point	Canal Refining Co.		N	1952	1.7	4.8
1962	Shreveport	Atlas Pro., later Pennzoil		N	1963	18.5	45.0
1967	Good Hope	Good Hope Refining			1968	6.5	70.0
1967	Venice	Gulf Oil			1968	20.3	28.7
1968	Convent	Texaco			1968	110.0	140.0
	St. James	S.W. Pallet, later North American Petroleum, now Louisiana Jet, Inc.			1969	3.0	15.0
1971	Lisbon	Claiborne Gas. Co.		N	1972	6.5	6.5
1971	Belle Chasse	Gulf Oil		N	1972	155.0	195.9
1972	Dubach	Kerr-McGee			1973	11.0	11.0
1974	Egan	Continental Oil			1975	15.0	15.0
1974	Port Allen	Toro, now Placid Refining Co.		N	1975	36.0	36.0
1976	Krotz Springs	Hill Petroleum, subsidiary of Goldking		N	1977	5.0	5.0
1976	Garyville	Marathon Oil			1977	200.0	200.0
			14	8		589.0	774.9

Year Constructed	Location	Refiners	Total for State	New Entrants Indicated by "N"	Initial Operating Year	Initial Operating Capacity (000 b/d)	1977 Operating Capacity (000 b/d)
	Michigan						
1959	Wyandotte	Wyandotte Chemicals		N	1960	2.4	-0-
1964	Rapid River	Upper Penninsula Refining, (later Gustafson Oil), then Delta Terminal		N	1965	1.5	-0-
1975	Marysville	Consumers Power Co.		N	1976	37.6	34.1
			3	3		41.5	34.1
	Minnesota						
1952	Wrenshall	International Refining Co., now Continental Oil		N	1953	10.0	23.5
1955	Pine Bend	Great Northern Oil Co., now Koch Refining		N	1956	22.2	127.3
			2	2		32.2	150.8
	Mississippi						
1957	Purvis	Pontiac Eastern Corp., later Gulf Oil, now Amerada Hess			1958	10.0	30.0
1963	Pascagoula	Standard Oil of Kentucky, now Standard of California			1964	100.0	280.0
1966	Lumberton	Lamar Refining Co., now Southland Oil		N	1967	1.0	5.8
			3	1		111.0	315.8
	Montana						
1953	Mosby	Jet Fuel Refining		N	1954	.4	-0-
1955	Chinook	Diamond Asphalt Co.		N	1956	1.5	-0-
1956	Lodge Grass	Texas Calgary Co.		N	1957	1.0	-0-
1961	Wolf Point	Petrofuels Refining (later Spruce Oil) now Tesoro		N	1962	1.0	2.5
			4	4		3.9	2.5
	Nevada						
1967	Ely	Nevada Refining Co.		N	1968	1.3	-0-
1969	Tonopah	Tonopah Refining (Division of Newton Petroleum) now Nevada Refining		N	1970	.8	-0-
			2	2		2.1	-0-

515

Year Constructed	Location	Refiners	Total for State	New Entrants Indicated by "N"	Initial Operating Capacity (000 b/d) Year / Capacity		1977 Operating Capacity (000 b/d)
	New Jersey						
1958	Ft. Reading	Amerada Hess Corp.		N	1959	45.0	-0-
1973	Bayonne	National Recovery Corp.		N	1974	3.0	-0-
			2	2		48.0	-0-
	New Hampshire						
1974	Newington	A. Johnson & Co.	1	N	1975	14.0	11.7
	New Mexico						
1955	Farmington	Four-States Western, later Beeline (division of Frontier Refining)			1956	1.0	-0-
1957	Ciniza	El Paso Natural Gas, now Shell Oil		N	1958	7.6	18.0
1961	Bloomfield	Plateau Inc.		N	1962	2.5	8.7
1964	Farmington	Caribou-Four Corners, Inc.		N	1965	1.0	2.5
1972	Bloomfield	Thriftway		N	1973	1.2	7.5
1974	Lovington	Famariss Oil, now Southern Union Refining			1975	28.0	36.1
1974	Farmington	Giant Industries		N	1975	5.9	8.8
			7	5		47.2	81.6
	North Carolina						
1973	Wilmington	Pace Oil, now Trans Ocean	1	N	1974	12.0	10.0
	North Dakota						
1954	Mandan	Standard of Indiana			1955	30.0	49.0
1954	Dickenson	Queen City Oil & Refining Co.		N	1955	2.5	-0-
1954	Williston	Williston Basin, now Westland Oil Co.		N	1955	1.5	4.7
1954	Dickenson	Northland Oil & Refining Co.		N	1975	5.0	5.3
			4	3		39.0	59.0
	Ohio						
1954	Cincinnati	Standard Oil of California	1		1955	10.0	11.5

Year Constructed	Location	Refiners	Total for State	New Entrants Indicated by "N"	Initial Operating Capacity (000 b/d) Year / Capacity		1977 Operating Capacity (000 b/d)
	Oklahoma		1				
1967	Arnett	Tonkawa Refining Co.		N	1968	7.5	4.0
	Texas						
1950	Alba	Alba Refining Co.			1951	1.5	-0-
1951	Kilgore	Petrolite Corp.		N	1952	.7	1.0
1951	Brownsville	Port Fuel Co., later McBride Refining		N	1952	4.0	-0-
1953	Corpus Christi	Suntide, now Sun		N	1954	25.0	57.0
1953	Corpus Christi	Corpus Christi Refining			1954	1.7	-0-
1956	Carrizo Springs	Texas Calgary Co., now Tesoro		N	1957	1.5	26.1
1956	Pasadena	Texas Asphalt & Refining		N	1957	5.0	-0-
1958	Winnie	Texas Gas, later Union Texas now Independent Refining					
1958	Odessa	El Paso Natural Gas, now Shell		N	1959	7.5	13.5
1959	Irving	Great Western Producers			1959	13.5	32.0
1960	Texas City	Monsanto		N	1960	2.3	-0-
1960	Waskom	Waskom Natural Gas			1961	3.0	8.5
1960	Abilene	Debco Corp., later Premier, Adobe, now Pride			1960	2.0	-0-
1962	San Antonio	Flint Chemical		N	1961	.8	36.5
1962	Chocolate Bayou	Monsanto			1963	.6	1.5
1965	Irving	Petroleum Industries, now Texas Asphalt & Refining			1963	5.9	-0-
1971	Quitman	Wood County, now Gulf States		N	1966	2.0	6.0
1972	Silsbee	South Hampton Co.		N	1972	3.0	4.5
1973	Lacoste	Texas Fuel & Asphalt Co.		N	1973	2.0	18.1
1973	Corpus Christi	Quitana/Howell		N	1974	1.5	-0-
1974	Nixon	Pioneer Refining Co.		N	1974	12.5	44.6
1975	Corpus Christi	Saber Petroleum		N	1975	2.0	2.2
1975	Graham	Thriftway Oil Co.			1976	10.0	10.0
1976	Hearne	Mid-Tex Refinery		N	1976	1.0	1.0
					1977	.9	.9
			24	18		109.9	263.4

Year Constructed	Location	Refiners	Total for State	New Entrants Indicated by "N"	Initial Operating Capacity (000 b/d) Year / Capacity		1977 Operating Capacity (000 b/d)
Utah							
1954	Woods Cross	Sure Seal Corp., later Petroflex			1964	.5	-0-
1962	Woods Cross	Caribou-Four Corners, Inc.		N	1963	1.5	7.5
1973	Woods Cross	Crown Refining, now Western Refining		N	1974	1.0	10.0
1973	Woods Cross	Morrison Petroleum Co.		N	1974	1.5	2.5
1973	Roosevelt	Major Oil Co., now Plateau Refining			1974	5.0	7.5
			5	3		9.5	27.5
Virginia							
1956	Yorktown	Standard Oil of Indiana	1		1957	33.0	53.0
Washington							
1951	Edmonds	Union Oil of California			1952	4.0	-0-
1954	Ferndale	Mobil Oil			1955	35.0	71.5
1955	Anacortes	Shell Oil			1956	50.0	91.0
1957	Tacoma	U. S. Oil & Refining Co.		N	1958	10.0	21.4
1958	Anacortes	Texaco			1959	45.0	78.0
1967	Tacoma	Sound Refining Co.		N	1968	4.5	4.5
1971	Ferndale	Atlantic Richfield			1972	98.0	96.0
1976	Tacoma	United Independent Oil Co.		N	1977	1.0	1.0
			8	3		247.5	363.4
West Virginia							
1972	Newell	Quaker State Oil	1		1973	9.7	9.7
Wisconsin							
1951	Superior	Lake Superior Refining Co., now Murphy Oil ***	1	N	1952	3.8	45.4

***This refinery was "re-erected."

Year Constructed	Location	Refiners	Total for State	New Entrants Indicated by "N"	Initial Operating Capacity (000 b/d) Year / Capacity		1977 Operating Capacity (000 b/d)
	Wyoming						
1960	Greybull	Gordon Refining Co., now Oriental Refining Co.			1961	.1	-0-
1964	Cowley	Sage Creek Refining Co.		N	1965	.5	1.2
1972	La Barge	Mountaineer Refining Co.		N	1973	.1	.3
1973	La Barge	Southwestern Refining Co.		N	1974	.6	.5
1976	Glenrock	V-1 Oil		N	1977	1.0	1.0
			5	5		2.3	3.0
		50 State Total	129	89		2,016.2	3,089.9

Source:

Derived from Bureau of Mines, U.S. Dept. of Interior, Mineral Industry Surveys, "Petroleum Refineries in the United States," 1950-1977.
Federal Energy Administration, "Trends in Refinery Capacity and Utilization," June 1977.
National Petroleum Refiners Association, "U.S. Refining Capacity," 1951, 1961-1977.
Oil and Gas Journal, "Refining Construction Projects," 1950-1977.

Figure 1
PIPELINE MARKET

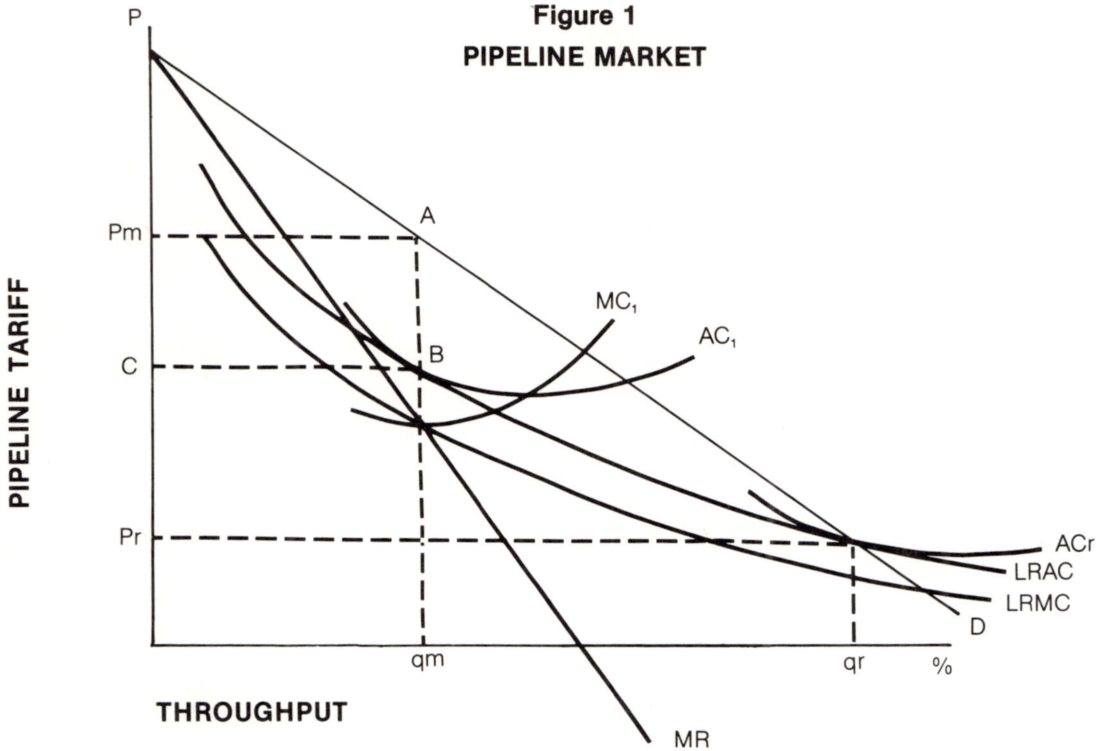

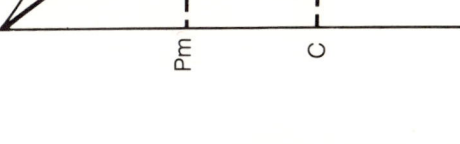

Demand and Cost Conditions Underlying the Department of Justice Theory of Pipeline Undersizing.
Source: Adapted from M. Canes & D. Norman, Pipelines and Antitrust (API Critique 005, Nov., 1978) Figure I

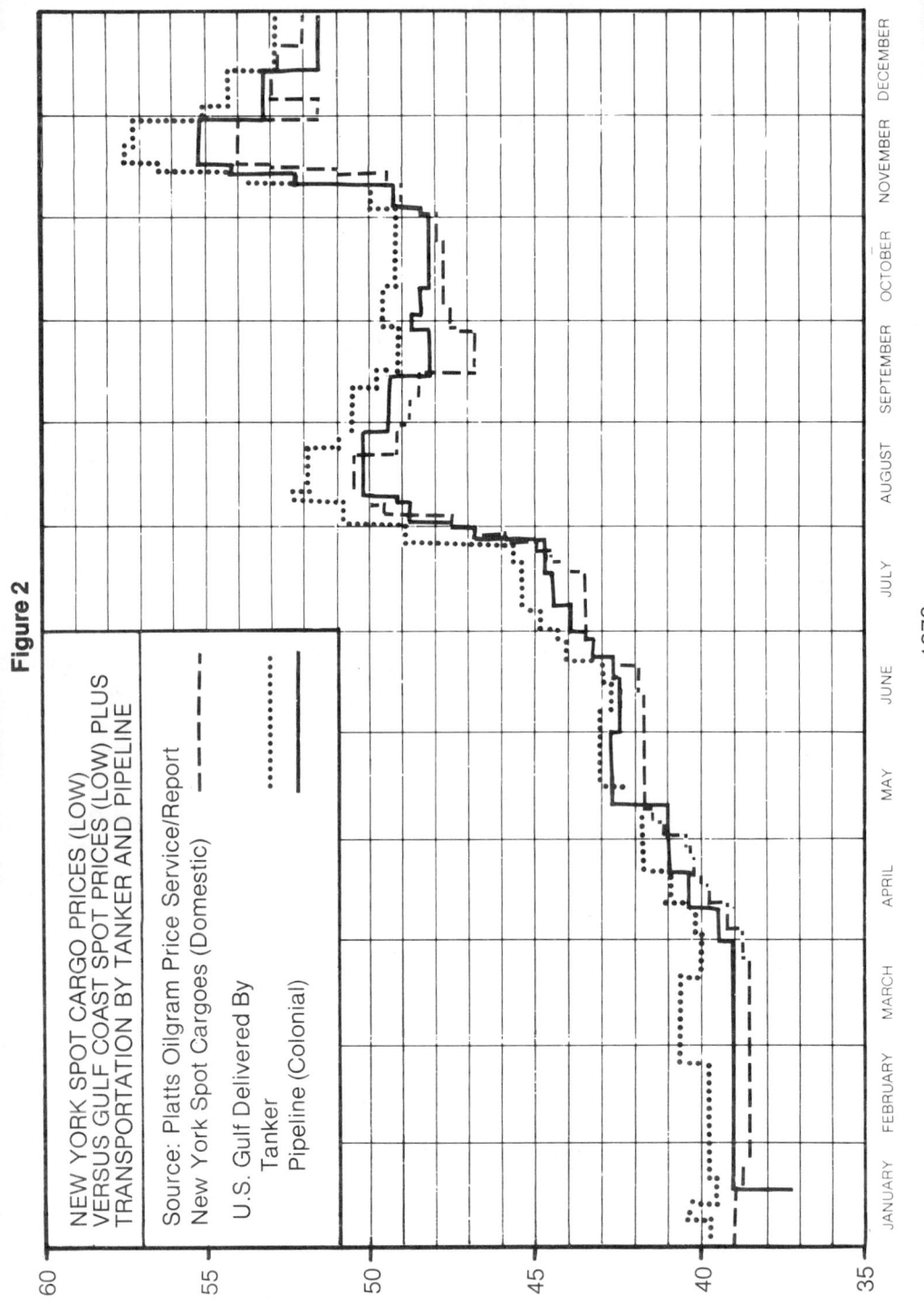

Figure 2

Figure 3

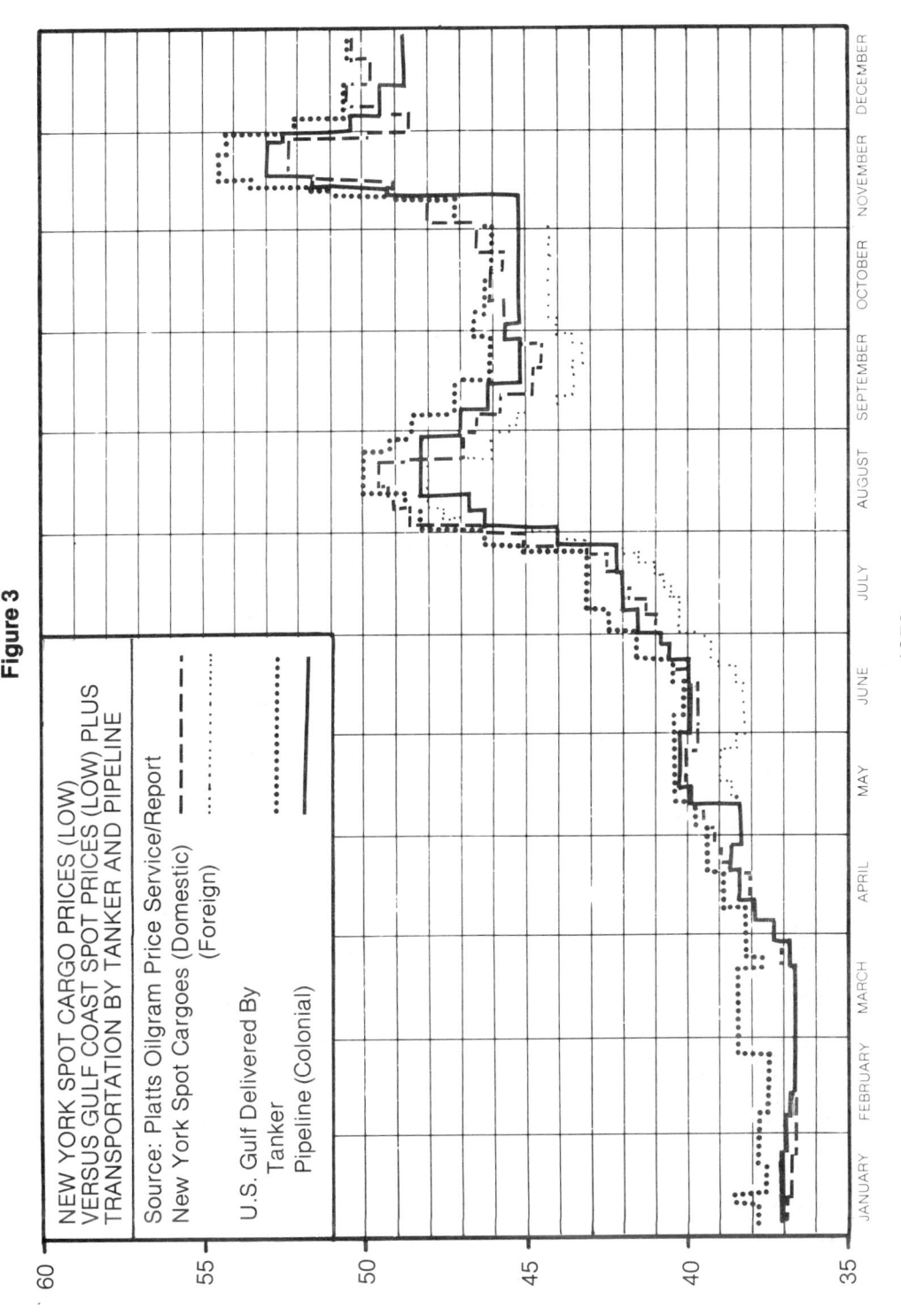

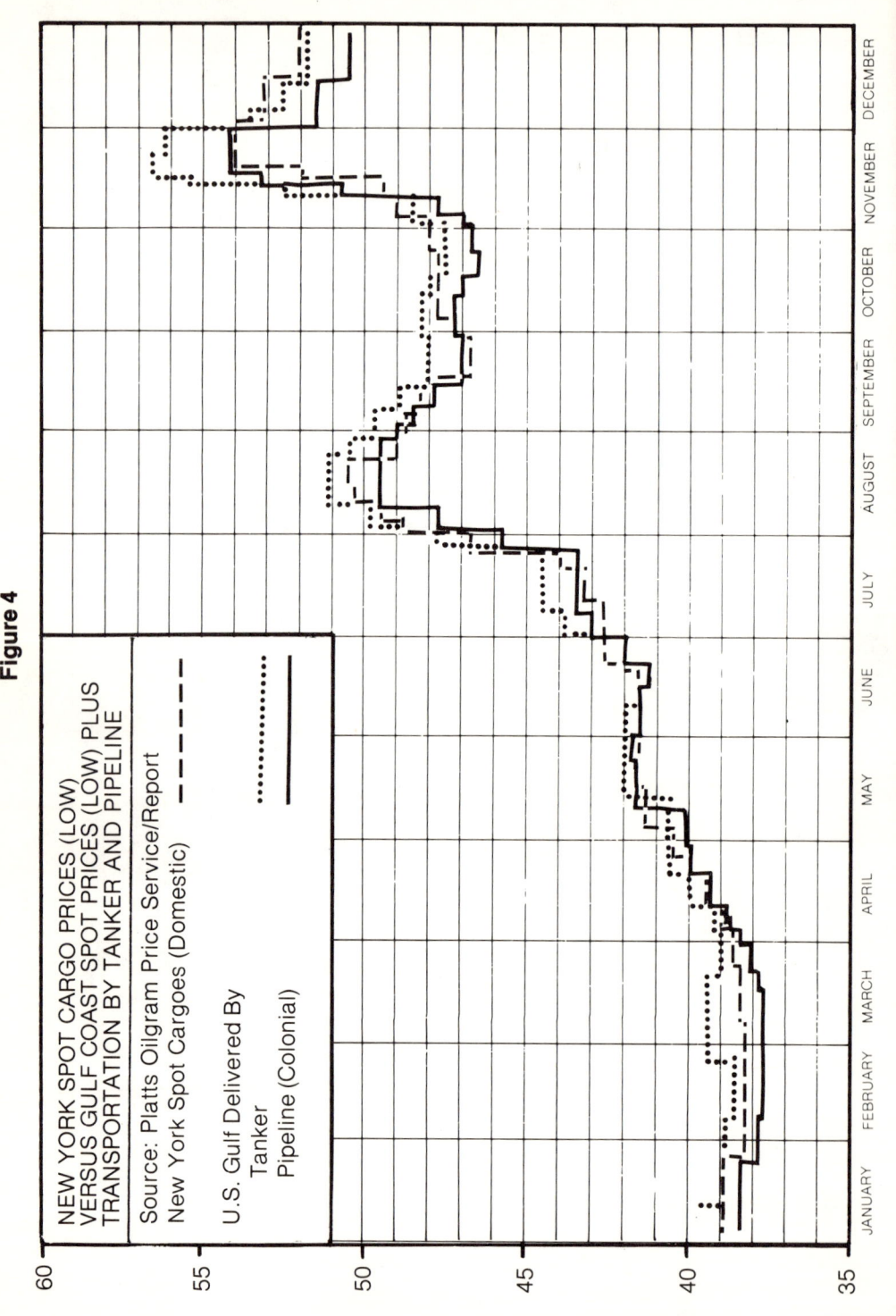

Figure 4

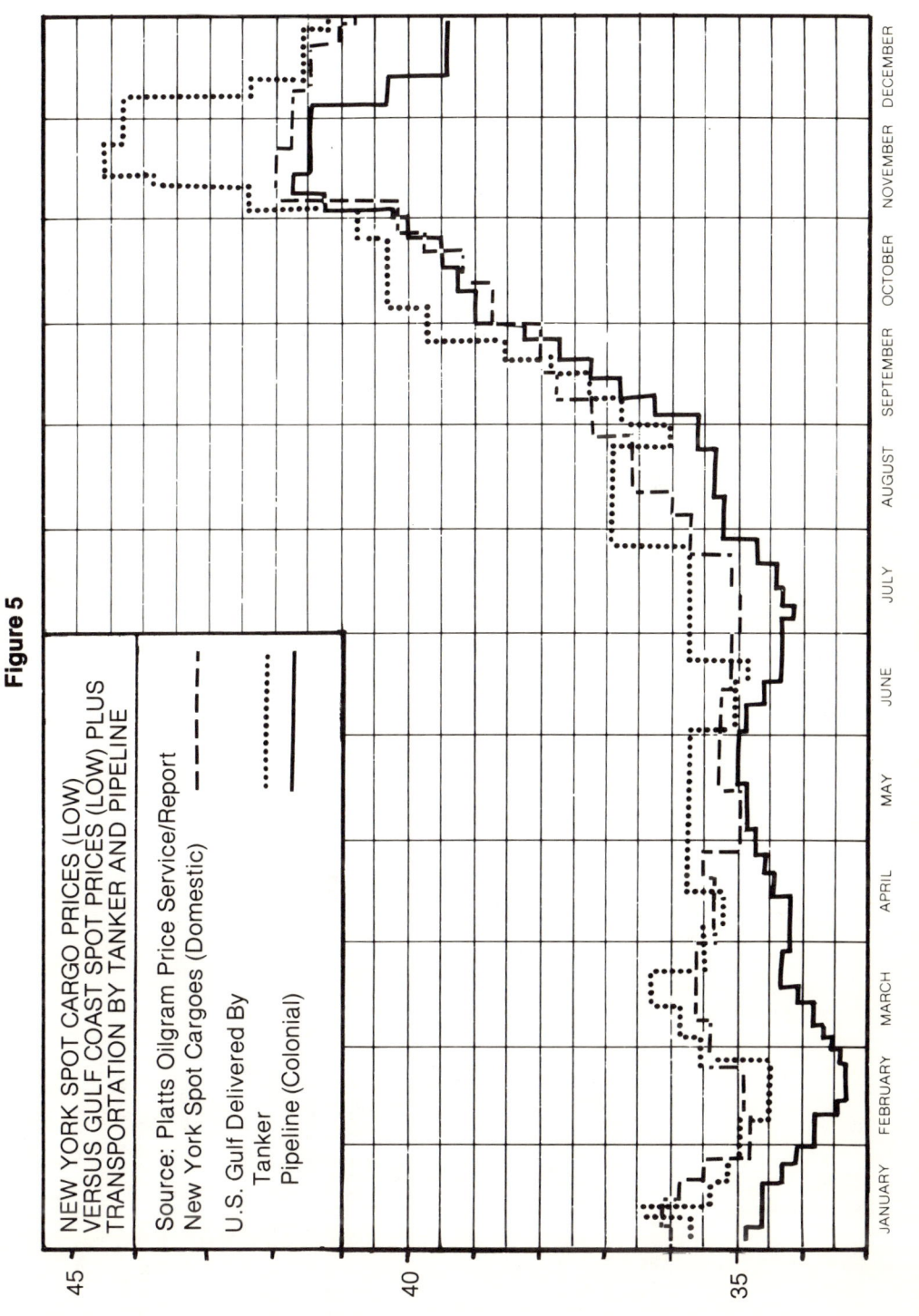

Figure 5

NEW YORK SPOT CARGO PRICES (LOW) VERSUS GULF COAST SPOT PRICES (LOW) PLUS TRANSPORTATION BY TANKER AND PIPELINE

Figure 6
CRUDE OIL NOMINATIONS AND PURCHASES
By Company
Average Barrels Daily

Nominator	May 1976 Field Nomination	May 1976 Additional Demand	June 1976 Field Nomination	June 1976 Additional Demand	Total Change May to June	March 1976 Field Nomination	March 1976 Additional Demand	March 1976 Field Purchases	Purchases Over/Under Nomination
Abney Oil Company, Inc.	300		300			300		323	+ 23
American Petrofina Co. of Tex.	39,645	82,000	39,695	82,800	+ 50	39,675	82,800	35,139	- 4,536
Amoco Production Company	345,000		345,000			350,000		343,190	- 6,810
Apco Oil Corporation	8,105		8,119		+ 14	8,020		8,101	+ 81
Ashland Oil, Inc.	49		48		- 1	55		49	- 6
Atlantic Richfield Company	178,000	2,000	178,000	2,000		177,500	2,500	175,364	- 2,136
Basin, Inc.	7,969		9,432		+ 1,463	9,406		9,124	- 282
Bigheart Pipe Line Corporation	4,133		4,133			4,308		3,662	- 646
Black Hills Oil Marketers Inc.	86		65		- 21			47	+ 47
Caddo Oil Company, Inc.	10		10			10		6	- 4
Cardinal Pipeline Corporation	115		115			115		77	- 38
Champlin Petroleum Co.	3,000		3,000			3,000		3,134	+ 134
Charter International Oil Co.	14,740	329	15,420	283	+ 680	15,339		15,137	- 202
Chevron Oil Company	73,800		73,800			73,800		70,852	- 2,948
Chevron Oil Co. The Calif. Co. Div.	900		900			900		527	- 373
Cities Service Oil Co.	112,000		112,000			112,000		109,536	- 2,464
Clark Oil & Refining Corporation	6,428		6,408		- 20	6,463		6,368	- 95
Coastal States Crude Gath. Co.	6,000	50,000	6,000	50,000		6,000	50,000	3,647	- 2,353
Coffield, H. H.	18		18			18			- 18
Coffield Oils, Inc.	133		133			133		28	- 105
Coffield Pipe Line Company	4,932		4,932			4,932		931	- 4,001
Compton Corporation	2,059		2,445		+ 386	1,861		2,295	+ 434
Continental Oil Company	54,500		54,500			54,500		51,311	- 3,189
Cosden Oil & Chemical Company	38,650	21,350	38,650	21,350		38,650	24,350	40,078	+ 1,428
Crown Central Petroleum Corporation	4,930	40,000	4,930	40,000		4,995	40,000	4,600	- 395
Diamond Shamrock Corporation	18,040	17,960	17,890	18,110	- 150	17,290	18,710	13,416	- 3,874
Exxon Corporation	532,000	200,000	532,000	200,000		536,000	200,000	540,412	+ 4,412

525

Nominator	May 1976 Field Nomination	May 1976 Additional Demand	June 1976 Field Nomination	June 1976 Additional Demand	Total Change May to June	March 1976 Field Nomination	March 1976 Additional Demand	March 1976 Field Purchases	Purchases Over/Under Nomination
Falco, Inc.	234		213			184		85	- 99
GEC Purchasing Company, Inc.	224		223		- 1	251		223	- 28
General Crude Oil Co.	20,875		21,500		+ 625	21,000		20,490	- 510
Getty Oil Co.	20,400		20,600		+ 200	20,400		20,894	+ 494
Graytex Corporation	831		808		- 23	862		778	- 84
Gulf Oil Corp.	155,400		156,400		+ 1,000	155,400		154,292	- 1,108
Harrdun & Harrdun	43		52		+ 9	48		18	- 30
Hill, A. G.	85		75		- 10	80		65	- 15
Hunt, H. L. Est. of	4,266		4,624		+ 358	4,466		4,489	+ 23
Koch Oil Company	66,353		73,557		+ 7,204	72,257		73,557	+ 1,300
Lacy, R., Inc.	250		254		+ 4	254		202	- 52
La Gloria Oil & Gas Co.	12,210	2,100	12,210	2,100		12,210	2,100	11,817	- 393
Marathon Oil Company	40		40			40		33	- 7
Matador Pipelines, Inc.	8,209		8,843		+ 634	8,192		8,843	+ 651
McCurdy, E. J., Jr.	55		55			55		42	- 13
Mesa Pipe Line Company	8,831		9,760		+ 929	10,050		9,760	- 290
Miller Oil Purchasing Company	35		35			63		32	- 31
Minerva Refining Company	144		144			144		61	- 83
Mobil Oil Corporation	281,100	63,900	280,100	64,900	- 1,000	284,300	60,700	280,232	- 4,068
OKC Corporation	2,750		2,750					1,985	+ 1,985
PML Corporation	547		567		+ 20	551		428	- 123
Peninsula Pipe Line Company	1,921		1,951		+ 30	1,836		2,018	+ 182
Permian Corporation, The	175,000		175,000			175,000		175,000	
Phillips Petroleum Company	92,364	22,636	93,855	21,145	+ 1,491	97,328	17,672	93,855	- 3,473
Pride Pipeline Company	1,776		2,054		+ 278	1,544		2,033	+ 489
Quintana-Howell Joint Venture	19,065		19,124		+ 59	18,079		19,173	+ 1,094
Richland Pipeline Company	175		175			175		139	- 36
Scurlock Oil Company	88,000		88,550		+ 550	87,000		89,965	+ 2,965
Shell Oil Company	300,500		300,500			300,500		297,446	- 3,054
Signor Refining Co.	500		500					132	+ 132
Skelly Oil Co.	9,523		9,516		- 7	9,846		9,511	- 335
Sohio Petroleum Co.	26,695	73,000	26,695	73,000		26,505	73,000	23,821	- 2,684
Stanmar Oil Corporation	2		2			4		5	+ 1
Sun Oil Company	178,000	47,000	177,100	47,900	- 900	177,100	47,900	165,482	-11,618

Nominator	May 1976 Field Nomination	May 1976 Additional Demand	June 1976 Field Nomination	June 1976 Additional Demand	Total Change May to June	March 1976 Field Nomination	March 1976 Additional Demand	March 1976 Field Purchases	Purchases Over/Under Nomination
Sunset Pipeline Corporation	1,000		1,000			1,000		600	- 400
Tesoro Crude Oil Company	15,816		16,546		+ 730	17,955		15,526	- 2,429
Texaco Inc.	165,000	58,000	165,000	58,000		170,000	53,000	158,415	-11,585
Texas City Refg. Inc.	14,755	3,000	14,723	3,000	- 32	14,661	3,000	13,438	- 1,223
Texas Petroleum Company	400		400			400		400	
Texpata Pipe Line Company	1,035		904		- 131	1,026		694	- 332
Three Rivers Refinery, Inc.						500			- 500
True, Ted Inc.	430		430			463		385	- 78
Union Oil Company of California	85,000		85,000			85,000		76,915	- 8,085
Vickers Petroleum Corp.	9,459	4,000	9,459	4,000		9,459	4,000	7,126	- 2,333
Waggoner, W. T., Estate	4,460		4,731		+ 271	4,706		3,116	- 1,590
Walker And Company	177		177			177		99	- 78
Weco Development Corp.	380		380			380		305	- 75
Western Crude Oil, Inc.	3,846		4,960		+ 1,114	3,867		3,763	- 104
TOTAL	3,233,703	658,075	3,249,485	688,588	+15,782	3,260,588	679,732	3,185,042	-75,546
Total Nominations and Additional Demand	3,921,778		3,938,073			3,940,320			

Source: Texas Railroad Commission

TABLE I

RATIO OF THROUGHPUT TO CAPACITY FOR SELECTED PIPELINE SYSTEMS

	Explorer		Platte	Wolverine	Yellowstone
	South Mainline	North Mainline			
1952			0.57		
1953			0.76		
1954			0.91		0.17
1955			0.88		
1956			0.82		
1957			0.81		
1958			0.75		
1959			0.81		
1960			0.83		0.67
1961			0.84		
1962			0.88		
1963			0.91		
1964			0.85		
1965			0.86		
1966			0.78		
1967			0.72		0.85
1968			0.84		0.89
1969			0.81		
1970			0.75		0.88
1971			0.69	0.74	
1972	0.31	0.22	0.73	0.72	0.92
1973	0.68	0.43	0.82	0.94	
1974	0.50	0.49	0.89	0.89	
1975	0.44	0.37	0.75	0.83	
1976	0.40	0.32	0.59	0.69	0.95
1977	0.70	0.84	0.52	0.81	
1978			0.45	0.82	

Note: 1.00 is ratio when throughput equals rated capacity under normal conditions.

Source: Donald Norman, "The Performance of Integrated Pipeline Companies: Evidence on the Undersizing and Access Issues," Policy Analysis Research Paper, American Petroleum Institute, (forthcoming). Data originally collected by the Association of Oil Pipe Lines.

TABLE II

RATIO OF SHIPMENTS TO CAPACITY FOR THE PLANTATION PIPELINE, 1950-1978; DESCRIPTION OF PRORATIONING ON PLANTATION

	T/C	Description
1950	1.04	Minimal proration
1951	0.69	Proration removed
1952	0.75	Prorated mainline east of Charlotte, N.C., during winter 1952/53; Macon/Columbus, Ga. 4" Line; and Tennessee line. All proration ended in 1953 by construction of new pipelines or stations.
1953	0.85	No proration
1954	0.93	No proration
1955	0.95	Prorated mainline east of Doraville, Ga. during winter of 1955/56.
1956	0.82	No proration
1957	0.88	No proration
1958	0.84	Proration in early part of year of Bremen/Macon-Columbus lines. Proration lifted with construction of Bremen/LaGrange 4" loop.
1959	0.87	Prorated 4" Helena/Montgomery, Ala. line.
1960	0.94	No proration. Proration lifted on Helen/Montgomery upon construction of new 8" line.
1961	0.95	Prorated mainline east of Doraville during winter of 1961/62
1962	1.01	Prorated mainline east of Doraville during winter of 1962/63.
1963	0.99	No proration
1964*	0.71	No proration
1965	0.86	Minimal proration east of Doraville due to tankage shortage at Greensboro, N.C.
1966	0.83	Prorationed east of Doraville during early part of the year until the Greensboro tank farm in operation; also Columbus, Ga. 4" line.
1967	0.92	No proration. Prorationing on Columbus line eliminated with 8" line construction.
1968	0.79	Prorated system early in the year until newly constructed 30"/26" lines operational.

1969	0.88	Minimal proration on mainlines during 3 winter months of 1969/70.
1970	0.95	Minimal mainline proration first month or so of the year.
1971	0.77	Prorated the Columbus and Birmingham lateral lines until new lines were added.
1972	0.85	Prorated east of Bremen as a result of manifold fire in September.
1973	0.89	Nominal proration on Roanoke lateral early in the year until Bremen manifold completed.
1974	0.87	No proration
1975	0.90	No proration
1976	0.78	No proration

Note: 1.00 is ratio when throughput equals rated capacity under normal conditions.

Source: Donald Norman, "The Peformance of Integrated Pipeline Companies: Evidence on the Undersizing and Access Issues," Policy Analysis Department Research Paper, American Petroleum Institute, (forthcoming). Data originally submitted by Plantation to the Association of Oil Pipe Lines.

*First full year Colonial was shipping product.

TABLE III

RATIO OF SHIPMENTS TO CAPACITY ON COLONIAL, 1965-1978; DESCRIPTION OF PRORATIONING ON COLONIAL

	Line 1: Houston to Greensboro, N.C.	Line 2: Houston to Atlanta	Line 3: Greensboro to Linden, N.J.	Description of Prorationing
1964	0.49		0.20	
1965	0.81		0.52	
1966	0.97		0.69	
1967	1.00		1.00	Line 1 Prorated starting in July
1968	1.00		0.92	Line 1 Prorated
1969	1.00		0.94	Line 1 Prorated
1970	1.00		0.92	Line 1 Prorated
1971	1.00		0.92	Line 1 Prorated
1972	0.87	0.50	1.00	Line 1 Prorated
1973	0.86	0.67	0.96	Line 1 Prorated
1974	0.87	0.67	0.85	Line 1 Prorated
1975	0.89	0.67	0.86	Line 1 Prorated
1976	1.00	0.68	0.90	Line 1 Prorated
1977	1.00	0.74	0.91	Line 1 Prorated
1978	1.00	0.80	0.92	Line 1 Prorated until Dec. 1, 1978. Starting Dec. 1, Line 3 began prorationing, while prorationing eliminated between Houston and Greensboro.

Note: All ratios based on approximate measures of shipments. 1.00 is ratio when throughput equals rated capacity under normal conditions.

Source: Donald Norman, "The Performance of Integrated Pipeline Companies: Evidence on the Undersizing and Access Issues," Policy Analysis Department, American Petroleum Institute, (forthcoming). Data originally submitted by Colonial to the Association of Oil Pipe Lines.

TABLE IV
"MAJORS" PIPELINE SYSTEMS

	Extent and Duration of Prorationing 1976-1978	Percent of Proration in Segment(s) Affected	Percent of Proration in Entire System	Response to Prorationing	Average Daily Unused Capacity, Entire System
01M	Lower 48: No Prorationing Alaska: 1978 - 8/21 - 11/12	0.62%	Alaska System: 0.62%	Negotiation underway for increase in system capacity.	1976: 40% 1977: 35% 1978: 44%
02M	No Prorationing				1976: 34.3% 1977: 34.6% 1978: 35.8%
03M	No Prorationing				1976: 29% 1977: 39% 1978: 44%
04M	Two (out of 34) Segments: 1976: Segment "A": entire year. Segment "B": Feb. 1977: Segment "A": Jan, Feb. Segment "B": Nov, Dec. 1978: Segment "B": entire year.	Segment "A": Jan 1976 through Feb 1977: 26.3% Segment "B," Feb 1976: 19.3%; Nov 1977-Dec 1978: 32.8%	1976: 1.40% 1977: 0.37% 1978: 1.34%	Segment 'A': Throughput on this segment has declined and, as a result, no expansion is necessary since prorationing was ended. Segment 'B': Economic evaluation for a means to alleviate this prorationing is in process and is expected to be completed shortly.	1976: 25.4% 1977: 26.3% 1978: 28.1%
05M	No Prorationing				1976: 19.1% 1977: 16.8% 1978: 16.8%
06M	No Prorationing				1976: 39.4% 1977: 41.0% 1978: 37.4%

531

	Extent and Duration of Prorationing 1976-1978	Percent of Proration in Segment(s) Affected	Percent of Proration in Entire System	Response to Prorationing	Average Daily Unused Capacity, Entire System
07M	Lower 48: Two Segments, Jan.-Feb. Alaska: 1978: 8/21-12/31	Lower 48: Segment "A": 3.9% Segment "B": 2.6% Alaska: Approx 5%	Lower 48: 1977: 0.025% Alaska: Approx 5%	Lower 48: Proration on the two segments was alleviated in late Feb. 1977, because of weather moderation which reduced fuel oil demand. Alaska: No final plans for alleviation have been developed.	Lower 48: 1976: 55.7% 1977: 53.1% 1978: 52.7% Alaska: 1976: 0% 1977: 0% 1978: 0%
08M	No Prorationing on crude lines. Prorationing on product lines has been confined to one segment from Feb. 1977 to Nov. 1977.	6%	0.1%	Proration ceased end of Nov. 1977 when additional horsepower was installed at new pump station.	Product Lines: 1976: 26.7% 1977: 25.4% 1978: 20.3% Crude Lines: 1976: 56.9% 1977: 53.5% 1978: 55.5%
09M	No Prorationing				1976: 30.0% 1977: 26.2% 1978: 26.3%

	Extent and Duration of Prorationing 1976-1978	Percent of Proration in Segment(s) Affected				Percent of Proration in Entire System	Response to Prorationing	Average Daily Unused Capacity, Entire System
			1976	1977	1978			
10M	Minimal and sporadic prorationing during past three years: Segment "A": Aug 1977, Oct-Dec 1977, Jan-May 1978. Segment "B": Aug, Sept 1978. Segment "C": June, Oct-Dec 1976, Jan, Feb, April-Dec 1977, Jan, Feb, April, July-Dec 1978.	"A": 0 "B": 0 "C": 3.6		7.8 0 10.6	6.7 2.0 11.2	1976: 0.31% 1977: 1.05% 1978: 1.11%	Segment "A": Two capacity expansions were made, the first being completed in Aug. 1977 and the second in Nov. 1977. Crude oil production handled by this segment peaked in Oct. 1977 and declined thereafter. After Nov. 1977 this segment had excess capacity but was prorated until June 1978 because of optimistic tenders. Segment "B": Intermittent proration continues and a possible expansion is being studied. Segment "C": No plans to expand this segment since capacity appears adequate.	1976: 25.7% 1977: 22.0% 1978: 23.1%
11M	No Prorationing							1976: 68% 1977: 66% 1978: 66%
12M	No Prorationing							1976: 54.8% 1977: 55.4% 1978: 58.8%
13M	No Prorationing.							1976: 54% 1977: 57% 1978: 59%
14M	No Prorationing.							1976: 60.0% 1977: 40.0% 1978: 34.3%

	Extent and Duration of Prorationing 1976-1978	Percent of Proration in Segment(s) Affected	Percent of Proration in Entire System	Response to Prorationing	Average Daily Unused Capacity, Entire System
15M	No prorationing on crude line system. Minimal prorationing on one segment of product line system in Midwest in 1977, but no prorationing on product line system in East.	Feb-Dec 1977: 26.6%	7.5% (of the mid-continent system)	A booster station was added on the prorationed segment in 1978.	Crude Lines: Mid-Cont 1976: 22.1% 1977: 28.7% 1978: 36.8% South West 1976: 36.8% 1977: 36.8% 1978: 36.8% Product Lines: Not Available
16M	No Prorationing				Not Available
17M	No Prorationing				Not Available
18M	Minimal prorationing -- out of 75 segments just one has been prorationed in last three years.	20%	Not Available	Segment was constructed before 1950 and the change in crude supply patterns in recent years has caused this problem.	1976-1978: 54%
19M	No Prorationing				1976: 36.8% 1977: 39.7% 1978: 39.3%

Source: Donald Norman, "The Performance of Integrated Pipeline Companies: Evidence on the Undersizing and Access Issues," Policy Analysis Department Research Paper, American Petroleum Institute, (forthcoming). Data originally collected by the Association of Oil Pipe Lines.

TABLE V
JOINTLY OWNED OR OPERATED PIPELINE SYSTEMS

	Extent and Duration of Prorationing 1976-1978	Percent of Proration in Segment(s) Affected	Percent of Proration in Entire System	Response to Prorationing	Average Daily Unused Capacity, Entire System
01MJ	1976: none 1977: Mainline System: Feb, Mar, Nov, Dec. 1978 Mainline System: Jan, Nov, Dec., plus one lateral and one spur line	1976: none 1977: 10% 1978: Mainline, 10% in Jan, 5% in Nov and Dec. Minimal proration on lateral and spur	1976: none 1977: Approx 1% 1978: Approx 0.6%	1978: On Mainline System constructed 72 miles of 26" line; proration removed in Dec. 1978 on spur and lateral lines. Studies underway to increase capacity.	1976: 8% 1977: 19% 1978: 15%
02MJ	1976: Mainline Segment: Jan-Dec. 1977: Mainline Segment: Jan-Dec. 1978: Mainline Segment: Jan-Dec.	1976: 17.5% 1977: 18.5% 1978: 30.5%	1976: 6.5% 1977: 3.7% 1978: 17.3%	1976: Line capacity increased from 500,000 BPD to 648,000 BPD by addition of three pump stations. 1978: Added 326 miles of 4" loop and constructed nine pump stations on Mainline System.	1976: 8.3% 1977: 10.8% 1978: 7.2%
03MJ	No Prorationing				1976: 18% 1977: 12% 1978: 14%
04MJ	1976: June-Sept. Seasonal prorationing related to tourist traffic 1977: June-Sept. 1978: June-Sept.	Not Available	Not Available	The situation is being monitored. Prorationing to date has been sporadic and, as far as company can tell, the shippers were able to meet marketing requirements.	1976: 33% 1977: 34% 1978: 32%
05MJ	No Prorationing				1976: 50% 1977: 50% 1978: 50%

	Extent and Duration of Prorationing 1976-1978	Percent of Proration in Segment(s) Affected	Percent of Proration in Entire System	Response to Prorationing	Average Daily Unused Capacity, Entire System
06MJ	No Prorationing				1976: 61.4% 1977: 23.3% 1978: 30.4%
07MJ	No Prorationing				1976: 31.1% 1977: 47.5% 1978: 52.1%
08MJ	No Prorationing				1976: 7% 1977: 9% 1978: 35%
09MJ	No Prorationing				1976: 22.6% 1977: 19.9% 1978: 41.1%
10MJ	No Prorationing				1976: 68.8% 1977: 69.6% 1978: 69.6%
11MJ	No Prorationing				1976: 36% 1977: 28% 1978: 26%
12MJ	No Prorationing				1976: 47.0% 1977: 47.5% 1978: 48.7%
13MJ	1976: none 1977: none 1978: Dec 1978 (expected to continue through 1st quarter 1979). Prorationing was limited to one of five segments.	Dec 1978: 1%	Dec. 1978: 0.5%	Prorationing in Dec. 1978 was the result of increased demand to meet supply shortages caused by the Iranian situation. Company has under consideration a plan to expand facilities on affected segment to increase capacity.	1976: 35.5% 1977: 28.9% 1978: 28.8%

	Extent and Duration of Prorationing 1976-1978	Percent of Proration in Segment(s) Affected	Percent of Proration in Entire System	Response to Prorationing	Average Daily Unused Capacity, Entire System
14MJ	No Prorationing				1976: 86% 1977: 82% 1978: 85%
15MJ	No Prorationing				1976: 12% 1977: 4% 1978: 1%
16MJ	1976: One Segment, Aug.-Nov. 1977: One Segment, Dec. 1978: none	1976: 11.8% 1977: 2.8%	1976: 11.8% 1977: 2.8%	Two new pump stations and additional horse power added to existing pump stations in 1978 to increase capacity.	Not Available
17MJ	No Prorationing				1976: 76% 1977: 82% 1978: 82%
18MJ	1976: Three Segments, Dec. 1977: Three Segments, Jan-Mar, Dec. 1978: Three Segments, Jan, Feb, Nov, Dec.	Not Available	Not Available	Pumping units added in 1979 to increase capacity.	1976: 41% 1977: 38% 1978: 34%
19MJ	No Prorationing				1976: 39% 1977: 43% 1978: 47%
20MJ	No Prorationing.				1977: 55% 1978: 50%

537

	Extent and Duration of Prorationing 1976-1978	Percent of Proration in Segment(s) Affected	Percent of Proration in Entire System	Response to Prorationing	Average Daily Unused Capacity, Entire System
21MJ	1976: none 1977: One Segment, Jan-Feb (3 weeks total) 1978: none	1977: 21.8%	1977: Approx 1.0%	Due to the seasonal demand, through-put fluctuates and at times pipeline operates at full capacity. However there is no demonstrated need for an expansion to date.	1976: 51.2% 1977: 41.8% 1978: 54.8%
22MJ	No Prorationing				1976: 55% 1977: 51% 1978: 54%
23MJ	Minimal prorationing on 6 of 22 segments during past three years.	Not Available	Not Available	Not Available	1976: 48% 1977: 39% 1978: 43%
24MJ	No Prorationing				1976: 52% 1977: 47% 1978: 53%
25MJ	No Prorationing				1976: 42.6% 1977: 41.3% 1978: 48.4%
26MJ	1976: Prorated full year 1977: Prorated full year 1978: Prorated full year	Not Available		Carrier has no plans for alleviating prorationing as such action is considered uneconomical due to small additional volumes, high costs, and low competing carrier tariffs. Carrier has expanded horsepower of system to maximum economical limit and any further expansion would require looping.	1976: 0% 1977: 0% 1978: 0%

	Extent and Duration of Prorationing 1976-1978	Percent of Proration in Segment(s) Affected	Percent of Proration in Entire System	Response to Prorationing	Average Daily Unused Capacity, Entire System
27MJ	1976: Jan 1-Feb 13: All terminals except one. June 26-Sept 13: two terminals. 1977: Nov 1976-Feb 1977: All terminals except one. Feb 9-Mar 1: all terminals. Mar 1-Mar 4: three segments. 1978: Jan 12-Feb 27: all terminals. Aug 15-Sept 28: one terminal. *Note:* The capacity at delivery terminals for LPG rather than pipeline was prorationed.	Not Available	Not available.	In Nov. 1977, carrier completed an expansion in pumping capacity resulting in an increase in total pipeline throughput capacity from 103,000 to 120,000 BPD on part of system. This expansion was designed to extend the period during which high seasonal demand can be met.	Based on June: 1976: 59% 1977: 49% 1978: 42%

Source: Donald Norman, "The Performance of Integrated Pipeline Companies: Evidence on the Undersizing and Access Issues," Policy Analysis Department Research Paper, American Petroleum Institute, (forthcoming). Data originally collected by the Association of Oil Pipe Lines.

TABLE VI
"INDEPENDENT" PIPELINE SYSTEMS

	Extent and Duration of Prorationing 1976-1978	Percent of Proration in Segment(s) Affected	Percent of Proration in Entire System	Response to Prorationing	Average Daily Unused Capacity, Entire System

011	1976: Oct, Dec. (Terminals only)
	 1977: Jan, Feb, Sept, Oct, Dec. (Terminals only)
	 1978: Jan-Mar, May, Oct. (Terminals only)
		1976: All Terminals: 6.8%
		 1977: 5 Terminals: 13.2%
		        2 Terminals: 15.4%
		 1978: 4 Terminals: 9.3%
		        1 Terminal: 17.3%
		        2 Terminals: 13.7%
			1976: 6.8%
			 1977: 14.0%
			 1978: 12.0%
				Company is in process of surveying its present shipper demand and potential new volumes to demonstrate that a capacity expansion is justifiable.	1976: 21.1%
					 1977: 15.8%
					 1978: 14.6%

021	No prorationing on trunk lines. Gathering systems have been prorated for short periods due to inclement weather, power failures, etc., but have never been continuously prorated.				1976: 49% 1977: 54% 1978: 56%
031	No Prorationing				1976: 15.8% 1977: 15.5% 1978: 10.1%
041	No Prorationing				1976: 19.2% 1977: 28.0% 1978: 20.0%
051	No Prorationing				1976: 17% 1977: 29% 1978: 53%

	Extent and Duration of Prorationing 1976-1978	Percent of Proration in Segment(s) Affected	Percent of Proration in Entire System	Response to Prorationing	Average Daily Unused Capacity, Entire System
06I	No Protationing				1976-1978: 79%
07I	Prorationing during past three years has been brief and limited to certain terminals. The need to proration was caused by seasonal demand for products.	Not Available	Not Available	Capacity was expanded in 1977 and 1978 to help alleviate the necessity to proration.	1976: 54% 1977: 51% 1978: 27%

Source: Donald Norman, "The Performance of Integrated Pipeline Companies: Evidence on the Undersizing and Access Issues," Policy Analysis Department Research Paper, American Petroleum Institute, (forthcoming). Data originally collected by the Association of Oil Pipe Lines.

May 25, 1978   **TABLE VII   SPOT PRICE COMPARISONS — CENTS PER GALLON**
Source: Platts Oilgram

Price Service	NEW YORK				U.S. GULF					
	Domestic		Foreign [1]		Waterborne [2]		Pipeline [3]			
	Low	High	Low	High	Low	High	Low	High		
Premium Gasoline										
Spot Cargo	41.750	42.000	—	—	41.000	41.500	41.500	41.625		
Transportation	—	—	—	—	1.900	1.900	1.330	1.330		
Total	41.750	42.000	—	—	42.900	43.400	42.830	42.955		
vs. Base	Base	.250			1.150	1.650	1.080	1.205		
Regular Gasoline										
Spot Cargo	40.000	40.125	—	—	38.500	39.000	39.000	39.125		
Transportation	—	—	39.000	39.250	1.900	1.900	1.330	1.330		
Total	40.000	40.125	39.000	39.250	40.400	40.900	40.330	40.455		
vs. Base	Base	.125	(1.000)	(.750)	.400	.900	.330	.455		
Unleaded Gasoline										
Spot Cargo	41.625	41.750	—	—	40.000	40.500	40.500	40.625		
Transportation	—	—	—	—	1.900	1.900	1.330	1.330		
Total	41.625	41.750	—	—	41.900	42.400	41.830	41.955		
vs. Base	Base	.125			.275	.775	.205	.330		
No. 2 Heating Oil										
Spot Cargo	35.375	35.500	—	—	33.500	34.000	33.750	34.000		
Transportation	—	—	—	—	2.170	2.170	1.330	1.330		
Total	35.375	35.500	—	—	35.670	36.170	35.080	35.330		
vs. Base	Base	.125			.295	.795	(.295)	(.045)		

[1] Foreign equivalent prices only quoted for regular gasoline.

[2] Tanker transportation based on "last paid spot tanker rates" using clean product rates from U.S. Gulf to New York. All grades of gasoline costed at mogas rate and No. 2 heating oil at the gas oil rate.

[3] Pipeline rates based on "East Coast pipeline rates" for Colonial to Newark, New Jersey. Rates volume weighted 75% from Pasadena and 25% from Lake Charles. This volume split is based on an estimate from National Energy Transportation System Map No. 12, "Total Petroleum Products Movement," accompanying NET. Colonial rates quoted are 1.35 cents/gallon from Pasadena and 1.26 cents/gallon from Lake Charles. Higher U.S. Gulf Coast spot cargo prices for pipeline shipments reflect the cost of moving from refinery to the Colonial entry point.

Index

A

AEI (American Enterprise Institute), 103
AEI Conference on Oil Pipelines and Public Policy, 103, 286, 304, 317, 323, 402, 407, 441, 452, 459, 476, 477
AOPL (Association of Oil Pipelines), 74, 77-78, 402, 405
API (American Petroleum Institute), 41, 81, 144, 284, 310, 387, 427, 431
ASA (American Standards Association), 42, 44
ASME (American Society of Mechanical Engineers), 44
Abourezk, James, Senator, 215
Acorn, 422
Admiral, 74
Adobe, 421
Air Force Pipeline, Inc., 172, 422
Aitchison, 312
Ajax Pipe Line Company, 18, 25, 118-119, 126, 131, 139, 191, 210, 273, 274, 313
Alaska Natural Gas Report, 286, 337
Alaska Gas Pipeline, 232
Alaska Pipeline Bill, 264, 269
Alaskan Senate and House Committees, 250
Alcoa decision, 433
Allegheny, 422
Allen, Jack M., 143, 193, 265
Alpar Resources, Inc., 265
Alton Railroad, 273
Amerada, 74
Amerada-Hess, 426-427, 430, 433
American Enterprise Institute's (AEI), Conference on Oil Pipelines and Public Policy, 103, 286, 304, 317, 323, 402, 407, 441, 452, 459, 476-477
American Maritime Association, 258
American National Standards Institute, 41
American Petrofina, 33, 66, 272, 449, 422, 426-427, 433
American Petroleum Institute (API), 41, 81, 144, 284, 310, 387, 427, 431
American Petroleum of Texas, 73
American Pipe Line Corporation, 251, 299
American Society of Mechanical Engineers (ASME), 44
American Standards Association (ASA), 41, 44
Amoco Oil Company, 187, 196-198, 222, 226

Amoco Pipeline Company, 61, 66, 70, 73, 74, 76, 83, 105, 116, 122, 126, 128-129, 169, 350, 400, 402, 439, 440
Anderson-Pritchard Oil Co., 172
Anhydrous ammonia, 173
Antitrust and Monopoly Subcommittee of the Senate Judiciary Committee, 159, 163, 177, 187, 192, 194, 215, 219-220, 249, 266, 292-294, 296, 298-299, 308, 329, 342, 348-349, 352, 354, 358-359, 413-414, 423-424, 426, 430-431, 434-435, 439, 442-444, 447, 449-450, 455, 467, 470
Antitrust Division, Department of Justice, 144, 258, 265
Apavins, Thomas C., 304
Apco Oil Corporation, 67, 172, 221, 226, 262-263, 413, 460
Apex Oil Company, 72, 343
Arab oil embargo, 156, 190, 257
Arapahoe
 case, 147, 316, 318
 decision, 316
 Pipe Line Company, 24, 61, 105, 117, 122, 126, 146, 284, 319, 422
Arco, 37, 70, 74, 83, 127-130, 304, 422, 424, 439-440, 421
Arctic Slope Regional Corporation, 275
Ashland Oil Company, 64, 66, 68-69, 73, 329, 335, 359
Ashland Pipeline Company, 69, 73-74, 197, 222, 328, 330, 331, 334-335, 344, 452, 426-427
Ashland Refinery, 34, 37
Association of Oil Pipe Lines (AOPL), 74, 77-78, 402, 405
Atlantic and Great Western Railroad, 3
Atlantic Refining Company, 16-17, 111, 196-197, 219
 Consent Decree, 111, 139, 157, 238, 284, 314
Atlantic Richfield, 429
Atlantic Seaboard, 5
Attorney General
 Alaska Natural Gas Report, 419-420
 Deepwater Port Report, 333
 First Report to Congress, 177
 Second Report on Interstate Oil Compact, 199
Augusta Pipe Line Company, 172
Aurora, 37
Averch, Johnson Theorem, 351

B

B&B, 66
B.P., 304
Badger
   Petroleum, 73, 75
   Pipeline Company, 25, 72, 75, 83, 129, 402
Bacigalupo, David, 472
Bain, Joe S., 160, 162–163, 179–180
Baker, Donald, 461, 470
Baltimore and Ohio Railroad, 3, 4
Bangert, Charles, 262
Bank Holding Company of 1956, 463, 464
Barge(s), 37, 132–133, 143, 177, 440
Barksdale, J. H., 72
Barnsdall, 210, 297
Bartlett, Senator, 163
Batch(es), 22–23, 79–81, 85, 89–91
Batching, 80–81, 83, 357
   Pigs, 87
   Sphere(s), 87
Bayh, Birch, Senator, 215–216, 316–317, 444–445, 453, 475
Bayou system, 21, 210
Beacon Oil Company, 413, 472
Belle Fourche Pipeline Company, 164, 172
Berge, Wendell, 239
Bergson letter, 319
Big Eight, 148, 181–182
Big Inch, 21, 37, 62, 91–92, 108, 211
Binstead, Charles, 262
Black, Justice, 147
Black Mesa, 173
Blue Bell case, 200
Brown shoe market foreclosure doctrine, 434
Broyhill, James T., 290
Brundred Brothers, 79, 80–81, 313
   Decision, 312
Buckeye Pipe Line Company, 5–7, 11, 22,24–25, 83, 110, 127, 170–172, 251
Buck vs. Bell, 281
Buffalo Pipe Line Corporation, 17, 37
Burns Terminal, 333
Butte Pipe Line, 24. 196, 223, 226

C

CRA Pipeline Company, 66, 73, 172, 221
Calder case, 472

Calhoon, Jesse M., 262
California Company, 199
California Standard, 12, 24
Calnev Pipe Line Company, 297, 422
Calvert, D. W., 296
Campbell, Philip P., 253
Capital Structure, 237
Capline System, 26, 81, 197, 221, 225, 336, 346, 406, 439, 448
Cardinac Petroleum, 74
Carter Oil Company, 191, 210
Celler
   Committee Consent Decree Hearings, 146
   Subcommittee, 263
Cenex, 66
Census, Bureau of, 161
Cerra, Arthur J., 259, 315
Certificates of Convenience and Necessity, 107
Central States Marketing Company, 64
Champlin Refining Company, 19, 74, 130, 173, 297, 449, 474, 421, 426–427
Charter International Oil Company, 64, 74, 344
Chase Manhattan Bank, 217, 221, 238
Cherokee Pipe Line Company, 25, 37, 83, 119, 129, 411, 422
Chevron (Standard of California), 74, 128, 222, 316, 421
Cheyenne Pipeline Company, 164, 172, 422
Chicago and Atlantic Railroad, 6
Circuit Court of Appeals, 10th, 200
Cities Service, 24–25, 67, 117, 209–210, 215, 220, 222, 426
Civil Process Act, 285
Clark Refinery, 33–34, 36, 74, 222, 449, 426, 433
Clarke, Owen, 263
Clayton (anti-merger) Act, 49
Clearwaters, Keith, Deputy Assistant Attorney General, 443
Coal
   Report, 286
   Slurry, 159
Coastal States, 66, 74, 444, 449, 426–427
Cole Act, 20
Coleman, William T., Secretary of Transportation, 41
Collins Pipeline, 344
Colonial Pipeline Company, 37–38, 44, 63, 65, 69, 78, 82–84, 89, 105, 127, 134, 172, 197, 206, 212–213, 215–220, 258–261, 284–285, 295, 297, 308,

Colonial Pipeline Company (Cont.)
314, 316-317, 330-331, 334, 338, 343-345, 349-352, 354, 359, 405, 407, 435, 446, 452, 467
Columbia Oil Company, 8
Commerce
Department of, 193, 460
Oil, 429
Commodities Clause, 254-256, 269
Common
Carrier Crude Lines, 125
Purchaser, 53-54, 282
Stream, 351
Competition
Bureau of, 424
in Oil Industry, 288
in the U.S. Energy Industry, 425
Competitive Rules, 150, 152
Completion Agreements, 124, 242-243
Congress, 107, 143, 148, 151-152
Congressional Research Service, 134, 248
Report on National Energy Transportation, 414, 469
Conoco, 221
Consumer Energy Acts Hearings, 74, 217, 261, 425, 430, 434, 443
Contes, 308
Continental
Illinois National Bank, 246
Oil Company, 24-25, 67, 73-74, 83, 129, 210, 221-222, 225-226, 297
Pipeline Company, 260, 297, 332, 426
Cookenboo, Leslie, 115, 125
Cook Inlet Pipeline, 81, 117, 423
Copper, Kennecott, 463
Corporate Accountability Research Group, 260
Cost of Capital, 229
Court of Appeals
District of Columbia, 140, 277
Fifth Circuit, 141, 276
Tenth Circuit, 199-200
Crescent Pipe Line Company, 9
Crown
-Rancho, 422
Central Petroleum Company, 64, 74, 176, 344
Crude
Company, 359
General, 450
Oil gathering lines, 144
Pipelines, 159

D

DOE (Department of Energy), 155, 255, 307, 316, 318
Organization Act, 141, 277, 280
DOT (Department of Transportation), 41, 44
Dead weight tonnage oil tankers, 144
DeChazeau & Kahn, 130, 180, 185, 189, 202
Debt Ratios, 240
Deepwater
Ports, 39, 151, 401
Port Act of 1974, 68, 150, 418
Port Report, Attorney General, 101, 286, 333, 335, 341, 345, 352, 401, 416, 441, 476, 497
Department of Justice (DOJ), report and recommendations on the Deepwater Ports Act, Licenses in 1976, 335
Delaware Corporation Law, 248
Delta Refining Company (Earth Resources Company), 195-196, 222
Delivery terminal, 60
Denver Oil Company vs. Platte Pipe Line Company, 353
Depreciation charges of carriers by pipe lines, 310
Derby, 66, 73
Diamond-Shamrock Oil Company, 128, 226, 501, 513
Dirksen, 72
Discounted Cash Flow (DCF) analysis, 157
District Court in the District of Columbia, 146
Divorcement, 294, 303
Dixie Pipeline Company, 238, 297, 359
di Zerega, Thomas, 263
Dorchester Refining Company, 66, 421
Dougherty, Alfred, 290
Douglas Oil, 74
Drake, Colonel Edwin L., 1
Dryer, Edwin J., 472
Dunn, 139

E

EPA (Environmental Protection Agency), 39, 84, 148
ERC (Energy Resources Council), 306, 431, 460, 464
Earth Resources Company, 195

Eastern
  States Petroleum Company, 257, 412
  Transmission Corporation, 297
Eastland, Senator, 260
Ecol, 430
Economies of Centralized Management, 185
El Paso Natural Gas Company, 121, 463
Elkins
  Act, 28, 145-147
  Amendment, 254, 284
  Consent Decree, 237, 247, 314
  Stephen, 254
Ellis, Otis, 444
Emergency Natural Gas Act of 1977, 270
Emery, Jr., Lewis, 8, 169
Empire Gas & Fuel Company, 209
Empire Transportation Company, 3-4
Energy
  and Natural Resources, Senate Committee on, 134
    Coop, 66
    Industry Hearings, 434
    Petroleum Allocation Act, 154
ERC (Energy Resources Council), 431, 460, 464
Engineering Committee, 227
Engman, Lewis, 290
EPA (Environmental Protection Agency), 39, 84, 148
Erie Railroad, 3-4, 8
Ernst & Ernst, 259
Esso Standard, 171
Eugene Island Pipeline System, 81, 117, 222, 227
Eureka and Southern Pipe Line, 23
Eureka Pipe Line Company, 170
Evangeline Pipe Line System, 25, 127
Evans, Joe, 264
"Excessive Rates" of Return, 299
Explorer Pipe Line Company, 37, 63, 66-72, 78, 83, 88, 95-96, 113, 126-130, 155-156, 172, 197, 213, 220, 223, 226, 240, 243, 249, 272, 294, 334, 339, 342-343, 347-348, 350-351, 406, 443-444
Explorer—Williams Tariff, 273
Export Administration Act of 1969, 149
Exxon, 37, 112, 192, 197, 200-201, 213, 249, 260, 287-288, 297, 304, 316, 348, 354, 422, 438, 450, 458-459, 421, 423-424
  Case, 287, 290-291

F

FEA (Federal Energy Administration), 155, 460, 462
  Regulations, 291
FERC (Federal Energy Regulatory Commission), 89, 107, 134, 140-143, 151, 164, 166, 206-207, 232-233, 237, 255, 267-269, 272, 277, 282, 301, 309, 314, 326, 355-357, 359, 467, 476-477
  Valuation of Common Carrier Pipelines proceedings, 265, 441
FPC (Federal Power Commission), 141, 144, 280-281, 299-300, 324, 418
FTC (Federal Trade Commission), 62, 75, 148, 150, 152, 181-182, 253, 284, 286-291, 294, 406, 418, 424, 429, 438, 442, 450, 453-454, 462, 471, 424, 427-428, 430, 432
  Act, 289
  Versus Exxon et al, 424, 430
  Report, 418
F&M Case, 290
Famariss, 74
Farmers Union Central Exchange, Inc., 140, 221, 278-279
  Versus Great Lakes Pipeline Company, 355
Farmland Industries, 172
Faulkner, Dawkins & Sullivan, Inc., 464
Feasibility Study Committee, 225
Federal
  District Court of Colorado, 400
  Energy Regulatory Commission (FERC), 89, 107, 134, 140-143, 151, 164, 166, 206-207, 232-233, 237, 255, 267-269, 272, 277, 282, 301, 309, 314, 326, 355-357, 359, 467, 476-477
  Government Mandatory Crude Oil Allocation Program, 459
  Manual for Complex Litigation, 287
  Oil and Gas Corporation (FOGCO), 475
Field tanks, 47, 49
Financial Accounting Standards (FASB), 325
Flexner, Donald (Deputy Assistant Attorney General, Antitrust Division), 103-104, 306, 459
Flint Ink, 421
Flow lines, 47

Flying switch, 88
Ford Foundation, 438
Foreign flag operations, 134
Forest Oil Company, 9-10
Four Corners Pipe Line Company, 24, 117, 422
Frasch process, 7
Free Pipe Line Law, 253
Friends of the Earth, 148
Frontier Refining Company, 199

G

GATX, 84, 196, 339, 344, 351
Galey, John, 9
Garfield, James, 253
   Report, 253
Gary, Raymond, B., 233, 237-238, 250, 466
Gasland, Inc., 196
Gateway, 226, 285
Gathering Systems, 202
   Line(s), 49-52, 55-56, 91-92, 202, 311
Gatherer(s), 50-56
Gauger, 52-53
Georgia Pacific, 463
Getty, 182, 422, 426, 433, 449-450, 455
Gillette, Senator Guy, 261
Glacier Pipe Line, 122, 260, 285
Gladieux Refinery, 34-35
Good Hope
   Industries, 196
   Refineries, 196
Goldstein, Samuel, 72
Goodman, David, 155
Government Operated Oil Function (GOOF), 474
Gravity bank(s), 81
Gravcap, 81
Great Lakes
   Order, 147
   Pipe Line Company, 18-19, 130, 173, 201, 210, 272, 274, 297-298
Great Northern, 73
Guffey and Galey Pipeline, 10
Guffey, James M., 9
Gulf, 262, 332-333, 420, 426
   Line, 226
   Oil, 11-12, 19, 24-25, 38, 66-67, 74, 110, 127
   Refining Company, 126, 210, 213, 222
Gustofson, 73

H

Haddock, Hoyt, 258-259
Hadlick, Paul E., 258
Hansen, Clifford P., 266, 280, 442
Harbor Pipe Line System, 24, 171
Hardin, E. P., 167, 328
Harshman, J. B., 297
Hart, Senator Phillip, 444-446
   Subcommittee on Antitrust and Monopoly of the Senate Judiciary Committee, 161, 165, 167
Haskell, Senator, 264
Hawaiian Independent Refinery, 429-430
"Hell or high water" clause, 168
Hepburn Act of 1906, 13, 253-254, 256, 309
   Amendment, 253, 256, 267
   Bill, 254-255
Hess Oil & Chemical (Amerada-Hess), 222, 422
Higgins, Patillo, 10
Hope Natural Gas Case, 141, 273-275, 277, 279-280, 324, 448
Hruska, Senator Roman, 464
Hudson, 66
Hulbert, Richard C., 296-297
Humboldt Refinery, 2
Humble (Exxon), 12, 70, 73, 74, 222, 226
Hunt Oil, 450
Husky Oil, 66, 172, 221
Hydraulic gradient, 59
Hydrocarbon Transportation, Inc., 73

I

ICA (Interstate Commerce Act), 13, 141, 143, 145, 215-216, 219, 246, 253-254, 261, 268-270, 274, 276, 279-280, 284, 295, 308-310, 351, 353, 400, 440
ICC (Interstate Commerce Commission), 13, 19, 26, 29, 43, 46, 54, 56, 75, 79, 81, 89, 91, 92, 107, 129, 139, 140-141, 144, 159-160, 164-165, 196-197, 219, 237, 246, 253, 255, 256, 261, 263, 267, 269, 271-280, 287-290, 298, 301, 307-315, 318-319, 324-326, 328, 330, 336-337, 341, 345, 350, 354-357, 359, 402-403, 413-415, 417, 419, 432, 434, 443-444, 452, 450, 459-460, 420
   (Now FERC), In the Matter of Pipelines, 309

Ickes, Secretary of Interior, 20
Ikard, Frank, 215, 316, 431, 475
Ileo, Michael J., 318
Illinois Pipeline Company, 118, 191, 210
Interstate
    Compact to Conserve Oil and Gas, 54, 177
    & Foreign Commerce, House Committee on, 279
    Oil Compact, Attorney General's Second Report, 199
    Oil Pipeline, 38, 54
Independent Petroleum Association of America), 143, 193, 264–265, 319, 412
Independent(s), 56–58, 164
    Gatherer(s), 54–55
    Producer, 194
Independent Refiners Association of America), 472, 424
Independent Tankers & Common Carrier Truck Rates, 259
    Terminal Operator's Association, 262
Indiana Pipeline Company, 7, 23, 170
"Industry lines," 211
Information Transmission, 188
Integrated Oil Operations of the Senate Committee on Interior and Insular Affairs, Special Subcommittee on, 301
Integrated pipeline, 55
Interior Department, 331
Intermediate products, 175
International Energy Agency, 317
Interprovincial Pipe Line, 119, 196, 225, 441
Interstate carriers, 55
Iowa Plan, 201
Iranian production, 183

Jackson, Henry, 265
Jacoby, Professor Neil H., 161–162, 202
Jeeping, 40
Jenkins, Thomas D., 316–317, 342
Jersey Standard, 210
Jet Lines, Inc., 172
Johnson, Arthur, 460
Johnson Oil Company, 194
Johnson, Ray M., 264
Jointly operated pipelines, 203, 211, 214, 218
Jones
    Act, 133–134
    David, 164–165, 171–172, 264
    L. Dan, 264
    Vernon T., 69–71, 129, 155, 249, 339, 348
    William K., 316, 318
Justice, Department of, 49, 64, 101–105, 142, 150, 152, 211, 253, 260–261, 266, 276, 283–286, 300, 317, 322–327, 331, 334, 340, 345, 402, 404, 407, 415, 418–419, 424, 427–428, 430, 432–434, 436, 441, 445–446, 448–449, 452, 458, 476, 477

K

Kaneb, 328
    Pipe Line Company, 83, 172, 174, 198, 251, 296, 298, 330, 348–349, 351, 422
Katy Railroad, 172
Kauper, Thomas B., Assistant Attorney General, 163, 446
Kaw Pipe Line Company, 73, 75, 422
Keech, Judge, 147
Kelley, J. Paul, 313–313
Kennedy
    E., Senator, 62, 68–69, 151, 289–290, 293, 441
    Robert, 258
    Staff Report, 163–165, 221, 253, 265, 272, 282, 287, 289, 293–299, 317, 326, 338, 341–342, 345, 347, 352, 354, 356, 400, 401, 424, 439, 449–451, 454
    Subcommittee, 103, 159, 265, 303–304, 315, 327, 421, 440, 459, 467, 476
Kerr-McGee, 73, 221–222, 449, 427, 433
Keystone Pipeline Company, 16–17, 19, 25, 37
Klausner, Joseph, 260
Knox, Navy Secretary, 20
Koch Industries, 56, 73, 83, 129, 176, 420–421, 430, 450
Kopolow, Meyer, 70–72
Korean War, 22

L

Labor Statistics, Bureau of, 45, 274
La Follette Committee Report, 298, 338
LaGloria, 66

Lakehead Pipe Line, 196, 411
Lamont, John, 261
Laurel Pipe Line Company, 25, 37, 83, 119, 127, 153, 423
Lead lines, 47
Lease Automatic Custody Transfer (LACT), 53
LeGrange, Ulyesse J., 304–307, 328, 452, 477
Leonard Pipe Line Company, 66, 74
Line fill, 59
Lion
  Monsanto Company, 64, 344
  Oil Company, 200–201, 221
Litt, Judge, 246
Little Big Inch, 21, 25, 37, 108, 131, 211, 260
Liquid Petroleum Gas (LPG), 25
  Pipelines, 159, 173
Lodge Amendment, 254
Lodge, Henry Cabot, 253
Long Beach Midland Line, 424
Long Beach suit, City of, 292–292
Loop, 286, 336, 340–341, 429
  The deep water unloading port, 26, 68–69, 133
  Incorporated, 151
Louisiana
  -Arkansas Refiners' Association, 311
  Pipeline Company, 12, 463
Lucas, Anthony, 10
Lundberg Survey, Inc., Report, 404, 415, 450
Lundberg Study, 220

M

MAPCO, 297, 328, 422
Martin Oil Company, 343
Maskin, Alfred, 258–260
MATCH (Mobil, Amoco, Texaco, Continental, and Humble [Exxon]), 226, 260, 285
MEBA (National Marine Engineers Beneficial Association), 218, 262
Magnuson-Moss Federal Trade Commission Improvement Act, 290
Magnuson, Senator Warren, 74
Magnolia, (Mobil), 12, 38
Madisonville Terminal Corporation, 222
Mencke, Professor Richard, 181
Mandon, 197–198
Mapco, Inc., 172

Marathon Oil Company, 67, 73–74, 170, 196, 222, 344, 357–358, 426, 458
Marine
  Engineer's Union, 262
  Petroleum Company, 70–71
Martin Oil, 71–72
Matador, 56, 420
McLaren, Richard, 260
McMann Oil Company, 12
McLean & Haigh, 191
MCN line, 332
Mead, Dr. Walter, 205, 211–212, 216, 218, 445–446
Medders, Tom B., Jr., 264
Medicine Bow, 128
Mellon, Andrew, 9
  William, 9
Metropolitan Petroleum Company, 64, 344, 428
Metzenbaum, Howard Senator, 317
Mid-
  American Pipe Line, 25
  Continent Petroleum, 117–118, 210, 297
  Valley Pipe Line
    Company, 66, 119, 126, 200, 222, 423
    System, The, 110, 212–213
  West Refineries, 116
Midland Cooperative, Inc. 73, 221
Mills, W. W., 9
Mines, Bureau of, 15, 443, 445, 456, 431
Minnelusa Oil Corporation vs. Continental Pipeline Company, 314
Minnesota, 422
  Pipeline Company, 56, 225
Missouri-Kansas-Texas (Katy) Railroad, 25
Mitchell, Professor Edward J., 177, 261, 266
Mobil Oil, 12, 38, 70, 73–74, 83, 126, 129, 209, 214, 221–222, 226, 249, 317, 430
Mobil
  Alaska Pipeline Company vs. United States, 141
  Pipe Line Company, 328, 332–333, 335–336, 439–440
MOIP (Mandatory Oil Import Program, 153
Moody's, 238
Moore, Beverly C., Jr., 260–261
Morgan Stanley & Co., Inc., 233, 250
Morison, H. Graham, 169–170
Mother Hubbard, 144, 146, 284
Motor carriers, 132
Murphy, Oil Corporation, 64, 73, 195–196, 226, 235, 344

Myers, Stewart, 237, 245, 467

N

Naph Sol Refiners, 37
National Congress of Petroleum Retailers, The, 262
National Environmental Policy Act of 1969 (NEPA), 123, 148–149
National
  Cooperative Refinery Association, (NCRA), 73, 129, 221
  Defense, Council of, 145
  Energy
    Board of Canada, 156
    Transportation System, 443
  Oil Jobbers Council, 333
  Petroleum Refiners' Association, 263, 289, 427
  Refining Company, 312
  Transit
    Company, 311, 341
    Pipe Line System, 5, 7, 23
    Pump & Machine Company, 23, 170–171, 222
  Transportation
    Policy 107, 132
    of 1940, 256
National
  Industrial Recovery Act (NIRA), 255
  Oil Marketers Association, 258
  Marine Engineers Beneficial Association (MEBA), 218, 262
Natural
  Gas Act, 280, 281
  Gas Transportation Act of 1976, Report on the, 246
Natural Monopoly Characteristics of Pipelines, 102
  Resources Defense Council, 148
Navarro Refining Company, 12
Newmont, 332
New York
  Central Railroad, 3–4
  Ontario and Western Railroad, 8
  Produce Exchange, 7
  Transit Company, 16, 170
  Transit System, 23
Nickers, 74
Nielson, 172
  Transit System, 23
Nickers, 74
Nielson, 172
Nonowner shippers, 64–66, 75–78

Northeast Petroleum, 429
Northern Pacific, 226
  Pipe Line Company, 7, 23
  Pipeline System, 170, 213
Northville Dock Corporation, 64
Northwest Improvement Company, 226
Northwestern, 463
  Refining Company, 222
NPRA, 449

O

Occidental, 164, 293, 429
Office of Defense Mobilization (ODM), 152–153, 170
Oceangoing ice-breaking tankers, vessels, 134
O'Hara, Donald, 449
Ohio Oil Company, 6, 23, 25, 37, 170
Oil and Gas Journal, 223
Oil, Chemical and Atomic Workers Union (OCAW), 427
Okie, Pipe Line Company, 56, 422
Oklahoma
  Pipe Line Company, 12, 118, 313
  Stripper Well Association, 264, 412
Oil Pipelines and Public Policy, Conference on, 103
Olympic, 77, 285, 336, 402, 444, 429
O'Mahoney
  Joseph, 316, 341
  Report, 72
  Senator, 48
OPEC, 151, 262, 288, 428, 451
Operating lines, 331
Origin station, 59
Outer Continental
  Shelf Lands Act, 331, 332
  Lands Act Amendments of 1978, 151, 476
Owen, Robert, 474
Ozark Pipeline, 38, 126

P

PAD (Petroleum Administration for Defense), 128, 425
  Districts, 165, 169, 222
Palomo, 422
PAW (Petroleum Administration for War), 20
PICA Report, 177, 197
Panama Canal, 121
Pan American Petroleum, 222

Pasotex, 38
Patterson, George, 23
Peabody Coal, 463
Peirson, Walter R., 187, 198, 202
Pennsylvania
  Free Pipeline Act of 1883, 8
  Railroad, 3-4, 9, 171-172
  Transportation Company, 3
Pennzoil, 262, 463
Permian, 73-74
  Corporation, 316, 342, 501
Peterson, Ronald J., 262
Petrofina, 426-427
Petroleum
  Industry Hearings, 163, 184, 214, 262, 435, 452, 442, 447, 471, 475
  Information Corporation, 421, 456, 459
  Rail Shippers Association, 79
    Decision, 355
  Refiners' Association case, 290
  Specialties, 37
Petrolia, 74
Pew, J. Howard, 119
Philadelphia and Reading Railroad, 4
Phillips, 297, 422
  Investment Company, 67, 74, 83
  Petroleum case, 281
  Petroleum Company, 17-18, 25, 210, 221, 226
  Pipe Line Company, 18, 25, 128-130
PICA Report, 263, 265, 296, 352, 424, 433
Pine, Judge, 147
Pioneer Pipeline Company, 297, 440
Pipeline
  Cases, The, 256
  Company, 272
  Companies, Annual Report for, 310
  Demurrage and Minimum Shipment, 355
  Financing, 241
Piper, J., 23
Placid Oil Company, 222, 226
Plantation Pipeline Company, 19-21, 37, 63-65, 78, 83, 86, 88, 105, 127, 197, 210, 212-213, 224, 259, 338, 348, 407
  Regulation, FERC Notice of a Proposed Statement of Policy on, 164
Platte Pipe Line, 24, 58, 63, 78, 105, 122, 126, 156, 196, 225, 353, 406, 439, 449
Platt's Oilgram Price Service, 443-444
Powder River Pipe Line Company, 359-400, 448
Prairie Oil and Gas Company, 10, 12, 16, 253

Pipe Line Company, 79, 116, 118, 311, 341
Presidential Advisory Committee on Energy Supplies & Resources Policy, 152
Producers and Refiners Oil Company, Limited, 8, 9, 12, 169
Producers Oil Company, Limited, 8-9
Product cycle(s), 90
Prorationing, 78, 408-410, 447
Proxmire Committee hearings, 260
Prudhoe Bay, 53
Public Utility Holding Company Act of 1935, 463-464
Pump & Machine Company, 171
Pumping stations, 59
Pure Oil Company, 9, 19, 24-25, 191, 210, 222, 239

Q

Quality banks, 81, 83
Quintana, 450
  -Howell Joint Venture, 421

R

Railroads, 132, 143
  Commission, 56
  Reading, 4-5, 8-9
  Reunitalization & Regulatory Reform Act, 279
Rail transportation, 134-135
Rancho System, System, The, 25, 209, 212-213, 346
Reduced Pipeline Rates and Gathering Charges proceeding, 314, 355, 451
Refiners' Association case, 290
Reynolds number, 86
Richfield, 24
Rock Island Refining Corporation, 73, 196, 221
Rodino Committee, 110, 403, 411
Roosevelt, President, 20, 253
Run ticket, 52
Ryan, Dr. John M., 134, 321

S

Salt Creek line, 400
Santa-Fe, 422
SEADOCK, Incorporated, 151, 286, 333, 340, 429, 469, 476

SEC (Securities and Exchage Commission), 61, 325, 461
Safety, 41-44
San Diego Pipeline, 173, 251
Santa Fe, 173, 251, 328
Scurlock, 56, 74, 450, 420-421
Seaboard, Incorporated, 151, 171
Seamless pipe, 108
Seaway, 26, 67, 81, 126, 130, 221, 226
Security Oil Company, 11
Seiberling, Congressman John, 151
Senate
    Commerce Committee, 443
    (S. 1167) Hearings, 447
    Interior Committee, 467
        Special Subcommittee of, 263-264
    Judiciary, Antitrust & Monopoly Subcommittee, 159, 163, 177, 187, 192, 194, 215, 219-220, 249, 266, 292-294, 296, 298-299, 308, 329, 342, 348-349, 352, 354, 358-359, 413-414, 423-424, 426, 430-431, 434-435, 439, 442-444, 447, 449-450, 455, 467, 470
    Antitrust Subcommittee on Oversight of Antitrust Enforcement, 419
        Committee, 446, 461
    Special Subcommittee on Integrated Oil Operations, 237, 245, 338, 345
Service Pipe Line Company (Amoco), 24, 61, 126, 147, 222, 225
Services, F. S., 73
Shallow
    River, 195
    Water Refining Company, 195
Shareholders agreement, 150
Shatford, John E., 311-312, 451
Shell
    Oil Company, 12-13, 16, 19, 25, 67, 73-74, 84, 126, 128, 131, 186, 196-197, 210, 213, 226, 239, 243
    Pipeline Company, 221-222, 239, 311-312, 316, 350, 438-439, 426, 429
    Shenefield, John, 62, 103, 164, 172, 265, 303, 315, 319, 327-329, 340, 406, 420-423, 428-429, 438-442, 450, 452-454, 458, 467, 470, 476
Sherman Act, 13
Shipbuilders Council of America, 259
Shipper-owned line, 214
Sierra Club, 148
Siess
    Charles P., Jr., 262-263, 460
    -Sun Pipe Line, 295

Signal Oil Company, 64, 74
Silvertip Pipeline, 260
Sinclair, 297
    Consolidated Oil, 209
    Oil and Refining Company, 11-12, 24-25
    Pipe Line Company, 127, 209-210, 222
Skelly, 73-74, 210, 221, 297, 455
Slick, W. T., 165, 192, 200, 202, 246, 426
Small Business Administration, 262
Small Business, House Select Committee on, 264, 315, 444
Smith-Conte
    Hearings, 258-260, 308, 315, 342-343, 404, 436
    Subcommittee of the House Small Business Committee, 329, 414
Smith
    Oil, 73
    Neal, 265, 297
Smyth vs. Ames, 277, 280
Socal, 74, 284, 421
Socony, Vacuum, 17
Sohio
    Pipe Line Company, 66, 73-74, 110, 116, 118, 121, 136, 191-192, 201, 210, 212, 222, 304, 312, 328-329, 465
    Report, 286
Solar Refining Company, 6
Solomon, Ezra, 318
Southcap, 222
Southeastern Pipe Line, 19-20, 210
Southern
    California Gas Company, 121
    Pacific, 328
        Pipelines, Inc., 25, 164, 170, 172-173, 198, 251, 422, 424
    Railroad, 64, 172
Southwest Pennsylvania Pipe Lines, 23, 170-171
South Hampton Company, 66
Southwestern
    Crude, 74
    Refining, 501
Spahr, Charles E., 110, 328, 403, 411
Spavins, Thomas, 105, 305-308, 328, 430, 441, 452
Spindletop, 10-11
Spur lines, 82
Stafford, George M., 263-264, 298, 345, 444
Standard & Poor's, 235, 238, 323

Standard Oil Company of
  California, 197, 199, 214, 329, 452
  Indiana, 7, 10, 147, 186, 209, 235, 426
  Kentucky, 19, 210
  Louisiana, 11, 118
  New Jersey, 13, 16, 19, 38, 118, 191, 210
  Ohio, 191
Standard Oil, 463
  decree, 23
  group, 4–12, 14–17, 170, 191, 251, 253
Stanford University Graduate School of Business, 318
Stanley, Morgan, 155, 466
Stanmar Oil Corporation, 4
Stanolind (Amoco), 311
State Department, 460
State of Texas Deepwater Port Authority, 469
Fred Steingraber, 259, 261, 308, 331
Sterling, 117
Stevenson, III, Adlai, Senator, 74, 264, 402, 444
Street, J. R., 70
Stripper-well, 142
Sun Oil Company, 11, 16, 37, 67, 73–74, 83, 110, 119, 127, 130, 201, 212, 221–222, 226, 262, 422
  Pipeline Company, 262–263, 423, 440
Sunder, Shyam, 305, 328
Suntide, 74
Superior, 24
Super tankers, 31
Supreme Court, United States, 141, 147, 255–256, 261, 268, 278, 310, 316, 319, 474
Surge tank, 53
Susquehanna-Sun Oil, 17, 19

T

TAA (Transportation Association of America), 29, 319, 415, 451
TNEC (Temporary National Economic Committee), 197, 326, 451
  Hearings, 257, 264, 288, 298, 316, 388, 412
TVA (Tennessee Valley Authority), 474
Tankers, 132–133, 143
Tank
  train, 135
  trucks, 135
TAPS (Trans Alaskan Pipe Line System), 26, 29, 40–41, 45, 53, 58, 81, 96, 106, 112, 119, 121, 124, 126, 134–137,

TAPS (Cont.)
  139–140, 142–143, 148–150, 155, 275–276, 304, 309, 318–319, 321, 329, 339, 346, 444, 451–452, 464, 466–467, 469
  Authorization Act, 149, 155
  Agreement and Grant of Right-of-Way for, 149
  Rates Cases, 150
  Stipulations for the Agreement and Grant of Right-of-Way for, 149
Tavoulareas, William P., 184, 317
Tecumseh Pipe Line, 119, 411, 422
Tender(s), 78–81, 84, 89
Tenneco Oil Company, 64, 73–74, 164, 293, 344, 433
Terra, Arthur J., 259
Tesoro, 35–36, 74
Texaco, 11–12, 24–25, 67, 209–210, 221–222, 226, 284, 297, 421, 426
  -Cities Service, 73–75, 402, 422, 439
Texas City, 501
  Refining Company, 64, 344
Texas
  Company, The, 11, 126, 209
  Deepwater Port Authority, 441, 467
  Eastern, 260, 440
    Transmission, 25, 37, 67, 83, 127–128, 131, 221, 226
  Empire Pipeline, 311
    Company
      Illinois, 18, 209, 311
      New Mexico, 74, 76, 117, 402, 411
  Pipe Line Co., The, 127, 147, 222
  Pacific, 450
  Railroad Commission, 48–49, 57, 79, 341–342, 450, 456–457, 459, 420–421
  Stanolind Pipeline, 311
  Western Oil Lines, Inc., 169
Texoma, 26, 63, 66–67, 81, 126, 130, 196, 221, 223, 226–227, 440, 449
Thompson, R. C., 243
Throughput and Deficiency Agreements(s), 97, 124, 168, 242–244
Thunderbird Petroleums, Inc., 200
Tidal Pipe Line Company, 147
Tidewater Pipe Line Company, 4–5, 24, 147
Tidioute Pipe Line Company, 3
Tierney, Paul J., 29, 45, 451, 464
Tosco Corp., 449, 426–427, 433
Trade Agreements Extension Act of 1955, 152
Transportation
  Act of 1920, 310

Transportation (Cont.)
  Association of America (TAA), 29, 319, 415, 451
  Department of, (DOT), 41, 44
  of Explosives Act, 43
  of Explosives and Other Dangerous Articles, 43
Trans-
  Western Oil Lines, Inc. 230
  Mountain Oil Pipeline Corporation, 164, 173
Treasury Department, 460
Triangle Refineries, 72
Trucks, 132, 143
True, Dave, 74, 172
Trunk line carrier, 55
Trunk line(s), 48-49, 53, 55, 58-59, 61, 78, 91-92
Tulsa County Bar Association, 103
Tulsa group, 18, 210
Tunney, Senator, 214, 297
Tusconana Pipe Line Company, 16-17, 19, 118, 130, 171

U

ULCC (Ultra Large Crude Carriers), 39, 133
Unconnected wells, 49
Uncle Sam Oil Company, 13
Undersizing theory, 436, 440, 442, 449, 452
Uniform System of Accounts for Pipeline Companies, 140
Union, 66, 74
  Pacific, 173, 221, 297
United
  Gas Pipeline, 262, 463
  Refining, 74, 221, 430
United States vs.
  American Petroleum Institute, 144
  Atlantic Refining Company, 146, 294, 316
United States
  Bureau of Corporations, 253
  District Court for the central district of California, 291
  Geological Survey, 227
  Pipe Line Company, 8-9, 169, 251
  Supreme Court, 13, 255-256
Unitized field, 53
University of
  Chicago Graduate School of Business, 305
  Pennsylvania Law Review, 350-351

V

VLCC (Very Large Crude Carriers), 39, 133, 151
Valuation
  Act, 278-279, 325
  Case, 272
Valuation of
  Common Carrier Pipelines, 300, 319, 326, 452, 467
    Proceedings Commission, 318
Valuation
  Act, Reports, 139
  Section, 280
Valley Towing Service, 195
Vandergrift and Forman, 3
Van Syckel, Samuel, 2-3, 30, 116
Vertical
  divesture, analysis of, 506
  divorcement, 14
  integration, 174-177, 179, 181-182, 187, 190, 194, 202
    in the oil industry, 257
  Support Members (VSM), 137-138
Vickers
  Energy Corporation, 413
  Petroleum, 221
Voluntary Import Program, 153

W

Wabash Pipe Line, 25, 128
Waidelich, 215-217, 220
Walsh, Louis J., 197
Warren Petroleum, 25
Webb, Charles, 263
Wertheim & Company, 171
West Coast
  case, 284
  Pipeline Company, 169, 250
Western
  and Shell Pipe Line Corporation, 400
  Crude Oil, 74, 221
  Oil Transportation, 400
West Shore, 76, 402
West Texas, Gulf Pipe Line Company, 66
Wherry Committee Hearings, 256
Whiting, Indiana, 7
Wilderness Society, 148-149
William Brothers, 73, 173
  Companies, 129, 164, 173, 201
  Energy, Exploration, 164

William Brothers (Cont.)
  Pipeline
    cases, 266, 280
      Company, 70, 72-73, 83-84, 128, 164, 173, 223, 249, 272-274, 277, 296-298, 307, 232, 347-352, 354, 403, 440, 456-457, 468
Wilson, Bruce, Deputy Assistant Attorney General, 64, 69, 446
Wilson, Wallace, 246
Winger, John, 217
Wolbert, Dr. George S., 299
Wolverine Pipe Line Company, 25, 63, 78, 238, 406, 449
Wood River Refinery, 131
Wooddy, L. D., 140
  Exxon Pipeline Company, 318
World War II, 19, 22, 25, 37, 88, 91
Wyco, 128

Y

Yancey, Robert E., 69, 197, 328, 334-335, 358-359, 452
Yellowstone, 63, 78, 128, 221, 406, 449

DATE DUE

GAYLORD			PRINTED IN U.S.A.